.

IN MY TIME

IN MY TIME

THE GREATEST CENTURY OF CHANGE IN THE HISTORY OF MAN

June 7, 2008

A Memoir

Charles E. Willingham

Charles E. Willingham

iUniverse, Inc.

New York Lincoln Shanghai

IN MY TIME
THE GREATEST CENTURY OF CHANGE IN THE HISTORY OF MAN

iUniverse books may be ordered through booksellers or by contacting:

iUniverse
2021 Pine Lake Road, Suite 100
Lincoln, NE 68512
www.iuniverse.com
1-800-Authors (1-800-288-4677)

ISBN: 978-0-595-42930-1 (pbk)
ISBN: 978-0-595-68196-9 (cloth)
ISBN: 978-0-595-87270-1 (ebk)

Printed in the United States of America

This book was written and rewritten over a long period of time, in different countries and on different continents. As I was writing, I sometimes relied on research to supplement my memories of people, places, and events, or to develop or better express my ideas. The extended time period and diverse geographical locations—plus the fact that I'm not a professional writer or historian—made it challenging for me to keep thorough records of my research. Moreover, as the writing process stretched into many months and years, my research would merge with my own ideas, and I'd forget what was based on memory and what was based on research. In addition, information in the text is occasionally derived from clippings, letters, brochures, notes, matchbook covers, and scraps of paper that I've saved for decades and that are impossible to credit or track. I've made every effort to acknowledge my sources, and I regret any failure to provide proper recognition or acknowledgment.

Cover: El Hombre (Man), by Rufino Tamayo, 1953, vinyl with pigment on panel, 216 x 126 in. (5 m 48.64 cm x 3 m 20.04 cm), Dallas Museum of Art, commissioned by the Dallas Art Association through Neiman-Marcus Exposition Funds.

I dedicate this book to my lifelong friend, William Allen. Without his friendship, partnership (in most of my life), encouragement, suggestions, and support, this book would not have been possible. I read all his books (The Fire in the Birdbath and Other Disturbances, To Tojo from Billy-Bob Jones, Walking Distance, and others), and was inspired by them. My style has been influenced by his writing, but it is not nearly as good. His classical university education in writing made his works equal to the very best, and head and shoulder above most. I'm proud to say I know him.

I'd also like to thank Dr. Kathryn Ward for all her time and effort in doing excellent editing and in providing help with the organization of the book. Without her expertise, In My Time would just be a collection of junk stories.

Contents

———————

Part Five: The Sailing Years

Part Six: The End of an Era

Preface:
A Personal Word from the Author

When I retired and had time to look back, I decided to set my memories down on paper in a sort of diary-after-the-fact. I wanted to make sure I had a record to look through when the memories began to fade. As I considered my life from 1939 to 2007, I realized I might have some stories my friends would appreciate, so I decided to share these.

When I thought about my earliest days, I realized much of what I thought I remembered wasn't really my first-hand memory at all, but what I'd been told by my parents and others when I was very young. I've included these second-hand memories. I also thought I should describe the world I was born into. So I've tried to describe my part of the world as it was before I was born.

After that, I just tried to describe the things I remember best. I began to see what these experiences said about me and how my life took shape—how I wound up where I am today. It was quite a revelation to see myself as part of a hugely important piece of both American and world history, and I decided my recollections were something to share with a much larger audience than just my friends. What these descriptions might yet turn into, I don't know; so for now, I'm offering this as a work-in-progress to readers who might remember or appreciate the times and places I describe.

I've tried to stick to a simple pseudo-chronology, but lots of things are simultaneous and complex, so the story doesn't always move forward as precisely as a calendar does. That's not the way life happens anyway. Some parts of this narrative are funny; some serious or scary; some classified as top secret by the government; some technical; and some talkative. That was the way my life was. As the book now stands, I hope you can pick this up, start reading at any point, and not have to worry about "Who's on first? What's on second? I don't know is on third." The anecdotes will, I hope, stand up on their own so you won't have to read from beginning to end. The chapter headings give approximate time frames or approximate lengths of time for the period during which the events took place.

Part One of this book is called "The Other Dallas." Every city has two sides to its story. Most people worldwide know J. R. Ewing and the TV version of Dallas. My story is about the other side of the tracks or the river or the interstate—every city has such an area. But even before the development of the interstate, there was an invisible division between the haves and the have-nots in Dallas. On the north side of town were Highland Park and University Park, the exclusive bastions of the haves; Southern Methodist University and many large churches surrounded the area, defining the WASP population. As you went farther south from Main Street (and it did not take too many blocks), the lower-income families bunched together, trying to scratch out a living.

My childhood friend, William Allen, grew up to write a book about Charles Starkweather, a teen serial killer who was our contemporary. Starkweather came from a socio-economic background very similar to Bill's and mine, and I think there must be a gene that made the difference between Starkweather's life and ours. With the human genome having been identified and mapped, there are great expectations that the next fifty years will bring amazing information about how DNA works, but gene or no gene, growing up in poverty does have an effect. You quickly learn how to survive, or you get left out. I come from a long line of survivors. In fact, my doctor told me once, "I have some good news or some bad news, depending on how you look at it. Sometime in your ancestry, like in the hunter-gatherer era, your forefathers were not very good hunters or gatherers. So they didn't eat every day. As a result, they developed a gene that you've inherited. It allows you to gain weight if you eat three meals a day. You can survive very nicely eating once every other day."

The title of Part Two, "The Military Years," is self-explanatory.

Part Three, "Living The American Dream," tells how I achieved the lifestyle to which I'd always aspired.

Part Four, "New Beginnings," is about starting over after my bubble burst.

Part Five, "The Sailing Years," covers my dream of owning a sailboat.

The final section, Part Six, is called "The End of an Era," and it describes my retirement and what I've learned in my time.

Part One:
The Other Dallas

Chapter One:

My Pitiful Lineage
(1920 to 1939)

Street-corner sidewalk preachers (Bible thumpers) and birds (grackles) are two things downtown Dallas used to be known for.

Aunt Effey owned a so-called boarding house—which was really a chicken ranch, massage parlor, or whatever else you want to call a whorehouse. Aunt Effey's whorehouse was located near McKinney Avenue, a few doors from Hibernia Street in Dallas. A Bible thumper on the corner of McKinney Avenue had been preaching hellfire and damnation against the place for some time. He said the Lord God would rain down hellfire and brimstone on the house like he did on Sodom and Gomorrah, and all the sinners should repent before the fast-approaching Judgment Day.

Meanwhile, bird droppings from grackles in the trees in front of the house were getting on the Johns' shoes and messing up Aunt Effey's nice red carpets. Aunt Effey told my dad she'd pay him a penny for each bird he killed.

Dad thought about it for a while and discussed it with his best friend Sam. Sam's dad worked at Texas Portland Cement and Lime Company in Cement City in West Dallas as a blaster in the limestone quarry.

Cement City was a small town in West Dallas created to support workers in the cement plants operated in the area. The town was situated on the Texas and Pacific Railroad line, about five miles west of the Dallas County Courthouse.

As a good father often teaches his son what he does, Sam's dad had taught him how to rig a blasting charge. Sam borrowed his father's keys to the explosive shack, and late one Friday night dad and Sam went out to Cement City. They stole a stick of dynamite, a blasting cap, some wire, and a charger. Saturday afternoon, they rigged the dynamite in the top of a tree outside Aunt Effey's boarding house and waited for dark, when the birds came home to roost. About 10:00 PM they set off the dynamite—and thus, modern technology found a way to impact even the oldest profession in the world.

It blew the top out of the tree and broke windows for blocks around. Aunt Effey's patrons—in various states of undress—jumped out second-story windows, and a few even broke legs or ankles. In the midst of mass confusion the police and fire departments arrived, barricaded the area, and surveyed the damage. After the chaos settled down, the police arrested the street preacher. His socially acceptable song of hellfire and brimstone suddenly seemed like a threat, so they charged him with trying to blow the place up.

Dad said he never did get paid for all the birds he killed. He couldn't find enough parts to make a complete bird, just a lot of feathers and pieces, but not enough to count.

My Dad: At Work and at Play

As a child, Dad was one of the best "top splitters" in Dallas. At that time (the 1920s), kids played with tops. They'd wind twine around them and throw them to the ground. As the twine unwound, it would spin the top and set up gyro momentum that kept the top balanced on its sharp point. The game developed to see who could throw a top hard enough and accurate enough to hit another spinning top in the center and split it into pieces. Dad split fifty-four tops in one year.

Dad was also a marble shooter. He was very good. He'd won some very nice "tolls," some "cat's eyes," some "monkey's fists," and other prized marbles. He had to win them, as he lacked the money to buy them.

He was good with yo-yos, too. Cheerio and Duncan made the best yo-yos. You could wax the string so when the yo-yo was down at the end, it would spin until you yanked the string and put a kink in the line. Then it would catch and wind the string around the center shaft and climb back up to your hand. Dad could do "walk the dog," "around the world," "over the waterfall," "rock the cradle," and several other yo-yo tricks.

Dad at work, however, was less determined. After high school, he was a painter and paperhanger for a while. Jobs were hard to come by in the 1930s, so he'd do any odd job. He worked a lot at Fair Park, building, repairing, and tearing down display booths for the many trade shows out there. Dad also did some bootlegging during Prohibition for his father, a Baptist Preacher who had a still in the basement of the church—anything to make a buck. Dad's father also owned a sawmill in East Texas near Troup.

Dad was a dancing fool and even managed to make money at it. During the Depression, Dad often entered marathon dancing contests. The sponsors always provided food and drink. And if you won, you could get $100, which was a lot of money in those days. Every time he'd hear of a dance marathon, he'd get there as fast as he could. He'd hop a freight train and go anywhere there was a marathon.

He became somewhat less industrious as time went by.

My Illegitimate Mother

My mother was born Hortense Edwards about January 10, 1908, near Belk in Lafayette County, Alabama. I say "about" because I've never found a birth certificate for her. She was the illegitimate daughter of Margaret Edwards and Lee Taylor. I really shouldn't say she was illegitimate because, since there's no birth certificate, there's no proof, and it's too late for a DNA test, seeing as how everyone involved is dead now. When Mother died, the doctor wanted to know the details of her parentage as he filled out her death certificate; when the situation was explained to him, he concluded he shouldn't list Lee Taylor as her father, so he left that part blank.

Margaret Edwards, her two brothers, and her daughter Hortense (my mother) moved to Shannon, in Lee County, Mississippi, before Hortense started grade school at age eight. Margaret worked at Vaughn Dairy in Mississippi. Mother and Margaret moved to Frost, Texas, when Mother was a teenager. Margaret went to work at Couch's Boarding House. Margaret liked to have a snort of Jim Beam every year at Christmas, but Mother never drank hard liquor. In Frost, they lived with the Yarboroughs, Margaret's cousins.

Mother graduated from high school on June 1, 1928. There were only twenty-two in her graduating class, yet it was the biggest class ever to graduate in Frost. Mother and Margaret moved to Adams Street, in front of Central High School, in Fort Worth that same year, and Mother went to work as a sales clerk in an F. W. Woolworth store. The pay was twenty-five cents an hour, and she worked fifty-six hours per week. She worked extremely hard all her life for very little money.

Mother's nickname was Frankie. She had a girlfriend named Lois McNut who worked at the other end of the counter at F. W. Woolworth. Mother worked in the school supplies section, and Lois had picnic supplies. Lois preferred to be called McNatt, for obvious reasons. McNatt's boyfriend was in the Army at Fort Bliss in El Paso, Texas. I mention that because one weekend he came home and brought a buddy with him named Harry Douglas Gilbert.

Harry was a very tall man at 6 feet 4 inches. Lois and her boyfriend had a Sunday party, and that's where Frankie (Mother) met Harry. Harry went back to Fort Bliss, and the two began corresponding. Mother's choice of men was not any better than her choice of work. She once said, "Harry didn't know how to do anything. He was just a mule skinner in the Army." His father had put him in a private school, but Harry didn't like it and left. After that, Harry's father ran him off from home even though he was a teenager. Harry, a rambling man, went to El Paso and joined the Army.

When Harry was discharged from the Army he came to Fort Worth, and he and my mother were married on September 1, 1930 (Labor Day), by Reverend Jess Skyles and his wife, Lenora. Harry went to work for Wolf Brand Chili Company in its bean factory. He handled the net full of beans that was lowered into a vat of boiling water to cook.

My Brothers and Other Minutiae

Dr. Grieve (a woman doctor) delivered both of my half brothers. Ralph Dean Gilbert was born February 6, 1932. Raymond Kelton Gilbert was born January 9, 1933. At around this time, Harry insisted that his mother, who didn't have any money, come and live with them. She earned some cash doing in-house clothes washing, and supplemented that with a small allowance from Harry's father. My grandmother Margaret didn't like Harry's mother but thought Harry's father was okay; "He kept to himself." Harry was not like his father; he always wanted to be in on everything.

Mother divorced Harry sometime early in 1937. The last I heard of Harry, he was buried somewhere in Connecticut.

The state welfare people wanted to put Ralph and Raymond in Buckner Orphan's Home in Dallas, but Mother absolutely refused. She insisted, "will keep them with me as long as I live. If I eat, they will eat. If I starve, they will starve." So the state welfare people arranged for Mother to get a job in a WPA garment factory in downtown Dallas, near the bus station.

Dad's mother, Grandmother Willingham, lived with Mr. and Mrs. Gant on Rutledge Street, on the other side of the Southern Pacific railroad tracks at Scyene Road and Second Avenue in South Dallas. Mr. Gant chewed tobacco and always had spit running out of both sides of his mouth. Grandmother Willingham would come over to Mother's place to baby-sit Ralph and Raymond. One day her son, Charles H. Willingham, Jr. (nicknamed Jot) came with her, and that's how my father and mother met. At that time, Dad was a painter and paper-hanger with real promise. Mom said, "He was very active with his dancing, also."

After several months they went over to the home of Dad's oldest sister, Sleta, picked up Grandmother Willingham, and drove to Waxahachie, Texas, where they were married on the lawn in front of the courthouse on June 19, 1937. It was the holiday known as "June 'Teenth," the anniversary of the day the Blacks were freed in Texas. Jot did not have enough money to pay the Justice of the Peace, so Mother paid the man his dollar.

Minutia: I, Charles Edward Willingham, was born on February 20, 1939.

Mother had an old-fashioned pedal-operated Singer sewing machine that she used to make quilts to sell. Jot worked odd jobs for the State Fair of Texas, and he brought home clothes, drapes, curtains, and rags from the trade-show booths. Mother used these to make quilts. Once he brought home some purple valor curtains from an automobile show, and Mom made me a pair of bib short pants out of them. That's when Dad started calling me "Sut Tattersall." We lived in a trailer that my father built on an old automobile chassis and parked behind a house at 800 Third Avenue and Ash Lane. Then things got better. Sometime in 1940 or early 1941 we moved into a duplex at 2210 Second Avenue.

Chapter Two:

The Second Avenue Years (1939 to 1945)

Our House

I was almost a year old when we moved into the duplex on Second Avenue. Our side had a living room, dining room, kitchen, bathroom, bedroom, and screened-in back porch. Mom and Dad slept in the bedroom, which had a double bed and a chest of drawers. We three boys (Ralph, Raymond, and myself) had bunk beds in the dining room during winter and on the screened-in porch during summer. We shared a chest of drawers in which each boy had his own drawer. This was quite a step up for us.

There was a natural-gas space heater in the living room, with six hollow sculptured ceramic bricks that radiated heat. I warmed my bottom at that heater on many a cold morning. The house lacked insulation, and the air was cold just a few feet from the heater. For summer we had an evaporative cooler, but it didn't work very well. It just kept the room damp and moldy-smelling as water was pumped and dripped and recycled endlessly through the system.

We had an old couch and an overstuffed chair in the living room, and a big old console radio that had a dial with a picture of the world. We'd all sit around the radio at night and listen to Gabriel Heater, Edward R. Morrow, or Walter Winchell give us the news. Our dining room had a large table with six chairs, but we used it only when we had company and only in summer, because in winter we switched the dining-room table with the bunk beds from the back porch.

We had one of those classic old iceboxes in the kitchen—I mean a real icebox with real ice to keep things cool, not an electric refrigerator. Mother couldn't afford to pay to have ice delivered, so every Saturday we'd trudge to the icehouse four blocks south on Second Avenue. My brother Ralph Dean had a red Radio

Flyer wagon that could hold a hundred-pound block of ice. The icehouse worker would load it in the wagon, and we'd take turns pulling it up Second Avenue. When we got home, Mother would go to work with her ice pick to cut it in two so she could lift the fifty-pound blocks one at a time into the top of the icebox. Those iceboxes were very well insulated, and the ice would last all week—if Mother could keep us kids from constantly opening the icebox door.

In the kitchen there was also a gas range and a small white table on which sat a clock radio. Mother liked to listen to the news on that radio in the morning and evening over a cup of coffee. It was her only time to herself. We ate most of our meals at that little table. We didn't have many dishes to put on the table, just a few plates and fruit jar water glasses, so we didn't need a lot of room.

The Other Side of the Duplex

The other side of the duplex was occupied by the Stovalls. Their son, Clarence, was in the penitentiary for stealing. Mother said he was the meanest kid in the world. Their oldest daughter, Marylee, took care of me sometimes. The last time Mother saw Marylee was when we lived in Rylie and she went to the washeteria near the Kaufman Pike drive-in theater: Marylee was there doing her laundry.

As I've said before, Mother's nickname was Frankie. Mr. Stovall came over drunk one day and asked, "Frankie, do you have a tater?" Mother gave him one, and Mr. Stovall went off grinning like a Cheshire cat. He sliced it, threw it in a pot, boiled it, and ate it. I don't think he even washed it. It was probably the only food he'd had in days.

Mrs. Stovall would come home, run into the house, sit down at the piano, and bang on it for hours. She wound up in Terrell's Mental Hospital.

What We Ate in Those Days

A nice old man had a store across the street from us, Rubin's Grocery Store. Old Man Rubin liked my brother Raymond, so Mother would give Raymond ten cents and send him to Rubin's to buy a soup bone. Old Man Rubin would give Raymond the biggest bone in the store, the one with the most meat on it. Mother would put it in the largest pot she had and boil it with some potatoes and carrots, a couple of tomatoes, and an onion or two. That would be all we'd get for the week, one bowl of that stew every day.

Our other main meal was pinto beans. Early Saturday morning I'd sit in the kitchen at the little white table and help Mother separate the beans from the rocks and dirt that were mixed in with them. While we worked, we listened to *Your Hit Parade* on the radio. It was sponsored by a jewelry store at 1924½ A Elm Street. The address—with that "½" and that "A"—always fascinated me, but I could never remember the name of the store. Anyway, a lot of the songs in those days were inspirational. These are the ones I remember best: "Accentuate the Positive," "Dream," "My Dreams Are Getting Better," "It Might As Well Be Spring," "'Til the End of Time," "High Hopes," "The Trolley Song," "Oh, What a Beautiful Morning," "Buttermilk Skies," and "Keep On The Sunny Side."

When we were done separating the beans, Mother would put the whole pound in her big soup pot, fill it with water, and let it cook all day on low heat. She'd add more water every couple of hours and stir it to keep the beans from burning. The next day was Sunday, and she'd put the pot back on to heat while she baked biscuits. Our big meal on Sunday was two biscuits cut open and covered with one cup of beans. That single meal would have to do for the whole day. On Monday, she'd add more water and heat the beans again. This went on all week until there were no more beans in the pot, just warm, brown, bean water and biscuits—which was usually our Friday-night supper. Sometimes, for variety, she baked cornbread instead of biscuits.

My dad liked to put hot pepper juice on his beans. Of course, when I saw him do this, I had to have some, too. I'd reach out for the jar, and Dad would put it down within my reach. As soon as I licked the top, I'd start to scream. My mother would take some butter on her finger and spread it on my lips. When the burning stopped, I'd reach for more. I still like the taste of that hot sauce.

Our one great treat per month came when Mother went down to the corner and bought four tamales for twenty cents from a Mexican street vendor with a pushcart. When she got home, she'd heat a can of Wolf Brand chili and spread it thinly over each of the tamales. Sometimes we'd get a biscuit with it, or maybe some cornbread. And if we were really lucky, we might get a slice of cheese. Those were the best tamales ever!

One time Mother went to get the tamales, but the Mexican was nowhere to be found. Mother asked around the neighborhood and learned that the Health Department had put him out of business. He was using the local dogs and cats in his tamales.

I don't care—those are still the best tamales I've ever had.

Fred and Eunice Williams

Fred and Eunice Williams lived in another duplex next door to ours. Fred drove a delivery truck for the local bakery. Sometimes he'd give us day-old cakes, donuts, and cookies left over from his route. Other times he'd catch a fish at the little rock bridge in Fair Park and give it to us with some day-old bread. We went through some hard times, and what we got from Fred was just about all that kept us alive one winter.

One day Fred took me out to a farm in North Texas. I don't remember much except the barn. It was huge (to a small kid, I guess any barn would be huge), and it had two side rooms and a loft. Pigs were kept on one side, and wouldn't you know it, to get to the loft steps you had to go through the pigpen. The farm boys told me to run to the steps or the hog would get me. I think that was just the country boys scaring the city-slicker kid. I don't think the hog gave a damn. But that loft was great with all the hay to play in. We spent most of the day up there. Had a lot of fun.

The people who lived behind Fred and Eunice had a little baby with a tiny head the size of an orange. The welfare people came and took it and put it in a home somewhere.

There was a tongue-tied girl that lived on the other side of Fred and Eunice. She couldn't pronounce my brother's name. Instead of "Ralph Dean" she'd say "Watt Dean." No one around there spoke proper English anyway, only "poor-white-trash Texan."

Dad and the 112th Cavalry

Dad joined the 112th Cavalry Division of the Texas National Guard on November 16, 1940, so he could get an extra thirty dollars per month. He reasoned that if there was a war, the National Guard would be the last to go. Also, if he did go to war he didn't want to walk, he wanted to ride—hence, the cavalry.

Well, the Texas National Guard was one of the first units to be activated. Dad was immediately sent to the South Pacific. Trucks and Jeeps were of little use in the jungles there, so they used the horses as pack mules to carry ammunition to the front. One day in a counterattack, for some reason the Japanese bypassed his squad, which then became separated from its unit. Dad was lost for twenty-one days in the jungle. He wound up eating his horse.

The Hospital

Mother used to tell a story of putting me in my bassinet in the trunk of the car, and driving around with the trunk lid open. I was delighted just watching the trees go by. As long as we were moving, I was happy as a clam at high tide.

Nevertheless, I was very sick during the first five years of my life. I had pneumonia seven times before I was five years old. It was wartime, and Children's Hospital had to give me sulfa drugs to fight off the pneumonia because all penicillin went to the military.

My most vivid memories from this time in my life are of the hospital. I remember how the hospital looked and smelled. The memory of the odors is very rich with the clean sheets and antiseptic smell with the hint of Lysol. I remember one time I decided to move my bed out of the dormitory. I did it by holding onto the wall of the cubical and pulling my bed into the middle of the room. Then, by holding onto other beds and cubicles, I moved my bed over to the door. Then I opened the door and pulled the bed into the hall.

That's when the doctors decided I was well enough to go home.

Another time I was back in the hospital and my bed was next to the corridor wall. The wall had sliding-glass windows so the nurses could see into the ward. I'd slide the glass back and yell down the hall, "Is dinner ready?" I received three hot meals a day in the hospital—a lot more than I ever got at home.

I remember lying in my hospital bed at night and listening to the trains as they approached Union Station. I could hear their whistles blowing for street crossings. The trains didn't go very fast in town, but there was a kind of Doppler effect. The whistle frequency would rise as a train approached a crossing—especially the one near the Children's Hospital at Oak Lawn and Maple—then the frequency decreased as the train passed the hospital and headed into town. I always wondered why the frequency went up and down. To this day, I get a strange, lonesome feeling when I hear a train blowing its whistle for a crossing.

I also remember lying in my hospital bed and seeing the flying red horse, or Pegasus, turning slowly around on top of the Magnolia Building. Later the Mercantile National Bank was built, and it had a searchlight that rotated around and over the flying red horse. There was a joke that there were two horses, one on each side of the pedestal, so Fort Worth wouldn't think Dallas was a one-horse town.

Every day the nurse would come in and get a blood sample from my finger for a red-blood-cell count. As I protested, she'd hold my hand and have the needle ready. Then she'd say, "See the pretty bird out the window." Being an animal lover, I'd always look and get pricked in the end of a finger. Soon all fingers on both hands were sore.

The Medical Arts Building

We went to the Medical Arts Building in downtown Dallas to see Felix F. Wilder, M.D., for radiation treatments. The hollow sounds of the hallway, and of the area where the elevators were, were like the stillness in an art museum.

The ride up the elevator to the doctor's office took my breath away. Sometimes the elevator made a funny clanking sound as it stopped and the doors opened on several floors on the way up or down. I always wondered what made the clanking sound and if it was going to let us fall. The wood-paneled halls and the hardwood doors with transoms made for a dark, spooky hallway. Looking out the window of the doctor's office, I was fascinated seeing how small the people and cars looked from that high up.

I remember being placed on a cold table and the nurse putting lead patches over my eyebrows and eyes for the radiation treatments.

More about the Hospital

One time in the hospital a doctor kept me in an oxygen tent. Every now and then he'd put his head in and listen to my chest with his stethoscope. He gave me a teddy bear to play with, and I have a picture of me with that teddy bear. I've sometimes wondered if the bear was a test to see if I was allergic to fuzz or something. I had a spelling board that had the letters in a circular slot around the outside, with a diagonal slot for creating words with the letters. The idea was to move the characters around the circle, pick out letters to spell a word, and line them up on the diagonal. It wasn't much of a success in my case. To this day, I still can't spell.

On one of my last trips to Children's Hospital, the doctor wanted to remove the lower part of my right lung. My mother refused. The doctor was very upset and tried to get me thrown out of the hospital. Mother complained to the hospital's management, and they let me stay. I sure thank her now for not letting them cut me up.

They performed a laparoscopy, running a tube down my nose into my right lung. They put some medicine directly on the infection in the lower right lobe. I still have a dark spot in the lower part of my right lung that shows up in X-rays. I think it's just gristle and a scar.

One time when I came home from the hospital, Mother asked what I wanted to eat. I said I wanted biscuits. She made a dozen, and I ate every one.

Streetcars

There were new streetcars with shiny two-tone green paint. When you climbed on, you dropped your coins into a contraption that looked like a glass lantern. They'd jingle down to the metal plate at the bottom, where the conductor could see through the glass to count them. If the change was correct, he'd press a long metal lever on the side of the lantern. That opened the tiny trap doors, and your coins would drop down into a box. There'd be an ominous grinding sound as if the coins were being eaten by some unseen monster. Then the conductor would push open the metal flap at the bottom, take the money, and deposit it in the money-changer hanging on a chain around the glass lantern.

He'd rotate the big rheostat arm in front of the change box, and the car would hum as it moved forward. The rheostat arm had notches in it, and it would click firmly into the next position as the humming increased. Walking back through the streetcar to take a seat, you could see long rows of yellow pine-ribbed seats. Each seat had a little metal sign attached to the top. One side of the metal sign said "White," and the other side said "Colored." The fifth from the last seat always said "Colored." So the Black (or "Colored") people always sat at the back of the car. If there were more Coloreds than Whites, the conductor would turn the sign over farther up the aisle.

The streetcar was a double-ender. That meant there was a conductor station at each end. When the car reached the end of the line, the conductor picked up his rheostat arm and moneychanger and went to the other end. As he went, he moved the seat backs to the other sides of the seats, so they'd always be facing forward. He'd go outside and pull down the rope on the spring-loaded arm that picked up electricity on the wire above the streetcar. At the other end, he let out the rope to engage the roller on the wire.

Taking the streetcar to town from 2210 Second Avenue, you went past Fair Park's Forest Avenue entrance, the Grand Avenue entrance, and the Music Hall. Then you made a right turn around the Music Hall to the Main Fair entrance on Parry Avenue. Then a left turn on Exposition Avenue. Then down to Main Street and another left turn. That's where you entered the "Colored," or "Negro," district known as Deep Ellum. There were shacks, beer joints, pawnshops, and honky-tonks with Blacks hanging around on the streets. There was always jazz and blues coming from those joints.

When you passed what was called the White-Faced Pawnshop on the corner of Harwood and Main, catty-corner from Heart's Furniture store, you were out of the Negro district. Titche-Goettinger Company—a department store—was on the next corner. Next was H. L. Green's at the corner of Ervay. Every year the March of Dimes had a booth in the street outside H. L. Green's, and it was fascinating to see all those shiny dimes lined up. You knew the Majestic Theater was in the same block on Elm Street. Neiman Marcus was on Main to Commerce Street on the other side. But we couldn't afford to buy anything they had and only went there at Christmas to see the decorations.

At S. H. Kress, we always went to the basement coffee shop. The food-service counter was next to the toy counter. I'd look at the toys while Mother would have a cup of coffee and rest.

The Palace Theater on Elm Street had a huge Wurlitzer pipe organ. During intermission and between shows, the organ console would rise up from under the stage with the organist playing and a spotlight on him. He'd play several numbers, then turn around with a great big smile and wait for the applause. Then he'd turn back around and start playing again as the console was lowered under the stage. It was quite a show.

The streetcar would continue past the F. W. Woolworth dime store, and past the A. Harris department store between Main and Elm. A block farther down was another store, Sangers, on the right side of Elm Street. Then came Houston Street, with the big red courthouse and Union Station just around the corner to the left.

The city really kept the streets clean. At night when you left the movies late, you'd see the city washing the streets with big water trucks. This also kept homeless people from sleeping on the street. You felt very safe waiting for a bus or streetcar in downtown Dallas late at night.

I remember riding the Jefferson/Tyler streetcar to see Aunt Omega, Dad's sister, who lived out near the end of the line. The streetcar went down to Houston Street, turned left past Union Station, then crossed the viaduct to Oak Cliff. The streetcar viaduct didn't have any sides on it, so you felt like you were suspended in space. As the streetcar click-clacked across the rail-section joints, it rocked from side to side, threatening to throw you into the Trinity River bottom—or worse, into the river itself. I always moved away from the windows and stood in the middle of the car so I wouldn't fall out.

Once across the river, we went past the Oak Farms milk factory, then past the streetcar barns, then by Burnett Field baseball stadium and down to Jefferson Avenue. We passed the Baptist church on the corner of Marsalis, across the street

from the Oak Cliff Carnegie Branch Library. There were used-car lots and a grocery store before we came to Beckley Avenue, with a drug store on the corner and
its soda fountain. Then down to Zangs Boulevard, which had Zales Jewelry Store
on the southeast corner.

We went past the Texas Theatre and up to Red Bryan's Smokehouse, a barbe-
cue place, across the street from Sears. After passing the diner on the left, we
turned left on Tyler Street, then right on Delaware. After a small dogleg, we were
on Burlington Boulevard. It had a wide field under high-voltage power lines,
with the streetcar track next to the road on the right.

Aunt Omega would always come whistling through the old house to greet us.
After our visit, when we were going home, it was lonesome. I was scared at night
waiting for the streetcar to turn around near Hampton Road. That was the end of
the Tyler line.

The North American Aircraft Factory

Mother went to work at the North American Aircraft factory in Grand Prairie.
To be hired, she had to show her birth certificate. But as I mentioned earlier, she
didn't have one. She wrote to some friends in Lafayette County, Alabama, and
got a signed statement about where and when she was born.

Mother rode to work with Fred and Eunice Williams' son Robert. She'd get
up at 3:00 AM, fix Ralph and Raymond their lunches, and then leave. She had to
be at work at 6:30 AM. They'd park across Jefferson Avenue in a huge parking lot,
and she had to walk across acres and acres of that parking lot to get to the bridge
over Jefferson Avenue.

They worked ten-hour days. She started at North American working for sixty
cents an hour on the assembly line, installing radios in the AT-6 Trainers. After a
year she received a raise to seventy five cents an hour. Two years later, when she
was laid off after the war was over, she was making ninety cents an hour.

She paid Robert $2.50 a week for transportation. When she missed her ride,
she had to get a streetcar downtown to the Westbrook Hotel on Jackson Street,
to take a bus to Grand Prairie. The bus was an old cattle truck. She would have to
stand up all the way. There were no seats, and passengers had to hang onto
leather cattle straps that hung from the top.

Grandmother Willingham

Grandma, whom I also called Mimi, had one of those old toasters that opened from the top on both sides like a double vee. The toasting elements were in the middle, so you had to turn the slices of bread over to toast them on the other side. When Grandma babysat me, she'd scratch my back to try to get me to take my nap so she could get some rest. At naptime I could often hear what sounded like a small airplane flying overhead. On hot summer days in Texas, without air conditioning, an oscillating fan would drone on and stir the air in an attempt to cool us.

One day while my grandmother was cooking she gave me a metal pie plate to play with. I threw the pie plate onto the hardwood floor with a bang. Mimi said, "That scared me!" and I replied, "It scared me, too."

There was a small Baptist church on the corner, and Grandma Willingham made sure we went on Sunday. Grandpa Willingham had been a Baptist preacher in Mixon, Texas, before he died. At church I'd go downstairs to Sunday school and play with the cut-outs of Jesus, the disciples, Joseph, Mary, the shepherds, and the sheep. I'd put them on the felt storyboard while the Sunday-school teacher told Bible stories.

I enjoyed going to Sunday school while Grandma was in church upstairs. One Sunday it was raining, and Grandma said that because the weather was bad she thought we'd just stay home. I told her it was very important to go to church. We had the radio on, and just at that moment, as if by divine intervention, the radio announcer said the same thing.

We went to church that day.

Higginbotham-Bailey-Logan Company

After the war, Mother went to work at Higginbotham-Bailey-Logan Company at 1100 Jackson Street. It was one of those garment-factory sweatshops that you hear stories about. Mother did piecework, sewing buttons at seven cents each and buttonholes at nine cents each. Pete, a repairman who was leaving the company, taught her to set up her machine and change it for the different-sized buttons and buttonholes.

One day she ran a sewing needle through her finger and had a hard time getting anybody to pull it out. They were too busy trying to keep up with the pace of work.

Mother dressed me up in an Army uniform with a Sam Browne belt. It consisted of a wide belt and a strap that crossed my chest and passed under the left epaulet. Then she took me downtown to have my picture taken. That day she took me into Higginbotham's to meet the people she worked with. We rode up in the employee elevator. The old gentleman elevator operator must've been eighty. He said he'd been running that elevator for over fifty years. Her boss kept asking me where my daddy was, and I said, "I told you two times, my daddy is in the Army." He gave me a ribbon for my uniform.

More Fun Times

One time we went to the barber shop, and the barber asked me how I wanted my hair cut. I replied, "Black and curly." I was a blond-headed kid.

There were some storm drains down by Mrs. Cole's place. We weren't supposed to go inside them, but you know how kids don't pay attention to rules. It was real spooky. The drainpipes were about six feet in diameter, but to a four-foot-tall kid they were enormous.

Second Avenue was the main U.S. Highway 175, running southeast to Athens and East Texas. There was a lot of traffic after the war, with soldiers returning and people moving. My brothers and I would sit on the front porch, counting the Red Ball motor freight trucks and Mayflower moving vans, and identifying the various types of cars as they passed.

Fair Park—Home of the State Fair of Texas

We went down to Fair Park during the off-season and walked around Leonhardt Lagoon. Sometimes Mom would give us a couple of stale slices of bread, and we'd feed the big old catfish from the little bridge near the Band Shell. The mouth on one of those fish was big enough to take a whole slice at once. Sometimes there were casting rings in the pond, so sport fisherman could try their skills at casting. Both rod-and-reel casting and fly-casting were tested. I don't know what the prize was, but the attendant and judge never missed a ring with either type of cast.

I liked to see the birds and insects, the plants and trees, and the fish and other wildlife that made their home at the lagoon. I always wanted to know more about the flora and fauna around the area. I learned about strange plants with exotic names, such as "Lizardtail," "Cypress Knees," and "Duckweed." I loved to observe this special habitat that was home to more than seventy species of birds, like the "Least Bittern" and the "Chimney Swift." I liked to watch the damselflies and other insects in the lagoon.

Most of the buildings in Fair Park were Art Deco in style and had opened in 1936 for the Texas Centennial celebration. There was a Texas Ranger station near the south end of Fair Park where we entered, with a bronze statue of a ranger in the covered driveway. The buildings were mostly made of logs, and the place was built like a small frontier town with wooden sidewalks. There was a replica of the Alamo down near the Texas Ranger station. I don't know what happened to it, but it's gone now. I guess they tore it down to make room for parking at the Star Plex.

I liked the Health Building with the clear plastic human where you could see all the blood vessels and internal body parts. I was fascinated with how the human body worked and how doctors and scientist figured things out. At that young age I was equally fascinated with the sperm-and-egg story of conception, although nobody ever explained how they got together.

The Aquarium

I loved the aquarium with its varied collection of six thousand aquatic animals. It included marine and freshwater fish, reptiles, amphibians, and invertebrates. It was cool and dark in the aquarium—a nice break from the Texas heat. The tanks had native flora and fauna and natural underwater habitats to make the fish to feel at home. The only light came from inside the tanks, and the fish couldn't see us watching as they swam by. There were large sturgeons, cod, barracudas, and sharks in the big tanks, as well as an alligator snapping turtle weighing 135 pounds, some venomous lion fish, and a six-foot alligator gar. There were colorful sunfish, angelfish, and other tropical varieties in the smaller tanks. There were also turtles, shellfish, octopi, water snakes, and eels. The five-foot electric eel was my favorite. I always wondered how it generated the shock.

The aquarium's Amazon Flood Forest exhibit showcased thirty species of fish from the Amazon River. The piranhas always scared me because of stories I'd heard. The World of Aquatic Diversity featured unusual and bizarre species, including jellyfish, poisonous stonefish, "fishing" anglerfish, and "luring" batfish. If these weren't bizarre enough, there were nine displays of seahorses and their unique relatives. Seahorses seemed strange because the males gave birth to babies from their pouches and spent their days entwined with their tankmates. Seahorse relatives on display included the monogamous blue-striped pipefish, the paper-thin shrimpfish, and the jumping alligator pipefish.

Fish feedings occurred daily at 2:30 PM. Sharks were fed on Sundays and Wednesdays; the moray eel ate on Mondays; piranhas dined Tuesdays and Saturdays; the alligator gar were fed on Thursdays; and the American alligator ate on Fridays.

The Art Museum

The best of all was the art museum. I always enjoyed seeing the beautiful paintings. When I think of the Dallas Museum of Art, I always remember a painting called *El Hombre* by the Mexican artist Rufino Tamayo. It showed a guy reaching for the stars while a dog went after a bone. I would sit for hours looking at that painting. It symbolized my search for knowledge and for myself. It was in front of the entrance to the planetarium, another of my favorite attractions and one I went to as often as I could afford.

The Natural History Building

The natural history building was full of stuffed animals in their natural environments. It was really educational.

The Hall of Texas Mammals presented wildlife dioramas of species native to Texas: badgers, beavers, coyotes, elk, gray foxes, mule deer, muskrats, ocelots, otters, prairie dogs, raccoons, ring-tailed deer, and white-tailed deer.

Of course, the Trinity River Mammoth predated those mammals. It lived 20,000 years ago and once grazed along the banks of the Trinity River.

Going farther back in time, a mural of prehistoric Big Bend provided a backdrop for a display of Texas dinosaurs. This showed that the scary Tyrannosaurus rex was not the only big toothy meat-eater in our state's past. There was also the Acrocanthosaurus, a huge, fierce dinosaur that once lived in North Texas and probably made some of the famous dinosaur footprints around Glen Rose, Texas. The Tenontosaurus was smaller and herbivorous. There was a primitive plant-eating ornithischian (or "bird-hipped") dinosaur that has no name and is from the Proctor Lake area, southwest of Dallas. Alamosaurus belonged to a group of animals called sauropods, long-necked dinosaurs that were among the biggest creatures ever to walk the earth. Sauropods were so large that scientists once thought they had to live in lakes and rivers so the water could help support their weight.

At the time I didn't know dinosaurs laid eggs, but there were several displays of nests full of dinosaur eggs.

The Hall of Texas Birds presented dioramas of avian species native to Texas, including the Attwater prairie chicken, wintering geese, and screech and barn owls. The Hall of Texas Wetlands exhibited the alligator, great blue heron, ivory-billed woodpecker, pileated woodpecker, reddish egret, spoonbill, sandhill crane, blackbird, and wood stork.

Bison Hall showed Texas around the turn of the twentieth century. I learned that bison played a major role in shaping the prairie and helped maintain the diversity of prairie grasses that supported other animals.

I was interested in the wild pigs and wild sheep. For example, there were the largely herbivorous peccaries; left alone, they're harmless to people, pets, and property. I found out that the distinctive feature of the pronghorn—its unique horn—consists of a permanent, bony core and a horny covering that is shed annually.

I learned about other animals, too: big cats and bears and wolves. The mountain lion formerly roamed much of the western hemisphere; once common in Texas, it was eventually eliminated from most of the state. The jaguar is the biggest American cat preying on large animals such as peccaries and deer; it was formerly found throughout the southern half of the state, but none have been seen for many years.

The black bear, once common in central and southeast Texas, was thought to have disappeared from the state, but it's now seen in the Davis Mountains. The only confirmed record of a grizzly bear in Texas was near Sawtooth Mountain in 1890. Gray wolves once roamed the western two-thirds of Texas, but the last verified gray wolves were killed in the Trans-Pecos in 1970.

Automobiles and Science

During the Texas State Fair, it was always exciting to visit the automobile exhibit and see the new cars and gadgets. The auto industry really put on a splash to introduce their new cars and get everyone excited about owning one. The big message was that the war was over and now was a good time to travel—by car, of course.

"See the USA in Your Chevrolet!"

But the most exciting time for me was spent in the science building. All the new developments in chemistry and physics were on display, such as nylon, plastics, and lava lamps, along with the snap of lightning from the Van de Graaff generator, the crackle of the Tesla coil, and the buzz of the Jacob's ladder.

There was so much knowledge to absorb! In the 1940s, there hadn't been any information on atomic energy until Hiroshima and Nagasaki (besides, I was too young to understand it), but after the war, atomic energy was a huge topic. Electrochemistry, the study of the interaction between electrical energy and chemical energy, fascinated me. The flashing pretty lights always caught my eye and sparked my interest in learning how things worked. Neon was a bright spot in the exhibits, combining the shapes of blown glass with different gases and electronic circuits. An endless variety of tube effects seemed possible. I was fascinated, observing the details of how things worked. For example, I always took twenty or thirty minutes to watch the Rube Goldberg exhibit.

I also liked to visit the WRR studio and see the announcers and disk jockeys working.

The Hall of State, the Railroad Museum, and the Farm and Livestock Shows

The entrance to the Hall of State featured a golden archer. Inside was the Portico Tejas room with the Hall of Heroes in back just before the Great Hall. On either side of the entrance were the North, East, South, and West Texas Rooms, which had various exhibits of documents, historical plaques, and portraits of important people such as politicians.

Nearby, the Age of Steam Railroad Museum displayed vintage railroad equipment. Exhibits included Pullman sleeping cars, cabooses, massive steam engines, and diesel and electric locomotives. The exhibit held one of the oldest and most comprehensive heavyweight passenger-car collections in the United States, with a complete pre-World War II passenger train—including steam locomotive, coal car, baggage car, coaches, lounge cars, Pullman sleeping cars, dining car, and steel caboose. In addition, historic structures such as a railway post office, plus signals and assorted small artifacts, made it one of the finest railroad museums in the Southwest. Union Pacific's steam locomotive, Kansas City Southern's boxcar, and a Lone Star Producing Company tank car made up a freight train. A Santa Fe Railway diesel electric locomotive was also on display.

We kids were allowed to climb all over the trains. I guess the museum people thought there was no way we could damage those big iron horses. We were also allowed to ring the big bell, but we couldn't blow the whistle—there was no steam.

I always enjoyed seeing the farm stuff. The tractor and farm-equipment exhibit was always interesting—with its tractors, cotton pickers, plows, rakes, and seeds—as was the livestock: horses, cows, swine, sheep, goats, rabbits, and even trained chickens. The rodeo was fun to watch, as was the judging of prize animals for the 4H and FFA. I spent hours watching these shows. And I loved visiting Elsie the Cow in the Borden exhibit.

A lot of these exhibits were free, but some things at the state fair required an entrance fee. They were just too expensive, and so I missed out. I always wanted to see the hot-dog stunt cars of Joey Chitwood, but never had the admission. Also, I always liked to watch roller-skating, but never could afford the ice-skating show until I was out of the service.

The Midway

The midway was where most of my time and money went.

It didn't take me long to figure out that I wasn't going to win anything at those game booths, so most of the money went for rides. The roller coaster was the most popular. Another fun ride was the Rotor. This was a large cylinder where you stood against the wall while the cylinder spun so fast you were held in place by centrifugal force. When the cylinder reached its top speed, the floor dropped down and left everyone hanging on the wall.

Of course, you had to have a corn dog with lots of mustard.

One year at the state fair, as Mother and I walked down the midway, she saw Minnie Pearl from the Grand Ole Opry just sitting there on a park bench. You could tell it was Minnie Pearl—she was wearing her signature hat with the price tag hanging down in front.

Cotton Bowl and Band Shell

The Cotton Bowl was the largest building at the fairgrounds. They always played the Texas-versus-Oklahoma University football game there during the fair, and that never failed to attract a crowd. But I didn't yet fully understand the game, so I was a little like Andy Griffith in his comedy routine *What It Was Was Football*:

I asked this feller that was a-settin' beside me, "Friend, what is it that they're a-hollerin' for?" Well, he whup me on the back an' he says, "Buddy, have a drink." "Well," I says, "I believe I will have another Big Orange." … It's some kindly of a contest where they see which bunchful of them men can take that punkin' an' run from one end of that cow pasture to the other, without either getting knocked down or steppin' in somethin'….

At night there was the light opera at the band shell. We'd sit outside and listen. I was fourteen or fifteen before Mom and I went to see *One Touch of Venus*. Mom said I was less interested in the performances and more interested in seeing how the stage crew did the scene changes in the dark. Anyway, I really liked the show.

There was a large above-ground swimming pool just inside the Grand Avenue entrance. It's gone now, as are the roller rink, the Alamo, and the Texas Ranger station.

Other Early Memories

Before Dad was discharged, he was sent to Hot Springs, Arkansas, for some "R & R" (rest and recuperation). Mother went to Hot Springs on the train to visit him.

Every morning before sunrise, Mom was up and off to work. Mrs. Cole, another baby-sitter, would come to the house around 7:30 or 8:00 AM. We could hear her heavy "Nanny" shoes clunk across the wooden porch. She would stop and put her umbrella down with a thump. Then she would fiddle with the door. During this time we'd jump back in bed, pull the covers over our heads, and pretend we were asleep. Mrs. Cole would come in and try to get us all up, then go make toast or cereal for breakfast.

One morning we heard the usual clomp, clomp, clomp across the porch and the thud at the door. Then the door opened—and Dad yelled, "Isn't anyone going to welcome me home?" All our feet hit the floor at the same time. He was home from the war, and we were all around his neck, screaming and hollering. Finally, he put his suitcase down and sat on the couch. We all begged for combat stories, but he was sick of the war. He showed us some coins and his medals and a few trinkets. But he was very tired and wanted to get some rest.

Not long after that, Dad took me fishing. He told me to be careful, because a really big fish might drag me into the lake. The next time Mother saw me, I was behind a tree, next to the water, reaching around both sides of the trunk to hold my fishing pole.

The first Christmas that I remember, I got a set of Lincoln Logs. I liked them so well that the next year I received Tinkertoys. Those were real special to me—now I could build more than just log cabins. The next year was an Erector Set, and boy, did I go to town with that! Now I could build a lot more things than with the Tinkertoys. But all of the toys seemed limited to me, because there were never enough pieces to build really large projects.

The last Christmas on Second Avenue, I received a Lionel electric train. But I didn't get to play with it much. My dad and uncle took over. They got down on the floor, and played with it like little kids—until they broke it.

I always wondered why it had three rails when the train tracks down the street only had two. That always bothered me.

Sleta's House

Aunt Sleta was another of Dad's sisters. One day I was over at Sleta's house on Morrel Street. My cousin Sylvia was going to the store, and I wanted to go. I ran into the back bedroom and asked Sleta if I could go. She said okay. I ran through the hall and into the front bedroom. The floor had just been waxed. When I tried to turn the corner to go out the door, down I went! My head hit the corner of the steel bed, which split it wide open just above the right eye. Needless to say, I didn't make the trip to the store. I still have the scar. It has moved up into my hairline but still reminds me of that day.

Still More Early Memories

One summer day in 1944, my brothers and I were going over to Grandma's house just for something to do. We had to go down Second Avenue past the icehouse to the Southern Pacific Railroad tracks at Scyene Road. There was a short-cut you could take to Rutledge Street by going down the railroad tracks a couple hundred yards. On the way, my brothers Ralph and Raymond stirred up a bees' nest, and I was hit just above the right eye again. I ran home screaming. Mother pulled the stinger out of my swollen eye, then she put some cigarette tobacco with spit on the sting. That took most of the pain away. But it was swollen for another day or so.

After he came home from the Army, Dad worked for the Greyhound Bus Company as a diesel mechanic. He'd had some diesel engine training when he returned from the South Pacific, before he was discharged. The Army had done away with horses, so the 112th Cavalry had become a mechanized cavalry. Dad had a shoulder patch from the Army that showed a Bengal tiger with a half-crushed tank in its mouth. Anyway, he was very good at his job and soon became one of the best mechanics at Greyhound. He was promoted to final check-out. He could tell what was wrong with one of those big bus engines just by listening to it.

While Dad was working at Greyhound, the drivers in Memphis went on strike. He was sent to Memphis to work on the buses and refill them so they could be driven to Texas. We took our first train trip to Memphis, Tennessee. One little girl on the train screamed and cried for over an hour as we left Dallas. Finally she went to sleep. She was asleep about a half hour when we went through Sulphur Springs. Her mother woke her up and said, "Wake up, Ruthie, and see the pretty lights"; the lights were made by gas burning off of the oil rigs. Ruthie started to scream again and continued for another hour.

The train's rhythmic click-clack of the steel wheels sounded like a heartbeat, and the gentle swaying back and forth rocked me to sleep.

When we arrived in Memphis, I had to keep quiet all day while Dad slept. He was working the night shift. Mother bought me a small pot-metal AT-6 airplane to play with. She took me to the park where we fed the squirrels. We also went to the Memphis Zoo, where we saw the bears, who waved at us.

One day we were going to Fort Worth for something, I don't remember what for, but as we went through Grand Prairie, near North American Aviation, we saw a huge flatbed truck. On it was a German V-2 rocket. It was heading west on Highway 80. I think they were taking it to White Sands Missile Range in New Mexico. They probably took it off a ship in Houston and brought it to North American Aviation so the engineers could take a look.

Chapter Three:

The Ramsey Years
(1945 to 1949)

The House and Yard

Growing up in Dallas's working-class suburb of Oak Cliff in the 1940s and 1950s was an experience I never got over.

Our one-bedroom house at 2327 Ramsey had a small living room, a kitchen/dining room, a tiny bathroom, and a screened-in back porch. We had the same old furniture as on Second Avenue, including the little white kitchen table and radio. The kids slept in bunk beds out on the back porch, and this time we didn't move inside when the cold weather came. In winter, Mom heated bricks in the oven, wrapped them in towels, and put them in between the sheets to warm the beds before we got in. "Bed warmers," she called them. We slept with a stack of heavy quilts on top of us, so it was difficult to turn over.

In summer, we slept in the backyard so we could get any little breeze that might stir during the hot nights. Often, as I lay looking up at the stars and moon, I wondered where they came from, how they got there, what kept them there, and if anyone would ever get to the moon. My brothers Ralph and Ray erroneously taught me to identify Venus as the evening and morning stars; they showed me the Big and Little Dippers, Orion, and the North Star. A telescope and a chemistry set were always on my Christmas list. As I lay in the backyard looking up at the stars I would sing to myself:

> The world outside
> Belongs to me
> If I learn the science
> I will rule the night
> And own the moon and

Tell the stars when to shine.

I'd changed the words from the *Warsaw Concerto*, written by British composer Richard Addinsell for the 1941 film *Suicide Squadron*, which I had seen.

There were two old peach trees in the backyard. My brothers Ralph and Ray sometimes brought baby squirrels back from their hunting trips, so they built a wire cage around the trees and put the babies in there. Of course, the squirrels became pets, and we'd bring them in the house to play with. One day one of the squirrels crawled up inside the console radio, and we pulled its tail off trying to get it out.

There was a mulberry tree on the south side fence near the alley, and we spent many hours picking mulberries and eating them. But I always had a hard time getting the mulberry stain off my hands, lips, and clothes. Still, I wish I had that old tree on my place at the lake where I live now.

There was no fence at the alley, so running up and down the alley was my first means of escape when I was in trouble with Mom or needed to get away from my brothers.

I loved to sit on the front porch in the morning before school. I liked feeling the warm sun on my face and listening to the neighborhood birds. I'd hear the mockingbirds sing. Well, maybe "sing" is the wrong word. The mockingbirds imitated every bird call they'd ever heard. Over and over. They screamed out these calls while jumping up and down like feathered maniacs in the top of a tree. There was also the gentle cooing of mourning doves, the chatter of sparrows, and the occasional shriek of a blue jay. At least twice a year our neighborhood would be invaded—either by blackbirds or crows, I never knew how to tell the difference. There were also endless bands of migrators, a hundred miles wide, stretching from horizon to horizon. I think these were starlings.

Organized Crime

One morning I was sitting there listening to the birds and drinking up the morning sun, when an explosion almost knocked me off the porch.

Mom and Dad came running out. "What happened!?"

We piled into the car and were driving north on Beckley when we saw smoke. At Conrad Street we could see a crowd gathered about a block away. When we arrived at the source of the smoke, we saw the burnt-out frame of a car.

It was a car bomb!

This is the report from *The Dallas Morning News*, Wednesday, November 30, 1949:

Gambler's Wife Killed as Bomb Blasts His Car

Mrs. Herbert Nobel, 36, wife of a Dallas gambler, was killed at 8:00 AM Tuesday by an exploding bomb affixed to the starter of her husband's car and apparently intended for him.

The Nobels lived at 311 Conrad, Beckleywood Addition, Oak Cliff.

In a last-minute change in plans, Nobel drove away in his wife's car, leaving the car ordinarily driven by him.

The death car, a late-model Mercury, was stripped to the bare wheels and chassis by the force of the blast.

Numerous attempts on the life of the former gambler have been made and, in one instance, an unknown marksman fired at Nobel while he was confined in a Dallas hospital.

A few months later, another murder attempt was made against Mr. Nobel. Here's the account in *The Dallas Morning News*:

Herbert Nobel was blasted as he walked out of the door of his home on Conrad St. in January 1950.

January 24, 1951 an assault charge was filed … [against] Louis Green, ex-gambler-associate and Dallas underworld leader.

There'd been stories in the newspapers about the Dallas police raiding illegal gambling halls and confiscating one-armed bandits. The police would take these to the Trinity River bottom and run over them with bulldozers. I think there was an unwritten agreement between the police and the underworld that the latter could keep their pinball machines and punchboards but not the one-armed bandits. The problem was that Herbert Nobel had nothing but one-armed bandits, and he wasn't about to part with them. So to keep peace with the police, the other gangsters decided Nobel had to go.

They finally got him. On August 7, 1951, *The Dallas Morning News* reported:

Herbert Nobel died when he stopped at the mailbox of his farm near Grapevine Tuesday morning and waiting assassins detonated an explosive charge placed in the ground.

Fun for Free

Severs Avenue, between Ramsey and Beckley, was a leafy tunnel running past tidy bungalows with well-kept lawns. The tunnel, created by trees and shaded from the sun, cooled the street, and the hot air at the end produced a draft that caused a welcome breeze on stifling summer days. It was a great place to skate or play. One year for Christmas I received a pair of steel roller skates with a skate key. The skates clamped onto the soles of my shoes and were adjusted and tightened with the key—a tiny wrench-like gadget that turned a screw on the side of each clamp. The metal skates made quite a racket as I rolled along concrete sidewalks and streets. But skating on the asphalt of Severs Avenue was a lot quieter and smoother. With the shade from the sun and the cool breeze, it was the ideal place to skate. And I never heard a word of complaint from the people who lived there.

Daugherty's Pharmacy, 2009 Beckley Avenue, was the area's only drugstore. It had a magazine rack where I read comic books I never could afford to buy. A couple of doors away was Perkins' Ice Cream Parlor. My brothers Ralph and Ray hung out there, and most of their friends did, too. The place had a couple of pinball machines, and Ray was great at pinball. He could easily rack up enough points to win himself a free game. I wasn't big enough or old enough to play, and I didn't go in there much as it was mostly older kids and I never had money for ice cream. But it was fun to go and watch him play pinball.

Down the street was the Beckley Theatre. If you had a quarter, you could get Saturday-matinee admission (nine cents), a bag of popcorn (a dime), a small coke (a nickel), and have a penny left over. Even at those low rates I couldn't often afford it, so my friends and I would stand out front and when folks started coming out, we'd try to sneak inside by walking in backwards.

The ushers almost always caught us.

But we had another trick. In the hottest part of the summer the theater doors were left open. One of the kids in my little gang would keep an eye on the theater employees and when they all had their backs turned, he'd give the signal to run. We'd charge through the lobby and into the theater and plop down in seats in the balcony.

One Halloween we saw a Frankenstein movie, and Frankenstein actually came into the theatre! Talk about scared. He sure was a big guy. Good thing he couldn't run in those big shoes.

If you headed south on Beckley toward Illinois, on your left there was an M. E. Moses dime store and a supermarket. The supermarket was air-conditioned and a cool place to play on hot summer days. We'd go in the supermarket, and one of us would sit in a shopping cart while the others pushed him around the store. Until the manager noticed. Then we'd be chased out.

There was an old stone filling station and garage on the corner of Beckley and Ohio.

Saner and Beckley Baseball Park was kind of the end of the world. On the south side of the park was a line of trees. After that, nothing until you got to Waxahachie. The ball diamonds were lit up with thousand-watt floodlights. We'd go there at night and watch the softball games and chase foul balls into the woods. About 1948, a developer put a housing development in from Saner for about ten blocks. It included a new grade school, Clinton P. Russell, which I attended for one year—but I'll talk more about this later.

Sometimes we fantasized that we were famous scientists, like Madame Curie or Louis Pasteur, engaged in brilliant research. On Wooden Avenue, near Daugherty's Pharmacy on Beckley, there was a medical clinic. This was in the days before the concepts of "biohazard" or "dangerous medical waste." My friend and I would raid the clinic's trashcans for empty medicine bottles. A special treasure we'd sometimes find was a glass tube with a rubber stopper in one end and a rubber insert in the other. We'd carry these things home, wash them in hot soapy water, and clean them up as best as we could. Then we'd bring them to the makeshift chemical laboratory in our clubhouse. We'd take the glass tube and shove its rubber insert inside with an ice-cream stick. In the heat of the sun, air pressure would build up and blast the stopper off the other end.

We called this our "Power-of-the-Sun Tube."

Church

I'd had so much fun at Sunday school on Second Avenue that I was really looking forward to finding a good church in Oak Cliff. But the first Baptist Church we attended was a real disappointment.

The preacher screamed, hollered, banged the lectern, and said we were all sinners. Born in sin and lived in sin and would die, go to hell, and burn forever—if we did not get saved and donate 10 percent of our assets to the church.

Well, I was only six years old. I didn't think I'd had time to commit that much sin and deserve that type of punishment. The worst thing I'd done was smoke, and most everyone I knew smoked. So I'd have a lot of company. My parents would be there, so it wouldn't be all that bad.

I thought the punishment for smoking—to burn for eternity—was a bit harsh. I didn't know how long eternity was, but the way this preacher said it, it sounded like a long, long time. I didn't like the way things were going at this church. Then, when it came time for communion, they gave us a tasteless cracker and a sip of Welch's grape juice. That didn't seem like very much to honor Jesus.

We tried several other Baptist churches in Oak Cliff, but they were all the same. Hellfire and damnation. They didn't have anything good to say or anything to lift your spirit. It was no fun at all. We stopped going. Years later I attended other churches in the area, and they were still much the same. I decided to wait until college and could reason for myself before trying again.

Riding a Bicycle

Learning to ride a bike is an educational event most people remember forever. True, there are a few kids with perfect balance who never experience the disappointing tumbles, the elbow-scraping crashes, and the eventual thrill of success. But otherwise, it's about the same for everyone; hence, the saying: "It's like riding a bicycle. Once you learn how, you never forget." The bicycle is the first symbol and tool of freedom that every boy and girl recognizes. Parents also. Moms and dads recognize the danger it represents, and they know that once bike-riding has been mastered, they've begun to lose control.

Once we'd all learned to bike-ride, my friends and I would head down to Wooden Park and ride around. It was better than peddling along in the streets with all those cars. Or so our parents thought. A creek ran through the park with ten-foot cliffs on both sides. The creek itself was only a couple of inches deep, and we'd try to get our bikes into the water. Splashing through the creek was a great way to cool off on summer days.

On the way to the park we could pick up speed heading downhill on Wooden Avenue. Then, as you entered the park from Wooden, an additional steep bank led down to the soccer field. At the bottom of this bank was a light pole with a heavy guy wire. As you sped past, you could grab the wire and pull yourself up and see how far your rider-less bike would travel across the soccer field before finally falling over. One time I was moving at breakneck speed when I grabbed that wire and let the bike run. It traveled the whole length of the field and into the woods on the other side. When we found my bike, it was down in the creek. It had plummeted over the ten-foot cliff and landed upside down. The right handlebar was bent in by about 30 degrees. My mother wouldn't pay to fix it, and I didn't have the money, of course. But that bent handlebar became my own personal badge of honor—a testament that I'd once gone fast enough to have my bike reach the creek.

Paper Routes

Ralph and Raymond had paper routes. Because the newspapers were so thick and heavy on Sunday morning, they'd enlist me as their helper. We'd get up at 3:00 AM and head out to Saner and Beckley, where the truck dropped off the papers. We'd fold them, put them in paper bags, and load them on our bicycles. It was all I could do to hold my bike up with that heavy load. Laden with papers, I'd peddle behind my brothers as they traveled their routes.

Later, I got my own paper route—about fifty customers in my neighborhood, down Ramsey and Alabama Street. Delivering was straightforward and easy. It was the collecting I didn't like. Sixty percent of the customers paid on time. The rest always had an excuse. It was hard getting money out of certain neighbors. Some would say to come back for the cash on a specific day. But then you'd go back, and they still wouldn't pay you. Some never had any money at all. South Oak Cliff was not the most affluent area. I lived there, so I knew this first-hand. We were all in the same boat. But some of us were working hard to get out of the damn boat. Others just seemed to let the boat sink.

Our Old Plymouth

Mother bought an old Plymouth because she thought she needed it to get to work. My dad was unreliable; he was drunk most of the time when he was home from the night shift. So Mom learned how to drive—sort of. It was always a struggle to the death when she was behind the wheel.

Today in the computer industry there's an expression for those who are phobic about computers, those who'd rather die than work at a computer terminal, those who're afraid they might hit the wrong key and cause the world to be sucked into a black hole. We say they have "terminal fear."

My mother had "terminal fear" of driving.

She'd grab the steering wheel in a death grip, both hands holding tight, both her thumbs turned under. Then she'd fight that wheel tooth and nail until we got to where we were going. By the time we arrived, her thumbs would be blue from lack of blood.

In addition to her terminal fear, she had to put up with wisecracks. One day as we were turning off Beckley Avenue onto Jefferson, Mother had her turn signal on. This smart ass on the corner yelled, "When did you get those blinkers? For Christmas?"

Eating Out and Playing Bingo

Occasionally we'd go out to eat. One place I remember was down the street on Oak Lawn past Lemon Avenue and across the street from Phil's Deli. Another was Jay's Marine Grill. It was off of Oak Lawn, and they served great hot rolls and butter before they brought your meal. Plus, their fish dinners were out of this world.

One summer we went out Greenville Avenue to Victory Park to play bingo. I remember it well because I won. My prize was nine pounds of bacon and an orange squeezer.

Radio

I'd enjoyed the radio on Second Avenue, and as I got older I enjoyed it even more. Every night I'd listen in.

Captain Midnight was my favorite. He had this secret decoder ring that he sent cryptic messages with. As a listener, you could send for your own decoder ring, and once you had it, you could decode special messages from him during the program. I had to have one, so talked my mother into sending away for it. Once I got it, I spent many a night decoding those precious messages.

Of course, there was *The Lone Ranger* with his theme song, "The William Tell Overture." There was *The Green Hornet* and his special assistant, Kato. There was *The Inner Sanctum*, and *Sergeant Preston of the Yukon* (with his theme song, Emil Nikolaus von Reznicek's "Donna Diana Overture"), and *The FBI in War and Peace*, and *Sky King*. With my mom, I'd listen to *Amos and Andy*, *Fibber McGee and Molly*, *Our Miss Brooks*, *The Great Gildersleeve*, *Jack Benny*, *Red Skelton*, and *Bob Hope*—from the latter I remember the "Call for Philip Morris" ad, with the "On the Trail" music from *The Grand Canyon Suite* by Ferde Grofé.

Ralph and Ray's Hunting Trips

Ralph and Raymond each received .22-caliber rifles one Christmas. Ralph had a single-shot, bolt-action Mosburg, and Raymond had a single-shot, bolt-action Remington. They'd go hunting for rabbits and squirrels out Illinois Avenue past Zang Boulevard, which marked the city limits. On the other side of Zang was a

40-acre field with a cow tank in the middle. We'd go skinny-dipping in that tank in the summer. Anyway, there were lots of rabbits in the field, so my brothers didn't have any problems getting game. Just the other side of the field was a small tree-lined creek with lots of squirrels.

That was great for the first couple of years, but then a developer bought the land and built Wynnewood Shopping Center. That ruined the hunting and our family's main source of meat.

One day Ralph and Raymond went hunting out toward Duncanville on Highway 67. They bagged four squirrels, and Ray tied the feet together and hung them over his shoulder to start the walk back home. A car came by, and the driver stopped and asked to buy a couple of squirrels. While he talked to these people, Raymond set his rifle down on his shoe, and when the car pulled away, Ray picked up his rifle. It was loaded, and it discharged into his foot. The couple in the car heard the rifle go off. They backed up, put Ray in the car, and drove to a clinic where doctors X-rayed and dressed the wound.

Everything turned out okay because, luckily, the round had gone between Ray's toes and hadn't hit any bones.

The Butt Rangers

My parents both smoked, so I couldn't wait to try smoking, too. But kids weren't allowed to buy cigarettes. So how would I get tobacco?

My friends and I raided the ashtrays we found and collected cigarettes butts. We formed a special club: "The Butt Rangers, a Secret Smoking Society." I was the president.

We had a clubhouse in a tree in the woods at the end of Illinois Avenue. We'd take our stolen cigarette butts to this clubhouse, where we created our own homemade corncob pipes. We'd hollow out a corncob and put a long tiger-lily stem in a hole at the bottom. Then we'd strip the tobacco from the butts, put it in our corncob pipes, and smoke it. With all the smoke coming out of our tree house, that tree must've looked like it was on fire.

The club lasted about three months, until one day we got sick from too much smoke. We didn't really inhale, but there was enough secondary smoke in the tree house to make anyone sick.

Games

For entertainment, we played "kick-the-can." It was a little like soccer, but with a can—because no one could afford a soccer ball.

Then we came up with a better game.

On the corner of Ohio and Ramsey was a tall cottonwood tree. We'd climb up in it and watch the cars go by. Cottonwood trees have these big hard balls of seeds that eventually burst and release a cloud of fine, hair-like gossamer sails, like the dandelion. We'd pull the hard balls off and throw them at cars as they passed.

We improved on this game when we devised what we called "car traps." A car trap was a rock with a board across it and a handful of mud, or a "mud ball," on one end. If a car ran over the other end, the board would throw the mud ball at the side of the car.

The ultimate mud-ball game was "Annie over." We'd stand behind the house with a pile of mud balls while the "spotter," one of our club members, would hide in a bush or up in the cottonwood tree. When a car approached, the spotter would holler "Annie over!" and we'd hurl mud balls over the house and try to hit the passing car.

Neighbors, and How to Deal with a Bully

Next door to us on Ramsey lived Roy Cannon, whose father was in the carpet business. One summer his father made a six-foot kite. Using a large spool of carpet twine, Mr. Cannon flew the kite on weekends, but Roy and all of us kids were too small to hold it. When we tried, it lifted us up off of the ground and Mr. Cannon had to hold us down.

Across the street from the Cannons lived Jessie and Harris Huggins. Harris Huggins fell off his top bunk one night and everyone in his family ran to the kitchen to see if the icebox had fallen over. Charlotte and Judy Blaha lived next door to Jessie and Harris, directly across the street from me. Mr. Blaha had a printing press in his garage, and on weekends he printed invitations and circulars. We always liked to watch. Bill Bostic lived on the corner of Ohio and Ramsey, next to Charlotte and Judy. Bill was a good friend of Ralph and Raymond, and they all ran around together. Sandra and Gene Murphree lived on the corner of Ohio and Alabama Avenue. Down the street lived Medford and Neal Cox. One of my best friends in school was Jerry Heath, who lived on Michigan Avenue. Every now and then, I'd walk or ride my bicycle over to his house for a visit.

There was a house next to the Cannons on the other side from us, but I never knew who lived there. The next house was home to Eugene Watt Jones, Jr.—*aka* Bud, the local bully.

Across the street from where Bud lived was Jimmy Hawkins. One day Jimmy, Jessie, Harris, and I were playing on the Huggins' front porch when Bud came over. Immediately Bud started picking on me because I was the smallest. I didn't like it, so I smacked him, jumped off the porch, and ran for home.

It took him by surprise, and he hesitated a second before starting his pursuit. This gave me a head start. But he wasn't going to let me get away with hitting him, and he came tearing after me. He was about to catch me when I reached my side of the street. At this point, the territorial imperative instinct took over: I was more confident on my own turf. Besides, I knew he'd beat the tar out of me if he caught me.

When I reached the curb on my side of the street, I planted my right foot on the curb, turned, and let him have it right in the face. Again, it took him by surprise. (He was not a fast learner.) He was running so fast that when his face hit my fist, his feet went out from under him. He hung in the air for a split second, then hit the street headfirst. I turned, slowly walked to my house, went inside, and locked the door. I peered out the window to see if he was going to come and get me. But he was knocked out cold. It was a good two minutes before he came to. It seemed like forever. I thought I might have killed him. Finally, he got up and ran home crying.

The next day when I saw him, he had a black eye, a swollen cheek, and a big knot on the back of his head. He told everyone that I'd hit him with a baseball bat.

He never bothered me again.

Homemade Trolley

One day Bill Bostic, the good friend of Ralph and Raymond's, rigged a long line of cable from a tree behind his garage to another tree across the backyard. He put a much shorter piece of cable, like a handle, on a garbage can and looped it over the much longer cable.

This was to be a trolley car, to ride from the top of the garage across the backyard.

He tested it with some dirt and bricks, and it seemed to work okay. So now it was time for the first human to cross the yard in the new trolley. Bud the Bully demanded to be the first passenger. He was very insistent, so Bill let him have the honor. Bud climbed into the trashcan, and Bill shoved it off the garage roof. The trees bent, the cable sagged, and the downward plunge was momentarily halted about three feet from the ground—until the cable handle broke.

The gondola crashed to the ground, and Bud tumbled out of the trashcan. The crash could be heard all over the neighborhood. Bud's screaming could also be heard all over the neighborhood. He was badly bruised and had a couple of cuts and scratches, but no broken bones. His honor and pride suffered the most damage. His mother, however, was furious and wanted Bill's head on a platter.

Milk-Bottle Stoppers and Ice-Cream Sticks

Playing with milk-bottle stoppers was a major form of entertainment. Milk was distributed in glass containers with round cardboard inserts that sealed the tops of the bottles. After the milk was gone, the stoppers were thrown away and the bottles recycled for more milk. We kids would collect milk-bottle stoppers, and when we'd all get together, someone would throw down a stopper and one of the other guys would throw and try to cover it with one of his. If it covered more than 50 percent, he won and got to keep both stoppers.

Ice-cream sticks were another favorite. They were very utilitarian. All kinds of things could be made with ice-cream sticks, such as log cabins. Admittedly, the logs were flat, but enough of them could still make a log cabin. They were also used for stirring stuff, and they could be laced over one another to make small mats. The ice-cream sticks were very useful in stimulating our young imaginations.

Bow and Arrow

One Christmas, I received a bow-and-arrow set. The arrow was a target-type that could punch holes in paper targets, not the hunter-type with the razor-sharp arrowhead.

One day we were playing cowboys and Indians. Naturally, I was an Indian—I had the bow. One of the cowboys was running around the house shooting and hooting and hollering, so I raised my bow and arrow to a 45-degree angle and let fly. The arrow arched up about ten feet then fell, hitting the cowboy squarely in the back. He screamed and dropped in the dirt with the arrow sticking out of his back. I thought I'd killed him.

As I ran over to him, he stood up. The arrow had gone down his back between his shirt and skin. The only damage was a long scratch down his back. But it had scared the hell out of me. It taught me a lesson—not to rely on the odds of hitting something, because the odds may be one to one.

Another time, I ran around the house and pretended I was an Indian circling a covered wagon. I fired off an arrow and shot a hole in the front-door window, for which I got a good licking.

The Woods

To get to school, I'd head north down Ramsey to Illinois and turn right to face the sun in the morning. Then it was down the hill to Alabama Street, where Illinois ended in a high white barricade of one-by-sixes.

There were two options. You could go right, over the sewer pipe. Or you could go left, over the ridge.

If you went right, you had to balance your way across a sewer pipe that was exposed for about a block as it spanned the creek. The pipe was eight feet above the ground, supported by concrete pillars. Growing up from below were sunflowers that almost reached the pipe. Under the sunflowers was an impassible thicket of vegetation that looked like swamp grass. Who knew what kinds of snakes and varmints lurked down there? I didn't want to find out. I'd hop onto the pipe, step carefully across the creek, and jump off at the railroad tracks on the other side. A short trip north on the tracks led to an opening in the woods and a path you could take to school.

If you went around the white barricade to the left, you could follow a trail along the top of a fifteen-foot ridge. On the left were trees with low limbs, so you had to duck your head every now and then. On your right was the bottomland, with more tall sunflower plants and another thicket of swamp grass. (Some really big trees towered up from the thicket and one of these trees, the one nearest the barricade, is where we built our tree house.) About three-fourths of a block down the ridge, you were parallel to the creek. You could only cross the creek in a couple of places before you were blocked by the railroad trestle. To get under the trestle, you had to get to the other side of the creek. This was because the creek turned north again and ran parallel to the railroad tracks, and there was no place to get across. Once you were on the opposite side of the creek and under the trestle, you were home free. After a quick climb up the trail through the woods, you were out on Michigan just a block from school.

I preferred to go left and take the trail along the ridge. There were usually birds—sparrows, robins, an occasional cardinal—in the trees. It gave me a peaceful feeling to hear them sing.

Going to the right was scary. I was crossing the pipe one day with my brothers when I fell into the thicket below. I just knew it was full of snakes, and I was frozen with fear. I screamed my head off until one of my brothers worked his way through the thicket and rescued me. Another time, we were playing under the north end of the trestle on the support beams, jumping from one beam to another. I jumped and missed, and my head hit the beam, splitting my forehead open just above my right eye. (The cut was in the same place where I'd hurt my head at Sleta's house when I slipped on the waxed floor and fell against the steel bed in her front bedroom.) I was rushed to the local clinic and patched up with several stitches.

Crawdad fishing at that creek was a favorite summer pastime. It was cool down there in the shade. I'd filch a slice of bacon from our fridge, grab a length of string, and head for the creek. You tied the string around a bit of bacon and tossed it in the creek. Then you waited. Soon a crawdad would creep over, grab onto the bacon, and start to eat. If you hoisted the string slowly, the idiot would hang on, even when you'd pulled him completely up and out of the water.

Trinity Heights School

At Trinity Heights School, there were several temporary or "temp" buildings on the north end of the schoolyard, with a covered walkway between them and the main school, to accommodate the expanding baby boomer population. There was a large covered area in back, behind the cafeteria, which was under the auditorium. This was used when it rained. Behind that was the swimming pool, used only during the summer. To the north was the soccer field, down a level from the temp buildings. To the south was the baseball field that I crossed every morning coming to school and every afternoon going home.

I didn't want to go to school. When my mother took me the first day, I cried for an hour. That had no effect, so I finally stopped and sat with locked jaws.

The smell of new blue jeans, new shirts, three-ring notebooks, pencils, paper, and erasers was always the same the first few years. Everything was new for the first day of school. It was an annual fall ritual. Every year before school started, you were marched down to the store and bought new blue jeans, shirts, shoes, underwear, socks, and a coat, if it was needed.

After a couple of months, my dad asked how I liked school. I said, "I can't read and can't write and they won't let me talk." I was not happy. The primary function of school seemed to be the crushing of enthusiasm, curiosity, and creativity by enforcing conformity and discipline.

Dyslexia

I had dyslexia. But I didn't know it. And no one else noticed it, either. If they did, they didn't do anything about it or try to help me.

It was years later, when I was in college, that I figured it out.

I didn't see words like everyone else did. I transposed letters. Words didn't make sense. The words looked completely different to me than to the teacher. She had to tell me what the word was supposed to be. When I came to a new word, it would look all jumbled up. Sometimes I'd lick my finger and try to rub it out.

I was a very slow reader. I had to read every word several times before it made sense. When the words finally did make sense, they were boring. The words would be repeated several times: "'Oh, oh, oh,' said Dick and Jane. 'See Spot run. Run, run, run.'"

Worse than reading was standing in front of the class to recite that sissy stuff—poetry. The embarrassment of reciting poetry and being bored to death by "Run, Spot, run" really hampered my learning of proper English. I remember a poem I memorized one time to fulfill a school assignment. It went like this:

> Fleas
>
> Adam
> Had 'em.

The teacher said it wasn't good enough. I had to do it again. So I made a slight alteration:

> Fleas
>
> Adam
> Had 'em.
> And still
> Has 'em.

But the teacher was not impressed. I had to try yet again. This time I chose a poem by Carl Sandburg:

> Fog

The fog comes
on little cat feet.

It sits looking
over harbor and city
on silent haunches
and then moves on.

The teacher approved.

One time the school had a contest in the auditorium. The new students were marched around the stage, one at a time. They called a couple of us back up to walk around the stage with girls. Finally they crowned me "Posture King." Stella May Gay was my queen. Stella was a pretty girl with curly blond hair like Shirley Temple. We were given Burger King-like cardboard crowns and asked to sit on the stage in high-backed chairs. That was okay, but then we, as king and queen, were asked to dance. I refused. I didn't know how to dance. So they danced without me: round dances and square dances and I don't know what all.

When I went home I showed Mom and Dad my crown, but they weren't impressed. No real moral support.

School Savings Plan

One of the few good things I can say about grade school was they began a savings program for me. First National Bank of Dallas wanted more capital to invest, so the bank started the "School Savings Plan." Every Tuesday was Savings Day. We'd all bring our coins and savings passbooks to school, and someone from the bank would be there to take our money and make an entry in each passbook. It started a habit I could use for the rest of my life.

Halloween Pranks

A man named Jeff Sweeney, custodian at Trinity Heights School, got caught in one of our pranks.

The school was having its annual Halloween party. My friends and I went to the party, and when it was over we decided to do a little trick-or-treating on our own. We found a paper bag and put some dog doo and toilet paper in it. We went across the street from the school, put it on someone's front porch, set it on fire, and rang the doorbell. Then we ran like crazy.

The idea was that when the people answered the doorbell, they'd see the fire on their porch. When they'd try to stomp it out, they'd get poop all over the porch and on their shoes.

Anyway, as we raced past the school this old guy, Jeff Sweeney, was standing out front. When we dashed by, he took off running with us.

We didn't hang around to see what happened to him.

Forty years later, I was having coffee with my good friend, Jack Krueger, and his wife, Jackie. We were talking about the good old times, and Jackie told me how her father, Jeff Sweeney, a custodian at Trinity Heights School in Oak Cliff in the forties, had gotten caught up in a Halloween prank.

We had a good laugh about that.

Art Class with Mrs. Eubanks

Art, with Mrs. Eubanks, was my most enjoyable class. I liked the creative freedom she gave us. She'd assign a general topic and let us draw anything we liked.

I remember Fire Prevention Week. We were supposed to draw something having to do with fire prevention. I drew a grass field with a power line through it. There was a hunter in the field, and he was trying to shoot a bird on the power line. Instead he shot the power line in two, and it fell and set the grass on fire.

Another drawing I remember was of a large tree. Mrs. Eubanks told the whole class that I'd drawn the most realistic-looking tree of anyone. It was almost a photograph. Mrs. Eubanks posted it on the classroom wall where it remained all year.

I must've had some talent. But if I did, it was soon crushed out of me.

From "TeleCollision" to Calculus

One thing I liked to do in my early years was create what I called "TeleCollision." I'd get some boxes that had wrappers around them and remove the wrappers. Then, on the plain brown surfaces, I'd draw dials, switches, and viewing windows like the TVs I'd seen pictures of.

My love for technology was showing, even in first and second grade.

Years later, when I was in college taking first-year calculus, I realized what had been missing in grade school—intelligent explanations of how things worked. One night, while doing calculus homework in the middle of the college semester, I was so happy that I was jumping up and down! This was what I'd always wanted to know—how to figure things out. In calculus, you learn how to come up with the formulas used in math. I kept yelling, "Why didn't they tell me this in second grade?!" It would have made high-school math so much easier.

"Straighten up or Go Home!"

I didn't do well in second grade and had to repeat the second half. That was torture.

We had regular fire drills and later, because it was the Cold War era, we had atomic drills where we ducked under our desks and covered our heads to protect against flying glass.

My friend Lawrence "Fritz" Von Hegel lived at 2030 Marsalis Avenue. Being a German during World War II, he was not very popular. Neither was I. So we teamed up.

Mr. Harold Budd was the school principal. One day Fritz and I were sent to his office for throwing erasers or something, and Mr. Budd shook me until my head nearly fell off. Then he commanded, "Straighten up or go home!"

Fritz and I went home.

That wasn't a very difficult decision. I didn't want to be in school anyway.

After we were thrown out of school by the principal, we decided to run away. We each packed a pillowcase with some clothes and food, and started off south on Highway 77 toward Waco. We got as far as Ann Arbor Street before I ran out of steam from carrying the heavy pillowcase. Plus I was plenty scared. I decided to turn back. Fritz reluctantly went along with me. It was getting dark by the time we got home, and Mother had been looking all over for me. I left my pillowcase in the alley and didn't tell her where I'd been. She fixed me a great dinner of all my favorite foods. I was really happy we'd turned back.

I didn't get the belt that time. But I wasn't a stranger to whippings. My dad had a two-inch-wide leather belt with brads on it. He'd double the belt with the brads on the inside, which made the belt very heavy. Compared to it, the shaving strap some kids got hit with was a mere switch. One time Dad took about ten swings and blistered my bottom good. I remember lying on my stomach afterward, because I couldn't sit down, and thinking, "I'm never going to have kids and put them through this kind of hell."

Saleta

Dad's oldest sister, Saleta (Aunt Sleta) Willingham, was the oldest girl in his family. Born August 3, 1904, Saleta married Martin H. Jenson. They were the richest members of our clan.

Mr. Jenson built the brick house they lived in, and Saleta lived in that house at 1418 Morrell Street for over seventy years, until she died in 1998. It was a nice, big, brick house, and it always smelled clean. A large porch ran across the front of the house, and a great porch swing hung at one end. I spent many hours just relaxing in that swing.

The house had a large living room with a fireplace. The fireplace had a clock on the mantel with Westminster chimes. It was so nice to hear that clock chime. Aunt Sleta had a big formal dining room, with a huge table that could easily seat twelve people for dinner. Behind the table was a huge china closet with enough china for a banquet.

Behind the formal dining room was a large, clean, white kitchen, with a breakfast nook where the morning sun would always come shining in. A hall ran down the middle of the house from front to back. On the other side of the hall were three bedrooms, with a bath between each one. The master bedroom and bath were in the back, next to the kitchen.

From the formal dining room, there was a staircase that went upstairs to two rooms and the attic. Saleta's son, my cousin Jimmy Jensen, had these two rooms—plus the attic!—all to himself. He always had a ton of toys for Christmases and birthdays. His sister and brother, Sylvia and Dan, got lots of toys also, and these were all stored in the attic.

I loved to visit my cousin Jimmy. My favorite part of each visit was going to the attic and playing with the toys.

Property near Red Bird Airport

Melody Ricesinger was a girl my brother Ralph dated for a time in high school. She lived on a ranch at the corner of Highway 77 and Loop 12. It was an enormous ranch with quite a few horses.

My parents bought a couple of acres off of Loop 12, near Old Hickory Trail and Mountain Springs Road, out near Red Bird Airport. They planned to build a three-bedroom, two-bath house on the property, and we'd move out there. But the city had other plans. The city claimed the property by imminent domain and gave my parents only half of what they'd paid for the land. It's now the location of the Air Force base exchange warehouse.

Someone made a bundle on that deal, and it wasn't us.

Uncle Fat

My father's brother was called Uncle Fat, though his real name was Leland Dobson Willingham. We'd go to my grandmother's place when he was there, and he'd be playing dominos with Grandmother and her little old lady friends. I'd watch as he nervously turned the dominoes over between his fingers. Some he would turn the long way, and others he would turn the short way.

Later he confessed that it was not a nervous habit. He was polishing the dominos so he could know which players had the important ones. He was cheating his own mother and her friends at dominos!

Years later I saw a PBS TV show that demonstrated how professional crooks cheat at cards. According to the show, when card crooks are around, as soon as the cards have been dealt more than three times they've been marked.

Uncle Fat wasn't there much. I'm not sure what he did for a living, but based on some of the stories I heard, I think he was a professional gambler. He traveled around to find big-money games—that's why he was gone so much. Later I heard he was an undercover agent for the FBI. They'd send him to get the names of major players in illegal gambling spots.

One summer some relatives came to our house with their kids. While the parents visited, we kids played hide and seek. One of the girls hid her face, and after we were all hidden she started looking. She found me, saying, "I see you behind the monkey suit"—which was Mother's fur coat hanging on the back of the door.

DDT Trucks

One thing the government did for us poor folks (living in "Hungry Heights") was spray DDT to kill mosquitoes. The DDT trucks would come through the neighborhood with these big clouds of white smoke spraying out from both sides of the truck. Not having much sense, we kids would ride our bikes right through the smoke. The driver didn't think it was harmful, but I've always wondered, especially after reading Rachael Carlson's *Silent Spring*, if the DDT had any effect on my brain.

Maybe my brain was dumb to begin with.

Ice Storm

We brought our natural-gas heater, the one with the six hollow and sculptured ceramic bricks, from Second Avenue to our house on Ramsey. I spent a lot of time playing in front of that heater during the winter. Those old houses were not well insulated, and except for within a few feet of the heater, it would get plenty cold in the house. Not cold enough to freeze pipes, but nippy enough that you'd need a coat and hat if you weren't right there by that heater.

In 1949 we had an ice storm. The temperature at ground level was in the twenties, and a warm front from the south was overriding the cold air. The warm air from the Gulf of Mexico had a lot of moisture in it, so it rained. When the rain hit the power lines and tree limbs, it froze. Many of power lines and tree limbs couldn't support so much extra weight, and they snapped. There were a several power outages that year. A tree next to our house bent completely over to the ground. Lots of tree limbs broke off, but I was surprised by how many survived when it thawed out.

Our Trip to Great Falls, Montana

One of my cousins was John Pinkney Yarbrough. He and his wife, Pearl, moved a lot, as he was in the Air Force. John was a master sergeant in the Air Force and in charge of the officer's mess. After he returned from Saudi Arabia, he was stationed at Great Falls, Montana, and that summer we drove up to visit. Later John moved to Perrin Air Force Base in Sherman, Texas, where we visited a number of times and went fishing at the Perrin recreational center on Lake Tahoma.

On the way to Great Falls, near Mousala, Montana, we saw the following Burma Shave sign:

> He saw it.
> And tried to
> Duck it.
> He kicked
> First the gas
> Then the bucket.
>   —Burma Shave.

I saw a PBS special on marketing that said Burma Shave had the best advertising ever devised. The ads were among the least expensive, and the brand became one of the best recognized products in the history of advertising.

We listened to radio station XERF out of Del Rio, Texas, as we drove. The XERF transmitter was located in Mexico, which is why it had Mexican call letters, but the offices and studios were in Del Rio. With the transmitter in Mexico, the station wasn't bound by U.S. restrictions that limited transmitting power to 50,000 watts. XERF had over 500,000 watts and could be heard anywhere in North America. This was especially true at night, when American stations were forced to reduce power because the radio frequency carried so much farther at night.

One thing I remember about XERF is they were selling autographed pictures of Jesus Christ.

Rednecks

Our community, on the hard-luck side of the Trinity River, was a redneck paradise.

At that time, racism was essential to the redneck mindset and its panoply of other boneheaded prejudices. The fear of race-mixing was a constant topic, because Oak Cliff's communal identity hinged on being white, conservative, "saved," and married with children. The absolute belief in white supremacy was reinforced by old-time religion and male chauvinist pig-ism.

My family were "purely Oak Cliff," of course. The number-one rule was "don't mess up." Above all, you had to "cut it": cut the yard, cut your hair, cut the mustard—and "cut the crap, boy" when you spoke out of line. Docility was preferred over intelligence, which guaranteed the dumbing down of the individual to fit a rigid social mold. This bred a grinding boredom and an economic poverty locked into place by spiritual poverty.

I lived on the outskirts of the family—craving acceptance, but shrinking from ties that didn't bind so much as strangle. Trying to reason with them was like slamming into a wall of soft mud. Religion created an intellectual vacuum accompanied by poverty. But their backwardness was something they couldn't help. They were just muddling through as best they could. Over time, I came to regard them as good people, trapped on a treadmill, working too long and hard to see that things weren't getting any better.

Chapter Four:

The Eisenhower Year (1949)

A New House

We moved to a house at 2867 Eisenhower, in the new part of town. It was built in 1948, as was the whole housing addition. Clinton P. Russell Grade School was at 3600 Beckley Avenue, at McVey, and it was built at the same time as the new housing addition. But the builders hadn't made it big enough to accommodate all the baby boomers. The baby boomers of the time were straining more than just the school facilities. They were described as "a pig going through an anaconda," and they stretched the whole American economy out of proportion. So Clinton P. Russell Grade School expanded into a number of temporary buildings. All I can remember about that school is sitting in a class in the morning sun and having a devil of a time keeping my eyes open. I don't remember what that class was about, but the sunshine was wonderful!

To get to our house you had to go down Brownlee Avenue to Eisenhower Drive, the last street in the addition. Our place backed up to Zang Boulevard. The house was a plain rectangle with asbestos siding. It didn't have a garage and was only a little bigger than the Ramsey house. It had a living room and a combination kitchen/dining room. But best of all there were *two* bedrooms with closets, separated by a bathroom. Now we boys finally had our own bedroom—we were no longer stuck out on a cold screened-in porch.

Behind our place was an overpass where Highway 67 merged with Highway 77, which today is Interstate 35E. We didn't live there very long, and the main thing I remember about that house is lying in bed at night and watching car lights flicker like a strobe as vehicles crossed that bridge on Highway 67.

The boy next door was a hemophiliac. He didn't go to school very often. He had a phone set up where he could listen to the teachers in the classes and ask questions. Then he would do his homework, and I'd take it to school and turn it in for him. Mother explained what a hemophiliac was and warned me not to play rough with him.

Ocala, Florida

On July 4, 1950, we went to Ocala, Florida, to see our cousins, Jetty and Hort Yarbrough. Hort was my Uncle Lamb's son by an Indian woman in Oklahoma. Hort and Jetty were second cousins but got married to each other anyway. Hort, the husband, was a drinker. They lived in the country outside Ocala in a log cabin.

In Florida we went to Marineland to see all the fishes. The experts there would take questions from the visitors and answer them over a public-address system. The question most often asked was, "If you stopped feeding the fish, which one would survive the longest?"

The answer was, "The barracuda."

Dad Builds a House in Rylie

After the city claimed the property my parents had purchased off of Loop 12 near Red Bird Airport, Mom and Dad bought some more land—eight acres in Rylie, Texas, on the corner of Alexander and Fish Roads.

Dad negotiated a deal where he would tear down three old houses in South Dallas for the lumber. He was careful to save as much of the wood as possible, then he hauled it all out to the property in Rylie.

We'd go out to the country to help Dad when he was building the house. To get there, we'd go down Lancaster, over the Corinth Street viaduct, and under the railroad bridge to Lamar Street. Across the corner was the Greyhound Bus repair barn where Dad had once worked. We'd turn right and go down Lamar, past Texas Kenworth, where Dad now worked. Then we'd go past the Sears catalog warehouse. Next was Procter & Gamble, which ran a stinky business rendering horse fat into soap. We could tell when you were getting close to Procter & Gamble because it smelled so bad. We'd turn left on Hatcher Street and right on Second Avenue. Then we'd head down Second Avenue and under the railroad bridge, past the Last Chance liquor store, and into the Trinity River bottom.

The Trinity River bottom was the lowest spot in Highway 175. It was about two miles long. There were several narrow bridges there, and they were not very high. Every time the Trinity River overflowed its banks, this part of the highway was under water. Another two lanes were later added, and these were about twenty feet higher. When the river rose and the outbound side was under water, traffic could be routed to the high side. This way, the state could keep this vital link to East Texas open all the time.

Dad, Ralph, and Raymond built our house that summer. I helped by holding the ends of boards while someone else hammered them in place. I also was responsible for cleanup.

Chapter Five:

The Rylie Years (1950 to 1954)

Beginning of Growing Up

Every trip must begin somewhere, and I think my journey into adulthood began on a hot dusty Texas chicken farm. The poverty and deprivation of our early years was so terrible that none of us ever spoke of it, but the bitter experiences were seared into our hearts and left scars that never healed. The effect of these years was to develop in me an insane niggardliness, a hatred of property, and a desire to escape as quickly as possible.

But it wasn't all bad. There were fun times, too.

Eight Acres

The eight acres in Rylie were equally divided: one half was scrub oak and the other was a very poor pasture—so poor it would hardly grow prickly-pear cactus.

Cactus plants hid among the tall weeds, and I usually found them, running barefoot through the pasture in the summer. I'd be picking those hairy cactus spines out of my feet for a week. They were really fine and the same color as skin. You could feel them better than see them. Sometimes I'd take a toenail clipper and just clip the skin from the area where the pain was. The bloody hole would heal, but those damn cactus spines would not. They'd just keep on hurting forever, it seemed like.

A shallow gully ran between the two halves of the eight acres. There was a small stand of trees in the pasture, near the corner of Alexandra and Fish Road. A 20' x 20' foot tar-paper shack sat just inside the scrub oaks tree line. In the summer of 1950, right after Ralph and Raymond graduated from Adamson High

School, we moved into that shack. The house Dad was building was not quite finished, but it would be by the time I started at Rylie School in September.

When the house was finally built, we moved in. We had the same old furniture from Second Avenue. My parents furnished the kitchen with a gas range top and refrigerator from Socco Company Unclaimed Freight on Industrial Boulevard. I felt like Second-Hand Rose.

Dad installed a large attic fan in the hall. At night, we could close the living room and kitchen doors, and the fan would draw air through the bedroom windows. This made for a nice breeze at night, good for sleeping.

Out back, they dug a 4' x 8' hole, about ten feet deep, for the two big concrete septic tanks. There was also a trench, a "leach field" it was called, that was lined with bricks and that ran straight back from the septic tanks for about a hundred feet. The overflowing sewer water would drain out into this leach field and fertilize and water the grass in the backyard.

"Witching for Water" and Digging a Well

Dad installed a large, corrugated, galvanized-steel cistern for harvesting rainwater. But during the hot summer we needed more water than the rain-harvest system could provide. So after we got settled in the house, Dad began searching for a place to sink a well. We talked to our neighbor, Mr. Hill, and he and some of his friends came over with peach-tree branches and "witched for water."

That was really interesting. The peach-tree branch was shaped like a Y. A man would hold the top two parts of the Y, one in each hand. Then he'd aim the bottom end of the Y up in the air. He'd walk around in a zigzag pattern, searching for water, until the end of the peach-tree branch turned and pointed down. It was the dandiest thing I'd ever seen. There was no way to stop the thing from turning down. When it did, you knew there must be water down there somewhere.

We got the same result in five places in the pasture, near the clump of trees. I think the trees also knew something about where to find water.

Dad created a special contraption to dig the well. He had a steel tube, six inches in diameter and five feet long, with a slot about halfway down one side. He filed the end on the inside to make a sharp cutting edge. He drilled a couple of holes in the other end, ran bolts through these holes, and attached a short loop of steel cable. At the top of the loop he had a bulldog clamp. To this, he attached a hundred feet of steel cable, and he ran the cable through a pulley supported by a tripod of ten-foot poles. He rigged an old truck tire rim to wind the cable on, and attached it to the outside of the rear driver's side wheel of his old Studebaker one-ton pickup truck. On the outside of this rim, he rigged a cam-and-lever sys-tem.

He'd put the truck in "granny gear," the lowest speed in the transmission. As the back wheels slowly turned, the cam arm would hit a bolt that ran through the opposite side of the rim, and would raise the six-inch-diameter steel tube two or three feet. As the wheel continued to turn, the cam would get past the center of the wheel's axle and fly loose, dropping the heavy steel tube into the well hole. This pile-driving motion would cut a six-inch-diameter hole in the ground. He'd do this ten or twelve times, then he'd stop the forward action.

Next he'd slip the cable off the lever arm, put it on the rim, and throw the truck into reverse. This wound the cable up on the rim and raised the steel tube out of the hole. When the tube was out of the ground, he'd swing it over to the side and let it down. Then he'd stick a crowbar through the slot in the steel tube and knock the dirt out. When all the dirt was out, he'd let it back down into the hole. When it reached the bottom, he'd adjust the cable length for an additional two or three feet, start in forward "granny gear," and repeat the process.

He dug five dry wells; each was ninety feet deep. There didn't seem to be any water, unless it was somewhere farther down in the earth. So he went to the low-est point on the property, near Alexander Road away from Fish Road. There he dug a well six feet in diameter and thirty feet deep, and he found enough seepage to create three or four feet of water. That was enough to supplement the cistern water, because the well would fill up again overnight. Dad installed a good on-demand electrical pump at the well, and dug a trench diagonally across the field to the house. He laid a water pipe in this trench. Now we didn't have to pray for rain on the hot, dry, summer days in Texas. Now we had enough water.

Dad sent water samples to the Department of Health in Austin for analysis. When the results came back, we learned it was okay to drink the water.

Chicken Farm

Dad had read some farm magazines about how easy it was to raise chickens. The magazines said it wouldn't cost a lot to get started, and the feeding process could be automated so it wouldn't be too time-consuming.

Well, that is exactly what Dad wanted to hear: *"It won't cost a lot, and it won't take up much time."*

He resolved to build a chicken house.

Dad cut down some trees just inside the wooded area, left of the tar-paper shack and slightly in the back, and that's where he built his chicken house. It could be reached easily from the driveway by driving in front of the tar-paper shack. It was 50 feet wide and over 150 feet long. There wasn't much to it, just a big corrugated tin roof supported on both sides and at both ends. The roof had about fifteen turban fans to let out the hot air during the summer. The sides were mostly chicken wire, to allow the air to circulate. In the winter, we boarded up the sides with 4' x 8' sheets of plywood.

In the front there was a ten-foot area screened off for chicken feed and the automated feed hopper. This hopper had a trough with a chain in it that circled the entire chicken house, about ten feet in from both sides and from the far end. Overhead were the lights and water pipes. Besides the lights, there were brooders: wide, shallow, upside-down tin cones with a light bulb in each one. When baby chicks arrived, these cones were lowered to within an inch or two of the ground. The chicks would be comfortable underneath, thinking they were under the mother hen's protection.

Also hanging from the rafters were the "automatic" water dispensers—plastic containers with water tubes. When the chicks drank enough water from the container to bring the water level below the tube, fresh water would flow in via gravity, and the container would be full again.

We arranged to get some laying hens and a few pullets. Of course, the old biddies—the hens—all had names and became my pets. This made it hard to choke down Old Gertie at the dinner table when it came her time to go. We engaged the services of a "Dominecker" rooster to take care of the hens. ("Dominecker" is a corruption of "Dominique," the name of a breed of chicken.) Our Dominecker, whom we named Dominick, did a good job with the hens. Dad later built some rabbit hutches and brought home some nice rabbits. These became pets also.

Most mornings I'd feed the chickens right after I got dressed. This meant going out to the chicken house and dumping a couple of fifty-pound bags of chicken feed in the hopper. The feeder was controlled by a timer. When the timer turned on, it would drag a conveyer belt around the trough that circumnavigated the inside of the chicken house.

The feeding system was noisy, and when new chickens arrived, the noise created by the automatic feeder scared the hell out of them. They stampeded to the far end of the coop, trampling each other, and smothering the ones that were on the bottom of the pile. We lost over two hundred chicks in the first panic. We learned to put the new arrivals into groups of not more than a hundred, using two-feet-high chicken-wire fences every five feet or so in the chicken house. After a while, the new chicks learned that the noise of the automatic feeder meant "*food!*" They'd dash to the conveyer like Pavlov's dogs.

The watering was automatic, also, but because it was controlled by gravity and air pressure, it didn't make any noise.

When new chicks arrived in their cardboard boxes, we'd put them under the metal skirt of the brooders that we lowered from the rafters. The chicks would huddle under these in groups of fifty and peek out from under the edge. As the chicks grew, the brooders could be raised an inch or so each week.

In every batch of five thousand chicks, there'd be one or two that were not quite right. When the other chicks got scared, they'd all run to the same place. But those one or two would just run around in circles. We'd pick these out, along with the hurt ones, take them into our house, put them in a box, and hand-raise them. When and if they got better, we'd put them out in the hen house and let the old sitting biddies take care of them.

As the chicks grew, they'd start to get what's called "pinfeathers." When the down fell out and real feathers grew in, the other chicks would see the shiny quills and peck at them. If they picked long enough and hard enough, the pecked chick would start to bleed. Then it was "Katy, bar the door!" For the other chicks, the red blood was like a red flag waved in front of a bull. A whole lot of chickens would start to peck, so when you saw a chick bleeding, you had to remove it and take it into the house until it healed.

My after-school chores included doing dishes, gathering eggs, loading up the chicken-feeder again, and feeding the rabbits. Each time the feeder was loaded, you had to make a pass through the chicken house to look for injured chicks that needed rescuing, or dead ones that needed throwing out. The old sitting hens didn't like you coming in and stealing the precious eggs that they'd worked so hard to lay. They'd peck the hell out of your hand. I guess I don't blame them, but it did hurt. Sometimes my hands would bleed. We put a glass egg in every nest, to make each old girl think there was always at least one egg she could keep.

When the chickens grew to be fryers, it was time to ship them off to market. Around midnight, a big flatbed truck would arrive, carrying wooden pens for transporting the chickens. Each wooden pen held about ten chickens. In the dark of night, while the chickens were asleep, we'd go in and grab them by the feet. Every time you grabbed one in the dark, she'd crap on you. I guess you could say we scared the crap out of them.

Anyway, after being pooped on by fifteen hundred chickens during the night, you can imagine how we smelled. After the truck left, before daybreak, a bath was in order for everyone. But even an hour-long bath with lots of scrubbing would not get the smell completely off of you. Some mornings I had to go to school smelling like chicken crap. Most of the other kids lived on farms. They knew immediately where you'd been all night, and that it wasn't at a social dance.

And the ordeal wasn't over yet. The chicken house still had to be cleaned and prepared for the next batch. After school I'd trudge to the chicken house with shovel and wheelbarrow. I'd shovel the floor hay—which was full of chicken crap, of course—and wheel it out to the pasture. Sometimes it took me four or five days just to haul all that shit out of the chicken house and dump it on the pasture. After the floor was cleaned and new hay was laid, all the equipment—the conveyer, the water bottles, and the brooders—had to be cleaned. Then new two-foot fences had to be installed for the new baby chicks. All this took about two weeks. Then we were ready for the next batch of five thousand chicks.

That farm-magazine article, the one about how easy it was to raise chickens, was right: "It won't take up much time." And it didn't take up much of Dad's time.

Because I did most of the work!

Dad helped out on weekends, but most of the labor was done by me, during the week after school.

Then the rains came. After the rains, the weeds in the field where we dumped the chicken dung really started to grow. Within two weeks, there were weeds three inches in diameter and fifteen feet tall. I had to use an axe to chop them down. After that, our neighbor, Mr. Hill, drove his tractor over, plowed it up, and turned it all under.

We planted a garden in that field. We had the best truck garden I've ever seen. We grew corn, tomatoes, onions, cucumbers, cantaloupes, carrots, watermelons, pumpkins, radishes, lettuce, and I don't remember what all. But we really ate well that fall with fresh fruit and vegetables, plus chickens, eggs, and rabbits. We were in high cotton! Speaking of cotton … well, that's another story. First, let me tell about the jack rabbits.

Chasing Jack Rabbits

When the garden started to grow, the varmints came around—especially jack rabbits. Most afternoons after school, I'd chase the jack rabbits out of the garden. One day I got close enough to hit one with a hoe. I thought I'd killed it, but when I picked it up I found it was only stunned. As it jumped to run away, a claw on its back foot caught my right arm at the wrist and gouged a deep gash all the way to my elbow. I still have the scar as just a little reminder of that wonderful time on the farm.

Needless to say, I enjoyed eating that old jack rabbit.

Mr. Hill and Dominick

Our neighbor, Old Man Hill, was a roofer, and he whistled all the time. He looked like Percy Killbright in *Ma and Pa Kettle on the Farm*. Whenever he came over to our place, he always took something home with him, even if it was only a stick or a board. This was his personal philosophy.

Meanwhile I had a feud going with our old rooster, Dominick. Every time Dominick caught me in the yard—*his* yard—he attacked me. He'd jump and peck at me and try to claw me with his sharp heel spurs. Of course, I couldn't let him get away with that. I'd try to kick the hell out of him, but he'd keep jumping on me until I left.

One day, Mr. Hill was in our backyard and Dominick was there. The rooster hit Mr. Hill in the middle of his back and pecked the hell out of him.

Dominick was a mean old bird, but he did a good job of protecting his flock. You had to admire him for that. So in spite of our feud, I really liked him.

Tony's Horses and Cows and Pigs

Tony Hill was Mr. Hill's son. I'd go over to Tony's place in the summer, and we'd try to ride his horses. They had a great big mare and an old swaybacked gelding. We'd chase the horses all over their thirty-six acres until we finally cornered them. Then we'd jump on, bare back, and hold onto their manes.

The first thing they'd do would be to set off for the barbed-wire fence and try to scrape us off. If that didn't work, they'd head for the trees and go under the lowest limbs to try to knock us off. If *that* didn't work, they'd stop at the stock tank for water. For the rider, when the horses put their heads down to get a drink, it was like being on a big slide. It was all you could do to stay on. Most of the time we wound up in the tank, which wasn't all that bad. Summers in Texas were hot, and chasing the horses was tiring. So the dunk in the tank was refreshing. If you managed to hang on during their drinking, you wound up back in the barn. Which again was not all bad: you'd gotten a free ride and ended up in the shade of the barn.

Bull-dogging the calves was another thing we did to expend our youthful energy. After a cow had a calf, we'd go out and try to separate the baby from its mother. This was quite a job. Momma cow was a rough customer when it came to her calf. But if we managed to get the calf isolated, we'd grab it by the ears and chin and try to bulldog it like we'd seen in the rodeo. This would work the first few times. But the calf grew up pretty fast. After four months, it was more than we could handle, and the calf got the best of the match. It would drag us around the pen until we got tired or fell off.

One of Tony's chores was slopping the hogs. Mr. Hill would bring home pickup truckloads of day-old bread he bought for a penny a loaf. Tony would break open about a dozen loaves and dump them in a fifty-gallon barrel. When I helped, I'd eat a slice out of each loaf I opened. I'd never seen that much bread in my life and wondered why they were wasting it on hogs. Tony would add some corn and some kind of hog meal to the barrel with the bread. He'd add water, stir the slop with a board, and pour it into the hog trough. The hogs would go nuts! They'd climb over each other to get to the slop.

It did look pretty good. And it didn't taste too bad, either.

More Fun

It was a five-mile walk down Peachtree Road to the nearest grocery store. Or you could cut across the Hill's thirty-six-acre pasture and another farm. During the summer, it was a good way to kill the day—go over to the store and get a soda.

I had fun playing next to the cow pond in the summer. There always seemed to be a nice breeze blowing through there. Another fun place to play was down Fish Road, where several big oak trees stood in a circle creating a beautiful canopy over the sandy dirt. One year for Christmas I received a couple of pig-iron cars. When the weather got warm, I played with the cars in a little "town" I built in the dirt and sand under those huge oak trees. There were a lot of birds in the area, and they liked the sandy dirt, too. They took sand baths to wash the mites off their feathers. Then I came along with my cars and played in the sand. I wound up scratching all day and all night. Seems I'd gotten mites from the birds.

4H and FFA

At school I joined the 4H club and later the Future Farmers of America (FFA). I picked up a lot of valuable husbandry tips.

Like how to castrate hogs and bulls.

How to kill rabbits by hitting them in the back of the neck.

Or how to wring a chicken's neck and pull its head off so it flops around on the ground and throws blood everywhere—this drains the blood and keeps the meat from tasting gamy.

Real useful stuff.

The Thing

One Saturday night in 1951 we went to the Kaufman Pike drive-in where I saw a movie called *The Thing*. It scared the hell out of me. I couldn't sleep for several nights. A whippoorwill in one of the trees near the house kept me awake. I'd crawl into my parents' room and get under their bed until Dad would wake up. Then he'd make me go back to my room. It took me a long time to get over that movie.

I saw a few more science fiction films, such as *Destination Moon*, *The Day the Earth Stood Still*, and *When Worlds Collide*. These didn't scare me, but they did stretch my imagination and intensify my interest in science.

Tar-Paper Science Lab

Since I'd left the tree house behind, I took over the tar-paper shack for my indoor activities—a necessity, since most of the time it was too hot or too cold to play outside.

When I received a chemistry set for Christmas, I immediately I set up a lab in the tar-paper shack. I had bottles filled with all kinds of stuff. The chemistry set had a microscope in it, and I'd look through it at everything I found. That was really educational, and it piqued my interest in science.

Around this time I heard about an experiment conducted at the University of Chicago by Stanley L. Miller and Harold C. Urey. (You can find a full description of the Miller/Urey experiment at the Web site for Duke University's chemistry department.) These two scientists tried to determine how life started and what kind of environment was necessary for life to begin. They created a cloud from several gases: methane (CH_4), ammonia (NH_3), hydrogen (H_2), and water (H_2O). They ran a continuous electric current through the cloud to simulate a lightning storm (probably present in the early days of earth), and analyzed the results. They found that 10 percent to 15 percent of the carbon had become organic compounds. In addition, 2 percent of the carbon had become amino acids—essential to cellular life. The experiment proved that organic compounds could have been created in the earth's earliest days, and this important research led to number of other investigations.

I wanted to be part of the world of science! I had many interesting, educational, and fun-filled hours with my chemistry set, but my lab was too small and limited for such earth-shaking scientific investigations as the Miller/Urey experiment.

Still, I could dream.

Grandmother gave me a Silvertone radio that had a short-wave band. One day I heard a ham operator on the air. As I listened, he gave his QTH (that's ham talk for address), and he lived just down Alexander Road from us, right on my way to school. I jumped on my bike and sped down there.

He was surprised when I knocked on his door. He was also glad to see me. We talked for a couple of hours about ham radio. He gave me some books, and I got some more books, and I built a crystal set. Using some instructions I found, I put together a "rusty-razor-blade" radio, and it worked—but not as good as the crystal set with its cat whisker and earphones. Later I bought a telegraph key and got a Model T spark coil that had been used to generate the spark for the car's ignition in the old days. It worked great, breaking up a twelve-volt battery DC voltage into a couple of thousand-volt sparks. With the telegraph key and the Model T spark coil, I started practicing code. I found an old ringer telephone and kept that, too, in my tar-paper science laboratory.

Chemistry set, microscope, radio, telegraph, telephone—this was stuff I could really sink my teeth into. I was learning how things worked.

I was hooked on science.

Everywhere I turned, I started noticing how science was affecting our daily life. The evening news on TV reported on March 8, 1952, that an artificial heart kept a patient alive for eighty minutes.

The news reported on November 6, 1952, that the first-ever hydrogen bomb was exploded at Eniwetok Island in the Pacific. The "H-bomb"—as it became known—was suspended from a tower. The explosion blew both the tower and the island out of existence, and left a crater in the ocean several hundred feet deep.

The H-bomb had an explosive power greater than five million pounds of TNT.

Washeteria

Saturday was laundry day. Mom gathered up the laundry and Dad dropped us off at the washeteria near Buckner Boulevard and the Kaufman Pike drive-in theater.

There were no automatic washing machines, just square galvanized tin tubs.

These tubs were placed four at a time around a swivel post with a hand-crank wringer. The tubs were filled from the black garden hose connected to the hot-water faucet. The first tub was for hot, soapy water. Mother washed the clothes in it, using a washboard. Then she cranked them through the wringer and dropped them in the rinse tub. Then they went through the hand-cranked wringer again and into the third tub, which had bluing in the water. Out of that, through the wringer again, and into the last rinse tub. Then a final trip through the wringer and into the laundry basket. The first load was done. There were usually four loads, each taking about an hour.

All-Weekend Poker Parties

The entertainment my parents enjoyed most was going to Ode and Bill More's place in North Mesquite. They'd spend the entire weekend, from Friday night to Monday morning, drinking, smoking, and playing poker. Ode and Bill had a boy and a girl that I could play with. Most of the time, we were sent outside to play. But near bedtime, we had to come in.

When the parents weren't looking, we'd steal the half-full glasses of booze and drink them. It didn't take much to knock me out. I learned to hate smoking, poker, and liquor.

And with good reason! Scientific investigations were going on in the early 1950s that suggested a link between smoking and cancer. Studies indicated that the risk of cancer was proportionate to the amount of tobacco smoked. Other research implied that big-city pollution was also linked to cancer.

Nevertheless, people were living longer than ever before. When Roosevelt signed the Social Security Act on August 14, 1935, the expected life span for the average American was 55 years. (Social Security entitlement did not start until age 65; I guess the idea was that most Americans would be dead by then, so there'd always be plenty of money in the system.) By 1953, thanks to science, the life expectancy in industrialized Western nations was rising dramatically. In the United States, it was 68.4 years, having increased 13 years since 1935 and 21 years since 1900.

School

The school bus turned around at the corner of Alexander and Fish Roads, so every morning I'd go down to the corner and wait. The kids in the Hill family and the others from down Fish Road all had to catch the school bus there. We usually were the first on the bus, and we'd almost always take the back seats.

The school was a brick building located along Highway 175. There was an auditorium in the middle and hallways on either side, with classrooms on both sides of the halls. There were also a lot of temporary buildings in the back that were about the same length as the main building. On the east side at the south end of another temporary building, next to the Baptist church, was a building called the "doll house." This contained school supplies and candy.

On the other side, where the school buses unloaded and loaded, were the outhouses. There were about ten holes for sitting and a metal trough for standing up. It was very smelly. Every month or so a truck would come by and put a pump down in the hole and pump it out. Boy, did that stink!

Skit

The school had a student's day where students could show their talents. Some sang, some danced, and some did silly things. My class came up with a skit in which I was the patient and they operated on me. There was a sheet between us and the audience, and our shadows were cast onto the sheet from a light behind us. The shadows made it look like they were taking my guts out. It was actually just a rope lying on the table beside me, but in the shadow it looked like they were really pulling out my intestines. It was fun.

Paddle

One teacher had a paddle about three feet long with holes in it.

If you misbehaved, he'd make you bend over his desk in front of the class. Then he'd give you three whacks with the paddle.

I had my share of whacks.

Joe Gordon

I'll never forget Joe Gordon. I envied him to a fault, even to a sin. He was just too smart to be in school. It was a waste of his time. Most of the time, he'd be reading a book he held down next to his desk so the teacher couldn't see it. If the teacher called on him to answer a question and join in the class discussion, he'd ask her to repeat the question. After she repeated the question, he'd give the correct answer and go back to reading his book. He never missed the correct answer.

Pickup Football

One day we were playing pickup football during recess. Being small, I was a safety. The other side had a couple of the biggest guys in school on their team. It was a power play, with the big guys on their line pushing everyone on our line back, while this giant ran through the hole. The giant was George, the biggest guy in school, and I was the last one he faced before he had a clear run for a touchdown.

I tackled him.

His knee banged me in the side of the head and nearly knocked me out. But I locked my arms around his legs. He hit the ground with a thud, and it knocked the breath out of him.

I was the hero that day. Everyone was talking about the great flying tackle I put on George. I never forgot it.

George never forgot it, either.

History Teacher

The fifth-grade history teacher asked three questions:

How can you tell if a boxcar is loaded? (Answer: Boxcars are inspected when they're loaded. After the inspector closes the door, he puts a seal on the door handle with his stamp on the seal to indicate that he was the last person who inventoried the shipment. So if there's a seal with a stamp on the door handle, the boxcar is loaded.)

How many sides does a circle have? (Answer: Two—an inside and an outside.)

What song did the British band play when British General Cornwallis surrendered to American General George Washington at Yorktown? (Answer: "The World Is Turned Upside-Down.")

The first one was easy. The second one was a little tricky. The third one, nobody could get. And the teacher never told us the answer. I didn't learn the name of the song until I got to college.

The TV news reported on March 17, 1953, that at Yucca Flats, Nevada, more than two thousand marines were given the order, seconds after a massive explosion, to charge across the desert to their "objectives"—the destroyed fortifications, trenches, and arsenals of an imaginary enemy.

Danger Signs

My homeroom teacher taught a class on danger signs. She told us to go home and look for danger signs and be prepared to discuss them the next day. I forgot all about the assignment when I went home.

The next day, she asked Joe Gordon what his danger sign was. He went to the blackboard and drew a jagged line down the board. The teacher asked Joe what it was. Joe said it was lightning, and if it struck you it could kill you.

Next she called Sue Jennings up to the front of the room. Sue drew a wavy line across the board. The teacher asked her what danger it was for. Sue said it was a snake, and if it bit you, you could die.

Then she called me up. But I hadn't studied the night before. I thought for a minute, then picked up the chalk and hit the blackboard with it, putting a dot on the board. The teacher wanted to know what kind of danger sign that was. I told the teacher I wasn't sure, but Tony's older sister had told her family at dinner the night before that she missed hers. His mother fainted, his daddy crapped his pants, and the boy down the street shot himself.

SOB

There was this one mean kid in school who liked to hit you in the back of the neck with a judo chop if you weren't watching.

It's a good thing we didn't have a pistol at home, because I'd have killed that SOB and been hauled off to jail.

Bombs

On March 1, 1954, the news reported that a Japanese fishing vessel, the *Fukuryu Maru* ("Lucky Dragon"), had made its way to the Marshall Islands in hopes of a better catch. That morning her twenty-three-man crew was fishing east of Bikini Atoll, twenty miles outside the area that had been declared a "danger zone" by the U.S. A vast explosion occurred, and they and their ship were covered with ash. The explosion was an American hydrogen bomb. The ash was radioactive. The hydrogen bomb tested at Bikini was more than five-hundred times more powerful than that dropped on Hiroshima nine years earlier.

A year later, Albert Einstein and Bertrand Russell were two of nine Nobel Prize winners who signed a statement that "a war with H-bombs might quite possibly put an end to the human race." This document came to be called the "Russell-Einstein Manifesto." The problem created by the existence of the atom and hydrogen bombs was called "the most serious that has ever confronted the human race."

Fishing

My dad was a *great* fisherman. I went fishing and camping with Dad and my cousin, John, to the Brazos River. They bought a thousand yards of 1/8-inch line, five hundred hooks, and some flax line for leaders.

When we got to the river, they made a trot line. There was a special knot Dad tied each hook with, so that if a fish didn't take the bait but just brushed up against the flax line, the flax line would stick to him and the special knot would always turn the barb toward the fish. Then, as the fish swam away, the hook would set in his side. If he flopped, the next hook would get him.

They wove the trotline back and forth from bank to bank across the river several times, like a shoelace. No fish could pass through that gauntlet without getting hooked. Then we all went back to camp and took a three-hour break. They drank a six-pack of beer each until it was time to go back and run the line. We always had fish to eat. Some had three or four hooks in them that Mother cut out before she cooked them.

It sure was good eating.

Perrin Air Force Base Recreation Center

We'd also go fishing with John at Lake Texhoma, at Perrin Air Force Base Recreation Center. During one trip, a couple of young airmen were there, taking advantage of the facilities. They checked out a fourteen-foot aluminum fishing boat, with an eighty-five-horsepower Mercury outboard. The motor was so large that the bow of the boat was out of the water. When we saw them, they were a few feet from the dock trying to get the outboard started. They'd yanked on the starter line a couple of times, but the engine hadn't fired up, so they sprayed some starter fluid in the intake.

This time it started with a roar. Unfortunately it was in gear. When it revved up, the bow came up out of the water, and the boat flipped over on top of them. It immediately sank. The two managed to get out of the boat and swim back to the dock. The center's operator snagged the boat with grappling lines and dragged it ashore. I don't know what they did with that motor.

The two airmen checked out another fourteen-foot boat with a smaller engine, got it started, and went out in the lake. Later that evening, they returned. They were beet red! They'd been out on the lake all day in the hot Texas sun without shirts on. That night the sunburn made them cold, so they put on tee-shirts. Next morning, they had to be rushed to the base hospital. Their sunburns had blistered during the night and the blisters had burst, gluing the tee-shirts to their bodies. The base hospital put them in the burn ward and soaked the tee-shirts off of them.

Years later, when I was in the Air Force, we were told it was a court-martial offense to get sunburned. I guess those two airmen got court-martialled—and probably discharged for stupidity. I'd lived in Texas all my life and knew what the Texas sun could do. I almost never went out in the summer without a long-sleeved shirt and a billed cap.

Possum Kingdom Lake

We also fished at Possum Kingdom Lake. One time when we were there, the ladies in the next camp began to scream bloody murder at 1:00 AM. Seems a large raccoon had invaded their camp. When he walked under the ladies' cots, his back rubbed against them and scared the hell out of them.

On another night, there was this awful commotion at the dock. Dad and John ran down to see if someone had fallen in the lake. When they got there they saw an old black man beating the water and the side of a boat with an oar. They asked him what the problem was, and he said, "I ain't gonna let that alligator gar get in the boat with me."

Hattiesburg, Mississippi

One summer we went to Hattiesburg, Mississippi, to visit Dad's youngest sister Omega and her husband Mac McSwain. Mac was a manager for City Gas, and they'd transferred him several times. On our way down, we stopped in Jackson, Mississippi, and had fried ice cream.

Mac prided himself on being a great bass fisherman, so I begged my parents to let me go fishing with him. To my surprise and delight they did. We drove to a lake where Mac often fished and rented a small boat. We went up the lake to the marshes, and Mac taught me how to cast to the edge of the tall grass growing out of the shallow water. We'd each caught a couple of small bass when I hooked a huge one. Mac coached me on how to play the line to keep it from breaking. It took fifteen minutes to get the bass to the boat. As I pulled it aboard, it gave a big flop and fell back in the water. It was large enough to still have its tail in the water while its head and gills were above the side of the boat.

I can't say exactly how huge it was, but the big one always gets away.

Meanwhile, a thunderstorm was approaching when Mac noticed a school of fish feeding at the surface. Every cast into that feeding frenzy hooked a fish. We had our limit in about fifteen minutes and headed for home. What a great afternoon!

The next day Mom and Dad decided to get in on the action, so we went back to the lake and rented two boats. Mom and I were in one boat, and Mac and Dad were in the other. Now that I'd proven I was a great fisherman, I talked Mom into letting me drive our boat. We were having such a good time and catching so many fish that we didn't see the thunderclouds gathering.

When the wind started to blow from the downdrafts, we headed back toward the boat-rental place. But the waves quickly became too big for our small boats. Mac yelled for me to turn and head toward a nearby island. Enormous waves were about to swamp the boat. Mother was completely white with fear. But I made it to the island and beached the vessel. Mac beached his next to mine, and we tied them both to a tree. It rained like crazy, but we were on land and safe. After an hour, we noticed a large motorboat traveling toward us. It was the marina operator. He was looking for his rented boats and was glad to see us, all wet on the island. We were happy to see him, too. The storm had mostly passed, but it was still raining. The marina operator loaded us all on his boat and tied the two smaller vessels in tow. We all made it back safely and had a lot of good fish to eat the next week.

Stamp Collecting

I started stamp collecting. I sent off for a couple of mail-order catalogs filled with beautiful pictures of their stamps. But I didn't have the money to buy any stamps. Once the catalog companies had my name and address, they began sending specimens for my collection. It was sort of a scam, because they were sending me stamps I hadn't ordered. I couldn't afford to buy them, but it cost money to mail them back. I used my lunch money to return the first few packages. But I was not going to forfeit my lunch just to keep these guys in business. So I stopped returning their damned unrequested stamps.

It wasn't long before they started sending me letters demanding payment. When they turned the matter over to a collection agency, I figured it was time to 'fess up to the problem. I didn't want some goon coming out and breaking my fingers or legs.

Boy, were my parents mad! I got a good licking and a lot of talking to. My parents took care of the situation. I was ordered to get a job and earn some money to reimburse them.

Picking Cotton

Tony Hill was going to pick cotton that summer to earn extra money, so I decided if he could do it, so could I.

Before sunup, Mr. Hill trucked us over to a cotton farm ten miles away. We were each given a huge bag about ten feet long. It had a large strap that went over your shoulder, and a big hole in the top of the bag under your arm. We got down on our hands and knees and dragged that damned bag along behind us.

When you pick cotton, you have to reach into the cotton plant, put your fingers around the cotton boll, and pull it out. The flower that surrounds the cotton has sharp edges and stickers all around it. After the first hour, my fingers were a bloody mess. They hurt more than they ever had from the needle pricks the nurses in the hospital had done to me.

By lunchtime, I was nearly dead.

I hurt all over, especially my hands, knees, and back. I was hot as hell, and tired. But there was no way to leave until Mr. Hill came with his truck to take Tony and me home. I don't think I could have walked home, but I considered it.

After lunch, it was all I could do to get back in that damned row on my damned hands and knees with that damned bag. I was really going slowly by now, falling way behind everyone else. At quitting time, just after sunset, I dragged the damned bag to the damned scales. I was so exhausted that I couldn't even lift it. When they weighed it, I found I'd picked less than half a bag, less than a hundred pounds. I earned fifty cents for the day's work.

I didn't go back the next day.

Lacing Leather

Bill Hunt lived just down Alexander Road near the Old Seagoville Highway. His mother made money lacing together pieces of leather to make wallets, billfolds, and purses. I went down to his house, and his mother taught me to lace. It was a lot better than picking cotton. But I was pretty slow, and it took me twice as long as anyone else to lace a billfold. Bill's mother paid us ten cents for a billfold and twenty-five cents for a purse. I could do a couple of billfolds in a day but only one purse.

Still, it was better than picking cotton.

Boy Scouts

Mr. Hunt was the local Boy Scout leader. He invited me to come to a meeting at his house one night. It was a lot of fun. All the guys seemed to be having a great time, and they invited me to join.

The meetings always began with a Pledge of Allegiance and the reciting of the Boy Scout Oath. Then they had award ceremonies and presented all the Scouts with their new ranks and merit badges. I got the impression that no matter what you did, you were recognized for your effort. They gave awards for community service and just plain citizenship.

After the awards, they broke into individual patrols. They were planning a camp out in a couple of weeks. They'd already had a couple of camp outs that summer, and everyone talked about how much fun they'd had sitting around the campfire at night and telling stories.

By now was I hooked on the Boy Scouts.

I talked my parents into letting me join. But here came that old bugger of a problem: money. Scouting was a very expensive venture. You had to buy a uniform—shirt, belt buckle, red scarf, boots, socks, and hat—and of course, it had to be an official Scout uniform. Then there were the patches, the knife, and the manuals. We were already up to a hundred dollars and hadn't even started on camping equipment: backpack, tent, cooking gear, canteen, axe, flashlight, first-aid kit, sleeping bag, etc.

The list was long and getting longer. If I wanted to be a Boy Scout, I'd have to economize and start out with the minimum. I got the shirt, red scarf, hat, patches, manuals, and most important of all, the official Boy Scout knife. No self-respecting Scout would be caught dead without that knife. All the rest would have to wait until I earned more money. A lot more money.

Mulligan Stew

Neither the Boy Scouts nor the Lions Club had a permanent place to meet, so they gathered in someone's home or at a church. There were no coffee shops or restaurants for them in Rylie, Texas. The two groups mounted a joint effort to get a place to meet. One of the local Lions members donated a couple of acres at the creek bottom on Alexander Road. He couldn't use the land as it was so thickly overgrown with underbrush and grapevines that even the rabbits had trouble running around in there.

The new meetinghouse was to be made of cinder blocks, and so to raise money, the groups "sold" cinder blocks for fifty cents each. They'd also planned a mulligan-stew dinner and sold tickets to that.

Our Boy Scout troop camped out on the property. We had to cut down all the underbrush and clear the lot before we could pitch our tents and build campfires. The campfires smelled of oak and mesquite wood. The food smelled good and tasted terrific—I don't know if it was the wood smoke or my voracious appetite from all the hard work. But in spite of the work, it was great fun.

The Lions members brought in three great big old iron wash pots with feet. They were three feet in diameter and two feet deep. The Scouts spent a whole day cleaning, scrubbing, and polishing those pots. They were shiny and immaculate when we finished.

The day of the mulligan-stew dinner, the cooks took each pot and threw in three whole chickens (cleaned and dressed, of course); several steaks; a couple pounds of hamburger; lots of potatoes, carrots, and tomatoes; and lots of salt, pepper, and garlic. The Scouts had to peel the potatoes, scrub the carrots, and build the fires under the pots. We were also responsible for keeping the fires going. The cooks stirred the pots with boat oars (washed and cleaned, of course). Hundreds of people showed up in the afternoon. All the Lions' families and Boy Scouts' families were there, and they brought *their* relatives and friends, most of whom had bought cinder blocks. The Lions had laid out the building strings where the clubhouse was to be built and held hourly tours.

It was a great success. We made enough money to build our clubhouse. Construction was started shortly after.

Camp Wisdom

One summer our Scout troop went to Camp Wisdom with other Circle Ten Council troops. It was fun seeing and visiting with other troops, most of whom had more money and better equipment than we did. One of the troops had real Indian teepees, headdresses, and costumes. At night, around the campfire, they performed Indian dances. The highlight was the eagle dance and the awarding of the latest Eagle Scout ranks.

Then all us first-timers gathered in a circle around the campfire to listen to "The Old Indian Story." When it was over, we were required to chant the Indian blessing—"O Wa, Tay Goo, Siam"—while repeatedly bowing to the Fire God. We finally figured out what we were saying: "Oh, what a goose I am." I laughed so hard when I figured it out that I fell over on my side and couldn't get up!

After sunset, we'd look up at the stars and learn about them so we'd be able to navigate at night. I learned that the Big and Little Dippers were called Ursa Major and Ursa Minor, and that the two stars of the cup in Ursa Major pointed to the North Star, Polaris, in the handle of Ursa Minor. The stars at night were big and bright, deep in the heart of Texas—just like the song said.

Since then, light pollution has dimmed the view.

Then came one of the rites of passage for a young boy into manhood: the snipe hunt.

For those who don't know the term, a snipe hunt is a type of inititiation or hazing for inexperienced campers. The experienced guys play a joke on them and send them on a wild-goose chase. They might tell the newcomers about a bird called a snipe, and describe a ridiculous way to catch it, like making stupid noises as you carry a gunnysack through the deepest part of the woods. You wind up alone in the woods at night, forced to overcome your fear as you struggle to use your wilderness-survival and night-navigation skills to return to camp.

Somehow, I made it back.

The Owl Tree

There was a huge oak tree in the back corner of our eight acres on the opposite corner from Alexander and Fish Roads. It was about three feet across at the base. Being the tallest tree in our woods, it had been hit by lightning, and the lightning had created a hole in the big trunk. Various creatures took advantage of the hollow place way up in the tree and made their nests in it. Originally it was squirrels. One day, I climbed up to see what lived in the nest now.

Surprise—there were baby screech owls! From then on, it was called "The Owl Tree." Of course, I took the baby owls back home. I know they should have been left alone so they'd grow up and catch mice. But I took them to our Scout meeting the next night. We named our patrol the Owl Patrol.

Merit Badges

I completed all the initial requirements and became a Tenderfoot Scout. That was really a big day in my life. Later I became a Second Class Scout—another big day. Then I started on the requirements for First Class Scout.

There was a whole book about merit badges. You had to complete various requirements to earn them. Again, they all cost a lot of money, some more than others, which limited my selection. The Camping merit badges were easy and inexpensive as I'd already done some of the requirements: building a campfire, pitching a tent, cooking a meal, and so on. But there were other requirements I hadn't fulfilled. You had to build a lean-to for protection from the weather. You had to clear the area where the campfire was to be built, then arrange rocks and stones so the wind would aid the flames and not put them out. You had to build the campfire using only what you could find and what you had on you—your uniform and Scout knife, no matches.

To build the fire, first you found suitable branches, especially one to make a bow with. You took the shoelace out of your boot to make the bow. You needed a straight smooth stick for spinning, and a larger hard stick to create friction against. In this hard stick, you dug out a small hole for the spinning stick. Using the bow, you spun the spinning stick to cause friction against the hard stick. When the hard stick began to smoke, you added dry grass until it caught fire. You fed in progressively larger twigs until you had a roaring fire.

For the Cooking merit badge, you had to prepare breakfast, lunch, and dinner for the scoutmaster and a panel of judges.

For other merit badges we learned wilderness-survival skills, such as how to track, how to set snares and traps, and how to skin and eat small animals.

Scouting was great fun! One thing it did was build self-confidence and give you the feeling that you could survive anywhere, if necessary.

Another merit badge involved learning International Morse Code. I'd already started to learn it, with assistance from my ham-radio friend who lived down Alexander Road. With his help and his code-practice set, I demonstrated my proficiency, earned that merit badge, and qualified for First Class Scout. That was another really big day in my life. You could tell how proud I was by how far my chest stuck out when they pinned that badge on my shirt pocket.

Shortly after becoming a First Class Scout, I was promoted to Owl Patrol Leader. A year later, the scoutmaster and scout council promoted me to Junior Assistant Scout Master.

State Fair Chickens

One time I went to Fair Park, to the State Fair of Texas, with Oscar Hoyt. While we were there I got a chameleon and a couple of colored chickens.

As if I needed another chicken.

Church and Vacation Bible School

Some people came around advertising vacation Bible school. So I joined the First Baptist Church, next to the school.

Vacation Bible school was okay, but the church was just like those back in Oak Cliff: hell fire and damnation and we all were all going to burn for eternity.

After Vacation Bible school was over, I didn't go back to that church.

Dad's Casket-Company Story

Willie went to work for Texas Kenworth, catty-corner across the street from the Sears Catalog Warehouse store on Lamar. Directly across the street, next to Sears, was a casket company. Dad would tell the following story he heard from the workers at the casket company:

There was this black guy that worked night shift, and every evening when the lunch break came, he'd sit on one of the caskets and eat his lunch. One night one of the workers knocked off a few minutes early and got into the casket. When lunch break came, the other workers sat on all the other caskets so that the one with the guy in it was the only one available for the black man. He came and sat on that casket and spread out his lunch. Before he could take a bite, the guy in the casket said, "Get off of me." The black guy hit the floor running, and the workers said they never saw him again. He didn't even come back for his pay.

Back to Town

Dad worked for the post office during the Christmas holidays. At the end of the summer Raymond joined the marines, and Ralph went to work at National Battery in Carrolton, Texas. Ralph and Dad came to an understanding that Ralph needed to get a place to live that was closer to his job. Mom and Dad couldn't drive him every day. Ralph moved to an apartment at the corner of Peak and Brian where he could catch a bus to work.

That left me with a room to myself.

I think those advertisements for chicken farming never mentioned how much—or how little—money you could make. I don't think we made any money at all raising chickens. Anyway, between that and my dad's drinking problem, my parents had to sell the eight-acre farm.

We moved back into town.

Chapter Six:

The High School Years (1954 to 1958)

Chinchillas, or The Million-Dollar Plan

My parents bought a two-bedroom, one-bath house at 3222 South Ewing Avenue, and we moved back to town. Dad left home shortly after we settled in. I think this was inevitable and was part of the reason my parents sold the farm in Rylie.

The previous owners of our new house had raised chinchillas in what they'd called their "million-dollar plan" to get rich. It must not have worked, since they'd had to sell their house in Hungry Heights. There were two buildings in the back. The small one, a little 10' x 10' cinder-block house, was the first one the previous owners built for their "product." Soon their furry little rats multiplied and exceeded the building's capacity. So on the other side of the yard they built a 25' x 20' building. It was very well insulated—it had to be, to keep the little critters cool so they wouldn't lose their silky coats. It was also air-conditioned against the hot Texas sun.

We made good use of the two buildings.

Frankenstein's Laboratory

I took over the small cinder-block building as my "ham shack." I filled it with electronic parts I took out of old radios I couldn't fix. I tried to build new ones from the parts with no success. I brought my Silvertone radio and crystal set from the farm. Then I bought a Heath radio kit and built a one-tube radio that worked even better than the crystal set. It had a tube that converted the radio frequency (RF) to audio frequency (AF) and amplified the signal, all in one tube.

I learned Ohm's law and the color code for resistors and capacitors. I acquired a cardboard calculator from the surplus electronic supply house; this told me the value of the resistor or capacitor when I turned the dials to put the colors in. I found an electronics book that had oscillator circuits in it, and I built a Hartley oscillator, a Colpitts oscillator, and a tuned-grid, tuned-plate oscillator. Next, I built a two-tube radio. One tube converted the RF to AF, and the other tube amplified the AF signal. After that, I built a three-tube radio. This was good stuff!

I had to save my lunch money to buy a soldering iron. It was only four or five dollars, but that was two weeks' lunch money. One night I was building one of my projects, and my elbow hit the soldering iron and knocked it off the workbench. I grabbed it before I had a chance to think about the danger, but it reminded me very quickly when it put second-degree burns on my palm. I rushed in the house and put burn medication on it, but it took a month for my hand to heal.

I still had the old Model T spark coil. I channeled its output into an old burned-out thousand-watt flood lamp I got from the ballpark maintenance people. The high-frequency sparks jumping around in the argon gas made beautiful purple arcs. I also built a spark board, which was made with gold paint. The arcs would jump from gold flake to gold flake and dance all around the board.

I found a book that had Jacob's ladders in it, the kind you see in the Frankenstein movies. I was able to create my own Jacob's ladder using the spark coil. It was great! The arc started at the bottom of the vee, where the rods were close together, and then it climbed up to the top as the air around the spark got hot. Late at night, with the beautiful purple flashes, the dancing spark board, and the Jacob's ladder, my ham shack really did look like Dr. Frankenstein's laboratory.

Of course, all this arcing and sparking caused some radio and TV interference in the neighborhood. I didn't score any points with the neighbors.

I talked my mother into letting me buy a ham radio receiver for $119. It was a Hallicrafters Model S-40 B. I ran a folded dipole antenna made of 300-Ohm twin lead TV lead-in wire from the top of the house, past the ham shack, to the telephone pole in the alley. It was just the right length. The exact center, where I needed to connect my radio, was right above my shack. I hooked up the antenna and spent three wonderful years listening to the world.

I spent 90 percent of my free time, day and night, in the shack. As soon as I came home from school, I'd throw my books down and head to the shack. Mother would have to drag me into the house for dinner or bring it out to me. She could almost never get me to do my chores. I'd stay in the shack, building electrical circuits or listening to Radio Australia.

Radio Australia had a kookaburra bird (*Dacelo novaeguineae*, the scientists call it) as one of its advertisers' symbols. The kookaburra's howling forced "laugh" is the most distinctive and famous bird call in Australia—perhaps in the entire world. It sounds like a screaming woman who then breaks into a hysterical laugh.

Every night at midnight I'd play the laugh at full volume—much to the neighbors' chagrin.

Sometimes I'd stay up all night, working on my projects and listening to the world. The long-distance signals came in best at night because the F layer in the stratosphere was not burned away by the sun's radiation. Once the sun went down, the radio signals readily traveled around the globe.

I was in *hog heaven*.

When I received my novice-class amateur radio license, I was assigned the identification code KN5ASE—Kilo November Five Alpha Sierra Echo. The use of radio frequencies is regulated by law, and I was allowed to use my voice only on the two-meter ham radio band, or 144 megahertz (MHz) frequency. I didn't have the money to buy or the expertise to build the high-frequency equipment required for voice communication. On the eighty-, forty-, twenty-, or ten-meter bands, I was limited to International Morse Code.

My ham friend, however, had a kilowatt station he built on the two-meter band. Normally when the skip is in, all you need is ten watts. One night I was at his ham shack in his garage, and he was talking to another ham in Fort Worth. This meant his multi-element beam antenna was pointed west. Suddenly another ham from California wanted to break into the conversation. It turned out the California ham was using a Gonset Communicator two-meter transceiver, with only ten watts of power. He had a very strong signal of five by nine, which is as good as it gets in signal strength (five) and in clearest voice clarity (nine). Naturally, my friend got a similar signal report with his one-kilowatt signal. He could have done a moon bounce, which a lot of hams participated in. My friend was in the National Guard and used his system to talk to the base. His call was Gold Eagle 255, and the base was Eagle Nest 54.

Witchcraft

I talked my mother into letting me buy a transmitter kit to build. I bought a Globe Scout fifty-watt transmitter kit. Once I was on the air with it, I'd tune the transmitter output to the antenna with a burned-out six-foot fluorescent light. When the transmitter was on, the radiated energy lit the fluorescent powder in the burned-out light. The glow was brightest at the maximum energy-level radiated from the antenna, and decreased as the energy decreased.

When a quarter-wave antenna is properly tuned, the maximum energy should be at the ends of the antenna, and the minimum energy should be zero at the middle where the transmitter and receiver are connected. At night, I'd walk up and down the length of my antenna from the telephone pole at the alley to the top of our house, and I'd check to see where the peaks and valleys of the radiated energy were. This way, I could tell if my transmitter was tuned properly.

With the kookaburra laughing full-blast at midnight and the colored lights flashing from my Frankenstein-esque lab, the neighbors already thought I was nuts. Now they'd see me out there at night, walking around the backyard with this fluorescent light in my hand. I'd hold it high in the air and it would glow, first brightly, then dimly. But it wasn't connected to anything.

They warned their kids not to play with me because I must be using witchcraft.

They just didn't understand.

The next year, I passed the requirements for a technician license and received the identification code K5ASE. Meanwhile, I could pass the written part of the general radio license test, but I couldn't handle the thirteen-words-per-minute code speed. I think my dyslexia was kicking in. There was just something awkward about the spacing. The dots and dashes weren't close enough to organize into characters before the next character's dots and dashes started to come in. The dots and dashes were too far apart, and the characters were too close together.

Later, when I tried again and got my amateur extra license, I went straight to twenty words per minute and made a hundred on the code test. Twenty words per minute was a lot easier to me than thirteen.

The Power of Life and Death

The Krogers lived in the corner house next to Graceland Street. Mr. Kroger was a very large man, but not fat. He had dark piercing eyes and dark eyebrows, and I thought he really looked mean. But his wife was beautiful. She looked like the TV star, Stephanie Kramer. She was friendly, and I always enjoyed talking to her. They had two kids. Glonda was the oldest and nicest. Jimmy was mean.

Jimmy was small and always wanted to fight to prove how big he was not. I was two years older and about a hundred pounds heavier. I could've easily whipped him. One day we were playing, and he wanted to fight. I tried to go do something else, but he kept following me around, hitting and shoving. We got into a wrestling match.

I'd taken him down three or four times easily, but he hadn't admitted I'd beaten him and could anytime I wanted to. He got mad and kept on hitting and struggling as if he thought he could win. It never occurred to him he couldn't. I took him down for the fifth time and had his head in an arm lock. I was sitting on his body and could've easily twisted his head until his neck broke, or put my arm around his neck and choked him to death.

I remember that moment vividly. It was as if God was talking to me. I thought, "I don't want to kill him, although I could easily do so. For that matter, I don't want to kill anyone." I turned him loose, jumped up, and ran into the house.

I don't think I ever played with him again.

It scared me, the thought that I had the power of life and death over someone, right there in my hands. It was an awesome responsibility, and I didn't know if I was up to it or not. What if I had to go to war sometime and be expected to kill people?

My First Date

One of the girls next door looked just like Jane Wyman, but my first date was with Jimmy's sister Glonda.

Glonda wanted to go to a party but didn't have a date. She asked me if I'd take her. I didn't have a car, or even a driver's license. No one at our house had a car except Dad, and by now he was gone. Glonda talked her grandmother into letting me drive her 1947 Chevrolet. All her grandmother wanted to know was, "Can he drive a stick shift?" Of course, I could. The stick-shift lever was on the steering-wheel column, one of those new inventions since the war.

We went over to Grandma's house about 6:00 PM, and I had to back out of a real small one-car garage. There were only three feet of space on each side. Two feet at the door. But for me, it was a piece of cake. I managed it, and so we were off to the party.

I don't remember much about the party—it was just a bunch of her friends. Afterwards, like normal teenagers, we went to Kiest Park for a little necking. We were making out pretty good when the whole world lit up. The park police had their huge spotlight on us. An officer was at the window with one of those five-battery flashlights shining in my face.

He wanted to know what we were doing in the park at that hour.

"Nothing, officer."

"Don't you think it's time you went home?"

"Yes sir, officer."

"Then be on your way."

I said, "Yes, sir," started up the car, and drove off. I sure was glad he didn't ask for my driver's license.

When we got to Grandma's house, I parked in the small garage without a scratch on the car.

Bryant's Groceries

Bill Hunt was my friend who had lived down Alexander Road from us in Rylie, the guy whose mother made money lacing the leather purses. Bill's grandmother had a house at 1311 Waverly Drive. When Bill came to town for the summer, he called me to come over and spend Saturday night. He had a job at Bryant's Groceries, on the corner of Rosemont Avenue and Twelfth Street, delivering groceries. They needed another delivery boy, so I talked it over with my mother. She was really for it. Bill's grandmother said I could stay at her house with Bill, as she had plenty of room in her big old house—I guess she was kind of lonely living there all by herself since her husband died.

This was great! I'd now have a real income with which to buy all the things I needed for my ham shack. The store was open from 8:00 AM until 9:00 PM. We'd work and get paid for twelve hours a day. We had to be at work at 7:30 AM to open up. That meant uncovering the vegetables, freshening them up, uncovering the meats for the butcher, and putting out all the outside stands that were in the aisles. At night, after 9:00 PM, we had to clean up: sweep the floors, cover the vegetables and meats, and bring in the outside stands. Outside was a storage area for boxes and one for pop bottles, and both had to be kept clean and tidy—this was in the days when you'd return your pop bottles for a refund on your deposit.

We'd get up at 6:00 AM and shower and get ready. Bill's grandmother would fix breakfast, and we'd be out the back door by 7:30. The grocery store was on the next street, so we'd just go down the alley and around the corner.

Bryant's was a local neighborhood store that catered to this upper-middle-class neighborhood. There were lots of big old houses there, with ten or even twenty rooms in them. Many of the homes were owned by little old ladies who'd come into Bryant's to shop. They did this because they knew the store would deliver the groceries for them and collect payment (plus a small tip). To transport their orders, we each had a bicycle with a small front wheel and a big wire basket to carry the groceries. Sometimes we even put the items away for them. Some customers would call in their orders to Mrs. Bryant. We'd fill the order—with Mrs. Bryant's supervision because she knew most of her customers by name, address, and telephone number, and knew what they liked and which brands they preferred. She was long-time friends with most of the customers, who'd been trading there for years.

The pay was fifty cents an hour for twelve hours—thirty-six dollars a week, plus tips. Sometimes I'd bring home as much as fifty bucks, which seemed like a fortune in those days. The average tip was twenty-five cents, but sometimes the customer would give you fifty cents. I even received a dollar a couple of times.

South Oak Cliff High School

Our house on Ewing was one block from Boude Story Junior High School. Both of my brothers had attended that school, and I was so much looking forward to going there and experiencing some of the fun things they'd talked about.

But no! The powers that be had other plans. The school boundaries were set at Kiest Boulevard, at the end of the block where our house was.

I'd have to go to the brand-new South Oak Cliff (SOC) High School, which was not even built when my brothers went to Boude Story and then graduated from Adamson High School. The older schools were full, but SOC needed students.

South Oak Cliff High School Faculty

Mr. Ben A. Matthews was the principal of South Oak Cliff High School. He was kind of a legend in Dallas—at least, he thought he was. He announced all the Cotton Bowl football games and was known as "the Voice of the Cotton Bowl." Every morning at school, he'd come on the public address system with "Good morning, SOC-ites!" Then he'd give the day's date, welcome us all, and introduce the student who was to give the morning prayer. After the prayer, he'd proceed with the day's announcements.

Mrs. Elizabeth Lockhart was our dean. She was what I'd call a handsome woman, with severely short and mannish silver-gray hair. She was tall and projected authority. But at the same time, as far as I know, she was fair and kind. I think her strong image helped her—we knew she'd wouldn't be a sucker for any of our lame excuses.

Mrs. Rae Files Still was my homeroom teacher.

New Friends

I started South Oak Cliff High School in the ninth grade. Most of the kids I'd gone to grade school with at Trinity Heights were there, but they were a grade higher than I was. In those days, some Texas school districts used a semester structure, where you could fail a grade for a whole year or just a half year. When I'd failed second grade, it was for only half a year. But when I transferred to Rylie, they didn't have the half-year system, so I was put back another half year. This meant I was now a whole year behind all my old friends.

I met new ones though.

Bill Allen

I met Bill Allen in Mrs. Rae Files Still's homeroom. He sat in the back of the room just behind me. Bill and I became friends pretty fast as we were both cut-ups.

Bill was a romantic dreamer, and he was smart, too. Bill was a reader of books. He'd read most of the classics by Shakespeare and the other great writers. He'd read Thomas Wolfe's *Look Homeward, Angel* and *You Can Never Go Home Again.* He'd read William Faulkner's *The Sound and The Fury* and *As I Lay Dying.* Also John Steinbeck's *Cannery Row, Grapes of Wrath, East of Eden, Of Mice and Men, The Red Pony,* and others. He'd read *Lord of the Flies* by William Golding**,** and *Animal Farm* and *1984* by George Orwell. He'd read Aldous Huxley, Fyodor Dostoyevsky, and Albert Camus.

Bill must've read a book every night. He planned to be a writer, so he read all the good books he could find. He'd tell me about them. Every day he'd have another story to tell me. His stories got me interested in literature, but I was a slow reader. I couldn't keep up with my homework, so forget about reading anything else. When I did have time to read, the books I liked best, of course, were science fiction, and Bill had read those, too: Ray Bradbury's *Dandelion Wine, October Country, Martian Chronicles,* and others; Isaac Asimov's *I, Robot* and about twenty others; Arthur C. Clark's *Childhood's End.*

During our get-to-know-each-other period, one day Bill was showing me how strong his left hand was. He said for me to put my hand on his desk and he'd crush it. Well, I wanted to show him how tough I was. I let him hit my hand, but first I cupped so it absorbed the blow. This didn't sit well with Bill. He dared me to put it back and give him another try. I put my hand back on the desk, and he half stood up and was going to really crush my hand. He raised his left arm high in the air and came down as hard as he could. But at the last second, I thought, "This guy might really hurt my hand, so why let him?" At the last possible moment I yanked my hand out of harm's way. Bill's fist slammed into the desk with a crash. He almost fell onto the floor with pain.

We learned a lot about each other those first few weeks.

Bill had a customized 1953 Ford. It was leaded-in on the hood and trunk, and he'd had Oldsmobile bullet taillights installed. It looked a lot like James Dean's Mercury in *Rebel Without a Cause.*

Bill was interested in astronomy and was building a four-inch reflector telescope from a kit. He was grinding the primary mirror, which took many hours, even weeks. When it was finished, he had it silverized. He assembled it with a linoleum tube and an eyepiece, and mounted it in his backyard on a fence post. One night he showed me the planets Venus, Mars, Jupiter, and Saturn. We had a lot of fun with his homemade telescope, and it worked very well.

He let me read his copies of *Sky and Telescope,* and we discussed the constellations. I'd always wondered what the universe was made of and what held it together. We discussed the Big Bang theory, the oscillating theory, the steady-state theory—but we didn't come to any definitive conclusions. We also discussed Einstein's general and special theories of relativity, and a lot of Einstein's thought experiments.

Exciting things were happening in science. On May 15, 1955, the last of fourteen nuclear explosions in Nevada took place at Yucca Flats. On July 17, 1955, the first atomically generated power station began operation in the U.S. On November 27, 1955, the Soviet government announced that it had successfully carried out a series of atom and hydrogen bomb tests.

Meanwhile, we saw all of the science fiction movies: *It Came from Outer Space, The War of the Worlds, Rocket to the Moon, The Creature from the Black Lagoon, This Island Earth, Forbidden Planet, Rocket Ship XM, The Time Machine, Twenty Million Miles to Earth, Invasion of the Body Snatchers, Them, The Blob, and 20,000 Leagues Under the Sea.*

Eddie Greding

I met Eddie Greding in homeroom, too. Bill and I hid out in the back row, but Eddie sat up front. He was always in the front of the class. Eddie was the studious one, the serious one. He was hungry for knowledge. He always did his homework and then some. He'd even do extra work for extra credit.

He once said, "I want to know all things."

His primary interest was biology, but he also liked physics and chemistry. He was usually too busy studying to play around with Bill and me, but I guess that's one thing we liked about him. He'd fixed up this room above his family's garage as a laboratory. He had all kinds of specimens in jars (some in formaldehyde and others just sitting there): snakes, lots of frogs, several types of spiders, rats, squirrels, rabbits, opossums, coons, dogs, cats, and other things he'd picked up dead off the road.

After seeing my Frankenstein lab, Eddie borrowed my Model A spark coil. He needed it to create a spark to simulate lighting in an experiment he was doing. He had a five-gallon glass water bottle. He put some water in it and pumped the air out by filling it with methane, hydrogen, and ammonia gas. Then he put the spark gap in the top and let it run for a week. The jug turned reddish-brown. He was excited. He had succeeded in creating a primordial soup. It contained several of the organic chemicals known as amino acids, the building blocks of proteins, and basic ingredients in all terrestrial life. However, after another week when nothing crawled out of his soup, he abandoned the project.

Bill wanted to put a couple of night-crawler worms in it to make Eddie think he had succeeded in creating life. I discouraged the idea, because I knew Eddie would immediately write a paper on the experiment and try to get it published. Then he would become the laughing stock of the entire scientific community.

Eddie had an X-ray tube. He did experiments with radiation on black widow spiders. As often as he did those experiments, I think mankind is lucky he never produced a mutation of some sort. After all, according to Carl Sagan, one lucky alpha particle could have hit on any one of the billions of protons, neutrons, or electrons in the DNA of one of those spiders. This was better odds than the Texas lottery. History might have been changed forever. Ignore the fact that scientists were spending billions of dollars to create huge accelerators to do this.

Eddie didn't want to create an atomic explosion; he just wanted to modify the spiders' DNA a little. Think of it—he might actually have created a new genus or species of spider. He might have produced a ten-foot arachnid the size of the ants in the movie *Them*. One lucky transmutation, that's all it would've taken, to secure the place of Eddie Greding in science books and history books forever.

I think the spiders all died from overexposure to radiation. It was the expected result, but at least he'd tried. I liked that.

Eddie also had a microscope. I could've spent hours in his lab looking through the microscope at his specimens. Like Bill, Eddie had a telescope, and the two of them taught me a lot about astronomy. They expanded my knowledge a thousandfold, identifying constellations such as Orion, Taurus, Sagittarius, Aries,

Canis Major (Sirius), and many others. Eddie explained terms and concepts, and talked about galaxies, globular clusters, nebulae, and star clusters. He taught me about the celestial sphere, about time and magnitude of brightness with the spectrum, and about the Doppler effect or Hubble's red shift. Eddie and I discussed the Big Bang theory and the steady-state theory, antimatter and dark matter. We didn't have any final answers.

The Wild One

I didn't dress like a science nerd.

I dressed like The Wild One.

The Wild One, with Marlon Brando riding a Harley, was the hottest movie of that era. I bought a black motorcycle jacket just like Brando's, along with black pants, black motorcycle boots with full metal taps, and a long white silk scarf. I even put black gloves in the jacket's epaulets. It was a lot of style but no substance—I had no motorcycle. However, it made a big impression at school. When I walked down the hall, everyone got out of my way when they heard those steel taps hit the floor.

I let my sideburns grow long like Elvis Presley's, and I kept that sullen look all the time.

Neisner's

One day Glonda Kroger—the girl from my first date—stopped me in the hall and said that the place where she worked, Neisner's (a new five-and-dime store in the A. Harris Shopping Center), had just let their two stock boys go because they were too young. If I wanted a job, I should go there after school and apply.

Boy, did I want a job near home! I was living at home by then but still working at Bryant's Grocery on Saturday, and getting there was a problem. I'd wake up early, ride the bus up to Jefferson, then catch the streetcar over to Waverly. A. Harris Shopping Center, on the other hand, was only a few blocks away, and I could easily walk there after school. Besides, if I had a job to go to, I could get out of classes two hours early.

I practically ran over there after school.

I filled out an application. The manager, Mr. Napier, interviewed me on the spot. He said he'd have to talk to my present employer and check out my references, but if there were no bad reports, the job was mine. I sat on pins and needles for two days. Then Glonda told me I'd gotten the job.

She said I'd need to wear a white, long-sleeved shirt, but that blue jeans would be okay. I nearly always wore a white, long-sleeved shirt under my black motorcycle jacket, so I was ready. Really ready.

After school I reported to work. Reverend Gay Weaver was the assistant manager. He was a preacher at a small Baptist church down near Waco. There was a floor manager, but I don't remember his name; he wasn't there very long. There were three floor ladies who supervised the ten teenage salesgirls. Ten teenage salesgirls! Boy, was I in high cotton. No, forget the cotton—my cotton-picking job had left me with an aversion to cotton. A fox in a henhouse was more like it.

Mr. Evans was stock manager; unfortunately, he was gay and liked young boys. I had to say no to him a lot of times while he worked there. Fortunately, he wasn't there very long. Lil was the marking lady and there was another lady temporarily helping her. There were two lunch-counter waitresses and a dishwasher. That just about rounded it out.

I had a very successful career at Neisner's. First I just replaced the two stock boys. Then I replaced the stock manager. I also had to support the dishwasher during peak lunch times—he was really slow.

One day Mr. Napier came down to the stock room to help move things around to make room for the incoming Christmas stock. He said, "Grab those manhole covers and put them over here."

I looked around where he was pointing, but I didn't see any manhole covers. So I asked him what he meant.

He said, "Those Kotexes."

I nearly fell over. I'd never heard them referred to that way.

After the store had been open for a year, Lil and the temporary marking lady left. I'd do the ordering, shipping, receiving, marking, and storing for all the new items. One day Mr. Napier was working with me. He was ordering Hula Hoops, and he asked me how many I thought he should order. I said to get a gross, but no more—I figured the Hula Hoop craze would be over soon. I was right. We sold our last Hula Hoop in about six months and never had a request for more.

The floor manager and the dishwasher left, and I took over their tasks. The counter girls did most of the dishwashing—I got called in when they were behind, which could be quite often if they were busy. As floor manager, I was the superior of the lady floor supervisors. One by one they started to leave—I guess they didn't like working for a young kid. Soon there was just Mr. Napier, assistant manager Reverend Weaver, me, eight teenage sales girls, and the counter help. This was our crew for about a year.

Then Reverend Weaver left. Mr. Napier wanted to make me the new assistant manager, but the company wouldn't let him. Assistant managers had to be at least twenty-one, and I was only eighteen.

Girls

I dated a few of the teenage salesgirls at Neisner's. Everyone knew the stockroom was my territory. I was like the old troll under the bridge in the fairy tales. The girls had to pay a toll to come down and get something. The toll was a hug. Most would ask my current girlfriend to retrieve things from the stockroom for them, so they wouldn't have to pay the old troll his toll. The girls I was dating didn't mind giving me a hug or a kiss to get whatever they were after.

My cutest girlfriend at that time was Pat McKissick. She was the 1950s dream girl: pony tail, pink cashmere sweater, full pink skirt with a white poodle on it, bobby socks, and black-and-white saddle shoes. She worked in the music department at Neisner's and played all the latest rock-and-roll and doo-wop songs: "All I Have To Do Is Dream," by the Everly Brothers; "Silhouettes on the Shade," by the Rays or the Diamonds; "Shangri-La," by the Four Coins; "In the Still of the Night," by Dion and the Belmonts; "Earth Angel," by the Penguins; "Oh, Donna," by Ritchie Valens; "Shaboom," by the Chords; "I Only Have Eyes for You," by the Flamingos; "Sincerely," by the McGuire Sisters; "You Are My Special Angel," by the Vogues; "And That Reminds Me," by Della Reese; "Smoke Gets in Your Eyes," "That Magic Touch," "Twilight Time," and "My Prayer," by the Platters.

And of course, everything by Elvis.

Pat lived on Terrace Drive not far down Beckley from the A. Harris Shopping Center. She went to Blessed Sacrament Catholic Church School at the corner of Eighth and Marsalis. One fall she hurt her back in gym class and was laid up in bed for several weeks. I'd go over to her house at night after work and keep her company as long as her dad would let me. She was the girl I took to my senior prom.

Werner Von Braun

At the Oak Cliff Carnegie Branch Library I found a book based on a 1952 *Collier's Magazine* article called "Man on the Moon—The Journey." The book was by Dr. Werner von Braun. Von Braun had been a key figure in Germany's rocket program during World War II. After the war, he came to the U.S. and became America's leading rocket scientist.

The book outlined the requirements to get to the moon, including the three-stage rocket with a reusable third stage. It described the amount of thrust necessary for each stage. His drawings showed that about thirty rocket engines were needed, based on the thrust available in each rocket engine at the time (1950). A rotating space station orbited the earth. And a shuttlecraft took the men from the space station to the moon.

I checked the book out and never returned it. I still have it, and I still read it often.

Another Move

When we went to town on the bus, we'd go down Ewing past Morrel Street with its Holy Rollers (or Pentecostal) Church. Then we'd cross Clarendon Drive, travel about one-half block, turn left on Thirteenth Street, and go past the clinic where I received my complete polio-shot series.

About this time, Raymond came home from the marines with his wife Lavonne and their two kids, Karen and Richard. (Lavonne had a brother named Marvin Harvey who married a girl named Sara Beth, or "Tuttie," Ferguson. Today they live in Duncanville, and Sara Beth has a dance studio.) Raymond was in training for a union plasterer's job. They didn't have a place to stay, so Mother let them move into the big chinchilla house in our backyard. I don't think Mother was able to afford the house payments after Dad left. So when Raymond started a paying job, Mom sold him our house.

We moved again, this time to a duplex down near Saner on Marsalis.

Meanwhile, back at school....

Test

Shortly after school started that first fall, each classroom of kids was herded into the library and given a standard skills test. The point was to see if you should be in high school or not. If your scores weren't high enough, I guess they sent you back to junior high.

When the results came back, I scored 6.5 for reading speed, but 14.5 for reading comprehension. In other words, I read as slowly as a sixth-grader, but understood as well as a sophomore college student. My math and science scores were both at a tenth-grade level.

Fortunately, they didn't send me back to junior high. I guess they thought I was smart enough, just a little slow.

Bonza Bottler Day!

Bonza Bottler Day is when the day, the month, and the year are numerically the same.

Bonza Bottler Day celebrations were begun by a woman named Elaine Fremont, a devout Baptist who wanted a holiday that could be celebrated without alcohol, drugs, stimulants, etc. There's a Bible passage—"So teach us to number our days"—in Psalms, and this was apparently her inspiration. She choose the name because "bonza" or "bonzer" means "great" in Australian slang, and "bottler" sort of means "to get wild" or to "loose control" in British slang (actually it means "to pee your pants," but Ms. Fremont probably didn't know that). Some say "bottler" also means "great" in Aussie slang. So Bonza Bottler means (or is intended to mean) "to go wild over things being so great" or a day that's "doubly great." I remember exactly where I was—in what room in the school and how the sun was shining in the window—when the teacher first pointed out the date: 5/5/55. There was a discussion about when it would occur again, in eleven years, one month, and one day—6/6/66—and what would we be doing on that day.

Mr. Houston's English Class

In 1956, in the tenth grade, Bill Allen and I were in Mr. Neil B. Houston's English class. I hope I can do him justice, because it was the *best* English class I've ever attended. This man should have been given the Teacher of the Year award several times over. Mr. Houston was only about eight or nine years older than we were, so he knew us pretty well. He was working on his master's degree, at SMU (Southern Methodist University), I think. Because he was still in school as we were, he had an affinity for us.

Bill and I tried all our practical jokes on him, but everything just rolled off him like water off a duck's back. He ignored most of the stuff we pulled. Other times he'd just say, "Not very original," and continue on with the class. But he also had a subtle way of challenging us to use our brains for something more than childish stuff—although he'd never chastise us or call us childish. He praised us for any little positive contribution to the class, and he let Bill write short stories instead of boring compositions.

One day he had us each stand up, one at a time, and talk about something we felt strongly about. I think it was his way of finding out what motivated each of us so he'd know how to encourage us.

When it came my turn, I stood up and talked about going to the moon and Dr. Von Braun's book. If you remember, in 1956 nothing human had ever been out of the atmosphere, so these ideas were pretty fantastic. I finished my talk by slamming my fist onto my desk and shouting, "We will go to the moon!" He really liked that. It was exactly what he was looking for—something emotional and motivating.

Bill had read a science-fiction book in which a couple moved into a new apartment and the woman had misgivings about the janitor. She told her husband she felt the janitor was looking at her from an eye in the back of his head. Then the couple found some strange equipment in the basement. They suspected the equipment might be rocket engines, and that the apartment complex might be a rocket ship. One day they heard this tremendous roar, like rocket engines blasting off, and the building began to shake. To escape the spacecraft, they ran out of their apartment—only to find that their entire block was part of a huge rocket ship! Our English classroom was over the school's heating and air-conditioning system, which was in the basement. Every time one of the compressors came on, the floor would shake. Bill and I would look at each other, thinking the same thought.

Then we'd look out the window as though to see if the ground were falling away.

Mr. Tonn's Electric Shop and Mechanical Drawing Class

A house in Kessler Park reminded me of Frank Lloyd Wright's house in Pennsylvania, the one named *Fallingwater*. I was taking architectural drawing and had won several honorable-mention certificates for my drawings in statewide competition. I almost became an architect. I made straight A's in mechanical drawing, and the teacher, Mr. James Tonn, was also my electric-shop teacher. One day I brought in the ham radio I'd built. I set it up in the tool room and contacted several other hams on the air. He was impressed, but I still had to have a class project. I made a floor lamp. It was easy—it only took me one class period to make it. Here again I made straight A's.

Mr. Tonn made me his assistant to help the other students with their projects.

Science Class

I made A's and B's in physics with Mr. William Garrison, and in chemistry with Mr. J. D. R. Thomas.

In chemistry class, our lab book had a warning in big red letters: "Do *not* boil concentrated sulfuric acid." It didn't give any explanation. This was a lab class, right, where you were supposed to have scientific curiosity, and experiment, and learn things? First chance I got, I took a test tube half-full of concentrated sulfuric acid and held it over the Bunsen burner. I wanted to see what would happen.

I found out.

It boiled violently. After a minute, there was one great big burp that ejected all the acid from the tube onto the lab tabletop. When that stuff hit the tabletop, it looked like a miniature atomic bomb. A huge vapor cloud shaped like a mushroom rose from the spot where it landed and splattered boiling sulfuric acid everywhere. Good thing we had on safety glasses. My lab partner and I immediately washed our hands and faces with cold water.

No serious damage was done. Thank God. I had to go to work at Neisner's after school. By the time I got off work six hours later, my white shirt had nearly fallen off it was so eaten up by the acid. I had to throw it away when I got home.

Speech Class

In speech class I gave a presentation about how teleportation could work.

A transmitter would scan the body and determine its molecular structure and DNA. It would record the brain waves, then transmit them across a long distance at the speed of light, using the decaying atomic structure of the body as a power source. A receiver would recreate the human body using the DNA sequence and energize it with the brain waves. Pat Golf said she really enjoyed being in the speech class with me.

Love from Afar

I had a crush on Nancy King, who had dark flashing eyes and beautiful curly black hair. I wanted to run my fingers through her hair while I kissed her beautiful lips. She had perfectly symmetrical facial features and an hour-glass figure with a .7 waist-to-hip ratio. She wore tight sweaters and tight belts to show off her waistline, along with tight skirts and bobby socks. She had a walk that made me weak. She was just what a bunch of adolescent boys needed—something to keep their minds off their studies. Just what the male of the human species unconsciously seeks for procreation. And when he sees it, the vision creates an uncontrolled flood of testosterone to prepare for the task.

Then there was Patsy Harper, who had the regal countenance of Grace Kelly, Princess of Monaco, and still does.

Barbara Dempsey—she was a grand vision of beauty, and still is. Bill Allen summed her up this way:

I met Barbara Dempsey as early as the third grade. She had the prettiest long brunette hair, which I got to look at up close because I sat behind her in our first period class at Lisbon Grade School. Barbara was an all-American girl, but with something exotic about her, too, a little Spanish influence perhaps. She smelled great from just a foot away, and I remember that she was always perfectly dressed. Barbara was one of the first girls to make an effort to be friendly toward me. She had such an open, pleasant, and spirited personality back then, and she never lost it. I think that she probably set the national standard for me as what the ideal girl should be like. She wasn't stuck up, wasn't shallow or an exaggerated female—just good-natured and kind and interested in other people. I was very shy back then, and I believe that her friendship in class gave me some confidence when I needed it. I'll always remember her smile and bright brown eyes when she looked back over her shoulder at me.

She was a standard for us all, I suspect.

These golden debs could lift my heart to the sky. They were goddesses marching by this mere mortal in hell. But they were all so far up the social ladder that they never even noticed a poor kid like me.

Future Engineers Club

At school I was a member of the Future Engineers Club. When I was a senior, one night I received a call from an engineer at the Chance-Vaught aircraft plant. He told me that, courtesy of Chance-Vaught and in cooperation with the Future Engineers Club, I was invited to a banquet downtown at one of the big hotels.

It was a great night for me. The engineer picked me up, and we talked all the way into town. This was 1957. We talked about the International Geophysical Year, the period from July 1957 to December 1958 that was intended for scientific observations of geophysical phenomena. We talked about the Vanguard rocket. We talked about rockets in general. He said he was working on a Mach

20 airplane, a sophisticated project in 1957. I told him about the Werner von Braun book I'd gotten from the library. By now, I'd almost memorized it.

The banquet was great! There must have been two hundred engineers and scientists there, along with their student partners. I felt as if I belonged.

In 1957, to counter the threat of a Soviet air or missile attack across the polar regions, a far northern radar defense line was completed, known as the Distant Early Warning (DEW) Line. The Line later became part of the North American Air Defense Command (NORAD), the headquarters of which were built deep inside Cheyenne Mountain in Colorado. The U.S. developed an even more sophisticated and expensive system, the Ballistic Missile Early Warning System (BMEWS).

Sputnik, Rockets, and Bombs

On October 4,1957, the Soviets launched the first artificial earth-generated satellite into space. The satellite was called *Sputnik*, and it circled the earth in the astonishingly short time of ninety-five minutes.

I was living in the duplex at 2630 South Marsalis Avenue, and mother and I were watching the 6:00 P.M. news when they announced the launch and played the "beep beep beep" signal from *Sputnik*. For two days straight, the media played that signal every hour on the hour and sometimes more. Everybody and his dog had something to say about the event. People were caught up in the frenzy.

Bill Allen and I were no exception, and we decided to sharpen our skills in rocketry. We didn't want to be left at the starting gate in the greatest race in history: the space race! We went to a chemical supply house near Irving Boulevard. It was on Hampton Road, or Inwood Road, or Sylvan Avenue—I don't remember which. We bought a pound of charcoal, a pound of sulfur, a half-pound of ammonia nitrate, a half-pound of potassium nitrate (saltpeter), a half-pound of sodium nitrate. And twenty feet of fuse. We scrounged a bunch of lipstick tubes from our mothers' cosmetic drawers, and we set up our rocket lab.

First we mixed the standard gunpowder combination of chemicals and sure enough, it exploded. That was *not* what we wanted. We wanted it just to burn rapidly and create exhaust gases. Not blow our missile apart. We started varying the amounts of charcoal, to see if we could slow the oxidation down. We didn't have much success, so we started using potassium nitrate. Not much better. Then we started varying the amount of sulfur to see if that would slow it down. Not much success there either.

We tried switching from potassium nitrate to sodium nitrate. That worked a little better, but not much. Each time we had a different mixture, we'd fill up a lipstick tube with it and crimp the open end to make a nozzle around the fuse. (We'd refine the nozzle characteristics later, when we had a good sustained burn.) Then we'd light the fuse and drop it in a steel tube we'd buried in the ground at an angle. Most of the time, the thing would just explode and shoot the lipstick tube to somewhere in the next block.

Technically, we were refining our rocket fuel. But practically, we had built a one-centimeter mortar and were shelling the houses on the next block with spent lipstick-tube rocket cases.

What we needed was a good burning fuel, so we went back to the chemistry books. We read that powdered zinc and sulfur could be combined and would burn smoothly. Not rapidly enough to explode, but smoothly. That was what we were looking for. We returned to the chemical supply store and bought a pound of powdered zinc. This new mixture worked much better. We built a number of rockets with it. We even had one come out of the tube and hover for a couple of seconds before it ran out of fuel. Eureka! We were finally getting somewhere.

Now we needed a way of making cartridges, so that recharging the rocket wouldn't take so much time, and so that we could make the cartridges in varying mixtures, to test the different nozzles, which was our next step. Well, we read in the chemistry book that powdered zinc and sulfur could be melted and poured into molds—like the Thiokol that was later used in the solid propellant boosters on the Space Shuttle. But remember, this was 1957. The Space Shuttle was still twenty years away, although the technology was already available.

One day I was trying to melt some zinc and sulfur in a small metal can on our kitchen stove when it flashed over and blackened the kitchen walls and ceiling. The chemistry book didn't say that it takes precise temperature control and an inert-atmosphere environment to properly melt powdered zinc and sulfur. When my mother came home, she grabbed the broom and hit me, screaming, "Clean up this mess! Get rid of those chemicals! *No more rockets!*"

And "If you don't do what I tell you, I'll call the fire department and the police."

Later she read in the newspaper that the Boy Scouts in Highland Park were building a rocket on the SMU campus, with technical assistance from the university's physics and chemistry departments. Mom said if I wanted to build rockets, I should go join that group.

Meanwhile, Bill and I discussed how to get rid of all the chemicals we'd bought. We considered putting them all together, lighting them, and watching them go up in smoke. That was too simple. We needed a more dramatic end to our experiments.

We decided to make a pipe bomb.

We had a stainless steel pipe that was two inches in diameter, two feet long, and threaded on both ends. We bought two end caps and drilled a hole in one of them for the fuse. Then we mixed the leftover chemicals, put them in the tube, and screwed the ends on. We wrapped the last ten feet of fuse around the pipe and taped it all together. Now that we had this bomb, what would we do with it? We'd created a big dilemma for ourselves. We didn't want to blow up anything, just get rid of the chemicals. But now that we'd built the bomb, we wondered, would it work?

We felt like the guys on the Manhattan Project, the ones who built the atomic bomb, must have felt.

We drove out Illinois Avenue toward Mountain Creek Lake. That was the closest unpopulated area. At Cockrell Hill Road, there was only one building on the corner. We hadn't passed any others since the little grass airport. We turned left and started south. We hadn't seen any signs of human life for a couple of miles when we came to a small bridge over this little dry creek. I lit the fuse and threw the bomb as hard as I could out the window. Bill hit the gas and we took off like a striped-ass ape.

We must have been close to ten miles away when it went off. Hanging out the window, I first saw a bright red-orange flash, then heard the great boom. Hurray! It worked!

We didn't go back to see what damage it did, though we certainly wanted to. We knew that criminals return to the scene of the crime and figured the police and or fire department would be watching to see who came by.

The next Tuesday we went to SMU to find the Boy Scouts who were building their rocket. They were meeting in the basement of the Methodist church on the SMU campus, on the corner of Hillcrest and Mockingbird Lane. We were too late to lend our expertise on propulsion, as they already had a team of engineers from Chance-Vaught, Ling-Temco-Vought (LTV), and a few other companies, plus a couple of college professors. It turned out that one of the Scouts had done some independent research on his own, similar to what we'd done.

He'd blown up his family's garage.

Since then, they'd had a strict rule: *no* independent propulsion experiments. All work of that type would be done by the professionals. The Scouts could watch and perhaps could assist in a minor way.

The group's design was for an eight-foot rocket, two inches in diameter. They'd already built it, fins and all, and were working on the nozzle design. They needed help with the recovery system, so Bill and I volunteered. We were given a budget and some materials they'd already purchased. A couple of the engineers helped us with some ideas and gave us some suggestions. But for the most part, everything was left to us. Of course, they'd review anything we came up with.

This was great stuff! We were members of a real rocket-design team. We took our materials and went home. For several days, or nights I should say, we discussed the plan and some possible designs. We had a lot of very thick plastic, and a lot of rope. The main problem, at first, was determining how much material would fit in the two feet of the rocket allocated for the recovery system. The available space limited the size of the parachute and its shroud lines. We came up with the largest size we thought would fit into the two feet.

Then we had to determine how much drag the weight of the empty rocket would produce. We built our first test case. To check it out, we bought a spring-operated fish scale. The weight of the empty rocket was only eighteen pounds. We mounted the scale on the back of Bill's car, and attached the parachute to the scales. I stationed myself on the bumper so I could take the readings. We took off. We measured the drag at 10, 20, 30, 40, and 50 miles per hour.

At 50 miles per hour, the plastic shredded to pieces. Back to the drawing board. We modified the design to have the shrouds loop over the top of the plastic parachute. This time we didn't go 50 miles per hour, in case the plastic wouldn't hold. We still had to test the opening-chute shock, but we had some good test data to present at the next meeting.

The engineers were impressed with our methodology and our data. We delivered some valuable information they wanted to use.

We went back to work to test the opening-chute shock. We scrounged a bag and filled it with eighteen pounds of dirt, the estimated weight of the empty rocket, and attached it to the shroud lines. Then all we needed was someplace high to throw it off of. There are not too many high bridges in the flat Dallas area. The only one we could find was on Hampton Road just north of Highway 30. The Texas and Pacific Railway Company had put a cut through the only hill in the area, near the limestone quarry of Cement City, so trains loaded with limestone wouldn't have to climb that little hill.

Next evening at 6:00 PM we drove there and parked just before the bridge. We lugged the parachute and test-weight bag out to the center. To passing cars, it must have looked like we were throwing a person off the bridge. But damn the torpedoes, full speed ahead. We had research to do.

We threw the test-weight bag off the bridge. The chute opened about ten feet before it hit bottom. The plastic ripped to pieces—the opening shock was too much for the thin material.

We had some very bad news to report at the next meeting: they'd have to find another material, such as silk, which real parachutes were made of. In fact, maybe they could cut a regular parachute down to fit in the available space. They needed a material that would stretch and not tear. This was welcomed as valuable data for the engineers.

On November 4,1957, scarcely a month after the first Soviet success with an orbiting satellite, there was a second Soviet achievement, the launching of *Sputnik 2*. This satellite was placed into orbit with a dog named Laika aboard. The presence of Laika would enable Soviet scientists to study living conditions in space.

ROTC

Every Tuesday morning before school the Reserve Officers' Training Corps (ROTC) department had a parade and inspection. The platoon leaders were instructed to put their best cadets in the front row, so that when the colonel asked questions, they'd give the correct answer loudly enough for the newcomers in the other rows to hear. This way, the new cadets would know what questions were being asked and what the correct answers were.

Master Sergeant Comer was my idol and father figure. My dad had always reminded me of the actor William Holden—or I should say, William Holden reminded me of my dad. Master Sergeant Comer was like the top sergeant in *From Here to Eternity*, played by another of my idols, Burt Lancaster. My dad was an alcoholic, and he was gone by now. Fortunately, Master Sergeant Comer provided a positive role model for me to follow when I desperately needed one.

Hall Guard

The school had a policy that no one was allowed in the halls during class, except those with written permission—called a hall pass—from a teacher. The school administration had placed desks in the hall at the four corners of the school on all three floors. That made twelve guard posts for the ROTC department to command during school.

I became a hall guard. This meant I got out of study hall (you had to report to the study-hall teacher before going to man the guard post). Any student coming down the hall had to stop at my guard post and sign a clipboard and fill in the name of teacher who'd authorized the hall pass. Then I'd verify the pass and let the student proceed.

This system worked okay until spring, when the girls started to wear loose-fitting, low-cut blouses. When they bent over from the waist, instead of bending at the knees, to sign the clipboard, I could see all the way down their blouses to their navels. Some girls would get mad if you looked. Some would get mad if you didn't. Some would just gaze up at you and smile. Some didn't even wear a bra—these were usually the ones that smiled.

It drove me mad. I'd have to walk to my next class with my books held in front of me.

Drills

Every year the ROTC department had a competition to see who'd done the best job of learning the drill procedures for parades. I won Best Drill Cadet in South Oak Cliff High School three years in a row (1956, 1957, 1958). I was good—really good. I was awarded a special white medal and ribbon for my uniform, and was promoted to another rank each time I won.

I also joined the South Oak Golden Grenadiers drill team. We placed first or second in citywide competition every year.

I could fieldstrip and reassemble the following weapons: an M1 Grand, an M1 carbine, a .45 caliber pistol, a Browning automatic rifle, .30 caliber and .60 caliber machine guns, and a .45 caliber burp gun. I could even do all of these blindfolded. I qualified as a sharp shooter at the underground .22 caliber rifle range at school. I memorized several army field manuals. When I graduated, I was promoted to ROTC second lieutenant and given a certificate that I'd start as a private first class if I joined the Army.

Woosencraft Drill Competition

Meanwhile, the place where Mom and I lived was scheduled to be torn down to make room for a filling station by the new freeway. So we moved yet again, to 818 South Beckley, at the corner of Yarmouth Street. It was another upstairs apartment, only three blocks from the Dallas Zoo. At night I could hear the lions roar. Sometimes I'd go down to the zoo at night and walk around and look at the cages. The zoo had a carousel with mirrors and a large drum that played. It was fun just to sit and watch the carousel—I couldn't afford to ride it. Not far away, on Clarendon Drive, there was this train that a man had in his yard. It was big enough to ride on and was powered by a car engine. He must've had an acre of land along the creek, and he'd built a couple of railroad bridges over the water, and a tunnel. He let kids ride the train for five cents. On Saturdays and Sundays when we'd go down Clarendon, we'd see him out riding his train.

The new apartment was nothing to talk about, but it was where I lived when the Woosencraft Drill Competition took place during my senior year. I expected to win. For three years in a row I'd beat my closest rival, Jim Alphan. But this time, one of the judges knocked a hundredth of a point off my score for something.

Jim Alphan won.

I was furious that I'd placed second. As soon as I got home, I smacked the glass pane on the front door. It broke and my hand went through it. I cut the skin between my index and middle fingers clear to the bone. Blood came pouring out. Using my Boy Scout first-aid training, I wrapped a handkerchief around my hand and held it together to control the bleeding.

Bill Allen was waiting downstairs for me to change out of my ROTC uniform. I showed him my bloody hand and asked him to drive me to the local clinic. When we got there, they wanted to know if I had insurance. I gave them my mother's phone number at work and told them to call and ask her. They called, but it still took a couple of hours before they treated me. By the time they finally took a look at my wound, they had to yank the handkerchief off, spread the index and middle fingers wide apart, and run a plastic-bristle brush into the wound to clean it. They put a couple of stitches in the loose skin and taped my two fingers together. Then they put a bandage over it all, and finished the process with a tetanus shot.

Combat Squad

Sergeant Comer organized a combat squad to go out to the woods near Mountain Creek Lake. He wanted us to practice the maneuvers we'd read about in our manuals. We set up skirmish lines and attacked several hills using hand signals and silent commands. Maintaining separation was one of the important things we needed to practice. In combat you're not supposed to bunch up, because if you do, a single grenade can get you all. We spent Saturday doing maneuvers and camped out Saturday night. Sunday, we did the same and went home about 5:00 PM. Boy, it was fun playing war games.

Michael Bishop

Michael Bishop was another guy I met in high school. Bill and I were members of a group Bill would later define as "Loners and Dreamers," and Bishop was a loner and a dreamer, too. Bill and I went over to his house sometimes after school. He had a drum set, and he'd let me bang on it a little.

There was a great Frank Sinatra movie out at that time called *The Manchurian Candidate*. It was about the brainwashing techniques used by the Communist in Korea. I didn't know it at the time, but I'd later learn about these techniques practically first-hand in the Air Force. Anyway, my friends and I were always discussing strange topics like mind control, and at Bishop's house one day, the conversation got around to hypnotism. Mike said he could hypnotize anyone, and I made a bet he couldn't hypnotize me. Mike took the dare.

Mike and I went into the bedroom, and I lay down on his bed. Bill stayed in the living room. We were in there almost an hour, with Mike chanting, "Sleep, sleep, sleep," and sometimes swinging a bob in front of my eyes. Nothing had any effect. Bill got tired of sitting in the living room, looking through magazines, not knowing what was going on in the bedroom. He started to sneak quietly over to the bedroom door. I could hear him coming to the door and could visualize him sneaking like Wile E. Coyote would do, trying to catch the Road Runner. When he started to turn the doorknob, I couldn't hold it any longer and burst out laughing.

Mike was bent all out of shape because he'd failed to hypnotize me. But Bill broke up laughing, also. That ended the hypnotism session.

After I joined the Air Force, Bill and Mike joined the Army together on the buddy system. Bill didn't want to spend any longer in the military than he had to, and service in the Army lasted only three years. Bill and Mike went through basic training together at Fort Chaffee, Arkansas. One of the first things you learn in the Army is that your rifle is your best friend. You do everything in your power to keep it clean and working properly. And you never ever let it out of your sight or let anyone take it away from you.

During basic training, they had to do a night march through the dense woods of Arkansas. Bill says that about halfway through the march they were allowed a five-minute rest, to sit and drink water from their canteens. When the sergeant yelled, "Off your ass and on your feet," Mike grabbed his backpack strap but not his rifle sling. He jumped up and held onto Bill so he wouldn't get lost in the dark. They'd marched for another fifteen minutes when Mike realized he didn't have his rifle. In sheer panic he broke rank and ran back toward where they'd all rested. But it was a pitch-black night, and he soon veered into the woods and got lost. The sergeant had to stop the march to go find him. But they didn't locate the rifle that night. I think they retrieved it the next day. Mike must be the only man in the army to lose his rifle on a night march.

Golden Bear Suit

Our school always had a pep rally on Friday during football season. One rainy Friday morning Bill and I were sitting in the auditorium before school, and one of the cheerleaders came over and asked me to try on this bear outfit. I did, and it fit perfectly. They wanted me to wear the golden bear suit and serve as team mascot for the pep rally.

Well, I did it. But I was scared to death. I didn't like being up there in front of the whole school, jumping around and hollering. Besides, the bear suit was really hot.

There was a fight cheer that went like this: "Two, four, six, eight—we don't want to integrate!" There were no Blacks in our school. In fact, there were almost no Blacks at all in Oak Cliff in the 1950s.

Kodak Movie Camera

One Christmas Bill received a Kodak 8mm movie camera. We had a great time with it.

I'd hold the camera and lean out the car door while Bill raced down the center of the highway. I'd get as close to the pavement as I could and film the front wheel as we sped down the road.

When we played chicken with a semi-truck by driving down the wrong side of the highway, I filmed the oncoming truck. I did it in telephoto mode, so the truck looked a lot closer than it really was.

Another time, I lay on the center stripe of the highway while Bill drove at me at sixty miles per hour, swerving at the last second and barely missing me. Again I had the camera in telephoto mode so the car would appear closer than it was. But I didn't need the telephoto mode—Bill came close enough. I thought for sure I was being run over! I even felt the wheel clip the side of my foot as he turned. It made for a great movie.

Bulldozer

One Friday night I went down to Bill's house to look through his telescope. But the sky was overcast, so we walked down Ann Arbor Avenue to Marsalis, looking for something to do. A small shopping center was under construction. In the middle of the construction site was a new freshly painted light tower about fifty feet tall. We climbed to the top and wrote our initials inside the light housing.

After that, we noticed a bulldozer parked near the street, so we climbed on. Soon it became boring just playing with the levers. We craved something more exciting. I noticed a starter switch but I didn't see any key, so I pushed the starter anyway. The diesel engine sprang to life with a roar! It scared the hell out of Bill and me. The engine hadn't even hit its third stroke when Bill and I jumped off of it and ran toward some buildings at the back of the site. There was a barbed-wire fence across the back, and I cleared it like a jumper clearing high hurdles, but Bill didn't make it. He got all tangled up. I didn't stop to help, and I could hear him cussing as I dashed to his house. A few minutes later he arrived, all scratched up and with his shirt and pants ripped. We've had quite a few laughs about that little stunt. On my way home, I could hear the bulldozer still running. I wished we had put it in gear.

Our Trip to Levelland, Home of the Texas UFO

Bill used to come over to our apartment in the morning and pick me up to go to school. It was only a few blocks down Marsalis, and I could've walked or ridden the bus, but it was more fun if Bill came by. On November 3, 1957, at breakfast I heard on the radio that an unidentified flying object (UFO) had been spotted in Levelland, Texas. When Bill showed up, I ran out to the car to tell him, and it turned out that he had just heard about it himself on the car radio. The newscast had said that not only had numerous people seen something, but that several vehicles had stalled, apparently from being in proximity to the sightings.

It was too much to ignore.

We stopped at a 7-Eleven and picked up a newspaper. The UFO was headline news nationwide, and the paper gave more details than the newscast.

The first reported sighting had occurred at about 11:00 PM. Two guys on a highway just west of town had called the police department to say they'd seen a bright, yellow-and-white torpedo-shaped object about two hundred feet long. It had risen from a nearby field and shot right over their truck, quickly reaching a speed of six hundred to eight hundred miles an hour before it disappeared. It shook the truck, they felt the heat from it, their headlights went out, and the truck's engine died. Then once the UFO faded into the distance, their lights came back on, and they were able to start their engine. Over the next two hours, there were a total of seven separate Levelland UFO incidents reported that temporarily disabled vehicles.

Bill and I looked at each other. No way we were going to go to school that day. We were within driving distance of the most significant UFO sighting since Roswell!

We found Levelland on the map and headed west to see our first spaceship.

On the way, we had to stop for gasoline, but were short on money. We bought what gas we could and tried to calculate how far we could go. It was going to be tight, but we decided we'd at least go as far as half our gas would take us. Hell, maybe we'd get out and hitchhike if we had to—that's how important it was.

"This may be the biggest thing that's ever happened," Bill intoned quietly.

Well, we didn't make it to Levelland. It was farther out in the West Texas Panhandle than we'd calculated. Also we got caught in a terrific thunderstorm, one so heavy that other cars had pulled off the road. The more tired we got, the more the odds seemed stacked against us. When the fuel gauge got to half empty, the rain was even worse. The idea of hitchhiking on a flooded two-lane highway was discouraging. Finally we decided to turn around.

On the way back, the fuel seemed to burn faster, and we used every conceivable method to conserve. We kept the windows rolled up to reduce wind resistance. At the tops of hills, we'd turn off the engine and coast as far as we could. Then we'd pop the clutch and hit the accelerator, so as not to waste fuel starting the engine. Even so, we only made it halfway between Dallas and Fort Worth on the turnpike when we ran out of gasoline. We coasted into a filling station. Bill didn't know this, but my Uncle Fat had once given me a silver dollar as a keepsake souvenir, and I always carried it in my pocket. I spent that silver dollar on gasoline to get us back home.

We caught hell from our folks and the school principle that day. But the sighting stayed on our minds for years.

That day would eventually turn out *not* to be the end of our Levelland story.

Airports

There were three local private-plane airports besides Red Bird Airport. There was one on Highway 77, there was Flyers Field at the corner of Hampton Road and

Illinois Avenue, and there was another little grass strip on the north side of Illinois Avenue, out near Cockrell Hill Road on the way to Mountain Creek Lake. Bill and I'd pool our cash, and if we could come up with five dollars we'd skip school, go to one of the airports, and see if we could get an hour flight in a light plane.

With our interest in rockets, one time Bill and I discussed what would happen if someone stole a jet fighter from the Naval Air Station, took it to forty thousand feet, and did a nose dive into the middle of downtown Dallas.

Little did we know we'd anticipated the plan of the September 11, 2001, terrorists by almost half a century.

Recreation

As I've mentioned, I was inspired by a painting—*El Hombre* by Rufino Tamayo—that hung in the Art Museum at Fair Park. Every year when we went to the State Fair of Texas, I'd sit for at least an hour staring at and dreaming about that painting.

A roller-skating rink sat catty-corner from the Dallas Sportatorium, and we went skating there a number of times. Later when I was in the service, it was turned into a bar called The Branch Office. My brother went to The Branch Office every evening after work.

Wee St. Andrews Miniature Golf Course was one of the few places teenagers could go on dates, besides the drive-ins. It was great fun, and I went there often.

Oak Cliff Park had a 100' x 200' swimming pool with three high-diving boards (three meters high) and four lower boards (one meter high). We'd roll our bathing suits up in a beach towel, take the bus to Fifth Street, then get off and walk to the pool. I think it was seventy-five cents admission. When you were inside, there was a large room with lockers all around and little benches in front of them. After you put your bathing suit on, you could lock your clothes in a locker. There were cold showers, and you had to wade through this cold, two-foot-deep pool of chlorinated water to disinfect your feet before you could get into the main pool.

Under the diving boards, the water was ten feet deep and it extended about twenty feet in front of the boards. There was a slope in the pool bottom, rising to a depth of five feet. A rope with floats stretched across the pool where the slope began, so everyone would know where the deep water was. There were plenty of grassy areas where you could spread out your towel and sunbathe. There was also a sandy beach area, but no wave action.

One day Danny Shreves and I took our dates swimming. We were standing in the five-foot section and necking, when his girlfriend yelled, "Get your hand out of my—Stop it!" Embarrassed, Danny dove underwater and hid.

Kidd Springs Park was another big swimming pool. But it wasn't on a bus line, so I didn't go there often.

Sybil's Drive-In

Sybil's Drive-In was like the one in the movie *American Graffiti*. It was located where the two main arteries in Oak Cliff came together, at West Davis and Fort Worth Avenue, and it was huge! The parking lot must've been two blocks wide. Five or six rows of cars could park around the central hamburger stand. Female carhops, in cute little short skirts and tiny paper hats, would roller-skate to your car to take and deliver your order.

The cool, macho guys in their "bad" cars would always park in the last row. If you were cool and had a bad car, you'd cruise around the last row racking your pipes to find someone willing to drag race. If someone was, he just revved his engine. Then you'd agree on a time and place—usually a city street nearby. But sometimes, if there were two *really* cool guys in *really* bad cars, say from across town, and someone wanted to put his money down, you needed a measured and timed quarter-mile track and a referee to name the winner.

Drag racing was illegal, of course. And dangerous, too. But kids with bad cars were going to do it anyway. The local police finally took a "if-you-can't-lick-'em-then-join-'em" position to keep drag racing off the streets. And so, the Yellow Belly Drag Strip was built down in the Trinity River bottom, just a couple miles from Sybil's, past Loop 12. That way, the police could supervise things and make the inevitable racing as safe as possible.

Fast Food

Bill and I hung out at the Dairy Queen at the corner of Illinois Avenue and Lancaster. Bill later wrote a story called "Bridey Murphy at the Dairy Queen," which describes some of the things we thought about at the time, things like reincarnation and ESP. There was another Dairy Queen on Lancaster, just across the street from the Veterans Hospital near Loop 12. We'd cruise between the two Dairy Queens on Friday and Saturday nights if we didn't have dates.

Next to the Dairy Queen was a McDonald's where we'd sneak out of school sometimes for a quick lunch. Hamburgers were nineteen cents and an order of fries was a dime. It was cheaper than the school lunchroom, but not as nourishing.

The Knife Fight with Owen Corley

We were very much inspired by movies. I've already mentioned *The Wild One* with Marlon Brando, and *The Manchurian Candidate* with Frank Sinatra. Another favorite was *The Blackboard Jungle*.

But *Rebel Without a Cause* with James Dean, Sal Mineo, and Natalie Wood—that was *our* movie!

A lot of teenagers of the time, Bill and I included, saw this movie as the story of our lives. Most of us identified with James Dean. Bill and I would go out in the country and practice rolling out of his car, like they did in the movie before the drag-racing cars ran off the cliff.

The knife fight between with Bill and Owen Corley was very much in character with our James Dean images.

One morning after seeing *The Blackboard Jungle*, Bill decided to play a trick on the young and gullible junior-high kids in the area. We talked it over and planned to stage a knife fight between Bill and Owen. We went to the 7-Eleven store on the corner of Edgefield Avenue and Twelfth Street, over near Grinner Junior High School, and waited. Owen had a reputation as a bully, and we knew he'd show up for a fight even if it were a fake.

Sure enough, Owen's car came screaming in. He slammed on his brakes and squealed to a stop in the parking lot. No one within a couple of blocks could've missed that entrance. He jumped out, flicked open his six-inch switchblade knife, and snarled, "Bill, I'm going to cut you open, you bastard."

Bill looked around and said, "Someone give me a knife." I reached under the front seat of Bill's car and retrieved a huge hunting knife we'd planted there just for this show. I threw it to Bill, saying, "Here, use this."

Bill deftly grabbed the knife out of the air and said, "Come on, you son of a bitch. Let's see what you got."

Owen gladly played along. Bill and Owen circled each other and took a couple of swings. They missed each other by more than a foot, but it sure looked real.

The 7-Eleven manager came out screaming, "I'm going to call the police."

Bill said to Owen, "I'll see you down the street, and we'll finish this."

They both jumped in their cars, and I got in with Bill. As we looked in the rear-view mirrors, we could see about thirty junior-high kids running down the street to see the bloody end of the knife fight. That number swelled as they passed Grinner Junior High and word of the fight got around.

But Bill and Owen had no intention of finishing the fight. Instead we all drove back to South Oak Cliff High School and had a great laugh.

Tony Green

One Saturday night, Tony Green and I were walking between houses when we heard a phone ring. We looked in the direction of the ringing phone, and through a window we could see the lady of the house taking a bath. She stood up, got out of the bathtub, and answered the phone—stark naked. We watched for a minute, then Tony started to scream. We ran in different directions. I searched everywhere for Tony, but he was nowhere to be found.

Next morning I saw Tony in church. He was right in the first row, praying like crazy. I went down and sat next to him and asked him what had happened.

He said, "My mother told me if I ever saw anything like that, I'd turn to stone. I thought it was really happening. Parts of me started to get hard."

My Brother in Jail

The efficiency of the Dallas Police Department left a lot to be desired.

When my brother was thrown in jail for DWI (driving while intoxicated), my mother had a hard time finding him because of a clerical error—he was never booked.

I can't remember how, but she suspected he'd been arrested, so she called the police station. No, they said, they didn't have him there. She called all the hospitals to see if he'd had an accident. None of the hospitals had any record of him. She started calling all of his friends that she had phone numbers for. Finally, one of them said he'd been arrested and was in the Dallas jail.

Mother called her attorney, Bill Utay, who phoned the police station. They located my brother in the drunk tank, but he hadn't been booked. She was able to post bail, and they let him out the next morning.

Some time later, I chug-a-lugged almost a fifth of vodka one night and ran around the drive-in theater to help it take effect. It is a miracle I didn't die. Bill took me home and opened the door and gave me a little shove. I hit the floor with a great crash, then crawled to the bathroom and threw up. Mother came in, and I told her it was something I ate. But she knew better, having been married to an alcoholic for so many years.

Senior Day

At South Oak Cliff High School, we had what was designated as "Senior Day." The seniors got to ignore the dress code and wear anything they liked. They also performed at a special assembly in the auditorium. Some put on skits, some sang, some danced. Others just clowned around.

The auditorium ceiling was very low in the back where you came in—only about seven feet high. But it rose sharply and the floor dropped steeply, so the people in the seats at the back could see over everyone's head. The effect was like that of a megaphone, and any sound carried very well. On Senior Day, Bill and I were in the next-to-last row of seats. Janette Jones was onstage performing. She was one of the most beautiful Latin girls in school, with long coal-black hair. She was doing a modern-dance routine in a black skin-tight leotard when I leaned over to Bill and said, "Can you image snuggling up next to her on a cold winter's night?"

The megaphone effect sent my words through the whole auditorium. Half the audience broke out laughing. The guys all laughed because they were thinking exactly the same thing.

Rejected by the Air Force

Mother wanted to move to California. But I didn't. I told her the only way I'd go was if she bought me a 1956 Chevy. Cars were expensive—the car would cost about five thousand dollars, which was her annual salary at the sweatshop. I still remember her answer:

"I can't afford to buy you a car. It won't be long until five thousand dollars will only buy a very old car. Someday gas will cost twenty-nine cents a gallon—we can't afford that. If things keep going the way they are, it'll be impossible to buy a week's groceries for twenty dollars. If cigarettes keep going up in price, I'm going to quit. A quarter a pack is ridiculous. And the post office is even thinking about charging a dime just to mail a letter."

Meanwhile, I was fed up. At school I was flunking English. And at work, I couldn't get a promotion. So Danny Shreves and I decided to run off and join the Air Force.

We went down to the Dallas recruiting station. Danny had a consent form from his dad. I was eighteen, so I didn't need my parents' consent. We signed up, and they sent us right over to have a physical examination. Danny passed, and they shipped him out right away.

But not me.

They called me into a room and said I'd failed the exam.

How could this be? I was in perfect health. Besides, I was the best drill cadet in South Oak Cliff. They gave me a note for my parents and told me to ask Mom to take me to our family doctor. I went home and gave my mother the note.

The next day, she took me to our local clinic on Lancaster and showed the people the note. It said I had albumin in my urine—whatever that meant. The young doctors began a series of tests. First they drew a lot of blood. Then they had me drink this great big glass of glucose. The stuff was really sweet, and I nearly gagged before I got it all down. Next they started drawing blood again and again, at thirty-minute intervals. My arms were sore from all the needles. After several hours, they decided to run yet another test. They put this big needle in my arm and injected some radioactive stuff at thirty-minute intervals, taking X-rays before they put in more. By the time we got home, I was sick from all the tests.

Mom told me later that they took her into a conference room and told her I only had six months to live. Well, she'd heard that before—remember, I'd had pneumonia seven times before I was five years old—and she wasn't ready to give up. We went home, and she called one of my old doctors from the Medical Arts Building and got an appointment for the next day.

We went downtown to the Medical Arts Building to see old Dr. Cominski. He listened to our tale of woe and said to me, "Open your mouth." He stuck in a tongue depressor and said, "Say 'Ahhhh.'" I did, and he said I needed my tonsils removed. He scheduled me for the operation at the Catholic hospital on Ross Avenue near Baylor Hospital.

I went in at the scheduled time, and they scrubbed me down and dressed me in one of those gowns without a rear end so your fanny gets cold. Then they put me on a gurney and gave me a shot to relax me. It made me high as a kite—man, was that ever good stuff! They pushed me down the hall and parked me outside the operating room. I was so loaded that I tried to talk to everybody who came by. I couldn't get anyone to talk to me, so the next time a nurse walked by, I patted her on the ass. Boy, was she surprised! Finally, they wheeled me into the operating room, gave me a shot of Sodium Pentothal, and told me to count backward from one hundred. I got as far as ninety-seven, then everything went black.

They took my tonsils out. The next thing I remember was waking up with the most God-awful hangover ever. My head was spinning something terrible. In fact, everything was spinning. I just wanted to put my hand on something that wasn't moving. My throat was really sore. They brought me some ice cream and said I could eat as much as I liked. By the next day, I was feeling much better. The hangover was gone, so they let me go home.

It was a couple of days before I could swallow. I'd try to eat ice cream and it would melt and run down my throat, but I couldn't swallow. I began to wonder if I'd ever be able to swallow again. Finally, after three days, things started to get back to normal, and I could eat Jell-O and swallow.

Disaster Dallas

We moved again, to 347 South Marsalis. That's where we lived when the tornado struck. It happened on April 2, 1957, and the news media called it "Disaster Dallas."

Our apartment was an upstairs unit in an old four-plex down near the Oak Cliff Carnegie Branch Library on Jefferson and Marsalis. I'd just come home from school and was in the kitchen. I was looking out the window over the sink when I saw the twister, ten miles away, in the sky above near where I used to work at Bryant's Grocery. I watched it go all the way across Oak Cliff, then it disappeared down in the Trinity River bottom.

I ran into the living room and turned on the TV. Sure enough, all three stations had mobile units following it. Everyone was running around Dallas trying to get exclusive shots. It was one of the most photographed tornados in history. And of course, it was what all the newspeople showed and talked about for a week.

Oak Cliff Carnegie Branch Library

Many times I walked to the intersection of Marsalis and Jefferson Boulevard, where the Oak Cliff Carnegie Branch Library stood. It was an elegant little sanctuary—one of the community's true gems. That old library had saved my life, or at least my sanity, a number of times. That's where I found the *Collier's Magazine* book about Dr. Werner Von Braun's plans for a trip to the moon. I found other sacred texts there, too, groping for direction in that period when the self doesn't really know what it is yet. A poor reader, I liked to look at the picture books and magazines. With no clear sense of vocation yet, I had the ghost of an ambition forming. I dreamed of attending MIT (Massachusetts Institute of Technology), Caltech (California Institute of Technology), Stanford University, or the Air Force Academy. I'd get my physics degree, and someday I would build and maybe even ride in a spacecraft.

But my English grades always punctured my dreams.

Predictably, my hard-shell relatives claimed I was ruining my mind by thinking too much. Doggedly, I trudged on in the intellectual vacuum of poverty. When the library was closed, I'd continue to walk down Jefferson Boulevard. I'd look at the used cars and do some window-shopping, as I didn't have any money and most of the stores were closed anyway. I'd walk all the way to Tyler Street and back, about seven or eight miles round trip—anything to dissipate my pent-up energy and frustration.

Cut My Thumb

I was walking across the railroad trestle at the chalk hill by Clarendon Drive, just past the zoo, when I nearly cut my thumb off.

I was whittling on a stick with my new pocketknife as I walked, and I'd put my thumb on the back of the blade to cut a bigger chunk off the stick. The knife slipped. It cut through my thumb. I looked and could see the bone.

I wrapped my handkerchief around the thumb and applied pressure with my forefinger to stop the bleeding. The knife was so sharp that the cut didn't hurt. I calmly walked home and found a sterile bandage and some medication. But by now, the bloody handkerchief was stuck tight to the wound. I took a deep breath, then yanked.

I nearly fainted. I was ready to throw up. But I was able to put the medication on the cut. I'd just managed to stick the bandage on top when my legs went numb. I staggered to my room and fell onto the bed, nearly passing out.

When the skin finally grew back there was a huge scar. I now had a new thumbprint. That was quite a hole I cut in my thumb.

Satellites

On January 31, 1958, the first U.S. satellite, *Explorer I*, was sent into orbit. An Army-launched satellite, it weighed far less than either of the Soviet *Sputniks*, but it penetrated deeper into space than either of them. Despite the failure of *Explorer 2* to get into orbit six weeks later, four more U.S. satellites were to be launched in the next six months.

On March 17, 1958, *Vanguard I* was launched to test solar cells; on March 26, *Explorer 3*, to study cosmic rays and meteors; on April 23, another satellite, to study radiation above six hundred miles, with a mouse on board; and on July 26, *Explorer 4*.

The Legend of "The Hand"

As mentioned before, Bill Allen and I were members of the group that Bill would later define as "Loners and Dreamers." Eddie Greding was another member. We knew that somehow, if we stayed together, we'd prevail in this life.

Eddie was the zealous scientist. I admired that and wanted to be part of his work, contribute some tidbit, so I could later say I'd helped with one of his inevitable discoveries. He had the methodical discipline of a great scientist: he studied hard, he worked hard, and he kept outstanding notes on his experiments. But not all advances in science come from discipline. I knew that instinctively in high school.

The processes by which great discoveries are made was explored years later by Dr. James Burke in his book and TV series *Connections*. According to Burke:

"The triggering factor is more often than not operating in an area entirely unconnected with the situation which is about to undergo change.… There are certain recurring factors at work in the process of change. The first is what one would expect: that an innovation occurs as the result of deliberate attempts to developed it. A second factor, which recurs frequently, is that the attempt to find one thing leads to the discovery of another. Accident and unforeseen circumstances play a leading role in innovation. Another factor is one in which unrelated developments have a decisive effect on the main event. Physical and climate conditions play their part."

Eddie had something extra that set him apart. Although he calculated risk-versus-reward and rarely took careless chances, he was fearless. Eddie would try things just to see what would happen. He'd attempt things that didn't have the slightest chance of happening, such as his experiments in genetic mutation using X-rays on the black widow spiders in his collection. He was, therefore, more than just a great scientist. He was very nearly the kind of fanatical *mad* scientist you've seen in the movies.

He had no fear of being called mad. He also had no fear of animals, man included, because he knew they all reacted from three basic instincts: food, defense, and procreation. He had respect for animals, yes. He knew the boundaries and knew not to push them. He'd handle rattlesnakes, water moccasins, black widows, tarantulas, and other poisonous creatures. In his laboratory, he had all these and more, both alive and dead—skeletons and skulls, teeth and bones, skin and hair—and he was always on the lookout for new additions.

I was green with envy. I could've spent weeks in his lab looking through his microscope at the different specimens in his collection. He'd picked up a lot of the animals as road kill; for Eddie, death presented the opportunity to learn. He'd bring home the carcasses, cut them open, put the innards in bottles of formaldehyde, boil down the rest to get the bones. Then he'd glue and wire the bones together to make a skeleton. He memorized the names of the various body parts—he already could recite from memory most of the bones and muscles in the human body.

I don't think Eddie feared death. I suspected he'd already sold his body to science for research. He was unafraid of the un-dead, as well. He'd read all the available occult books and knew about the silver bullet for the werewolves, the wooden stake in the heart of a vampire, the garlic and the cross. He knew a few incantations and magical formulas, also: "An eye of a newt …" He must have one around the lab somewhere. "The hair of a frog …" He had hundreds of frogs—there must be one with a hair on it. "The ear of a bat …" Wait, he had no bat.…

Eddie needed a bat.

Meanwhile, Bill Allen was a persuader. He could get drones like me to do things. He was smart, like a martial arts master or a Shaolin Kung Fu priest. He'd use his opponents' strength against them. He was able to get things done his way.

Bill was a communicator. I needed that expertise. Ignorance of English and lack of command of the language was my downfall. English was like a foreign tongue to me. I spoke "poor-white-trash Texan." This was before bi-lingual education, diversity, and Ebonics. You did it their way, or they beat it into you with a wooden paddle, later replaced with detentions. But Bill knew English. He'd read books. He'd drunk from the fountain of wisdom. He'd shared with Eve an apple from the Tree of Knowledge.

Granted, I was good in math, science, mechanical drawing, and electric shop. But with only one of the "three R's,"—'rithmetic—I felt like an idiot savant, a 1950s version of Raymond Babbitt in *Rain Man*.

Bill and I were cruising in his 1955 Chevrolet one Sunday, down Loop 12 south of Dallas.

Loop 12 was one of the first highways to circle Dallas. Located on the city's south side, it was about halfway between Five Mile Creek and Ten Mile Creek (those names came from the distance to the creeks from the Dallas city limits at the time). Loop 12 was about midway between Dallas and Duncanville to the south. At the time, there wasn't much out there on the chalk hills of Loop 12, just scraggly cedar trees and some scrub oaks.

That Sunday we were coming back the long way from Mountain Creek Lake, west of Dallas at the end of Illinois Avenue, when I noticed a rock house, set back from the road. It looked abandoned. I pointed it out to Bill, and he said, "Let's go see it." So we turned up the dirt driveway and drove about fifty yards to the house.

It was set into a hillside. It had no windows, doors, or roof. It consisted of one room with a hole in each wall for a window. A rock stairway led to a cellar. We went down to see what was there. The cellar was just a single room. Three of its four walls were set into the hill; the fourth wall had two more large holes for windows. By the stairway there was a small room, like a closet, and inside the closet was a large wooden box and some scrap lumber. At one end of the closet was a foot-square hole that gave access to the stairs.

This house was the only sign of civilization for miles around. Other houses and streetlights hadn't yet made it out this far. By now it was afternoon, but we could tell the place would be very dark at night.

On the way home, we discussed possible uses for our new-found miniature castle. Eddie immediately came to mind. Bill said Eddie had spoken several times about needing bats for his lab. He wanted them alive, if possible. He'd wanted to go with Bill down near Austin, to a cave he knew about. But he and Bill could never get together.

This looked like a perfect opportunity to play a trick on Eddie.

Bill would call him on Monday and tell him about the rock house and the possibility of finding a bat there. Possibility, mind you, not that we had seen any, just the possibility.

Eddie swallowed the bait.

He was hooked, he wanted to go immediately, but Bill had to play the fish a little. Bill hesitated, said he had to work. That was okay, it would take Eddie a few hours to get his gear together. He wanted to make a net and some bags to put the bats in, and get something for them to eat. He'd be over to meet Bill when Bill got off work at 10:00 PM. I had to work at Neisner's until 9:00, so Bill took a break from work, picked me up at 9:00, and drove me out to the rock house. Then he went back to his job.

So there I was at our little castle, and everything was going as planned. I had some time to kill, so I went downstairs and arranged the wooden box to look like a coffin. I put some of the lumber on top to make a cover. That didn't take long, and I went back out near the road so I could see them coming. While I was waiting, I noticed a chalky mud puddle near the rock I was sitting on. I took off my class ring and watch and put them in my pocket. I pushed the sleeve of my black motorcycle jacket up over my elbow. Then plastered my hand and arm with chalky mud and let it dry. When it dried, it cracked when I moved my hand.

In the light of my flashlight, it looked real ugly and scary.

I was sitting about twenty feet back from the road. Bill and Eddie couldn't see me, but I saw them as they went by. That first drive-by was my signal to go get in the casket. I ran downstairs and climbed in the box. I pulled the loose lumber on top to cover me, but left an opening so Eddie would see my cracked chalk-white hand and arm.

I heard Bill and Eddie enter the house. There was some shuffling around upstairs as they searched for bats. It didn't take long for Eddie to see there were no bats, as there was no roof. Almost immediately, he started down the stairs. I noticed the glow of the flashlight as it reflected off the ceiling.

When they reached the cellar, Bill said, "Eddie, did you see something?" Then the light shone on my hand.

Eddie said, "Bill, don't move. There's a hand in that box."

Knowing Eddie would want the hand for his collection, I had to think fast, before he discovered our little ruse. I had taken a deep breath when Eddie came down the stairs. As he approached the box, I held my breath so I wouldn't be moving if he looked inside.

I now let out with the loudest screeching growl I could muster with my commanding military voice.

At the same time, I thrust my cracked, white, gnarled hand toward Eddie in hopes of grabbing him.

I think my hand almost hit him in the face as he reached down for his prize. Pieces of boards and lumber came apart and crashed around in the small closet as I jumped to my feet. The unholy noise was deafening as it echoed around the stone and concrete basement. When I got to my feet, I reached through the small opening to the stairs and tried to grab Eddie's leg. I managed to get part of his pant leg, just enough so he knew he wasn't safe yet.

Within five seconds, all was quiet. Eddie and Bill were nowhere in sight. I climbed the stairs, went outside, and began walking down the dirt driveway. When I reached the highway, Eddie's car was already quite a way down the road. Still in reverse without any lights on, it was going backwards at an impressive speed.

Afterward, Bill and I laughed all the way back home. Eddie just kept saying, "Thank you. Thank you. If it hadn't been for you two, I might never have known true fear.

True to his scientific bent, he went straight home and spent the rest of the night writing the following description of the events of that night:

The Hand

I can remember quite clearly all that happened on this night of March 23, 1958. You will probably not believe me now, but I am confident that you will when I have finished penning this fantastic story.

For quite some time now, I have had a consuming interest in astronomy and biology, as well as the history of magic etcetera. Among my studies of the biological sciences I have had a very great interest in bats, and their connection to legend. It is well known that all throughout history bats have been greatly feared and mistrusted because of their nocturnal characteristics. It is a known fact, however, that bats are for the most part entirely harmless to man. As a matter of fact, they greatly benefit mankind by destroying many insect pests.

Wishing to secure a few live specimens for study, I, and my friend Bill Allen, set out for an old house made out of stone which Bill knew of out in the suburbs of Dallas, Texas. Bats live in old deserted houses quiet often, and we hoped to find a few still in hibernation. At 10:02 PM, I picked Bill up from work, and we set out, our equipment consisting of two flashlights, and a polyethylene net, which we had constructed previously for the purpose of capturing the bats, which we hoped to find in the old stone house. We had some difficulty in finding the house, having passed it up once, but finally located it while coming back. We pulled up a bumpy old driveway, and parked, facing the house. Quitting the car, we advanced slowly to the door. The night was, it seemed to me, unusually dark, there being no moon in the sky, and the stars obscured by heavy clouds. Entering the house cautiously, we searched diligently for bats in the front, and finding none, proceeded toward the back, in hopes of finding some in the darker recesses of the old house.

Seeing some old steps leading to a lower level, we descended into a stone room, I going first, and Bill following closely behind me. My first impression upon entering the room was one of gloom and oppression. Two large windows looked out to the south, and a large pine tree was directly in front of one, weaving to and fro in the wind. Looking around the room with the aid of my flashlight, I saw standing at an angle in a niche in the wall a wooden box, about 6' x 1.5' x 1.5' in exterior dimensions. I couldn't help noticing that it looked remarkably like a coffin. Dismissing so naive a thought, I glanced across the room at the windows, and looked out apprehensively into the darkness.

Suddenly Bill said: "Eddie, did you see something?" I detected a note of alarm in his voice, and so turning said: "No, I didn't—did you?" What I now saw as I looked at the box filled my heart with terror, and the thought of it defies my very sanity—for there in the crack in the box was a yellow, cracked, seemingly dead hand. I said, quite calmly, "Bill, don't move. There's a hand in that box!" The hand began to move slowly forward, exposing an arm. [In reality, the hand was moving very fast, Eddie was experiencing a time dilation from all the adrenalin flooding in his body.]

Here I shall pause to attempt to describe my feelings at this moment. When I first beheld the dreadful hand, I was so utterly surprised that I was unable to speak. It all seemed as though it were a terrible dream. All at once, my knowledge of the horrible legends of the vampires and werewolves returned to me in a rush, and terror seized my heart in a grip of iron. We had absolutely nothing with which to defend ourselves, save our flashlights and our net, and a feeling of utter helplessness came over me. I felt rooted to the spot, and simply stared, fascinated, at the terrible hand, slowly moving out of the box. Believe it or not, my scientific curiosity overcame my fear, and I would have liked to stay and determine what the creature was, but Bill suddenly ran for the door. I was certainly not about to be left alone with the creature, and so I lost no time in following Bill out of the house. As I ran up the stone stairs, I heard the monster scream behind me, and I could hear it getting out of the box. Surely, nothing could possibly be more horrible than the dreadful knowledge that there were only a few feet of space separating me and this fiend of the darkness, and I expected at any second to feel a cold hand pulling me back down into the yawning blackness, to my death—or worse. I, therefore, determined to place a few feet more of space between me and my pursuer, and so I ran wildly up the stairs, and out of the house, passing Bill up! I didn't even take the time to look back to see if it was following until I was at the car. Never would I have thought myself capable of running so fast, but under the influence of so horrible a thing, amazing things can be done.

On my rather undignified exit from the house, I deliberated upon whether to use the car to flee, or whether to not take even this much time (perhaps, I thought, time enough for the monster to ascend the stairs and emerge also) and just "take the road." Bill settled the question by getting into the car—for I couldn't leave him to the mercy of whatever was in the house. Thank God, the door (for once) opened, and leaping into the car, I started it. I backed out into the road, at the same time, saying to Bill: "This is our only chance!" I didn't take even the time to shift gears, but simply backed out into the road. I backed down the road and didn't stop until we were a good two hundred feet away from the house. Then, turning on the lights, I recognized my "friend," Charles Willingham. My two "friends" Bill and Charles had arranged the whole thing as a joke on me, and we laughed over it all the way home, and for many days thereafter.

An interesting project would be a mathematical comparison of our velocity into the house compared to the velocity coming out. It would probably come out in the ratio of about one thousand to one! Nonetheless, I doubt if ever I shall forget my feeling at being confronted with what I though to be a monster from the world of darkness.

—Edward J. Greding Jr.

Carswell AFB and Being Locked Up

I was scheduled to go to Bill Hunt's wedding at 6:00 PM, when Danny Shreve came home on leave from the Air Force and caught me before school. We hadn't seen each other since he joined. We had a lot of catching up to do, so I skipped school. It was uniform day, and I was wearing my master sergeant ROTC uniform. Danny said he needed to go to Carswell Air Force Base (AFB) in Fort Worth to pick up his pay, so off we went, chatting all the way. He told me about his Air Force training. He was going to be a B-47 crew chief when he got to his new duty station. About 10:00 AM, we arrived at the main gate. The guards wanted to see his leave orders. He didn't have them with him. Then, seeing my uniform, they wanted to know where *my* leave papers were. I tried to explain that I was not in the Army, just ROTC.

They marched us inside and threw us in the brig. They were going to hold us until they could prove Airman Danny Shreve was not AWOL (away without leave). I might as well be there also, because it was Danny's car we were in and I couldn't have gone home if I wanted to.

They called some Air Force person in Dallas and sent him to Danny's house to get a copy of his leave papers. The experience took up the entire day. About 7:30 PM the Air Force guy finally called from Dallas saying he'd seen a copy of Danny's papers and Danny was authorized to be on leave. So they let us go.

But it was too late. I didn't make it to Bill Hunt's wedding. I didn't even get to call him.

Graduation and Typing Class

I had problems in high school because of English. In some of my earlier English classes, the teachers passed me just because I was older than everyone else. But in high school I'd failed a couple of English courses, and I'd had to take make-up classes in summer school. By the time senior year came around, I was reluctant to take English for fear of failing and not graduating.

I was doing okay in math and science. And in ROTC, I'd won Best Cadet in South Oak Cliff High School three years in a row, *and* I was named second-best cadet in the city of Dallas my senior year.

English was my Achilles' heel.

Bill was taking a typing class because it was supposed to be easy and he needed it to become a writer. He talked me into taking the class with him. If I passed, I'd have just enough credits to graduate.

Dyslexia was my downfall in typing class. Something weird went on in my brain. If I was typing with my left hand and needed an "i," I'd type an "e." If I was typing with my right hand and needed an "e," I'd type an "i." Mixed dominance? Who knows? Maybe it was related to my difficulties with eye-hand coordination, which had always kept me from doing well in sports. Anyway, reaching forty words per minute without errors seemed impossible. I could type fast enough, but when the teacher factored in my errors, I wound up with only fifteen or twenty words per minute. If I typed slower and made every keystroke count I could be error-free, but I couldn't make forty words per minute.

By the time final exams rolled around, I'd managed to get through typing with just the minimum passing grade, so I had to pass the exam. The day of the test, Bill was sitting next to me when a thought occurred to him. Just as everyone was ready to begin, he leaned over and whispered, "Hey Charlie, did you ever stop to think how many times your thumb hits the space bar?"

At that moment, the teacher, Ms. Lura Black, said, "Go!"

We started the test. As long as I was conscious of trying to do it, I could hardly hit the space bar with my thumb at all. Finally I was actually trying to hit it with my fist.

You guessed it. I flunked the typing test.

And the typing class.

I was not going to graduate.

Poor Ms. Lura Black, bless her heart, felt so bad. She went to the principal and requested a make-up test. Not wanting me to be the only twenty-one-year-old student in school the next year, he agreed.

The test was scheduled for 8:00 AM on graduation day. Hoping for the best, I was all dressed up in a suit and tie. I could hear "Pomp and Circumstance" playing in the auditorium when the teacher said, "Go!" I typed as slowly and deliberately as possible to avoid mistakes. Finally she called, "Time." It was all over.

In the auditorium Bill, having the last name of Allen, was already marching up to receive his diploma. The teacher very carefully checked my errors. She counted them up, subtracted them out, and said, "Wonderful! You *passed*, with 40.1 words per minute." She grabbed the test, and we ran screaming down the hall to the principal's office. Everyone checked and double-checked the results. Then they started jumping up and down as they finished the final paperwork.

I rushed to the auditorium just in time to take my place in line—being Willingham, I was naturally at the end. I marched down to get my diploma. School was finally over.

After graduation, I went back to the recruiting office to try to join the Air Force again. This time I passed the physical with flying colors.

I was off to see the world.

Part Two:
The Military Years

Chapter Seven:

As a Cold War Spy (1958 to 1963)

Selective Service System

In the post-Korean War era, men were required to register with the Selective Service System when they reached their eighteenth birthday.

This conscription was a general policy of involuntary servitude under which young and otherwise free American men were forced to serve in the Armed Forces. The policy was given various names and was colloquially known as "the draft."

With the draft hanging over our heads and the unsettled state of the world, many in my graduating class opted to join and complete their service obligation before war was again declared. There were also benefits, such as job training, as well as the opportunity to use the GI Bill go to college, which many of us in Hungry Heights could not afford. The military was a way to escape the bondage of poverty.

I don't know how many of our class joined, but there were several I know of: Bill Allen, Mike Bishop, Wray Harrison, Eddie Greding, and Mike Duckworth.

I suspect there were many more, but I've lost contact with them.

Some got married, found jobs, and had kids. The number of guys in our class who went into law enforcement—and the number who wound up on the wrong side of the law—struck Bill and me as ironic. Bill later wrote a book about Charles Starkweather, one of the first modern day spree killers, and contrasted our life with his.

My choice of the Air Force was one of the best decisions I made in my life.

The Beginning

My military years began unceremoniously. My best friend, Bill Allen, and my girlfriend, Pat McKissick, took me to the station and said goodbye as I boarded a train.

The train was supposed to leave at 4:00 PM. But nothing happened until after midnight, when there was some bumping as we were hooked to a switch engine and moved to a siding. Sometime in the early morning we were attached to the southbound milk run. What should have been a three-hour ride to San Antonio took all day. We stopped at every little town along the way, arriving at about 4:00 PM ("1600 hours" in military parlance) the next day.

The sergeants who met the train were very nice. They guided us into blue Air Force buses and shuttled us off to Lackland Air Force Base (AFB). They showed us our new barracks and told us each to pick a bunk. They called us "rainbows" because we were all dressed differently and in colors. That would soon change. A sergeant escorted us to the mess hall where we all had a good supper. Then he took us back to the barracks and explained that our training would start the next day. He said to get a good night's sleep. "Lights out" was at 2200 hours. We heard "Taps," the lights went out, and everyone went to bed.

All hell broke loose at 1201.

One sergeant was running a coke bottle around the inside of a corrugated tin trash can and screaming, "Get up and sit on the floor!" Three other sergeants were yelling and turning over the bunks of recruits who hadn't got their wits about them. They had someone stack two footlockers on top of each other to create a makeshift lectern. We all had to sit in our underwear ("skivvies") at attention while they lectured us on fire drills and how fast the barracks would burn. If they ran a fire drill, we were to grab our blankets and be outside in our skivvies in less than thirty seconds.

Our training had begun.

At 1230 we were sent back to bed.

At 0100 there was a fire drill.

There was screaming and hollering as they herded us into the street. Needless to say, we didn't make it out of the barracks in thirty seconds. So there was another lecture, with us sitting in our skivvies at attention again, about how fast the wooden barracks would go up in smoke.

Then it was back to bed at 0130.

And another fire drill at 0200.

This went on until 0500 when they sounded reveille. Some guys showed up outside in their skivvies with their blankets when they should have been dressed. Boy, did they get chewed out!

That morning began with physical training (PT), then breakfast, then off to the quartermaster for uniforms. They had us strip and put everything we owned in a basket, with small items such as pocket change going in an envelope. These would all be locked away until we left the base. We wouldn't need them anyway, as the Air Force provided everything. The quartermaster issued each of us two boots, one pair of dress shoes, two pairs of fatigues, khakis, belts, fatigue hats, several pair of socks, underwear, tee shirts, and one dress-blue uniform with billed hat. We put on our fatigue uniforms and took the rest to the barracks.

Next we were marched to the paymaster's office and issued our first month's pay: $143. A sergeant was waiting at the end of the pay line. He had two clipboards. One was for the extortion money he collected, for this or that, and the other was the kitchen patrol (KP) duty roster. You had to sign one or the other. We were given a list of items to purchase at the base exchange (BX). These were basic toiletries—toothbrush, toothpaste, soap, shaving cream, razor, comb, etc.—plus the insignia for your collar and hat. The BX had most of the things pre-bundled. Interestingly, the final cost of all the items at the BX was equal to what we'd received from the pay line. But we didn't need money, because we wouldn't be allowed outside the barracks unless we were being marched someplace. A base liberty was weeks away. We were marched to get military haircuts, and then military IDs. They sent our pictures to the local newspapers with a short note about us being in the Air Force—for their own little public relations purposes. Ha!

Classification

Basic training at Lackland was essentially a classification period to see what kind of work each man was qualified for. There were outdoor exercises and indoor training classes. It was June, and the weather was so hot that we were encouraged to take two salt tablets per day because of the heat. If they put "Maggie's drawers"—a red flag—on the flagpole, the temperature was over 95 degrees and we couldn't do outside exercises, only indoor training.

Most of the indoor work involved testing. During the tests, if I saw a question that looked like it would lead to a position as a clerk typist or a supply grunt, I'd mark the wrong answer on purpose. I didn't want to be stuck in a supply depot or worse yet, in an office in some God-forsaken place as a clerk.

One of the other major indoor training classes was about venereal disease (VD). They showed several very graphic movies about the effects of every kind of VD there was. I didn't know there were so many types and how bad and nasty they were. The instructors preached abstention, but just in case, they also discussed all the preventive measures: condoms, jells, hygiene, and such. Just to reenforce the issue, they showed and discussed the medical procedures with big square needles and razor-sharp instruments that were used on your private parts to try and cure the problem.

The drill instructors (DIs) noticed I did all my commands perfectly and signaled me out as a leader. They asked if I'd had any drill training. I said, "Yes, three years of high-school ROTC." They said, "Great," gave me a three-stripe armband for airman first class (A1C), and put me in charge. That gave me a feeling of empowerment, and in that sense, it changed my life. I was not the same kid I'd been in high school. I was no longer an aimless follower but a leader of men. The discipline in the Air Force helped focus my abilities on my areas of interest. I started to develop a better sense of purpose. It was a way of channeling my energies to suit my temperament.

I marched my flight to and from PT, chow, and classes.

After a month of evaluation, I was advised that I'd be good in radar, and as an airborne electric counter-measures repairman, and in something else electronic, I can't remember. I knew what radar was. It might be interesting, but I was afraid I'd get sent to Alaska, Canada, or Thule (Qaanaag), Greenland on the Distant Early Warning (DEW) line. I said, "Let's try airborne electronic counter-measures repairman" (ECM 30153A). I had no idea what that was, but the key words were *airborne* and *electronic*. I'd always wanted to do something with airplanes.

AWOL

After the third week of confinement to the base, I was getting a little stir-crazy. So when we had base liberty and one of the other airmen asked if I wanted to go into town, I immediately said yes.

Base liberty meant we had to stay on base. But he had a friend named Bob who lived in San Antonio and could come on and off the base without problems. He arranged for Bob to pick us up. Bob brought civilian clothes ("civvies") for us to wear. We changed into civvies, and Bob drove fifty miles north of San Antonio to a swimming pool. We swam all day.

I was a little nervous when we came back, because technically we'd been AWOL. But we had no problem getting back on base. We changed back into our fatigues and were safe.

Shots

One of the first things that happens when you go in the service is they give you a physical exam and shots.

By now, I'd had my fill of shots. I'd been poked and prodded as a kid. I'd had the complete polio shot series. I'd had a tetanus shot after I put my hand through the glass that time when I placed second in the drill competition. I'd had blood drawn at thirty-minute intervals *plus* been injected with radioactive stuff at thirty-minute intervals when they'd found albumin in my urine. I'd had a shot to relax me when I had my tonsils out.

I'd had enough shots to last a lifetime!

So in basic training at Lackland, when the time came for me to get my shots, I asked what they were for. They said one was for polio. I told the doctor (or "corpsman" or "pecker checker," whatever you want to call him) that I'd already had my polio shots.

He just said, "Prove it," and gave me the shot.

There would be more polio shots to come. Eventually, I went to the trouble to get proof that I'd had the complete polio series and didn't need any.

But more about that later.

While I was going through basic training other services were making great leaps in technology. On July 23, 1958, the USS Nautilus left Pearl Harbor; passing through the Bering Straits it made its way under the pack ice on August 1, sailed underneath the North Pole three days later, and emerged from the ice east of Greenland on August 5. In its ninety-six-hour journey under the ice it had covered 1,830 miles. Eight days later it docked at Portland, on the south coast of Britain. There would be many such breakthroughs in the coming decades as part of the greatest century of change in the history of man.

Keesler AFB, Biloxi, Mississippi

After basic training, we were sent off to Keesler AFB in Biloxi, Mississippi. At Keesler, all the basic trainees were billeted across the runway in nice, two-men-to-a-room, two-story masonry dorms. We had desks for studying. You really needed them—there was a lot of homework. For the first nine months, for eight hours a day from 0800 to 1630 with thirty minutes for lunch, I took a course in basic electronic circuits. It covered everything you needed to know about DC and AC circuits, Ohm's law, Kerchoff's law, and all the other laws.

It started out very simply, with learning the color code. I'd already learned the color code, but not to this degree. I didn't have my little cardboard calculator to figure it out, either. It was hard because you had to do the calculations in your head.

Luckily, to help you remember things they had these little ditties. The one for the color code was: "Bad boys rape our young girls but Violet gives willingly." The first letter of each word stood for a color, which also had a corresponding number: B for black, which was zero; B for brown, which was one; R for red, two; O for orange, three; and so on. There was also: "Eli the Iceman." In an AC circuit, "E" stands for voltage, and "I" stands for current. "L" is inductive circuit and "C" is capacitive circuit. In an inductive (L) circuit, the voltage (E) leads the current (I), and in a capacitive circuit (C), the current (I) leads the voltage (E).

My friend Tony Villars and I were always first in our class. We studied together and usually had everything down by the next day.

Every morning we were up before dawn at 0500, did our PT, and went to breakfast. At 0700 we marched to the runway and down it to pass in review before the officer of the day. One day a hurricane was coming ashore about the time we started across the runway. The runway had an elevation of about two feet, and the water on the runway was a foot deep. The wind and rain were blowing at sixty knots. I was carrying the guidon flag, and I almost had to put it down to keep from being blown off the runway. When we got to school, the air conditioner was on, and we almost froze in our wet fatigues. Fortunately they let us go back to the barracks at lunch to change.

October 11, 1958, the U.S. launched its first moon rocket, *Pioneer*.

Debbie Reynolds

After a few months we were allowed to visit the service club on base. One weekend Debbie Reynolds was there to entertain the troops. It was a madhouse—five thousand servicemen and one woman. She had to have the air police protect her. She gave a great show, but got mobbed afterward, passed out, and had to be carried out of the service club.

Ebb Tide

After the first semester, we were given weekend passes to go into town. There was a great little nightclub on the main highway called the Ebb Tide. It had a small combo, with musicians playing piano, drums, base, and guitar. Sometimes a clarinet or horn player would sit in. They played a lot of great music. I liked to sit and drink and listen.

One of my favorite songs at the time was "Patricia" by Perez Prado.

One Saturday night I'd had too much to drink and my buddies wanted to go bar-hopping. Well, I could hardly stand, much less hop. When we got outside, I told them I was going to lie down behind the hedge next to the building, and for them to come by and get me when they started back to the base. It was sure nice in that hedge.

Another Saturday night, I came back to the barracks half polluted. As I was undressing for bed, Frankie Avalon came on the radio in my metal wall locker singing "Venus." I didn't know the radio was on, and the sound seemed ethereal, coming from inside the metal locker. I thought I had died and gone to heaven.

Girls!—From Gulf Park Junior College

One weekend we went to Gulf Park and stopped at a Denny's on the pier. There was a group of nice young ladies there. Naturally, we struck up conversations and tried to get dates.

They all went to Gulf Park Junior College, a private school for girls. To get a date, you had to go see the principal and dean of women and get on the approved calling list. The next day, dressed in my finest, I went to the college. To get on the approved calling list, you had to submit three character-reference letters: one from your pastor, one from the principal of your last school, and one from a business or professional associate, such as a doctor or lawyer. I wrote to Reverend Gay Weaver, whom I'd worked with at Neisner's; and to Mrs. Elizabeth Lockhart, dean of South Oak Cliff High School; and to my former boss Mr. Napier, manager of Neisner's. I explained the situation and asked them to send the appropriate letters.

Time went by, and I didn't hear back from Gulf Park Junior College. I figured I hadn't been approved. But I was later to learn something entirely different.

December 18, 1958, the U.S. launched its fifth satellite into space, an experimental radio relay station, which successfully recorded and then rebroadcast back to earth a Christmas message from President Eisenhower.

Trying to Outfox the Military

Once we'd completed the preliminary training, we began a four-month course in advanced electronics, eight hours per day. Advanced electronics went into detail about radios: amplitude modulation (AM), frequency modulation (FM), pulse-control modulation (PCM), radar, direction finding (DF), modulation/demodulation, radio frequency (RF), intermediate frequency (IF), audio frequency (AF), wire/tape recorders, cathode ray tubes (CRTs), signal/pulse generators, and detectors. This was the real stuff, where we put all the basics we'd studied into practical applications.

It was great! I could hardly go to sleep at night I was so excited. It explained all the questions I'd had when I was playing around with radios in my ham shack and getting my amateur radio license. At that time, I'd had more questions than answers, and now, here were the answers—and they all made sense.

As we approached the end of school, it was rumored that the top student would get the base of his choice. I went to see the first sergeant to find out if this was true. Yes, it was true, he said, but Air Force need had priority. So if you were first in your class and wanted one base and the Air Force urgently needed you somewhere else, you went where they needed you. I managed to see the records of the past six years and learned that *no one* had been sent to the base of his choice. Air Force needs had *always* taken priority. When it came time to fill out our request, I thought, "Where do I *not* want to go?" I wanted to see the world, so I wanted to go any place but home. I put down Carswell AFB in Fort Worth as my first choice, and either Europe or Asia as my second.

Well, you guessed it. I finished top in my class, and they sent me to Carswell AFB. I was the first guy in seven years to get his base of choice.

You just cannot outfox the military.

More Satellites and More Rockets

On January 1, 1959, there were six satellites in orbit around the earth. These included the *Atlas* satellites and *Sputnik 3*. On February 17, 1959, *Vanguard 2* was launched, followed by *Vanguard 3* on September 18, 1959. These weighed twenty-one and fifty pounds respectively, and they were expected to be in orbit between 301 and 1,790 miles above the earth's surface for fifty years. The aim of *Vanguard 3* was to send back photographs of the earth's surface, but this failed as a result of the satellite's irregular motions.

In addition, on March 3, 1959, a moon probe, *Pioneer 4*, was launched, and it passed within 37,000 miles of the moon. Scientific curiosity, competition with the Soviet Union, and the intelligence-gathering aspect of our defense policy were all in harmony. Then, on May 28, 1959, the U.S. launched two monkeys into space. The rocket in which they were traveling reached a height of three hundred miles, and then returned to earth—with the monkeys alive. There was a sudden realization around the globe that it might be possible to put a man on the moon.

Another Shot?!

Before we were released we had to have another physical and more shots. *Another* polio shot?! If you didn't have proof you'd had the series, you had to have the shot.

I took that one, but I was determined not to get that damned needle stuck in me again, not for polio, at least. So on my way to Carswell AFB, armed with my military shot record, I stopped to see my mother and asked her where I'd received my polio shots. She said, "The clinic on Marsalis, near the Dallas Zoo." I made a beeline for that clinic. They found my name in their files and filled out my military shot record: "Polio Series Complete."

The doctor signed it.

Now I had proof.

But would it be enough?

Carswell AFB, Fort Worth, Texas

When I reported to Carswell AFB, I learned I'd be working outside my career field. The equipment they had there was all "active jamming." I'd studied nothing about this in my classes at Keesler AFB, so I hit the books. Luckily Airman First Class Hagstrom, nicknamed "Doc," took me under his wing and helped. He was probably the smartest one there. Thank goodness he liked me!

We worked on B-52s, E and F, which had fourteen transmitters and power supplies, plus four-hundred pounds of chaff to jam signals.

Chaff was like aluminum foil cut into strips. There were some really long strips, as if a tube of aluminum foil had been cut across in quarter-inch slices. When thrown from a plane, the long chaff would unroll as it fell, and randomly long lengths of metal foil would descend slowly from the aircraft. There were also shorter "radar lengths" of chaff, cut to correspond in centimeters to the electrical length of the radar frequency. A common radar frequency was ten centimeters, or approximately four inches. These strips were pencil-thin and heavier on one side so they would fall more or less horizontally, but would flutter and tumble as they fell. Each one—and there were hundreds of thousands, even millions in four-hundred pounds—was intended to mimic a B-52 and confuse the enemy's radar. If the wind was just right, this stuff could stay up for hours and create a cloud-cover for many miles in all directions.

During July and August the temperature on the flight line—concrete that ran for miles—would be 102 degrees in the shade under the wing. That was cool compared to the 150 degrees inside, in the back of the B52s where the transmitters, power supplies, and chaff were. Up front, in the crew compartment, there was air conditioning, but not in the back. Because it was so hot, we'd take off our fatigues, stripping down to our skivvies and jump boots to work.

We always had to have two men on the airplane at any one time, in case of an accident. One would get into the tail hatch, which was normally seven feet off the ground, and catch the equipment as the other one "cleaned and jerked then pressed" each piece of equipment over his head like a weightlifter. Each transmitter weighed eighty-five pounds, and each power supply weighed fifty-five pounds. It was really an Olympic event, unloading and loading fourteen transmitters and power supplies and four-hundred pounds of chaff. We worked in five-minute shifts.

From that job, I remember there was this master sergeant who always carried a clipboard under his arm and was always hurrying to places unknown. He never seemed to have time to stop and answer any questions, much less engage in idle conservation. I never did see him actually do anything.

Sometime in May, two months after leaving Keesler, I received a formal letter from the dean of women at Gulf Park Junior College. It informed me that I'd been accepted and placed on the approved calling list. There was to be a formal party for all the new members on the list, so the young ladies could meet us and we could formally introduce ourselves. An RSVP was requested. Too late! But it was good to know they approved of my character. I showed the invitation to everyone who'd written recommendations for me, and I thanked them for taking the time to send the letters.

Carnival

In June, the base was flying B-52s continually to use up their fuel allotment so the Air Force would give them the same amount at the beginning of the fiscal year that started July 1. Every day, as the planes came in for refueling, the base would require a complete change of equipment to "configure for a different attack scenario." That meant taking all fourteen transmitters and power supplies off and putting a different set back on, and loading another four hundred pounds of chaff. Talk about a work out!

While this was going on, the base ran out of money. How could that happen? They said they didn't have enough to pay miscellaneous expenses until after the end of the fiscal year. The base commanders decided to have a carnival to make some extra money to carry us over until the new fiscal year began.

Each airman was issued twenty dollars' worth of tickets to the carnival and told to go sell them to the local civilians. We went out all week and tried to sell tickets. Selling is not one of my best talents. I managed to sell two in a week. On Friday, when we handed over the money we'd raised, we tried to turn in the unsold tickets, but they wouldn't take them back. All they wanted was twenty dollars. A noticed was posted on the bulletin board at 1400 saying there'd be a commander's call at 1600 for those who hadn't sold all their tickets. At 1600, a group of us went to the commander's call. The commander was not there, but the executive officer was. He informed us that when we'd joined the Air Force, we'd signed up to work seven days a week, twenty-four hours a day, with five minutes off each hour. He said we'd do so until the tickets were sold.

Well, that rubbed me the wrong way. My roommate and I discussed the issue and decided to go see the inspector general (IG) and let him know what was going on. On the way to see the IG, we stopped by the chaplain's office and explained the problem. The chaplain listened very closely and explained, "I've been an enlisted man and an officer. Based on my experience, I'd say they'll do more to you two for complaining than they'll do about this issue. My recommendation is that you pay the twenty dollars, go to the carnival, and drink it up in beer." After hearing that from the chaplain, I decided discretion is the better part of valor. So I paid the twenty bucks and had a good time drinking beer at the carnival.

Beatniks

My roommate and I became beatniks. We found a couple of coffee houses in Fort Worth and Dallas—The Eighth Day, The Coffin, The Rubaiyat, and The Cellar—where we could drink expensive espresso and listen to poetry. I bought a pair of bongo drums, and sometimes they'd let me play. Ha!

"Quiet Village" was one of the popular songs of the day. I had a radio and kept it tuned to KOA Denver. I remember listening to the "Procession of the White Elephant" from *The King and I* as I fell asleep one night around 0200.

My roommate had a 1958 Ford convertible and was dating a girl from Abilene, where he was from. She was a student at Texas Christian University in Fort Worth. We'd go to her dorm to pick her up, and sometimes she'd bring a friend along for me. Maybe this was for her own protection! Anyway, we always had a great time. The only problem was the girls had to be back in the dorm before midnight, because the house mother locked the doors for bed check.

One weekend I took a couple of rolls of that chaff with me when we went to the local park on University Boulevard and Camp Bowie Avenue. As we drove through the park, I stood up in the back seat and strung the chaff through the low-hanging tree branches. One branch was a little out of my reach, so I jumped up to hang the chaff on it. When I came back down there was no car under me. I hit the road and rolled for a few feet. No major damage, nothing broken, just injured pride and a lot of scratches and bruises. Boy, was I lucky we weren't going over five miles per hour.

A "Brush" with Amputation

I was in the best physical condition of my life, having lost twenty pounds. I'd gone from 175 down to 155, mostly from working in that hot Texas sun. I feared I was about to lose a few more pounds when, after a minor accident, the quacks at the base hospital threatened to amputate my foot.

This episode began one Sunday when my roommate and I picked up our girl-friends and went to the beach at Lake Worth, north of Fort Worth. About 1600 I was horsing around, chasing one of the girls down the beach, when I stepped on a rusty upturned bottle cap. Its sharp edges took off most of the under part my right big toe. To stop the bleeding, I tied a sock around it. That seemed to work. I didn't want to leave the party. I figured I could go to sick call on Monday. (Boy, how stupid young men are!) We had a great time at the beach and returned the girls to the dorm at 11:59 PM.

Next morning, I went to the base hospital. It was run by a bunch of sadists. When the hospital guy heard my story and saw the sock, a grin spread over his face.

He said I was lucky my foot hadn't turned green and wouldn't have to be cut off. I was even luckier that my jaw hadn't locked up. Of course, they gave me a shot—a tetanus shot. Someone yanked the sock off, while someone else brought out a pan of scalding antiseptic water to soak my foot in. They jammed my foot in the near-boiling water. Then they took it out and scraped it clear to the bone with a stiff-bristled brush. They called this procedure "cleaning the wound" and continued until there was very little skin left on the toe.

I sat there for an hour, soaking my foot while they prepared a bandage and did the paperwork. They said they'd need to see me again in a couple of days to check for gangrene, in which case they'd cut off my foot. But my foot healed nicely.

I'm sure they were disappointed that they didn't get to amputate!

World Series

September was World Series time. Almost everyone was watching the game in air-conditioned offices one day, when the sergeant told me to go check out some problems on one of the airplanes. It was hot outside! My partner and I went out to the aircraft. Two other crews were working on other planes. We hooked up the air conditioner, but it was still plenty hot in the crew compartment as the air conditioner hadn't yet had time to cool it down.

I was up in the electric counter-measures (ECM) operator's seat, tuning through the frequencies and checking out the receiver, when I heard the ball game on the radio. Listening to it as I checked out the transmitters, I got a fiendish idea. I tuned one of the transmitters to the TV-station frequency. I waited until a couple of men were on base and it was a three-two pitch. Then I switched on the transmitter and blanked out Dallas and Fort Worth TV with a megawatt of jamming power.

Everybody was pissed! What I had was a hundred times more powerful than the 100,000-watt TV signal. The squeals, clicks, and frequency-pulse modulation used for jamming gave it away as one of our jamming transmitters. So the FCC sent a tracking truck to the base to investigate where that signal was coming from. They knew it was Carswell, but they didn't know which airplane it was.

My partner didn't know what I'd done. I tuned the transmitter off of the TV station frequency and switched it off. I played around with some of the other equipment to use up some time—to make it look like we hadn't been in the airplane when the problem occurred. After a while, we buttoned up the airplane, turned off the air conditioner, went back to the office, and waited.

Everyone in Dallas and Fort Worth was complaining about missing that pitch. Six airmen were out on airplanes that day, so no one knew who'd blanked out the game.

Everything was quiet for about a week. Then one day, when I was out on another plane checking the equipment, I was told the first sergeant wanted to see me. Right away. I was to drop everything and go *now*!

I figured they'd somehow found out who had done the dastardly deed. It was time to face the music. I was resigned to my fate, as I trudged into the first sergeant's office. But I was safe. The first sergeant simply informed me that the Air Force had an urgent request for someone with my training. How did I feel about going to Rhein, Main, something or other?

I thought he said Maine, as in the state. "This is fantastic," I said to myself. "I've always wanted to go to New England."

Then I heard him say, "Germany."

Wow!

Background Check

Was I excited? You bet. I wanted to leave right away. But first I had to fill out this DD219 security background investigation form in triplicate. Then the FBI would do a background check. The process would take several weeks.

"Let's do it," I said.

I filled out the form listing all my relatives, and all the places I'd ever lived or worked. I think I had to put down the name of everyone I knew. My mother had to help me get some of the information. What a form!

Meanwhile, I was going home almost every weekend and visiting most of the people I'd known from before I joined the Air Force. I nearly always went to Neisner's where I'd worked. Most of my friends were there, and some of my old girlfriends, too.

One Saturday as I sat at the Neisner's lunch counter, Mr. Napier came up. He sat down next to me and turned his seat to face me. In a serious voice he asked if I was in any kind of trouble.

I said, "No, why?"

He told me the FBI had been to see him and Reverend Gay Weaver. The FBI had asked them both a lot of questions about me. He wanted to make sure everything was in order and that I wasn't in any trouble.

"Oh, that was for my background investigation to get my top secret security clearance," I said. "I'm being sent overseas!"

The Big Apple

On my way to Germany, I took a thirty-day leave to use up some of the leave time I'd accumulated. I spent two weeks at home and then two weeks in the Big Apple—New York City. I arrived in New York on November 1, 1959. They are very good to servicemen in New York. At the USO nearly everything is half price, or free.

I got free tickets to the top of the Empire State building, and to Radio City Music Hall, where I saw the premier of *A Summer's Place* with Sandra Dee. I also saw the Rockettes and many more sights. I stayed at the YMCA downtown for a dollar a night. I rose real early one morning and walked from Central Park to Greenwich Village. I stopped at a pizza place at noon and had a pepperoni and mushroom pizza and a beer. The walk took all day, but I was in no hurry. When I arrived in the Village, I found a coffee house and spent the evening there. I took the subway back uptown.

While in New York City I also visited the United Nations Building and the Museum of Modern Art, where I saw some fantastic art pieces. I remember one in particular. It was an infinite kaleidoscope of color and design in continuous motion.

Speaking of motion, it wasn't long until my New York City vacation was over and I was off to McGuire AFB, Mobilization Assistance Team (MAT) headquarters, for transport to Frankfurt, Germany.

Germany

I arrived at Rhein-Main AFB in Frankfurt am Main (Frankfurt a.M.), Germany, on November 19, 1959.

I was processed through customs, then I reported to my new duty station at the 6916th Radio Squadron Mobile in barracks building number 286. My background security check hadn't been completed, so I worked in the first sergeant's office for a little over a month. After work each day, we'd go to the airman's club and have a couple of those good German beers. The airman's club was named the Rocket Club. It was across the street from the Gateway Service Club, where Miss Leilue Remain and Miss Eleaim worked as hostesses, and the base library.

My new group was anxious to get to know me, so some of the non-commissioned officers (NCOs) that I'd be working for invited me to their houses for Thanksgiving, then for Christmas open houses, and then for New Year's parties. After New Year's, it was Fasching. Fasching originated in the Alps to chase the winter away. It's the German Mardi Gras. Fasching is a time when the Germans take off their wedding rings and everybody gets drunk and has a lot of fun. The favorite saying was "Macht nichts, ist Fasching." That means, "It makes no difference, it's party time."

We nearly always had a beer with lunch. You can image, after drinking only 3.2 percent in the states, and very little of that, how drunk I'd get on the good German 12 percent beer. I awoke one morning in March with an awful hangover and decided I'd better stop drinking that good German beer and get my shit together. I wrote to Bill Allen and asked him to send me a list of good books to read, so I could increase my reading speed. When the letter with the list in it arrived, I immediately went to the library and checked out one of the books. I'd go into the listening room and put on a good classical record and read. This went on for several months. This is where I found that a lot of the popular music I liked was based on classical music, such as the *Warsaw Concerto* by British composer Richard Addinsell, written for the 1941 film *Dangerous Moonlight* (also known as *Suicide Squadron*), a piece of music in the style of Rachmaninoff's *Piano Concerto No. 2*. There were many more pieces of classical music that I liked, especially the ones on the radio shows I listened to at night.

Diplomatic Passport

When my security clearance came through, I was taken to the American Embassy in downtown Frankfurt by one of the NCOs I'd be working for. At the embassy, I was issued a diplomatic passport. Regular passports were green with the United States seal on the cover. The diplomatic passport was a little larger and had a maroon cover. Inside was my passport picture, in which I wore civilian clothes. I'd had to borrow a suit coat from one of the other airmen, as I didn't have any civilian dress clothes. The passport simply stated that I was "abroad on official U.S. government business." Enough said.

Frankfurt am Main

On one of my first free weekends, I went to downtown Frankfurt and walked around. It's a beautiful old city, rebuilt well in the old style. When I got to the center of town, though, I was really disappointed. There, in the middle of these beautiful old buildings, was an F. W. Woolworth store. Sure enough, it was exactly like the ones in the states. Boy, was I bummed out. Here I'd come 15,000 miles and was seeing the same thing I could see at home.

I continued to walk around and found an old bombed-out church. It was the only building left after one of our bombing raids during World War II. The story I heard was that the Nazis noticed that the Allies tried not to bomb churches, so the SS or Gestapo had moved its offices into the church. When our side found this out, the church was put on the bombing list. A lot of nuns and priests were killed in the next raid. The bombed-out church was left there as a reminder.

I went out to Frankfurt Grueneburg Park, where the 5th Army Headquarters were in the I. G. Farben Poelzig building and where the Army exchange and commissary were located. The old I. G. Farbin building had been Nazi headquarters during the war.

Then, as it was getting late, I went back near the hauptbahnhof (the main train station) and found one of those German oom-pa-pa places, the Mari Gusti, on Kaiser Strasse. I went in for a beer. While I was looking around for a table, I ran into Wray Harrison, a guy I knew from South Oak Cliff High School. He was there by himself having a beer, and I sat down with him. It was the second time I'd run into him since graduation.

Dancing with a Dream

I was sitting in the service club reading a book at 1930 hours one day when I heard the brisk heel-taps—click, click, click—of a woman in high heels coming across the hardwood floor. Looking over the top of the book, I saw this beautiful lady, Lynn Wescott, heading for the back room in the club. She set up a record player, and she put on some old 1940s dance music. When I asked what she was doing, she said she was starting a ballroom dance class for one dollar per lesson. "I'll take a dollar's worth of that," I thought.

I was her first student.

The class grew very quickly to twenty or more. It was mostly airmen, but a few wives joined, too. I was pretty nervous the first couple of times I danced with Lynn. I was even shaking. I'd never in my life been that close to such a beautiful woman. It was plenty scary, but I finally overcame my jitters.

Dancing with Lynn was like dancing with a dream. She could anticipate exactly what you were going to do, unless you screwed up. Even then, she was so light on her feet and easy to lead, it looked like you'd done what you were supposed to do. She was very good at covering up mistakes. After just three lessons, I felt like a pro. Especially when I danced with her.

It was another matter when I danced with one of the women who were just learning. That was the blind leading the blind. You really had to know how to lead them. It was good training. We learned the fox trot, the waltz, the rumba, the mambo, the cha-cha-cha, the swing, the samba, and the tango.

I wasn't going to let this good thing end. I took the intermediate class, and the advanced class, too. In the advanced class, Lynn would sometimes ask me to help demonstrate the basic steps. When I finished the advanced class, I came back for a refresher course—ha! That's when she asked me to be her partner and help demonstrate dance steps to her new group. After that class, she asked if I'd come to one of her other classes in Wiesbaden.

Boy, would I!

This was opening up a lot of new possibilities: more time to spend dancing with her, plus a chance to meet and dance with some new women. It turned out even better than I imagined. Wiesbaden was Air Force Headquarters, and they had an entire hotel, the Amelia Earhart, that housed all the American secretaries who were there working for our government. There must have been five hundred of them, and fifty took the dancing lessons. After class, we always went into the ballroom and danced to a live band. Showtime!

Lynn invited me to go with her to downtown Frankfurt, to the Army's 5th Corps Headquarters Officer's Club where she was giving dancing lessons. I went with her and walked into that Officer's Club with her just like I belonged there. They thought I was a civilian like she was, and never asked any questions, and I sure wasn't going to say anything. I danced with a lot of the officer's wives and even the general staff officers. Sometimes I'd have to dance with the general himself to demonstrate the correct steps and the motions he needed to know to guide his wife around the dance floor. If they'd known I was a lowly airman, I don't know what they'd have done. They'd probably have court-martialled me.

Then Lynn asked me to go with her to Heidelberg, U.S. Army Headquarters, European Command and the 7th Corps Headquarters Officer's Club, another place where she was giving dancing lessons. Again I had to dance with the general.

After I'd spent two years of helping her with the classes, she booked two classes on the same day and time and asked if I'd take one and teach it myself. She said all I had to do was pick someone out of the class, like she did me, to assist. So I did. I became a ballroom dance instructor.

It really was showtime!

More Rockets and Other Items of Interest

January 29, 1960, an "Honest John" artillery rocket, on its way to a target and capable of carrying a nuclear warhead, was blown up by a Hawk anti-aircraft missile. The rocket warhead never reached its target. Experiments were immediately begun to increase the range of the anti-aircraft missile so that it could intercept and destroy any incoming intercontinental ballistic missiles. The new defensive missile was given the name *Nike Zeus*.

In February 1960, a nuclear-powered submarine, *Triton*, sought to circumnavigate the globe underwater without once rising to the surface. It was successful, completing its 41,519-mile circumnavigation in eighty-four days.

On March 11, 1960, the American space probe *Pioneer 5* was launched into orbit around the sun.

On April 1, 1960, *TIROS 1* was launched to photograph the world's weather conditions from space. TIROS stood for television and infrared observation satellite.

On April 27, 1960, the nuclear-powered submarine *Tullibee* was launched in the U.S. with long-range sonar.

Birthday Party

On April 30, 1960, I attended a party for my friends Shelp and Paige at The Countries, a guest house in Aschaffenburg, just outside the main gate. They served goat sandwiches. Moodie, the owner, was the pump frau. There was only one bathroom, and that was reserved for the ladies. The guys had to go outside and see how high they could pee up the wall. Chief Master Sergeant Stover's son, Robert Stover Jr., fell in the mud next to the wall, and we had to prop him up on the stairs to dry out. I still have a photo of him on the stairs. My friend Ski got so drunk he had trouble getting up. Other friends were there, too: Hoch, Pranuche, Lee, Lowe, Woods, Carol, Strain, Ramero, Middie, Stephens, Bennett, Rita, Britt, Cross, Acton, Notzon, and Raymond Padgett. Howe Callers was a photographer and took a group shot of all of us. When we returned to the barracks, Chen and Reardon had to roll our friend Barn up the stairs. He was a happy sit-down drunk.

Ice Skating

There was an ice skating rink just behind the barracks. I bought a pair of ice skates and went skating every night that I wasn't busy with dance class. Saturday and Sundays were also good days for ice skating. It was a lot like roller skating, which I did a lot of in high school. I put the roller skating training to use in ice skating for moves such as the figure eight, being "on an edge," holding the body and arms straight to keep from turning, or turning the shoulders to make a turn. I used a lot of the roller skating dance steps on ice also. It was fairly easy, once I'd achieved the strength in my ankles to keep the blade straight.

Schwimbad, or "Call Me Schwartz"

One Saturday a friend and I decided to go to a schwimbad in downtown Frankfurt. A schwimbad is an indoor heated public swimming pool. For the price of a German mark (twenty-five cents in those days) you could swim for an hour. On your wrist they put a colored band like the one you get at a hospital, and they used the color to tell when your time was up. My wristband was black. We were swimming for over an hour when the lifeguard started whistling at me and shouting, "Schwartz!"

I didn't understand—my name wasn't Schwartz.

One of the German swimmers could speak a little English, and he explained that "schwartz" in German means "black." The lifeguard was telling me that since I was wearing a black wristband, my time was up and I had to get out of the water. My friend and I left.

6916th Radio Squadron Mobile

The 6916th Radio Squadron Mobile was made up primarily of linguists who'd gone to the Presidio Defense Language Institute (DLI) in Monterey, California, or to Syracuse University in New York. They were predominantly trained in Russian, but many of them were multilingual and could speak several languages fluently. Most knew German, and a few knew Serbo-Croatian. They told a story about a professor who jumped out of the class window when DDT trucks came by spraying for mosquitoes. He thought for sure it was his betrayed country trying to gas him.

The eavesdropping system was called ELINT (electronic intelligence), the most familiar form of which was airborne eavesdropping conducted by planes cruising Russia's borders and recording all radio-communication traffic.

Maintenance Shop

Chief Master Sergeant Stover was in charge of the maintenance shop. Senior Master Sergeant McDougle was second-in-command of the maintenance shop and probably the smartest of the bunch. He had a little drinking problem, though. Or maybe it was Parkinson's disease, I don't know, but he had the shakes. He couldn't put a screwdriver on a screw or tune a variable resistor unless he could pick up both pieces. That way they both moved with the same motion, and he could hit the slot. He knew his electronics, though.

Our maintenance group had twenty-one NCOs and eight airmen. Because our airmen were frequently away or taking R&R (rest and recreation), typically only four airman were in the shop at any one time. The NCOs would take turns supervising us airmen. One would supervise from 0800 to first coffee break, and then another would supervise from first coffee break until lunch. After lunch, another sergeant would supervise us until second coffee break, and then another one from second coffee break until quitting time. Most of the NCOs had been in the Air Force a long time and had been cooks or air policemen, the bottom of the intelligence ladder, before cross-training into electronics. Most couldn't find their butts with both hands. It was best that they left the equipment maintenance to the airmen.

We had three black airmen. They thought the NCOs were always picking on them. The NCOs really weren't, but the black airmen thought so. One day Airman Campbell was mad and yelled, "You are always picking on me, motherfucking Silcott and Smith!"

We were instructed always to have our flight bags packed. We were to be ready to go anywhere in the world for two weeks on a two-hour notice.

And we had to have shots.

Yes, for a new duty station, they needed to give me more shots. We might be going to some God-forsaken, disease-ridden place, so I needed a lot of shots.

I was sent to the base hospital to get my shot record updated and to get all my shots for flying status. I asked the doctor, who was a first lieutenant, what each shot was for. When he came to the polio shot, I told him I didn't need it. I said my polio series was complete and that I'd had several extra polio shots that year and didn't need another one.

He said, "Are you refusing to take this shot?"

"Yes, sir, I am."

"You are refusing a direct order from an officer."

I said, "I know it, but I am not taking that shot."

He said okay, and I went straight to the first sergeant and told him what had happened. He just roared! He said, "I bet no one's ever told him they weren't going to take a shot." He said not to worry, that he'd take care of everything. A couple of months later, when I saw the first sergeant again, he said that being on flying status meant I had to have a polio booster shot every year. So next year, I'd have to get the shot again.

I knew they'd figure out a way to give me that damned shot.

C-130

The C-130 was the Air Force's workhorse cargo aircraft. It could fly long hours and do short and soft field take-offs and landings. It could easily fly on two of its four engines. It could also do what most aircraft cannot do: it could back up. The turbo props could be reversed to give forward thrust to slow the aircraft down on short field landings, and when parking, it could back up. This universal workhorse was ideal for our missions.

The aircrafts were part of the 7406th Material Command. They were maintained and flown by aircrews from the 7406th. The flight crew would be briefed on what we were doing, but they weren't allowed in the operations rooms during missions—they didn't need to know how our operations were performed. Most of the time, when we landed in our normal airfields it was no big deal, we just taxied up to our place on the ramp or hangar and our own crews took care of the aircraft. The aircraft carried a "general staff officers" call sign, however, so at the non-routine airfields, the transit crews would be on their toes trying to take care of the general and his aircraft. There would sometimes be some confusion when transit crews were not allowed on or in or near the aircraft. But we got VIP treatment everywhere we went.

Our aircrafts were outfitted with extra fuel tanks, not the tanks hanging under the wing. When we were on station and flying our orbit, the pilots would put the aircraft in what was known as "max endurance." The turbo props ran at a constant speed, and the velocity of the aircraft was controlled by the pitch of the props. The aircraft commander would turn the props so the engines used a minimum of fuel. This allowed us to stay on station for about eight hours, depending on where we were. Our flights usually lasted ten hours each with plenty of reserve fuel. If it was absolutely necessary, we could stay up for twelve hours. But that would cause us to run short on the minimum reserve fuel that enabled us to get to an alternate airfield if Rhein-Main was closed for any reason.

Exterior

Our planes were cargo aircraft modified by the LTV Aerospace and Defense Company in Greenville, Texas, to have fiberglass panels put on the sides. This would enable our antennas to receive radio signals without being obvious. Two dummy fiberglass fuel tanks, mounted under the outboard wings, housed a set of phased direction-finding antennas. A long wire antenna ran from forward of amidships to the top of the vertical stabilizer. We also had some stub antennas, about the size of king-size cigarettes, sticking out of the bottom of the fuselage. Otherwise, you couldn't tell our aircraft from any other C-130 cargo plane in the Air Force.

Interior

Inside, it was a different matter. Under the flight crew's deck, we had six inverters, which converted the aircraft's 28-volt DC power to 110 volts AC for our equipment in the back. There was an almost soundproof three-room compartment that housed the linguists, who were known as Ravens.

First Compartment

The first compartment had four operators, each at his own station, with two facing forward and two backward.

In the first compartment, Stations One, Two, and Three were alike. Each had two radios and two Ampex 600 tape recorders. The operators would tune one of the radios to a frequency they were supposed to monitor, and turn on the tape recorder. The tape recorders were voice activated, so anytime a voice or a signal came on, it was recorded. Each station had a digital signal piped in that automatically recorded, on one of the stereo tracks of the tape recorder, the day, date, time, frequency, and some other information. The operator used the other radio, the one not being automatically recorded, to scan and search for other signals of importance. He could copy signals on a second recorder, if necessary. Most of the time, however, the second recorder was kept loaded and ready so that when the first recorder ran out of tape, the operator could simply switch recorders while he reloaded new tape on the first.

Station Three was like Stations One and Two, except that it also had a direction-finding (DF) capability in one of the radios. These radios were connected to the two dummy fiberglass fuel tanks that housed the set of phased direction-finding antennas. As the aircraft flew along, the operator could take a DF track on the signal he was monitoring, and get the angle from the aircraft from which the signal was coming. At twenty-to-thirty-minute intervals, this would give intersecting lines, via triangulation, that established the location of the signal.

Station Four was the International Morse Code station. We had two Collins Radio J-51 receivers connected to the long wire antenna, and two Ampex 600 tape recorders. The operator would copy FOX-type five-letter code groups up to one hundred words per minute while carrying on a conversation. He would monitor the Russian radar nets and plot our position, as well as the position of all aircraft in the area that the Russians were seeing on their radar. He would plot all this on a map he had in his lap with a grid overlay. The grid overlay had to be changed every time the Russians changed their codes.

Between the two sets of operators, situated on the opposite side from the combination-lock door, was a fourteen-channel high-speed tape recorder. With that we could record some of the signals directly.

Second Compartment

Like the first compartment, the second had four stations—Five, Six, Seven, and Eight—for four operators, with two facing forward and two backward. Stations Five and Six were the equivalent of One and Two. Station Seven was the equivalent of Three, with direction-finding (DF) capability in one of the radios. Station Eight was like Station One. A high-speed video recorder with a three-inch-wide tape and a spinning recording head was positioned between the two sets of stations. It was state-of-the-art for copying video-type signals right from the air.

Third Compartment

The third compartment had two operators facing backward. One of these was the operation-control supervisor, known as the admiral. The third compartment was for overall operations management. Station Nine had a wide-band spectrum analyzer that could recognize when new stations came on the air. When this happened, the operator would tune down to the new station and see if it was of any importance. This operator would monitor the other operators, and a lot of the Russian command and control communications, to see if our country was in for any kind of trouble.

Maintenance Compartment

At the back of the compartments was a galley with a refrigerator, stoves, and ovens for heating our in-flight meals, hot and cold water, and two five-gallon coffee pots. There was also a head (restroom). There were the two cargo doors or crew-jump doors on the sides, and a fold-down bunk. Then came the maintenance compartment. It was located over the drop-down cargo door in the rear. After about five feet, a little behind the workbench, the floor inclined up toward the loading ramp. This was real nice (although cold) because you could put your feet up on the inclined floor while working on the equipment.

The maintenance compartment had a spare for every piece of equipment in the compartments, except for the fourteen-channel recorder and the video recorder; there were two spare Ampex tape recorders. On the workbench we had a superior Tektronix oscilloscope, some HP signal generators, an HP audio generator, a pulse generator, a signal analyzer, a spectrum analyzer, a tube tester, and a vacuum-tube voltmeter (VTVM). In the parts drawer, we had all the standard resistors, capacitors, tubes, transistors, nuvistors, and fuses; a complete set of bolts, nuts, screws, and washers; and all the hardware needed to fix any problem that came up.

I also carried some special parts in my toolbox, like the ball bearings for the motors of the tape recorders. Sometimes the operators would call us in and complain that the tape recorders sounded like Mickey Mouse. The technical term was "flutter and wow." I'd replace the recorder, and take the one with "flutter and wow" back to the maintenance compartment and work on it. I'd completely disassemble it, even disassembling the motor (which I don't think was done at depot-level maintenance), and replace the ball-bearing on the capstan. A flat place on the capstan ball-bearing was what caused the flutter and wow. So the ball bearings were important for ensuring that the tape recorders worked properly.

When one of the operators called us, we'd go in and trouble-shoot the problem and isolate it to a piece of equipment. We'd remove that piece, replace it with one of the spares, and get the operator back in action. Then we'd take the bad piece of equipment to the maintenance compartment and work on it. We were supposed to fix anything that went wrong while we were in the air. There was *no* excuse to abort a mission for equipment problems.

Heating System—Based on a Design in Hell

I don't know who designed the heating system in those RC-130s, but I hope he is dead and burning in hell.

If he's in hell, he knows what it was like for us working in the compartments.

In the winter, and it was not much better in the summer, the outside temperature at altitude was 33 degrees below zero. Air was piped in around each engine's exhaust to heat it, then directed into the compartments at the top. As any second grader knows, hot air rises and cold air falls. The SOB who designed the system must've failed thermodynamics or didn't take the course at all. The air temperature in the compartments would be 120 degrees at the top and 30 degrees below zero on the floor. About four feet off of the floor there was a temperate zone about two feet wide where the temperature was bearable. The maintenance compartment workbench was about thirty-six inches off the floor, so it was at the bottom of the temperate zone. In order to work and be fairly comfortable, I'd have to lie on the workbench and slide back toward the warm test equipment. Otherwise, if I stood or sat, my head would be on fire and my feet would be freezing.

If I was in my comfortably warm reclining position when the aircraft commander pulled back to "max endurance," I'd be put to sleep by the drone of the engines as they tried to maintain synchronization. The plane would tilt up from the horizontal by about 25 degrees and barely hang in the air.

Why I Was There

Eventually I found out why I'd been so urgently sent to Germany. One of the 7406th RC-130As (Flight 60528), with our crew aboard, had been shot down by the Russians near the Turkey-Russian border. All hands were lost.

The *Stars and Stripes* military newspaper printed a report of the incident on Saturday, July 16, 1960. Under the heading "Not Unusual Procedure" it said: "Electronic intelligence gathering of this kind has been carried on for years by America, Russia, and other countries from the ground, sea, and air. It is done all along the 4,000 mile Iron Curtain from Norway to Turkey. The U.S. dramatized the technique in February 1959, when it made public tape-recorded conversations of Russian fighter pilots shooting down an unarmed U.S. transport plane that wandered across the Turkish border into Soviet Armenia."

When we were on station in the Black Sea, we'd orbit between Samsun and Trabzon, in Turkey. There were radio beacons at both sites. The navigator would fly until he was perpendicular to one of the beacons. Then he would turn away from Russia and fly back to the other beacon. The Russians had seen us do that many times, so they created a decoy beacon on the same frequency and situated it just inside Soviet Armenia near the Turkish border. In the Flight 60528 incident, the navigator missed his normal turnaround because he was monitoring the Russian decoy beacon. The aircraft flew into Soviet Armenia's territory. The Russians wanted to get their hand on one of our planes and its crew, but the flight crew had instructions never to let that happen. When the Russians sent up a couple of MIG-17s to force the aircraft down, it wouldn't go. The MIG-17s, with pilots Gavrilov, Ivanov, Kucheryaev, and Lopatkov, shot it down with the loss of all aboard.

It crashed near Sasnashen, thirty-four miles northwest of Armenia's capital, Yerevan.

The remains of John E. Simpson, Rudy J. Swiestra, Edward J. Jeruss, and Ricardo M. Vallareal were returned to the U.S. on September 24, 1958. The remains of the other crewmembers, Paul E. Duncan, George P. Petrochilos, Arthur L. Mello, Leroy Price, Robert J. Oshinskie, Archie T. Bourg Jr., James E. Ferguson, Joel H. Fields, Harold T. Kamps, Gerald C. Maggiacomo, Clement O. Mankins, Gerald H. Medeiros, and Robert H. Moore, were recovered in 1998.

I was the replacement maintenance technician on the new crew. On the evening before a flight, you had to go to the orderly room and register with the sergeant of the guard and sign up for a wake-up call—usually a 0300 wake-up, because most of the flights took off at 0500.

I'd always have a restless sleep before a flight. The guy in the next bunk snored loudly. I'd usually be half awake when the guard came to wake me. I could hear his footsteps coming down the quiet hall. He'd shake me and shine his flashlight in my face to make sure I was getting up. I'd get up in the dark and take my clothes to the shower room and get dressed in there. The chow hall was always open to flight crews, so I'd walk down there and have breakfast, speaking in hushed tones. Then I'd walk to the shop. Just inside the gate, I'd have to run over to the door to get the eight-key combination lock open before the MP came by with his German shepherd guard dog. On cold mornings it would really be a race. Once inside, I'd get my toolbox and go out to the aircraft. I'd "preflight" the equipment to ensure everything was working, and check the maintenance log book for any write-ups. Next I'd go in for the briefing. After the briefing, we'd all lie around on the ramp and wait for the air crew to preflight and get their weather briefing. We'd all board and button up. The air crew would begin the startup checklist.

No matter what the weather was, we'd fly. No minimums for us. We'd go no matter what. We had to go; we were the ears of the Joint Chiefs of Staff.

If the fog was too thick for the flight crew to taxi, the ground crew towed us to the end of the runway and lined us up on the center stripe. As soon as the pilot could see the center stripe, we took off. If we had mechanical failure on takeoff, we'd go to an alternate air base. There were several fighter bases nearby that were higher up the mountain, usually above any fog. We could not return to Rhein-Main

We reported in every thirty minutes. If we didn't report in, the Joint Chiefs of Staff would know something was amiss.

Going North

When we'd "go north," we'd fly over Hamburg to Holland. Turning east, we'd head out over the Baltic Sea, then we'd turn back north. We'd fly toward Stockholm on the mainland and just outside of the island of Gotland in international waters. We'd orbit from just below Stockholm down to near Kalmar. This would give us good line of sight to Leningrad and Moscow and points north. We were given special "over the water" survival training. I don't really understand why. The water in the Baltic and North Seas was so cold that you'd only survive fifteen or twenty minutes.

One day when we were going north, the lights of the APS-89 radar detection system came on—and stayed on. This meant that some gun-tracking radar or missile system had locked on our position and was ready to shoot us down. We were watching Russian MIGs out our windows and could see their contrails in East Germany. After about fifteen minutes, a U.S. Navy RB-66 passed us. It was up there doing its thing and was on the same flight path as we were.

Escape, Evasion, and How To Perform a Lobotomy

Shortly after going on flying status, my group, about thirty of us, was sent to "escape and evasion" training. The Air Force was going to teach us to survive off the land, in case we were ever shot down or crashed somewhere. I'd had survival training in the Boy Scouts, so I thought this would be a cinch.

During the first two days of the two-week course, they taught us how to avoid getting captured; how to eat bugs, snakes, and edible plants; how to ration your consumables; how to navigate to a place so you could be rescued; how to get help; what code signals to use on the radio, if you had one; if you didn't, how to use a signal mirror to get the attention of a pilot.

I'd already learned most of this stuff, but the course was a good refresher, and it added to my information about how the military worked.

On the third day, we were given maps and a compass, and briefed on where we were to go, from Point A to Point B. They told us it was essential not to bunch up. If you were alone, it was easier to get through the woods without getting caught. They also told us the Army would be patrolling the area we were to cross. The soldiers would receive a weekend pass for each one of us they caught, so they had an incentive to find us. The Army guys were pretty good at this exercise, and they'd had plenty of practice.

We had a week to get to Point B, which was ten miles away. This sounded easy to me. Less than two miles a day. Piece of cake. Find a good hiding place, hole up for three or four days, then run the last few miles.

We were taken out to Point A, a spot on a road by a woods, and let out. Everyone else immediately went into the woods and hid. I decided to go down the road a ways just inside the edge of the woods, because I figured the Army guys knew where Point A was and would be waiting just around the next tree. Sure enough, a lot of the guys got caught the first day. Meanwhile, I'd found a nice, dry, cozy little culvert under the road. I crawled in and waited until night. After that, I came out of hiding only at night and traveled only in the dark. I figured it was harder to see where I was going, but it was better than getting caught. I managed to stay out three days and get about halfway there before they caught me. But even if I'd gone the whole distance, and it wouldn't have mattered. What they didn't tell us was that Point B was a simulated POW compound.

The Army was training interrogators how to question prisoners, and they were going to practice on us.

We were about to learn what to expect if we ever got caught by the enemy.

When they threw us in the compound, the first thing they did was line us up and have us strip butt-naked. Then they washed us down with a water hose. The air temperature was about 35 degrees, and we were freezing. They gave us a towel, some pajamas, a pair of shower thongs, and a number. You'd better not lose your number, for that was sure cause for a beating. The number was your only identification. What they were doing was removing any kind of self-identity.

In the simulated POW compound, our "captors" used a wide range of techniques for psychological warfare and brainwashing. These guys had learned most of their methods from American POWs who'd returned from brainwashing camps in Korea—and they'd learned well. They were really good at inflicting psychological and physical pain.

They told us that because we carried diplomatic passports instead of military identification, we'd given up all rights under the Geneva Convention and would be treated as spies. They said we were to be stood up against a wall and shot.

They put us in a little ten-foot-square room. All thirty of us. There was not enough room for thirty people, so your personal space was infringed on—more psychological warfare. There wasn't enough room for everyone to stretch out on the floor at the same time, so you had to sleep in shifts. We'd take turns lying down like sardines, your face next to someone's feet. But we had to try to sleep, because not sleeping at all would be really bad for your stamina and morale. The enemy would have liked that, because it would help to wear you down.

Throughout the day and night, they'd periodically come and get someone and take him for interrogation. While he was being interrogated, they'd come and get another prisoner and punish him with either cold water or a beating with rolled-up newspaper. This punishment was administered publicly, so all could see. The guard would say it was being done because the guy they'd just taken for interrogation had said so-and-so about you. The intention was to sow distrust among the prisoners. The guy inside had probably said nothing.

When I was called in for questioning, they made me stand at attention in front of this big desk, with my arms tied behind my back in two places. Just like in the movies, they held a bright light in my face. Two big guards stood next to me. The guy at the desk had his head down and was reading something—probably information they'd gathered on me from my 201 Service Record and DD219 DOD Request for Background Investigation. He was probably considering how to use that data to interrogate me. I must've stood there over an hour—at least, it seemed that long—before the interrogator finally said something. He still didn't look up when he spoke. This was to show non-recognition of me as a human being; I was just a thing and didn't deserve recognition. When the guy asked me the question, I gave my name, rank, and serial number.

Wrong thing to do.

The guard nearly knocked my head off with his rolled-up newspaper.

The interrogator told the guard not to hit me, that I was just doing what I was told to do. There was another long wait. Then the guy asked another question. This time, I didn't answer at all. I didn't want to get hit again. About a minute went by, and then the guard hit me again anyway. Once more the interrogator took my side and tried to keep the guard from punishing me.

This went on for what seemed like hours. The interrogator would repeat my name, rank, and serial number, and ask me if the information was correct. That was to start me saying "yes." Then he'd repeat another thing from his file, and ask if that was correct. "Yes." After a lot of yes answers on my part, he'd offer me a drink of water, a cup of coffee, or a cigarette. Then he'd change a phrase by one word to mean something completely different, to get me to say "yes" accidentally. Meanwhile, I was sure they were punishing one or two of the other prisoners back at our cell for things I'd supposedly said.

After a while, you began to really hate this goon next to you who kept knocking you in the head. But you started to like the guy behind the desk, because he kept taking your side. It's an interrogation technique called the "threat-and-rescue" method. It's similar to the "good cop, bad cop" approach used by police with civilian prisoners. It was certainly effective. The direct punishment from the guard just infuriated you, and made you determined to get even. But the overall "threat-and-rescue" method won your confidence and made you want to help the interrogator.

The next time I was questioned, there was a different interrogator. This guy was in my face screaming and hollering and jumping up and down while the goons were punishing me. They'd whack me in the head; pinch or hit my muscles to create charley horses; bend my arms, hands, wrists, and fingers until I screamed. They made me do a lot of physical exercise, too: squats, push-ups on my fingertips, standing on tip-toes until my calf muscles cramped and I fell down—and with my hands tied, there was no way to break the fall. Nothing was serious enough to do permanent damage, but it sure was painful. I later found out that these guys were doctors who knew how to really hurt you without causing any lasting injuries. They didn't want your family writing to a congressman, telling him how they broke something. That was not the point of the exercise.

The mean interrogator made you like the nice interrogator even more, so that the next time he showed up, you were happy to see him.

Another interrogator tried to make you think you were gay, or that you liked little boys. He'd accuse you of all kinds of things to see what your hot buttons were. When he found one, it was "Katy, bar the door." By the time you left that interrogation, you didn't know which way was up. Anything you said, he'd turn around against you to make it sound bad.

They tried a lot of other things, too, like threatening to hurt your kids, wife, or parents. They had a very complete file on everyone, even had current pictures of everyone's family at home, at the store, at school. They'd threaten to harm them. Or they'd threaten to send phony letters saying you were gay, you were married to another woman, you wanted a divorce—any number of things just to get at your family. They tried to get you to sign the letters or confessions of spying, biological warfare, or chemical warfare. They'd found out from everyone's commander or supervisor what everyone's hot buttons were, and they used them against us.

Boy, were they good at this stuff.

At one point you had to sit (not really the right word to describe the position) in a box that was much too small for you. The cramped position caused your body to work against itself. What you needed to do was relax and let all the tension out of your muscles. But that was very difficult, and the more it hurt, the more your muscles wanted to tense up. Which was the wrong thing to do in the cramped space. I remember seeing a variety show on TV, I think it was *The Ed Sullivan Show*, where this Hindu got into a 2' x 3' x 3' clear plastic box. He spent an hour in it during the show and then got out in good condition. Because he'd had yoga training, he was able to relax, control his breathing, and not have his muscles work against him.

One cup of slop per day was nutritious but terrible to choke down. It was made up of purée raw vegetables, coffee grounds, raw eggs with shells and green food coloring. It was really slimy and when you started to drink it, it all went down at one time.

There were always doctors around, disguised as guards, watching you, so you didn't really get hurt. But they did want you to feel the pain, so you'd know what to expect if you ever were really captured by the enemy. It was the most educational experience I'd ever had. The training identified your strong and weak points.

After a week, they returned our clothes, took us to a classroom, and debriefed us with an explanation of what they'd been doing.

In the classroom, they taught us how to perform a frontal lobotomy in case one of our comrades went completely mad from the psychological and physical torture. You take a sharp stick, raise the eyelid, and slip the stick in above the eyeball. Then you wiggle it back and forth over each eye to separate the frontal lobe from the rest of the brain.

This will calm the person down.

Feather in Your Cap

When an aircraft engine has to be shut down or quits, it's necessary to "feather" the propeller—that is, to stop the propeller from turning as the plane flies. This is required because when the propeller is turning but the engine is not running, extra drag is created, reducing overall performance. And of course, when flying with one or more engines incapacitated, you need all the performance you can get from the remaining engines. Sometimes, we'd return with one engine feathered. Sometimes, we'd come back with two engines feathered.

I once came back with three engines feathered.

It's possible for a C-130, which has four engines, to fly on only two engines and maintain altitude. In fact, the standard test-hop maintenance required after replacing an engine is to go to altitude, cut the two outboard engines, and fly on only the two inboard engines for an hour. Then you start the outboard engines and cut the inboard engines and fly for another hour on only the outboard engines. This aircraft can fly on only one engine, but it needs a second engine to maintain altitude.

On this particular occasion, to help balance the power for landing, the air crew started the least-problematical engine just before we began our final approach. On our way back, with only one engine we were losing altitude. But because we were coming down anyway, it didn't matter. We managed to land without crashing.

When you make it back safely with three engines feathered, it's a feather in your cap.

Going South

When we'd "go south," we'd fly over Munich, Germany; Innsbruck, Austria; Bolzano, Chioggia, and Brindisi, Italy. We'd orbit between Chioggia and Brindisi for about eight hours while fuel lasted. Then we'd land in Aviano, Italy for the night.

There wasn't much in Aviano. The airbase was small, and we were housed in Quonset huts. The town was tiny, with only a few buildings around a plaza with a fountain in the middle. It had only one bar, so we'd usually just have a couple of glasses of Italian wine and come back to the barracks.

In Aviano, I liked hearing the local Italian stations on our radio. One of the popular tunes in America at the time was "Al Di La," a song also played on Italian radio. It was great to be hearing it in Italy.

Turkey

"Going south" could also mean turning east after Brindisi; flying over Alexandroupolis, Greece; going on to Istanbul and Ankara, Turkey; and landing at Incirlik AFB in Adana, Turkey. The Turks hired a lot of foreigners to work for them. In the mess halls, all the cooks were German. To ensure good in-flight lunches, we'd always take a case or two of good German beer to give to the cooks. We'd keep the beer in one of the unheated parts of the aircraft so it'd be good and cold when we landed. The first thing on the agenda after landing was to take those cases of cold beer over to the mess hall. We always received great treatment from the cooks.

When I first started to going to Turkey, we were housed in Quonset huts. Talk about hot! There was a window air conditioner, but it didn't do much good. There were also huge mosquitoes. They'd keep me awake at night, so I started a campaign of revenge. I'd try to catch them instead of squashing them. Then I'd stick a pin through one of them, and stick the pin in the wall to show the others what I was going to do to them. Of course, the mosquitoes didn't care. But it made me feel better.

There wasn't much to do at Incirlik AFB. You could go to the airman's club, the base library, the bowling alley, the swimming pool, or the movie theater. That's about all there was except for the base exchange and the commissary. Every day, the snack truck (we called it the "roach coach") would come by with sandwiches, cold drinks, ice cream, and candy. If you asked the Turkish driver what he had, he'd say "Coke, root beer, grape, orange"—the only four words of English he knew. If you saw something you wanted, you just pointed.

We all spent a lot of time in the Quonset hut. Usually the guys would play hearts, double-deck pinochle, or poker. I never liked poker. One day I was playing solitaire, and a couple of guys came up and wanted to play poker. I told them I'd deal a few hands, but not bet any money. That was okay—they just needed something to do. We played a few hands and, as usually happens, I lost most of the hands. Then one of the guys said, "Let's play low ball." The object of the game is to get the lowest cards. And aces are high. He dealt me four aces. I threw down the damned cards, said, "That's why I don't play poker for money," and left.

I was reading *Hamlet* at the time, and would walk around the base memorizing Polonius's speech to Laertes ("And these few precepts in thy memory ...") and the famous "To be, or not to be, that is the question" soliloquy. I read most of the works by Shakespeare, including *King Lear* and *MacBeth*, and I enjoyed

The Taming of the Shrew, especially the part about "Spit in the hole, man, and tune again."

One thing that kept coming to mind was the painting *El Hombre,* by Rufino Tamayo, from the Dallas Museum of Art.

The first time we touched down at Incirlik AFB, I noticed several RB-47s on the ramp next to our hangar. One of the RB-47's had a picture of Charlie Brown painted on it. Another one had Lucy, and another had Snoopy. They were in the same business as we were, but with some kind of different scenario. I don't know what all they were doing—I didn't have what the military calls a "need to know." The navy had several of their RB-66 aircraft stationed at Incirlik also, doing their thing. There were a couple of RB-57 Cambaras, too. These had long wings like the U2 but had two engines. The engines were started with some kind of black powder charge. When an engine started, there'd be a loud bang, and a giant cloud of black smoke would pour out of the engine. I guess this was for remote locations that didn't have engine-starting equipment on site.

Then there was the U2. It was kept in the hangar until it needed to take off. They would quickly open the hangar doors and taxi quickly out to the end of the runway. They never stopped to run up their engines. They just ran out on the runway and took off. And what a takeoff it was! The U2 would go down the runway about a quarter of the way and then rise at better than a forty-five degree angle and go out of sight very quickly. Our cover story was that we were doing weather observations. I think that is what the CIA used also.

Why were we in Turkey, and why did we have so many orbiting stations? The Russian missile sites were in a line of sight from the Black Sea. We'd copy their telemetry data code, named "Mercury Grass," as it was transmitted from their rockets to their ground stations. Then we'd send it back to the States so our side would know how the Russian missile program was going. There was a radar bomb-scoring site there. Russian Bear and Bison bombers would take off from bases near Moscow and Leningrad and fly around Russia for a while to simulate flying to the U.S., then they'd come down to the bomb range. They'd simulate bomb runs against U.S. targets, then they'd return to their bases. We'd intercept the telemetry data with the bomb scores in it.

We knew how well the bombardiers performed before they did! They had to wait until they got back to their home base and had a debriefing to get their scores.

Adana

After several trips to Turkey, one of the guys wanted to go into Adana to buy some Meerschaum pipes. I decided to go with him and see what the town was like. Well, I don't have many good things to say about that part of Turkey. In fact, my impression of the entire Middle East is that God sure sent his son to the right place. They still need him there today.

A river runs through Adana, and it's spanned by a bridge built by the Romans. When I saw it in the 1960s, it had been repaired a couple of times, but the original seven arches were still there. It was amazing to see that old bridge and how well built it was. It clashed with the surrounding hovels and emphasized the inhabitants' lack of ambition. Their lifestyle seemed still to be that of pre-Roman times. The people had only progressed a notch or two beyond Stone Age man.

Francis Gary Powers and his U2 Flight

Meanwhile, on May 1, 1960, we were on our usual orbit between Samsun and Trabzon when the operators started saying the Russians had shot down a U.S. airplane inside Russia. No one understood exactly what was going on, but there was a lot of Russian communications traffic between aircraft and ground about the incident. We thought perhaps they'd shot down one of their own aircraft masquerading as American in a simulated air-combat training exercise. But that didn't make sense because we'd monitored a lot of these exercises, and this was the first one in which an aircraft was actually shot down.

A couple of days later we read in the *Stars and Stripes* that the Russians had downed one of our U2 aircraft over Russia. President Eisenhower said it was on a weather flight and must have strayed off course. Boy, was the Soviet Premier, Nikita S. Khrushchev, mad! They had the pilot, Francis Gary Powers, and they denounced him as a spy. They also had the downed U2.

Powers was sentenced to ten years' imprisonment for espionage.

Taking over the Compound after the CIA's Departure

The CIA folks in the hangar where the U2 was normally kept—along with all their equipment and everyone in their compound—were gone within a day. They'd packed up everything during the night and flown it out by transport aircraft.

A couple of weeks later our detachment commander, Captain David W. Weiland, received permission to take over the CIA hangar and compound for our operations. The CIA's area was more secure than the area we were in at the time. I was given a tour of the compound and asked to determine the power requirements, cabling configurations, and placements for our equipment.

I spent the next week calculating power and arrangements for our equipment. Thank goodness I went back to Germany and didn't have to do all the work. They had base maintenance people come in, redo the power in the building, and run the new cables we received from the States.

The next time I went south everything had been taken care of, and we were operating out of our new secure area.

Midas

May 24, 1960, the U.S. space satellite program launched *Midas 2*. It was instantly known as the "Spy in the Sky" satellite. Weighing two and a half tons, *Midas 2* carried 3,600 pounds of instruments. Its aim was to test the viability of an orbiting-satellite network that could send instantaneous warning of a ballistic missile attack.

RB-47s with Snoopy

In 1960 the Soviet Union began a series of tests in the Artic, off Novaya Zemlya. The hydrogen bomb, which was exploded in the atmosphere, was part of this series. It was estimated to be almost as powerful as the most powerful U.S. bomb on May 1, 1954—fifteen megatons.

There was a further explosion, and this one was so powerful that ground and air shockwaves were felt in Sweden, Denmark, Norway, France, and Germany. The British atomic research establishment at Aldermaston described it as "significantly larger than any previously recorded." It was a fifty-megaton bomb.

On July 1, 1960, we went north and were on station outside the island of Gotland near Sweden. We were there to monitor the Russians' test of their first one-hundred megaton nuclear blast on Novaya Zemlya, near the Artic Circle. (A PBS TV show once said this blast took place in September 1961, but I think this is incorrect and that history has been rewritten for security reasons.)

One of our RB-47s (ERB-47H (534281) of the 38th SRS, 55th SRW—with Snoopy painted on it!—was to fly over the Arctic Circle and through the radioactive cloud to get samples of its radioactive particles. These would help our scientists determine what the device was made of.

Well, the Russians didn't like our guys flying near the test site and ordered them to land in Russia. Of course, our guys refused. When the refusal to land continued, the Russian MIG-15, under the Soviet pilot Vasili Poliakov, shot them down at exactly 1803 hours, Moscow time. Of the six men on board only First Lieutenants John R. McKone and Freeman B. Olmstead survived. They happened to land in the middle of a fishing fleet in the Barents Sea near the Kola Peninsula, south of the Arctic Circle. They were immediately picked out of the cold Arctic water by the Russians, tried for spying, and sent to a Soviet prison. They were later traded for a Russian spy at Checkpoint Charlie, the same way Francis Gary Powers was traded on January 25, 1961. The story was printed in the July 1962 issue of *Reader's Digest*. I think the title was "Little Toy Dog," referring to the fact that the children of our pilot, Captain Willard G. Palm, had given him a plush toy of Snoopy, and he always had it in his airplane. This pilot's body was returned to Rhine-Main Air Base aboard a C-47 on Monday, July 25, 1960. The three other men, missing and presumed dead, were Captain Eugene Posa and First Lieutenants Dean B. Phillips and Oscar L. Goforth.

Paige, known as "The Crushed Cricket," was our company clerk. He was one of those who can't walk and chew gum at the same time. One summer he wanted to "go south" to see what all the rest of us were talking about. ("Going south" meant flying over Germany and Austria and into southern Italy. Sometimes we'd also go to Greece and Turkey.) Paige took a two-week leave and flew to Turkey with some friends. While they were there, the guys decided to go to Beruit, Lebanon, for a little sightseeing. Well, Paige was Jewish, but he still wanted to go, so they decided to let him. Our diplomatic passports didn't specify religion, so he had no trouble getting into Lebanon. The guys made him buy all the drinks at the bar—they threatened that if he didn't, they'd tell everyone there he was Jewish. They all had a lot of fun, and Paige didn't mind the kidding.

George Washington and "the Pill"

On July 20, 1960, the first nuclear-powered ballistic missile submarine, *George Washington*, which had not yet gone out on patrol beyond U.S. territorial waters, fired a Polaris missile from under water. The missile covered a distance of 1,100 miles.

A few months later, on September 24, 1960, an atomic-powered aircraft carrier, *Enterprise*, was launched at Newport News, Virginia. She had the capacity to sail for several years without refueling.

That year, 1960, was also a year of medical progress. There was the introduction of an oral contraceptive tablet for women, "The Pill." Surgeons also developed a mechanical pacemaker for the heart.

"Nyet"

The Russians were building a new surface-to-air missile (SAM) site near Sevastopol. Sevastopol was like the French Riviera to the Russians. It was where all their bigwigs went for vacation, so the Russian government wanted to protect them.

The U.S. wanted to know what the latest SAM sites looked like, so they sent an RF-101 Voodoo photo-reconnaissance plane over to Sevastopol to get pictures. At the time, our RC-130 was in the air doing our usual orbit between Samsun and Trabzon. The Voodoo plane passed under us, and the pilot was flying only about fifty feet above the water. He did this so the Russian radar wouldn't pick him up until it was too late and he was going 1400 miles per hour. About fifty miles out, the Russian radar did pick him up, and the Soviets immediately scrambled their MIGs fighter jets. He popped up to about 2,000 feet, flew over the site, and took his pictures. Then he pulled it back into a vertical climb toward Turkey and climbed to about 45,000 feet when he passed over us.

The Russian MIGs were mad as hornets. They were coming upon him as fast as their afterburners would carry them. There was one big problem—we were between the MIGs and the photo-reconnaissance plane. The admiral told the aircraft commander, "Get the hell out of here." The aircraft commander put the RC-130 in a power dive toward Turkey.

The fastest a C-130 can go is usually about 350 miles per hour. We were doing close to 400 miles per hour and the plane was shaking violently. But that was not nearly fast enough to outrun the MIGs. I was standing at the jump door ready to open it and parachute out.

Meanwhile the photo-reconnaissance plane was long gone, already back in Turkey by now. Our admiral was monitoring one MIG pilot's radio conversation, and he put it on the intercom so everyone could hear.

As the MIG pilot radioed his base, the admiral translated the message for us:

"I have the plane in my gun sights and locked on. Request permission to shoot."

The plane he was referring to was us!

After a very, very long minute, the response to the MIG pilot came back from his base:

"Nyet."

There was a loud sigh of relief from everyone, but we weren't out of the woods yet. We still had fifty miles to go to get to Turkey. The MIG flew around us. It delivered some cannon fire right in front of us. I guess he was hoping we'd turn back to Russia. No way. We were now in Turkish airspace. The Turks started to scramble their fighters, and the Russians headed home.

One of the guys in our plane said he wasn't scared. But it was ten minutes before he noticed his parachute's straps were so tight they were cutting off the circulation to his legs.

Water in the Wings

We often flew that route, orbiting between Samsun and Trabzon. One time, as we made our normal turn away from Russia, the two engines on the high side of our aircraft began to slow as if starved for fuel. When the aircraft commander (AC) tried to level the plane out, it overcorrected because all the power was on one wing and both engines on that wing were at full power. As the full-power wing got higher than the slowing-down wing, the full-power engines started to slow, but the first two started to run better.

We were in trouble.

The AC sent out a Mayday distress call and headed for Turkey. The nearest airfield was at our radar site on top of a mountain at Samsun. It had a short, soft field used only for pickup and delivery of personnel, supplies, and mail. Our RC-130 was designed to handle those types of airstrips, but under optimum conditions, with everything functioning normally. We didn't know if our engines would be working by the time we got there—or even if we *could* get there.

A quick calculation by the navigator indicated we could just make it. But he didn't know if we'd have sufficient reverse thrust to land on the short field. If the plane went off the mountain, our top-secret documents would be scattered all over the hillside.

The admiral began emergency destruction procedures for all classified material aboard. He put some of the code books in a wastebasket and set them on fire. Smoke started to fill the aircraft. The crew chief in the back radioed the AC that the plane was on fire.

As if he didn't have enough problems just trying to keep it in the air, now it was on fire!

He was about to give the "abandon ship" command when the admiral informed him that he was just burning classified documents. Okay, the aircraft was not on fire, so maybe the AC could get it on the ground.

But by that time, we might all be dead from smoke inhalation.

After we got down below 15,000 feet and depressurized, the crew chief opened one of the jump doors a little ways, and the copilot opened his window. This produced a draft through the plane to get the smoke out.

By this time, the engines were working just well enough to stop the airplane from going off of the mountain at Samsun.

We landed safely.

The subsequent accident investigation identified a problem with our destruction of classified documents. The burning of paper in a crippled aircraft was not recommended. A decision was made that from then on, water-soluble paper would be used for all classified material. That way, in an emergency the documents could simply be dissolved with water or coffee. Several months later, all of our document started to come to us on water-soluble paper.

But there was a bigger problem. The accident investigation revealed we had water in our fuel tanks. Lots of water. The crew chief is supposed to check for water in the fuel with a probe before each flight. There's typically a drop or two of water condensation, and it's easy to identify the different color of clear water on the bottom of the probe with the fuel above it. But in the dark, it's hard to see the color of the fuel in the probe, especially if it's all water, which this was. When we got to our cruising altitude, where the air temperature is 33 degrees below zero, the water froze. The remaining fuel was just running over the top of the ice. That's why the high-side engines were starved for fuel.

Water had been put in our fuel tanks instead of JP-4. When we took off, the fuel tanks were half full of water.

It was sabotage.

Mount Ararat and Noah's Ark

One of our regular orbits was from Trabzon down over Lake Van. We'd fly from Incirlik AFB over Diyarbakir to Van and turn north. From Lake Van, you could look east to the Russian border of Armenia. Just past that border is Mount Ararat.

I remembered reading about a couple of explorers who'd been up on that mountain. They claimed that they'd seen Noah's Ark, or something resembling

it. For years, there was a lot of discussion about this by historians and Bible scholars, but nothing ever came of the story.

So one day, while we were flying over Lake Van, I looked to try and locate
Noah's Ark. All I could see, for hundreds of miles in every direction, was treacherous mountain terrain. I didn't find Noah's Ark, but I knew that if we ever went
down in that area, there was no way we'd escape—unless one of the area's nomad
tribes carried us out in a camel caravan.

Mersin, Swimming in the Mediterranean, and "The Sheik of Araby"

Once during a routine southern flight, a couple of Turkish F-100s flew up beside
us and escorted us back to our base in Adana. There had been a revolution in
Turkey. We were told we wouldn't fly the southern route again until the U.S.
established diplomatic relations with the new government of Turkey.

It was like being laid off from a job. We didn't have much to do. So several of
us decided to go to the Mediterranean and try snorkeling.

We went to the motor pool and signed out a pickup truck. Then it was over to
the recreation hall for some swim fins, masks, snorkels, a spear gun, and a cooler.
Then we found the "roach coach" and bought sandwiches, cokes, and candy.
Then off to the mess hall for ice and C rations, in case of emergency. We took
several blankets from the barracks, and off we went.

Our destination on the Mediterranean was Mersin, Turkey. There was nothing between Adana and Mersin except desert. When we arrived in Mersin, we
stopped at the local marketplace to buy a couple of melons. We saw skinned
goats and chickens hanging in the open-air market with flies crawling all over
them.

I guess that's how they tenderize the meat over there.

We got back in the truck, went down to the dock, turned west, and drove
onto the sand by the sea. We must have driven along the beach for a little over an
hour. We didn't see a soul for fifty miles. There was nothing as far as the eye
could see. So we stopped and took out our blankets and spread them on the sand.
Then we took a dip in the Mediterranean.

It was beautiful. Really clean water. You could see the bottom clearly in fifty
feet of water. There were a couple of small reefs, about three quarters of a mile or
so offshore, with lots of fish for good spearfishing.

After we'd been in the water about an hour, we noticed a bunch of kids on top
of the hill overlooking our camp. I don't know where they came from. There was

nothing around for miles, but there they were. Some of the big kids had donkeys. We decided a couple of us should stay in camp, in case the kids got "sticky fingers." That was a good idea. An hour later, they were all practically sitting on our blankets. They just kept inching closer. By now, it was time to break camp, anyway, and start back.

As we prepared to leave, those kids grabbed anything we threw away. A tin can was a real prize—now one of them had a drinking cup. He was the Sheik of Araby, he had a cup and the others didn't.

The Bay of Pigs and a Man on the Moon—Soon?

On April 12, 1961, a Soviet air force pilot, Major Yuri Gagarin, made an orbit of the earth in an artificial satellite.

Five days later, on April 17, 1961, a group of 1,400 Cuban exiles, who had been living in the U.S., invaded Cuba at the Bay of Pigs.

On May 5, 1961, Commander Alan B. Shepard was catapulted into space in a Mercury space capsule on top of a Redstone rocket. Unlike Gagarin's flight, Shepard's was sub-orbital and short: fifteen minutes in time and 115 miles in distance.

On May 25, 1961, President Kennedy said in his State of the Union, "The United States must take a clearly leading role in space achievements, which in many ways may hold the key to our future on earth,' so much so that it would become 'a major national commitment' to put a man on the moon, and to bring him safely back to earth—and to do so before the end of the decade."

In August 1961, Soviet pilot Major G. S.Titov completed seventeen orbits around the earth in just over twenty-five hours.

Climbing the Hill

A couple of days later, we were really going nuts looking for something to do. We could see a little butte off in the distance, several miles away. Someone suggested we catch the bus that went out there and go climb that little butte.

The next day we packed a lunch, got our canteens, and headed out to the road at about 0900 hours. We flagged down a local bus and climbed aboard. I really mean "climb"—there were people all over that bus, on top and everywhere. They even had their farm animals with them, chickens and goats. One of the goats sneezed and stuck a good lunger on Joe's arm. Anyway, we paid the driver, and he took off. About forty-five minutes later we were almost even with the butte, so we stopped the bus and got off.

We walked about a mile over to the butte. It was a pretty rugged climb, about 2,500 feet, almost straight up. There was a lot of underbrush and rocks, but we thought we could make it. We started up this little ravine. We climbed and climbed and climbed. At about 1600 hours, we finally arrived at the top. When we did, we found this goat herder who'd been up there watching us all the time. A little road led down the back side of the butte to the highway. Good thing too, because I was too tired to climb back down. After an exhausting thirty-minute walk to the road, we flagged down a bus to take us back to Adana.

Saving Master Sergeant McDougle

I liked going to the swimming pool. It was cool there, and sometimes I'd see a good-looking girl. Like most pools, this one had a deep end and a shallow end. But there wasn't a rope across the pool marking the deep-end drop-off. One day I was hanging out in the pool, holding onto the side in about five feet of water just before the drop-off, when an attractive young lady came out with a bunch of kids. She took them to the opposite side and began teaching them how to swim.

She was in a bathing suit, of course, so I was keeping her under surveillance. But then this guy slowly moved in between us, in the middle of the pool, about ten feet away. All of a sudden, he slipped off the shallow part and into the deep end. He started thrashing around. It was kind of annoying because he was splashing a lot of water in the air, which blocked my view of the good-looking babe.

But he was flailing around so much, I decided he might be in trouble.

I swam out, grabbed him, and dragged him back to the side. Some others who'd noticed his frantic splashing helped me get him out of the water. They hauled him up the side of the pool and he lay there, spewing water out of his lungs. He seemed to be okay, but someone took him to the base hospital for a check.

I found out his name was Master Sergeant McDougle. After he left, one of the guys who'd helped get him out of the water asked me my name and outfit. I told him, and a month later I got a letter of commendation for saving the man's life.

Permission to Pilfer—Getting the Ground Equipment Fixed

A new sergeant was transferred to Turkey as a permanent-duty assignment. He must have volunteered. No one wanted to go to Turkey to stay, but we found out later he had good reason. He was selling his young wife to the lonely G.I.s stuck there. She could make ten times his income. His new job would be to maintain equipment in the new compound we'd taken over from the CIA after the U2 incident.

After he'd been there six months, the ground equipment was in such bad shape that the operators could barely get their work done.

The sergeant's pimp job was interfering with his maintenance work.

The detachment officer, Captain David W. Weiland, requested technical assistance to get the shop up and running again. Chief Master Sergeant Stover asked me if I'd go down for an extended temporary duty to help get the ground equipment back in shape. How could I refuse?

I took the next flight down, and sure enough, there were problems on every station. Nothing serious, but enough to hamper the operators from doing their jobs. I went to one station at a time. I'd cannibalize good equipment from the worst station, in order to get the first station completely operational. Then I'd start on the next station, and then the next. Out of eight stations, I'd typically end up with five good ones, two partially operating ones, and three that were completely down. I talked to Captain Weiland about the situation. He asked Chief Master Sergeant Stover for permission to pilfer good equipment from a returning aircraft, and replace it with bad equipment. The bad equipment would then be repaired at our home base. Everyone thought that was a smart idea, because it would enable us to get our stations operational as soon as possible.

The plan worked great. Within three weeks, all the ground equipment was operational, and I returned home.

I received a letter of commendation for that little trip.

Test Hop

When an aircraft engine is replaced, the maintenance crew is required to do a test flight, called a "test hop." During a test hop they completely check out the aircraft, determine that everything is working, and ensure that nothing was damaged during the repair. One day the flight crew asked if anyone from our group wanted to join them on a test hop. A test hop is usually pretty boring, but I decided to go.

We took off and climbed to altitude. The crew cut the two outboard engines, and ran for an hour on the two inboard engines. Then they started the two outboard engines, cut the two inboard engines, and ran for another hour. Everything was fine, so they started the two inboards. We still had five hours of fuel left, so the AC said, "Let's go sightseeing." He put the plane in a steep dive, and we plummeted from 25,000 feet to 5,000 feet. What a roller-coaster ride!

We were flying about 500 feet off the ground as we traveled over the desert, when a bump appeared on the horizon. The AC said, "Let's go see what that is." The bump turned out to be a hill about 800 feet high. There was an old castle on top. The AC banked the plane, and we flew around the bump and looked up at the castle. This was great fun.

After we'd seen enough of the castle, someone said, "Let's go to the coast and fly along the Mediterranean." So we did. We flew down the beach at about 500 feet. The water was so clear, looking at it was like gazing into a fish bowl. After a while, we came to some mountains that ran into the Mediterranean Sea.

A castle! Pristine water! Mountains! The area we saw on that test hop was absolutely gorgeous. It could easily become another Riviera. I don't understand why someone in Turkey doesn't develop it.

What Mountains?

Speaking of mountains....

We were on another orbit in the Black Sea near Romania. As usual, the AC was asleep with his feet up on the instruments, and the navigator was flying the airplane by looking into the radar scope. The crew chief came up to the flight deck from the galley. When he put on his headset and looked out the pilot's window, he asked the navigator, "What are those mountains?"

That shocked the AC awake. He steered the aircraft into an immediate high bank, full power, turning left toward Turkey. It nearly threw all of us out of our seats.

There were not supposed to be any mountains in the area of the Black Sea where we were flying. The navigator must have missed his west-end turn beacon.

It was another narrow escape. We'd been about to fly into Soviet Romania's airspace. That would surely have gotten us shot down. The U.S. had lost an RC-130A in a similar incident. In February 1959 it was shot down at the other end of the Black Sea, in Soviet Armenia. Everyone was killed.

Service Club Tours

The service club offered several Rhine area wine-tasting tours, and I took most of them. The tours offered wine tasting in the delightful village of Ruedesheim in the Monastery Eberbach, Hotel Krone Assmannshausen, Hotel Historisches Altes Haus, and Hotel Lamm. It was great. We visited many of the welcoming bars along the famed Drosselgasse and sampled glasses of the famous local vintage.

Ruedesheim is in the middle of the Rhine river valley at the confluence of the Rhine and Moselle rivers in the Rhine Gorge. We took the ferry over to Moselle and sampled their renowned Moselle wine also. We tried to drink the Moselle and Ruedesheim wineries dry.

We took the chair-lift from Assmannshausen to the Niederwald Memorial atop the mountain overlooking the Rhine. This is where Elvis Presley filmed part of his movie *GI Blues*. We also visited a number of castles along the Rhein from Koblenz to Bingen to Mainz to Cologne. Some I remember are Godesburg, Schloss Rheineck, and Drachenfels Castle. On each of the castle tours we always stopped in Ruedesheim and tried to finish drinking the wineries dry.

We took trips to the spa towns of Bad Schwalbach, Bad Kreuznach, Bad Ems, and Lahnstein. I never went to a single spa! But the towns were beautiful, and we took a lot of pictures.

And drank a lot.

Nuerburgring

On May 28, 1960, I went to Nuerburgring, the famous racetrack in the heart of the German countryside. I wanted to see Sterling Moss and Dan Gurney race.

All the celebrity drivers were there, with the most powerful sports cars of the day. There were Ferraris, Porches, Maseratis, Lotuses, and many others. The race was five hundred miles, and the track was ten miles around, so the cars had to make fifty laps.

It was a great race. I took a lot of photographs that I later developed myself. On the base there was a photo lab run by the service club. They'd teach you the processes, and then they'd let you develop your own film and make your own prints.

Three months later, on August 28, 1960, I took a service club trip to Burg Frankenstein and saw Frankenstein's Castle.

Cold Cash and Hot Cars

It was February, 1961. Lynn had just asked me to help teach dancing in Wiesbaden. I needed to get a car.

Before this, to get to class I simply walked down to the Gateway Service Club. Now I had to get to Wiesbaden. I'd been talking to our technical representative (or "tech rep"), who worked for LTV, about buying his Austin Healey. Some of my friends and I had gone to Wiesbaden a few times with some friends to look at cars, especially the E-Type Jaguar. We'd also checked out the new Triumph TR-5 that the car dealer had on display in the showroom. This car was so new that they only had the one on display. If you wanted to buy one, they had to order it from England, and there was a two-month wait for delivery.

Ever since I'd sobered up the previous March, I'd been sending most of my money home for my mother to deposit in my saving account for college. With the money I'd saved through the First National Bank of Dallas's "School Savings Plan" in grade school, plus what I'd squirreled away while working at Neisner's, I had nearly $4,000.

The LTV tech rep agreed to sell me his Austin Healey. So I wrote to my mother and asked her to send me my money. She didn't think buying a sports car was a very good use of my college savings, so she took her time withdrawing my funds. Two months passed before she sent me an American Express check for $4,000. I went to the tech rep, but he'd already sold his car.

I was bummed out!

I decided to go to Wiesbaden and see if I could get the car dealers to sell me the TR-5 demo model for cash. My friend drove me there, and darned if they hadn't just sold the demo. They wouldn't get another one for a couple of months. Now I wasn't just bummed out, I was pissed off.

I said, "Let's go see the Jaguar dealer."

The Jag was twice the price of the TR-5: $8,000. I'd have to finance the other half. There was one model on the showroom floor, and we looked it over for two hours. The salesman wanted me to take it for a ride. He knew—and I knew—that if I test drove it, I'd buy it.

I *really* wanted that car.

But I realized that if I bought it, I'd be dead in six months. I'd constantly be tempted to see how fast it would go. And driving on the German freeway or "autobahn" with the other sports cars—Porsches, Mercedes, Maseratis, and others—it wouldn't be long until I'd be enticed into a race.

Finally we left the showroom. I deposited my $4,000 in a savings account on the base. A week later, I bought an Opal Record for $1,500 from a GI going back to the states.

Selective Reenlistment

In June 1961, I was scheduled to be discharged from the military in a year.

The Air Force had what they called a "selective re-enlistment" program. The commander would talk to airmen who were within a year of their discharge, to find out what plans they had for when they got out. If the airmen were undesirable, they'd be encouraged to leave, and they might even be put on permanent mess-hall duty until then. But if the Air Force wanted to keep you, the commander would give you the "re-enlistment speech."

The day the commander spoke with me, I told him I'd been saving most of my pay. I said I had several thousand dollars in the bank for my education. I explained that when I got out of the service, I planned to go to college, get an engineering degree, then work for a big electronics company.

The commander said a lot of airman had wives and kids. These men couldn't afford to miss a paycheck, and they needed the medical care and other benefits the service offered. He said they usually didn't have any real plan for what they'd do when they got out. They also hadn't saved any money. With no plan and no bank account, he said he recommended re-enlistment for these guys. But this didn't apply to me.

He said he thought I had an excellent plan and should follow through with it.

He left, and I went back to work, confident that my military service would be over in a year. A week later, President Kennedy signed an executive order. It extended, for a minimum of one year, the enlistment of all critical career field personnel—which is what I was.

I'd be in the service for yet another year, whether I liked it or not.

"Knees Together"—My Ski Trip to Garmisch-Partenkirchen

The Gateway Service Club offered a ten-day skiing trip to Garmisch-Partenkirchen. It cost only a hundred dollars, which included the hotel. At ten dollars a day, it was not a bad deal. No meals were included, but the hundred bucks did cover ski rental and a ski instructor. I signed up. I had to provide my own transportation to and from Garmisch-Partenkirchen, but by now I had the used Opal Record. It had a standard 1953 Chevrolet engine and transmission, and it was a good little car.

I took a two-week vacation and was off! I drove down the autobahn to Munich the first night, got a room, then drove around town and did some sight-seeing. Next morning, I headed for Garmisch-Partenkirchen, site of the 1936 Winter Olympics.

We were to ski on the slopes of the Zugspitze, an extinct volcano that was the tallest mountain in Germany. It had erupted millions of years ago, creating a huge crater in its surface. Covered with snow, the crater was now a wonderful sloping bowl with one side washed out. There was every kind of skiing you could think of. The slopes went all the way from vertical to flat, and there was a cross-country route over the washed-out rim through the woods and down the side of the mountain twenty-three miles into Garmisch. There were a dozen chair lifts down in the bowl, and a few rope lifts up the steep parts. Another chair lift went from the hotel and ski shop at the top to a level spot at the bottom of the bowl, where beginners could learn to ski.

The service club had given us a map to the Eibsee Hotel, which was halfway up the mountain on the shores of beautiful Lake Eibsee. The Eibsee Hotel was a gorgeous Swiss-chalet-type hotel that the Germans had used for their senior staff officers' recreation during World War II. The U.S. had taken it over for our military's rest-and-recreation program. That's why it was so inexpensive.

A cog train came twenty miles from Garmisch-Partenkirchen to the Eibsee Hotel, then went into a tunnel for another six miles and up to the top of the Zug-spitze. At the summit there was another hotel, restaurant, and ski shop inside the old volcano. Some skiers would ride the cog train to the top in the morning and ski down to Garmisch for lunch. Then they'd take the cog train back up to the top and ski down again for dinner.

It was marvelous. The view of the lake from the dining room in the Eibsee Hotel was breathtaking—picture-postcard perfect. One night while I was having dinner, a huge buck deer came out on the frozen lake and looked around. When he was satisfied that it was safe to traverse the ice, he started across. Ten does came out of the woods behind him and followed him over the frozen lake. There were deer feeders in the woods not far from the hotel; I could see the feeders from the cog train, just before it went into the tunnel.

The first night, I settled into my room. There was a featherbed with a down comforter. You'd sink into the feather bed, and it nearly covered you. Then you'd pull up the comforter, and you were snug as a bug in a rug.

The first day, they rolled us out before sunup. After breakfast, they introduced us to Klaus, our ski instructor. He took us to the ski shop and had us fitted for skis and boots. Then we caught the cog train to the top.

I'd expected him to have us take the chair lift down to the flat part at the bottom of the bowl, where we'd learn the basics of skiing. But *no*. Klaus had us start at the top.

It was incredibly steep, about a 75 degree angle. I could stick out my arm, directly from my side, and touch the snow on the side of the slope. Klaus began teaching us to traverse. He demonstrated by pointing the toe of his ski across the mountain and sliding about thirty feet. Then he stopped, did a kick turn, and traversed back to just slightly below us. He told us to keep our skis flat and our weight on the uphill side of the ski so the edge would cut into the snow. We all skied over to where he'd turned around. Then he came and showed us how to do the kick turn.

To do the kick, you had to twist slightly and look downhill. Boy, was that scary! It must have been a mile or more, almost straight down, to the flat place. If you fell, you wouldn't stop rolling until you got there. If you didn't hit a tree, you might even keep rolling right over the washed-out side and twenty-three miles into Garmisch.

Klaus showed us how to put our ski poles under our arms and stick them into the snow behind us—to support us and keep us from falling—while we kicked the down-hill ski into the air and turned it around. Oh, great, now you had the left ski going one way and the right ski going the other way. It was a strain on your crotch! Then you'd kick the uphill ski into the air and turn it around. So now you were pointed back to where you came from and could traverse in the other direction.

Actually, it worked pretty well. Klaus had us do it about twenty-five times. By then we were starving, and the last traverse brought us to the chair lift so we could ride up and have lunch. After lunch, Klaus took us to the flat place and taught us to do the snowplow turn. We spent the rest of the afternoon doing ever-steeper snowplow turns. At the end of the day, we went back to our hotel and had dinner. I was exhausted. I sure was glad I'd been ice-skating or dancing every night, because I can't imagine what it would have been like if I'd been out of shape.

The next two days were a lot of the same, except the slope kept getting steeper. Klaus would holler, "Knees *zusammen*!" That's the German word for "together."

One time, when we were nearly all at the end of our traverse, this big captain skied over and fell down. When he hit the snow, he farted, and it echoed all over the bowl. You know how sound carries over snow. He smiled and said, "Whoops, sat on a frog!" Everyone just cracked up. I laughed so hard I was rolling in the snow, holding my sides, and thinking I might roll down the slope.

On the fourth day, Klaus taught us how to do a parallel turn. You keep turning slightly more downhill, and when your skis have passed the center line—an imaginary line running straight down the slope—you swing around, plant the upper edge of the downhill ski in the snow, and traverse in the opposite direction. I didn't like the part when you got to the center line. When this happened, both your skis were pointed straight downhill. There was nothing holding you back. You'd wind up speeding into Garmisch at a hundred miles an hour if you messed up. Klaus did remind us to fall down if anything like this happened. Falling down was one of the finer points to remember.

The fifth and sixth days were more of the same. We'd had two feet of snow on the fifth day, and the sun came out and melted the top couple of inches so there was a slight crust. It snowed another six inches the night before the sixth day, so it was really good skiing.

By now I was getting the hang of the parallel turn. I could traverse, parallel turn, ski right back, and do another parallel turn. I was getting pretty good. Until one time when I started my parallel turn, I jumped up a little too high, and when I came back down, I broke through the crust of day five's snow. My skis stopped, but my body momentum kept me going. I went through the air downhill about ten feet, in perfect ski position with my arms and ski poles beside me and my knees together. I landed in the snow on my head, and it broke through the crust also. My ski trail stopped and all you could see was the bottom of my parallel skis. I was upside down in the snow.

I was laughing so hard I could barely move. Someone rushed over and asked if I was hurt. When I stopped laughing, I told them no, to go ahead and ski and I'd dig myself out.

Well, that was easier said than done. My arms were pinned against my sides with the poles still in ski position. I finally freed one hand and started digging in the downhill direction. When the hole was finally big enough, my feet fell over and I could stand up. Now I was stuck in this hole that was about four feet deep. Getting out was a real struggle, mostly because I was laughing so hard. I finally extracted myself by rolling downhill. I put my skis back on and caught up with the group.

We spent the next couple of days practicing what we'd learned. On the last day, the ski school had planned a slalom downhill ski race. There were three categories: beginners, which I was in; intermediate; and advanced. Each level started a little farther up the slope. When it came my turn, Klaus moved me from the beginners' category to the intermediates' starting point. I put everything I had into my parallel turns and flew down the course. I made every gate perfectly, and my time was better than any of the intermediates. That night they had a banquet for the ski class and awarded trophies. They gave me the slalom cup for Best Intermediate Skier in the class. Boy, was I proud.

I'd become a speed demon on the slopes, and it carried over into my driving. The next day I started the trip back to Rhein-Main in my Opal. On a winding road in the mountains, I got stuck behind a really slow car. Finally we came to a straight part of the road. There was no oncoming traffic, so I pulled out to pass. I hadn't seen the crossroad—or the German Polizei. As I flew by, the German policeman just shook his finger at me. He didn't try to follow. Fortunately, they usually didn't give out tickets unless there was an accident.

I made it all the way back to the base that night.

E-Type Jaguar

One of my friends, an airmen named Dave, had brought his 1959 Chevrolet Impala to Germany with him. We used to drive into Heidelberg to go to the A&W Root Beer stand for hamburgers and root beer. One day we decided to take the autobahn to Wiesbaden. We were still on the city streets when this guy in an E-Type Jaguar pulled up next to us. Dave burned out when the light turned green and beat him to the next signal. Because we'd outrun him from the last light, when we got to the autobahn we took off as fast as the Chevy would go. There was no speed limit on the German freeway. The Jaguar came up behind us, and he blinked his headlights, indicating he was going to pass. Sure enough, he flew by like we were sitting still, and we watched his taillights disappear down the highway. He must've been going 150 miles per hour.

Environmental Movement

I loved the beauty of the environment in the German countryside. Speaking of the environment, in 1962 Rachel Carson published *Silent Spring*. With that book, she planted the seeds of an environmental movement that was to germinate and sprout in the 1970s into a mainstream political issue. Using evidence of environmental damage done by chemical pesticides, she exposed the folly of those who sought to conquer nature, those who disregarded the integrity of the natural world and the interdependence of all living things.

Sunrise in Paris

I'd been in Germany for two years, and I wanted to see more of Europe. A friend and I decided to go to Paris. We signed up for a two-week leave, and on March 24, 1962, we took off for Brussels, Belgium. On the way we stopped in Luxembourg, but there wasn't much to see there, only a church or two. In Brussels, we went to the site of the 1958 World's Fair and took a tour of the Atomium. It was a fascinating exhibit on nuclear energy.

We spent the night in Brussels and drove to Paris the next day. It's true that Paris is beautiful any time of the year. We first went to the USO to see what was available. They booked us in a hotel for ten dollars per night for a room with a bath. It was just off of the Avenue des Champs Elysees, the fifth street around the Arc de Triomphe, one block down on the Avenue Foch. This was great!

Driving in Paris is very exciting, especially around the Arc de Triomphe. It is a game of bluff. If you look at the other car, you have to stop. If you don't look at the other car he has to stop or give way. The circle around the Arc de Triomphe is four or five cars wide. Sometimes six if the cars are small. If you are going only one or two streets, stay in the outside lane. If you are going three or four streets, get into the second lane in. If you are going five or six streets, go to the third lane in. If going more, then go to the inside lane. We went to the fifth street most of the time but a couple of time I went to the inside lane and made several trips around the circle. It was great fun.

We walked down the Champs Elysees to the Place de la Concorde. Then we strolled along the Seine to the Eiffel Tower and took the tour of the tower. The next day, we visited Notre Dame, and in the afternoon stopped at a sidewalk cafe. We spent the following day at the Louvre.

One night we went to the Moulin Rouge.

Another night we went to see the show at the *Folies Bergeres*.

We also went to the Lido, one of Paris's most famous cabarets.

Afterward we went back to the USO. We talked to a couple of hostesses and got dates with them. We asked them to show us some of the non-touristy stuff. For the next two nights, they guided us around the out-of-the-way attractions of Paris. One of the places was down in the market area. It was real scary down there at night. We drove several blocks down this really dark street with big warehouses on both sides. Then, right in the middle of one of the blocks, there was a little red neon sign that said "Cabaret." Inside, we went down three flights of stairs and came out inside this wine cellar with high, arched ceilings. There was this mob of young people dancing to one of the best bands I've ever heard. It was party time!

We danced and drank wine until 3:00 AM. Then we drove up to the Bastille and watched the sun rise over Paris.

General Douglas MacArthur's Speech

On May 12, 1962, I was driving back from Frankfurt listening to the car radio when I heard General Douglas MacArthur's "Duty, Honor, Country" retirement speech, delivered at West Point. I shall never forget hearing it.

The Crisis

One thing I was told when I first got to Germany was to keep a flight bag packed for a two-week stay, and to be prepared to leave on two hours' notice for an unknown destination. Early in June 1962, a call came down to the shop at 0830 hours to get a plane and crew ready for a trip anywhere in the world. At the briefing, we'd find out where.

This had happened before. When people were needed for these trips, the non-commissioned officers always had something else to do—such as go to the dentist or take the kids or wife to the hospital—anything to keep from going on one of these fly-by-night jaunts. I'd always be the one elected—Shanghai'ed would be a better word. Plus, we never went anywhere nice, like Paris, Rome, or London.

So I figured that, once again, I'd be sent to someplace like Bum-Fuck, Egypt.

I went to the barracks, picked up my flight bag, packed a few extra things just in case, and reported to the shop. I loaded my flight bag and toolbox on the aircraft and did my preflight check-out. By then, it was time for the briefing.

We all sat down, and the officer informed us we'd be going to MacDill AFB in Tampa, Florida—in the good old USA.

I couldn't believe my good luck!

He also said we'd be under tighter control with closer monitoring than normal. We normally reported our position every thirty minutes, but on this flight we were to report in every fifteen minutes. They would provide more information when we arrived. Until then, we were to say nothing to anyone.

Boy, this was a mystery! Why would the Air Force want a bunch of Russian and Spanish linguists in Florida? And why the close monitoring? I knew that Russian trawlers regularly patrolled the Florida coast off Cape Canaveral—had the navy nabbed some who'd been intercepting U.S. missile signals, and they needed us to help with interrogations? We'd find out soon enough.

We took off at about 1000 hours and headed west. Over the Atlantic we came to a decision point, should we go north to Nova Scotia and refuel in St. Johns, or should we go south and refuel in Bermuda. The flight crew decided it was faster to take the southern route. So they headed south. Soon after that we reached the point of no return—and wouldn't you know it, we picked up a stiff headwind. Immediately, the navigator began estimating our fuel consumption to determine how much we had left. It would be close. We might have to use our reserve fuel. We always carried thirty extra minutes of fuel in case we had to land at an alternate airfield. But there wasn't another airfield within a thousand miles, so we didn't need the extra fuel.

Our refueling destination was a tiny island out in the middle of nowhere. I was standing behind the flight engineer looking over the aircraft commander's shoulder thinking, "There's sure was a lot of water out there in the Atlantic." There was no land in sight. I listened as the navigator kept calling out the remaining minutes of fuel left. The number got closer and closer to zero. Finally, "Land ho!" Right in front of us was this little speck of an island. It was Bermuda. Was I ever glad to see it!

We had sufficient fuel to land. It was sundown, and we were just in time for chow. The flight crew would need eight hours' rest before continuing, so we'd spend the night on the island.

After chow, several of us decided to go into town. We stopped one of those Bermuda jitneys—gold carts with fringe on top—and had him drive us around. We wound up all the way down on the south end of the island, at the Elbow Beach Surf Club. It was a real fancy place. One guy going in for dinner was wearing a white dinner jacket and black bow tie with his plaid Bermuda shorts, plaid socks, and white buckskin shoes. We weren't dressed for the Surf Club, so we walked around on the beach. There were a couple of secretaries there from New York, and we messed around and talked to them until 0100 hours. Then we headed back to the base.

Next morning, we were up early and took off about 0600. We landed at Mac-Dill around 1000. Because we carried the call sign of a general staff officer, the local transit crew was ready and waiting to service our aircraft. In addition, three staff cars were there to meet us; one had a brigadier general aboard, and the other two had full colonels. As the crew door was lowered, the transit crew was surprised to be met by an airman with his .45 caliber pistol drawn. He informed them that we had our own ground crew, and that they were not to go near our plane. Above all, they should never try to board it. All we needed from them was a fuel truck to top off our fuel tanks.

"Yes, sir," they said, and left.

The general and the two colonels came over and welcomed us. Our flight crew got into the staff car with the general, and the rest of us piled into the other two cars with the colonels. Our ground crew stayed with the aircraft.

The three cars drove us to Strike Command Headquarters, where we were issued temporary badges and escorted inside. We took the elevator down five floors to the briefing room. It was just like you see in the movies: stair-stepped rows of desks banked around a stage with maps, projector screens, and information boards. In the back were projection booths for showing movies and slides. A telephone sat on each of the desks. In the very front row, there was a red phone marked JCS. I think it was a direct line to the Joint Chiefs of Staff in the Pentagon, and I'm sure there was a similar phone in the general's office with a direct line to the Oval Office. Other phones were connected to the local police and fire departments, the mayor's office, several churches, and the coroner's office.

Hanging above the middle of the stage was an enormous map of Cuba.

After we all were seated, a general told us of reports that the Russians were building missile bases in Cuba. Our job was to find out where they were being built, so U.S. planes could fly over and take pictures. The Pentagon had put the Armed Forces on alert and was moving surface-to-air-missiles into the area. Tomorrow morning we'd to fly to the Gulf of Mexico and into the Eglin AFB missile test range. Then we'd head for Cuba, to try to locate the missile construction sites. We were to observe strict radio silence at all times. We were to talk to no one except the MacDill tower for landing. The Air Defense Command was being told that one of our C-130 aircraft would be there, and they'd be given our flight path to MacDill. We'd be under the direct control of the Joint Chiefs of Staff (JCS), and if we needed anything we only had to ask. Major Harper, the strike command security officer, would be our liaison.

It was the Cuban Missile Crisis.

The whole country ready to shoot down anything coming from Cuba. And here we were, not allowed to talk to anyone.

I didn't like the sound of this.

Flying around Cuba

Next morning we took off at 0600 hours, flew into the Gulf of Mexico, and turned south in the Eglin AFB missile test range. We flew around Cuba for about nine hours before starting back. When we entered the Air Defense Command's defense zone, we heard them call to us.

"Unknown aircraft, entering the U.S. at latitude 80 degrees west and longitude 25 degrees north, please identify yourself."

We were forbidden to reply. We'd been ordered by the general to observe strict radio silence at all times. So we said nothing.

Again came the call: "Unknown aircraft, entering the U.S. at latitude 80 degrees west and longitude 25 degrees north, identify yourself."

Again nothing.

When they came on the air the third time, two F-4 Phantom jets were flying on our wings. They could barely travel as slowly as we were going. They were just able to hang in the air, and they had to let their flaps and gear down in order to move so slowly.

We still didn't reply.

It was scary. I could see the pilots in the F-4 Phantoms.

But we knew they had our flight path and would follow us until we landed at MacDill. If we'd deviated from that flight path, they'd immediately have shot us down.

They turned and headed back to Homestead AFB at the tip of Florida.

As soon as we landed, staff cars were there to take us back to Strike Command Headquarters for our debriefing. We told them we had some good direction-finding (DF) cuts on a number of possible missile-construction sites. They took our tapes and notes, and ran them out to a waiting F-100 fighter sitting on the ramp with its engine running. It took off and flew the information to Washington.

They were in a hurry to get that information.

Top-Secret Repair

One day when we returned from a mission, I had to report, in our debriefing, the failure of one of our DF antennas in the fiberglass pod.

I was asked what I needed to fix the antenna.

I explained that first, we'd need a secure hangar in which to disassemble the pod. It was Top Secret, and we couldn't work on it out in the open on the flight line.

"Okay," they said, "We'll find you a hangar. What else?"

I said I'd never had any experience with the antennas or phase-shifter in the pods. That work was always done back at LTV in Greenville, Texas.

They said, "How about a new phase-shifter and an engineer from LTV to remove it, install it, and align it?"

Perfect. I said I'd prepare the aircraft so the engineer would need only a minimum amount of time to install and align the phase-shifter.

"Okay, done," they said. "We'll call and let you know which hangar and what time the engineer will be here."

They called about an hour later, told me to use Hangar 27, and said the phase shifter and engineer would be there at 2100 hours.

This was about 1600. I went to Hangar 27 and located the officer in charge. I explained that I needed his hangar and that he'd have to take his aircraft out so I could put mine in.

"What?! I'm not taking my aircraft out. We're in the middle of a hundred-hour engine check."

I told him to call Major Harper, and I gave him Major Harper's phone number.

He called. I could hear him say, "Yes, sir. Yes, sir. Yes, sir!" He hung up. With much grumbling, he called in the crew chief and told him to remove the aircraft and put it on the flight line.

I called the air police and explained that I needed to secure the hangar for "top secret." I said I'd need at least six air policemen.

Okay, they'd send over a detail.

By 1800 hours we had the other aircraft out and ours in. The air police arrived and asked what I needed.

I told them to get everyone out of the hangar and then secure it, to post a guard at all four corners and one at each door, and to allow no one in or out without my permission.

It took them just fifteen minutes to secure the hangar.

Once it was secure, I went in and began to unbutton the fiberglass pod to prepare it for the LTV engineer. About 2030, I heard a knock on the door. It was the sergeant of the guard detail. Two men lay spread-eagled on the ramp in front of the hangar. He wanted to know what to do with them.

I asked if they were from LTV.

By checking their badges, the air police verified that the two men were indeed from LTV.

The two engineers got up, and I let them inside the hangar. They carried in the phase shifter and immediately went to work. By 2330, they were done. We had the pod buttoned up and ready to go by midnight.

I told the air police detail they could go home, and I told our crew chief he could put our airplane back out on the ramp. The LTV engineers flew their twin-engine plane back to Greenville, Texas.

And I had a good night's sleep.

History, Rewritten

Meanwhile, the JCS had authorized photo-reconnaissance missions over Cuba to determine if missile construction was going on where we thought it was. They started by sending U2 aircraft from Laughlin AFB. The U2s flew high over Cuba to avoid attracting attention.

The Air Force had been running routine U2 flights out of Laughlin for some time; it was one of their primary U2 sites for photographing Honduras and Nicaragua. Then one day, one of the U2s crashed in Cuba. We intercepted the radio message from the Cuban jeep driver to Havana that the pilot, Major Rudolf Anderson, was killed in the crash—we knew he was dead before Castro did. However, for security purposes, the JCS waited about two months to tell his parents he was dead. Until that time, they kept listing him as missing over the Gulf of Mexico on a routine weather mission.

I've seen several TV specials on the Cuban Missile Crisis, and they've all said that the U2 shooting occurred during the blockade in October 1962 that aggravated an already tense situation and that nearly provoked World War III. For example, a PBS TV show said Major Anderson's plane was shot down on October 14, 1962. But August 14, 1962, is a more accurate date. Again, the JCS rewrote history and gave an inaccurate date, for reasons of national security.

After several more U2 flights, in order to get closer pictures with better resolution, JCS began sending in RF-101 Voodoo photo-reconnaissance planes. They got some pretty good pictures of the missile sites we'd identified. *Aviation Week*, *Time*, and *Newsweek* all published missile-site photos released by the President.

A Beach Bum in Clearwater

At MacDill in Florida, we flew only every other day, so I was off on alternate days. One time, I rented a car from the local Hertz and drove to Clearwater Beach. I bought a pair of beachcomber pants with a rope belt at the local beach store. I put these on, along with a white shirt. I tied the bottom of the shirt together around my waist. I had my Air Force sunglasses and shower clogs on. I really looked like a beach bum. I was having fun!

Boy, there were a lot of girls there. I picked up a date for the night and needed to go back to the barracks and change. When I arrived at the main guard gate, I didn't have my diplomatic passport. For that matter, I didn't have any identification. As the guard on duty was about to haul me off to the brig, I gave him Major Harper's number. He called. "Yes, sir. Yes, sir. Yes, sir." I could go.

I jumped in the car, drove to the barracks, changed, and picked up my passport. I had a great evening.

Freedom!

Greenville

The aircraft required some maintenance and needed to go to LTV in Greenville, Texas. So I went along. In Greenville, they put us up in a local motel. Bill Allen, my high-school friend, was going to East Texas State College in Commerce, Texas, so I gave him a call. He came right over. We sat around the swimming pool and talked, but for security reasons, I couldn't tell him what I was doing in Greenville.

Between June and October 1962, our aircraft made about forty-five flights around Cuba. Then our President called Russia's hand. With our identification of the sites, and the pictures the photo-reconnaissance planes had taken, President Kennedy held all the cards.

When the Cuban Missile Crisis ended, Ken Nunn came over on the replacement aircraft and I was sent back to Frankfurt. We flew to the Azores, refueled, spent the night there, and the next day we reached to Frankfurt.

We spent the next two months, November and December, flying our regular missions. I received a letter of commendation for my efforts in Operation Quick Fox, which is what the JCS called the Cuban Missile Crisis. Then, in January 1963, I was told I could go home.

On January 15, I flew to McGuire AFB. There I underwent processing for my discharge. It was quite an ordeal. They needed something for me to do for two weeks while my paperwork went through, so they put me on KP. I worked in the mess hall on the serving line. But I had to go over to the processing center about every other day for something. I tried to turn in my diplomatic passport, but no one would touch it. I still have it, although, of course, it has expired. When the final mustering outbriefing came, they told me I'd still be in the military for twenty-four hours after my discharge. I guess they didn't want me to celebrate too much that night. I went straight to the airport and caught the first plane home that afternoon.

It was a *wonderful* feeling to be free.

Part Three:
Living The American Dream

Chapter Eight:

The Home Again Years
(1963 to 1968)

Back to Dallas—A New Man

Now that the military had taught me some discipline, I had an objective and was focused on success. I would do what I told my commander in the selective re-enlistment interview: go to college, get my engineering degree, and go to work for a big electronic company.

After my release from the Air Force early in 1963, I returned to Dallas and moved in with my mother. She had moved to 1018 North Marsalis Avenue. It was the first building you came to in Oak Cliff, after crossing the Houston Street Viaduct, and it was right on the bus line. She'd moved as close to downtown as possible, without actually being downtown, to shorten her ride to and from work. Right behind Mom's house, at 612 First Street, was the VFW (Veterans of Foreign Wars). My dad often came over there with his new wife, so I got to see him when I wanted.

Once settled in at Mother's, I headed straight to the art museum at Fair Park. During my years in the military, I'd longed to sit in the museum and gaze at that Rufino Tamayo painting, *El Hombre*. But when I arrived at Fair Park, the Museum was shut down. A simple note was taped to the door: "Closed."

I was stunned. I'd so looked forward to seeing the painting.

I asked around and learned the museum had been relocated downtown, in the newly established Arts District, and was next to the new symphony house. I made a beeline downtown. All the way there, I silently prayed the painting hadn't been put in storage. When I arrived, I asked about it, and sure enough, it was hanging on a wall around the corner from the main entrance. I sat in front of it for over an hour, remembering how I'd felt when I first saw it. Later, in the new museum's gift shop, I bought a postcard of the painting.

That postcard now sits on a small easel in my bookcase. I see it every time I go into or out of my home office.

Arlington State College

After I'd been home a few days, I went out to Arlington and enrolled in electrical engineering at Arlington State College for the upcoming spring semester.

I'd sold my car before leaving Germany and was without transportation. Arlington was quite a way from home, well beyond the municipal bus routes. But Greyhound service went down Highway 80 West. The Greyhound bus came across the Houston Street viaduct, just past my mother's house, and up Zangs Boulevard to Davis Avenue, which was Highway 80 west. All I had to do was walk one block down and flag the Greyhound. That bus would pick me up, then drop me off three blocks from the college. It didn't cost much to travel this way, and I figured I could read some of my homework en route.

You had to stand in a long line to register for each class. Sometimes the class you wanted—on a specific day at a specific time—was full, so you'd have to settle for a different day and time. After registering, I was given the outlines for each course and a list of the books I'd need. I went over to the bookstore and picked up all the books—fortunately, I was able to buy used books, which saved money.

It was only a week until classes started, so when I got home I read ahead as much as possible. The freshman curriculum was standard for everyone: math, history, and English.

The math was trigonometry, which was a repeat for me, although having been out of school for five years I needed to bone up. And history would be okay.

But English—I dreaded English.

I knew there were problems on the horizon. In high school I'd only achieved an eighth-grade competency, or maybe it was even sixth-grade. Whatever level I was at, I wasn't ready for college English.

Trig

I was in Professor Lloyd Lassen's trigonometry class. On the first day he strolled in, perched on his desk, and explained his grading system. It was a bell curve. Most of us would fall into the first standard deviation in the middle, at about a C level, but some wouldn't make it.

That sounded a little ominous to me.

Next he asked if there were any questions on chapter one. Questions on chapter one? Most hadn't even picked up their books yet. There was no response from the class.

He continued, "No questions? Any questions on chapter two?

There was still no response.

"Okay, then, chapter three? No? How about chapter four?

Still no response.

"All right, then. Next time, we'll begin with chapter five. Thank you."

And he left!

College was sure going to be different than high school. A couple of students began to cry. Boy, was I glad I'd read the first three chapters and worked the problems.

Next time, as soon as Professor Lassen came in the room, someone asked why he was going so fast.

He said this was a repeat course. He said we'd had most of this stuff in high school, which was true, in my case at least.

When he asked, "Are there any questions?" my hand shot up. I asked about some points of trig, and he explained each one. Other students asked questions also, and we had a full class session.

When I arrived for the third trig class, the classroom was less crowded. There were fewer than half the original number of students. Some had transferred to another professor, but most had just dropped trigonometry.

History

History class went smoothly. I liked the fact that it went into detail about things we'd just skimmed over in high school.

For example, freshman history was where I learned the answer to one of the questions my fifth grade teacher in Rylie had asked and never answered: What song was the British band playing when Cornwallis surrendered to Washington at Yorktown? "The World Is Turned Upside-Down."

English

I'd read all the books on the reading list that my friend Bill Allen had sent me in the Air Force. This practice had improved my reading speed and my reading comprehension. But I was still way behind. On the first day of my college English class, I was so nervous my stomach was growling!

Freshman English was my hardest class. I knew that to pass this one, I'd need all the points I could get. I asked Professor Carter if she could assist me with some of the problems, perhaps even give me additional assignments for extra credit. Fortunately she was very helpful and spent an hour, three times a week, helping me with my writing.

The professor was great, but the class didn't impress me. Freshman English is mostly composition, and I remember that one of the assignments was to write a 150-word paragraph on comparison and contrast. We were to create a topic sentence, and then write about both sides of the topic.

A paragraph of 150 words? I had trouble getting even fifty words down on paper. I don't like flowery, fluffed-up writing, and it's my nature to use as few words as possible to get my point across.

Professor Carter and I had a long talk about what I could say about my topic sentence. When I left that meeting, I was still basically lost, but I had to turn in the assignment. So I went home and wrote a bunch of just plain bullshit. It was nonsense! I simply made it all up. It was as if I was running off at the mouth like an idiot, except I was doing it in writing. Most of it was not worth the paper it was written on.

I turned in my lousy paragraph, and was afraid to show up for the next class and get the paper back with my grade on it. But I mustered my courage and went. Professor Carter came into the room and began passing back our papers. My hands were shaking as I took mine. I didn't look at the grade right away. When I did, there it was, written large at the top of the page.

I couldn't believe my eyes.

For that bunch of bullshit, I'd gotten a B.

Texas Instruments

That first summer I took a job at Texas Instruments—or TI. It was only meant to be a summer job, but as it turned out, I worked there from June 1963 through March 1966. I applied at the location on Lemmon Avenue, passed their technician test, and was told to report on Monday.

With a typical starting pay of two dollars per hour, TI was one of the lowest-paying companies in the Dallas-Fort Worth area. But I thought it would be a good place to work until I got through college. Plus, one benefit of working at TI was the cheap price of food in the company cafeteria.

After processing through personnel on Lemmon Avenue, I was sent over to the location at Corporate Drive and Regal Row, in the Stemmons Industrial Park area.

My Involvement with a Criminal Enterprise

I'd gone to work at Texas Instruments on their Airport Surveillance Radar (ASR-5) system. At that time, J. Erick Jonsson was president of TI. J. Fred Bucy was my division manager, and Ed Frost was manager of the ASR-5 project.

There's an ASR-5 system at every major airport—it's the primary radar used by air traffic controllers to manage airport traffic. The ASR-5 was a critical element in the Federal Aviation Administration's (FAA's) flight-control system, so there were special requirement for the parts that we used. The guys at the FAA, to ensure they were getting the quality parts they paid for, required a paper trail tracing all components back to the manufacturers. Because so many lives depended on this system, if any part failed, they wanted to go to the supplier and find out what had happened.

As we neared completion of our ASR-5 contract, General Electric stopped manufacturing the power-supply tubes we used. GE decided that, since TI was the only customer for this tube, it was insufficiently profitable. We needed only a dozen tubes to finish production, but GE wouldn't budge. As TI tried futilely to negotiate with GE, our purchasing agents scrambled. They contacted Sam's Surplus, what we called an FBN (fly-by-night) operation in Massachusetts. Sam said he'd ship us six tubes.

When the box from Sam's Surplus arrived, it came with two surprises. First, there were only five tubes in the box. Second, an FBI agent accompanied the box.

The FBI wanted to know if TI was a regular customer of Sam's and part of its criminal enterprise—fencing parts stolen from the U.S. government.

Seems that Sam had contacts at the Westover Air Force Base supply depot in Massachusetts. These contacts were stealing parts from the base and selling them through Sam. Our company convinced the FBI that this was a one-time buy, and that TI would never do it again. The FBI left, and everyone was holding their breath, hoping the FBI wouldn't tell the FAA what had happened.

Somehow we finished the contract with exactly the right number of tubes—I still don't know where we got them.

Keeping Busy

I memorized the test procedure for each new subsystem and was a damn good troubleshooter. I was able to check out subsystems faster than anyone else on the production line. After a year, I was promoted to working foreman. That meant I still did my share of production, but I directed the work of all the other technicians, to keep production on schedule.

My commute to work—from Mom's home at 1018 North Marsalis to Corporate Drive and Regal Row in the Stemmons Industrial Park—wasn't bad. But that didn't last long. TI opened a new building in the Richardson/Dallas main TI complex, and the ASR-5 program was one of the first to move in. The place was huge! The workers literally wore roller skates to get around in the building as they set up the new assembly lines.

The drive to my new workplace, through downtown Dallas and up the North Central Expressway, was a real pain. So in January of 1964, I moved into a studio at Spring Brook Apartments, at 158 Highway 155 in Plano, Texas. I lived there for six months, and it eliminated that awful drive.

On Saturday night after work, I would stop by the local grocery store and have the butcher cut me two two-inch steaks from his best cut of prime beef. I picked up three potatoes for baking and a bottle of wine. Then I drove up to North Texas State University in Denton. By now Bill Allen was married to a girl named Leslie, so I'd drive to Bill and Leslie's where Leslie would cook the steaks and potatoes for us. We'd have a little get together and talk about things. Bill introduced me to several of his writer friends, Buddy Carter and his wife Frances, who was a nurse. Also Frank Frickie and several others. They would talk about college and writing. They had a pretty easy life, I thought. Bill told them how much I was working, what I was doing at TI, and the courses I was taking at SMU. This astonished the student artists, and their reaction inspired Bill to write a poem about it:

The Factory

> Don't think you'll ever be free of the Belt
> Don't think it for a minute
> From the moment we pop out of the slot marked IN
> Until we pop into the slot marked OUT
> We'll all ride that precarious conveyer
> Snatching what we can along the way
> But that's not what really gets you
> Nor is the fact that it's narrow and swaying

Always threatening to dump us into the Works
Or that we travel as much in night as in light
Or even that strange assorted instruments
Pick and poke at us all the while
No the big gripe is the thing keeps picking up speed
Making it all you can do to hang on
Much less take time off to figure out
What the hell's being produced

He wrote it on the spot in about twenty minutes and gave me the only copy, which I found in my personal collection of his unpublished works over forty years later.

Learning on the Job

Meanwhile, I'd taken a "work simplification training" class. It taught that, since we had two hands, we shouldn't let one sit idle while the other was doing something. This helped employees be more productive.

Now I took a course in transistor circuit design. We learned how transistors were made and classified, how the substrate was handled, and how each level of atoms was doped and deposited on the substrate. If the layer was to be positive (P), having fewer electrons, arsenic was added to the silicon or germanium; this combined with the free electrons and gave the silicon a positive charge. If the layer was to be negative (N), having a lot of free electrons, iodine was added to the silicon or germanium, to give it a negative charge. There was a mask deposited over each layer, and the substrate was sent through a photo-etch operation and an acid bath, to etch off the unwanted silicon leaving the proper geometric design of the transistor. For an NPN (negative, positive, negative) transistor, the first layer deposited was negative. Then it was etched. The second layer was positive. Then it was etched. And the third layer was negative. For a PNP transistor, it was just the opposite.

Hundreds of transistors were all made exactly alike on the same substrate, but they didn't all perform the same. Minute differences, at the molecular level, made each one perform differently. Testing was the only way to tell the difference. So after the transistors were made, they all went through a testing cycle. If the transistor had a high beta (β), it was given a high 2N2710 or a high number. If the transistor had a medium β, it was given a 2N2705 or a medium number. And if it had a low β, it was given a low 2N2701 number. The company charged a high price for a high β, a medium price for a medium β and a low price for a low β. Sometimes, they charged the same price for all the transistors in a lot.

We were taught how to design all kinds of transistor circuits with alpha (α) and β currents in transistors. The circuits were the same as what I'd learned in basic electronics in the military, but they used current instead of voltage-gain parameters.

More Science

On October 12, 1964, the seventh Soviet-manned spaceship was put into orbit. There were three people on board: the pilot, a doctor, and a scientist. The space-ship was brought back safely to earth after twenty-four hours.

October 16, 1964, the Chinese government exploded a nuclear device in the atmosphere.

Medical science in 1964 continued to make strides. The ability to save and prolong life offered the prospect of a larger, healthier, and longer-living popula-tion, not only in the affluent societies, but around the globe. One area of medical science in which marked progress was made was that of organ grafting. Drugs were being developed that could prevent tissue or organ grafts from being attacked and destroyed by the body's own immune system.

Computer technology was evolving and was coming to the aid of medical sci-ence.

The Nobel Prize for Chemistry was won by Dorothy Hodgkin of Oxford University, who had been able to determine the structure of biologically impor-tant chemicals, including cholesterol, vitamin B12, and a new antibiotic, cepha-losporin.

On March 18, 1965, a Soviet astronaut, Lieutenant-Colonel Alexei Leonov, emerged from his capsule for a few minutes while it was in orbit and became the first person to walk in space. Major Edward White repeated this feat within three months, leaving his space capsule and drifting alongside it, weightless in space, for twenty minutes.

On March 23, 1965, a manned spacecraft, *Gemini 3*, was sent into orbit; it was the first of five to be launched that year. Virgil Grissom and John Young were the first to maneuver a space ship while outside the earth's gravitational pull. On the following day an unmanned spacecraft sent television pictures back to earth from a mere 1,300 miles above the surface of the moon.

On April 6, 1965, the world's first commercial communications satellite, *Early Bird,* was put into orbit 22,000 miles above the earth's surface. It could transmit telephone calls—up to 240 simultaneous two-way conservations—and television pictures.

On December 4, 1965, *Gemini 4* was launched. Ten days later it carried out a rendezvous in space with another spacecraft, *Gemini 6*. Both returned safely to earth.

Shrike Missile—or "What the Hell Is a Twiget?"

After another six months, I was promoted from working foreman to foreman.

One of the technicians who worked for me was drafted, so I gave him my copy of *From Here To Eternity* by James Jones and told him to read it before he reported for duty. It would give him some idea of what he was getting into. Bill had planned to write a book about his army experience and call it *The Olive Drab Cage*, but he abandoned the idea after reading *From Here To Eternity*. According to Bill, James Jones had said it all.

When ASR-5 production was finished, I was assigned to the new Shrike missile production program. Shrike is an anti-radiation missile with a homing capability against radar installations. It's used by tactical military fighter planes. Shrike was being developed with engineering and manufacturing support from TI.

I set up and laid out the production test stations and tested the first one hundred systems that came in. The Air Force later renamed the missile AGM-88 HARM (high-speed antiradiation missile). TI won a quality award for the reliability of the missile. The AGM-88 can detect, attack, and destroy a target with minimum aircrew input. Using a fixed antenna and a seeker head in its nose, the missile's proportional guidance system homes in on enemy radar emissions. A smokeless, solid-propellant, dual-thrust rocket motor propels the missile.

This was a new system, so for all failures a failure-analysis card had to be filled out. When there was a problem you'd fill out the card, send it to the failure-analysis lab, and the engineers would get back to you and explain what had caused the problem. We were having some difficulties with the missile's cold cycle. There'd be a ringing or oscillation in the middle of the pulse. We didn't know what to call this problem to describe it to the engineers on the failure-analysis card, so I named it a "Twiget."

I eventually found that by replacing a specific transistor, the ringing or oscillation would go away.

We kept sending cards to the failure-analysis lab but never heard back from the engineers. We knew the lab was backed up and hadn't had time to analyze our problem. About six months into the contract, an engineer came out one day and asked me, "What the hell is a Twiget?"

I could show him but could not explain it.

Time to Go

After my second year at TI, I was promoted to night foreman over three programs: the Shrike missile system, the APQ-99 terrain-following radar system for the F-111 bomber, and the APS-88 infrared video mapping system.

When I'd begun at TI, the cafeteria prices were low, but now they seemed high. A Luby's restaurant had opened across the street and had all-you-can-eat deals for ninety-nine cents. The same meal in TI's cafeteria now cost $2.79. This made no sense, especially considering that the TI cafeteria was a benefit for employees, while Luby's was a profit-making business. One day I asked my division manager, J. Fred Bucy, why the cafeteria prices were so high. He said he'd look into it. A month later, he called me into his office and explained that the cafeteria was expected to make a profit that would pay for plant overhead expenses, such as lights and air conditioning.

I didn't like hearing that. It was the beginning of a growing resentment on my part against TI.

Meanwhile, nearby in Fort Worth, General Dynamics (GD) was tooling up to build a new line of F-16 aircraft. Like wildfire, the news spread around TI that GD was hiring technicians at $3.50 per hour—$3.50! I was making $2.53, including my ten cents per hour foreman's differential. My pay now seemed low to me, especially considering I was "walking on water" at TI—that is, I'd been promoted four times in three years.

The month GD began hiring, I lost twenty five of the 150 technicians I supervised. The second month, I lost another twenty. That was one third of my work force. To compensate for this loss of manpower, the company was having me hire new technicians off the street for $3.00 per hour. Guys with no TI experience, who were to be supervised by me, were walking in the door and getting more than I was making!

This called for action.

I went to see J. Fred Bucy and asked for a salary adjustment. I told him I'd lost thirty percent of my trained technicians, and thirty percent more were busy training the newly hired guys. Production was really hurting. Could he do something to stop the outflow?

No, he said, there was nothing he could do.

I gave him my resignation.

I'd demonstrated my technical troubleshooting skills and my organizational and management skills, and now I was ready to get into research and development to further my hardware-design expertise. I found a job with Collins Radio in research and development—making $3.75 per hour.

How I Met My First Wife

I was still quite a dancer, and I loved music. My favorite song at that time was "The Girl from Ipanema," with a melody by the Brazilian composer Antonio Carlos Jobim, who pioneered the *bossa nova* musical style of the early 1960s.

I met Joan Olin at a YMCA dance party. She was there with her daughter, Martha, who was closer to my age. I danced with Martha first. Everyone told me later that Martha went back to Joan all excited and told Joan what a great dancer I was. Next I asked Joan for a dance. After one dance, choosing between mother and daughter, there was no question who was my choice. It was the mother. It was Joan.

Joan had been previously married to Reverend Harry Olin and had helped him establish three Nazarene churches in southern Illinois. After twenty years of very strict religious living—with someone who wouldn't change, or couldn't—Joan had decided it was time to move on. So she went to work for the National Cash Register Company, NCR.

After a year as an office manager at NCR in Little Rock, she was transferred to Dallas. She now worked as a salesperson and a trainer in the accounting-machine and data-processing department. She worked on class 42 machines and was training seventy-five operators for a bank in downtown Dallas.

Joan and Martha lived on Gaston Avenue. It wasn't long until I moved from Plano to Gaston Avenue, just to be close to them.

Gaston Avenue Apartment Complex

There were sometimes weird things going on in my new apartment complex. For example, some guys lived there who attended mortuary school at Baylor. For an assignment they were having to build heads, ears, and noses. They'd bring their artwork out on the public patio and swimming pool area to show it off—or to scare the hell out of the girls.

One morning as I was leaving for work, someone had put Nazi literature on each door, brochures titled "One-Way Tickets for Blacks To Go Back to Africa."

Brinks Coffee Shop

Not far from the apartment complex, also on Gaston Avenue, was Brinks Coffee Shop where we often went for coffee. Martha got a job there as a waitress. The first week, she had to buy her uniform. That same week, being new, she broke a lot of dishes, which she had to pay for. When she received her first pay check, it was for only $12—for forty hours of work. The poor girl broke down and cried.

Joan and I occasionally talked to the owners, Norman Brinker and his wife Nancy. They later sold Brinks Coffee shop and started Steak and Ale, which became a huge success. Today it's a huge chain, with nearly twenty locations in Texas alone. Then Mr. Brinker formed Brinker International, and sold Steak and Ale. Eventually, he became a leader in the casual-dining industry. Today Brinker International is listed on the New York Stock Exchange and either owns, operates, franchises, or is involved in the ownership of hundreds of restaurants.

I'm sure you'll recognize some of the names: Chili's (which has nearly 1000 locations), Romano's Macaroni Grill, On The Border Mexican Grill & Cantina, Cozymel's, Maggiano's Little Italy, Corner Bakery Cafe, Big Bowl, Rockfish Seafood Grill, and EatZi's Market and Bakery.

Religion

Several times Joan and I visited Dr. Bowlin at the Presbyterian Church near White Rock Lake. He gave very inspirational sermons.

I still remembered the fun I'd had as a child, going to Sunday school while Grandma was in church upstairs—playing with the cut-outs of Jesus, Joseph, Mary, the shepherds, and the sheep; putting them on the felt storyboard while the Sunday-school teacher told Bible stories. But I hadn't yet forgotten some of the other churches I'd attended, where the preachers had banged on the lectern and said we were all sinners who would die, go to hell, and burn forever—unless we got saved and donated ten percent of our assets to the church.

At about this time, I read *Waiting for Godot*, a play by Samuel Beckett. The basic idea is that these guys spend the whole play waiting for someone named Godot, who never shows up. Some people say the play is an atheist drama and that Godot stands for God.

I had mixed feelings about religion. For example, at that time, selling liquor over the bar was illegal in Texas, because the churches were able to lobby politician into passing anti-liquor laws. The laws were silly, because you could be served drinks if you were a member of a private club. To skirt the laws, all the bars in Texas claimed to be private clubs, and you could buy a membership in any one of them for five dollars. I had a "membership" in one of these "private clubs"—the Century Club on Greenville Avenue, where I met two of my friends at the time, Caruth Junior and Jimmy the bartender.

Anyway, Joan—who was, you'll remember, a former preacher's wife—and I joined the twenty and thirty (T-N-T) singles Sunday-school class, sponsored by Momma and Papa Brown. It was at the Highland Park Methodist Church on the SMU campus. The T-N-T class met every Sunday morning in the same basement room where Bill Allen and I had gone in high school for our rocket-development project with the Boy Scouts. One of the other members was Joe Willman, who was trying to start an upscale ladies' boutique in Highland Park. One time after Sunday school, Joan and I went over to help him remodel the place and get his dresses all hung up. He was going to call his boutique Umphrese, which was his middle name.

Big Bang

In 1965, two young astronomers, Arno Penzias and Robert Wilson, accidentally discovered "cosmic microwave background" (CMB) radiation. Their discovery was used as evidence to support the "hot Big Bang" theory of the origin of the universe. (This theory is explained at the website for the University of British Columbia.) According to this theory, CMB radiation is a remant from the earliest days of the universe, when a huge, very hot, and very dense fireball exploded and created everything. As the universe expanded and cooled, the radiation decoupled from matter and cooled.

T-N-T

The T-N-T classes that Joan and I attended were interesting. The group invited Dr. Albert C. Outler, the Protestant representative to the Vatican's Second Ecumenical Counsel (Vatican II), to give us some insight into the Catholic Church. Dr. Outler was a professor at the Perkins School of Theology at SMU. He was an official observer at Vatican II, and on several occasions had been a representative and spokesperson for publications such as *Tradition* and *The Unity We Seek*. He also wrote books and articles, and he provided the response to the Dogmatic Constitution on the Church.

He presented our T-N-T group with a series of six lectures on *The Documents Of Vatican II*. He told of the division in the Catholic Church between two groups, the "Immobilisti" (or "Unmovable Ones" in Italian) and the "Reformers." The Immobilisti did not want any change in the church whatsoever. The Reformers were the new, younger priests who wanted to bring the church into the twentieth century. For example, they wanted to say the Mass in English or whatever the local language was.

As Dr. Outler described it, the political in-fighting in the Catholic Church was worse than in the U.S. Congress. The priests would fight for weeks over a single word or a phrase, and would usually end up in a deadlock. Then they'd adjourn, and go lobby for support of their position and for more votes for their position in the next session.

It was interesting to hear about, but I didn't think politicking and in-fighting were appropriate behaviors for priests.

Conspiracy Theories

Much to my chagrin, Joan was dating other guys in addition to me, and one of them was a lawyer. One night, Joan went with him to Kirby's Charcoal Steakhouse on Greenville Avenue. After they were seated, her date noticed that Judge Joe B. Brown was also there.

Brown was the judge in the murder trial of nightclub owner Jack Ruby—the man who'd killed Lee Harvey Oswald two days after the Kennedy assassination.

Joan's lawyer friend went over and told Judge Brown that he was representing a stripper from Jack Ruby's Carrousel Club. He said he had seen Ruby and Oswald together at the club. He told the judge that he would be willing to testify to this fact in court.

Judge Brown said this was not necessary, that they had all the information they needed.

Meanwhile, news had come out about a man named Major General Edwin Walker. He was a right-wing political activist who'd had been fired at in his living room on April 10, 1963, in Dallas. The news was that bullet fragments taken his home may have matched those fired by Lee Harvey Oswald.

Space Activity

On February 3, 1966, the Soviet spacecraft *Luna 9* made the first soft landing—as opposed to crash landing—on the surface of the moon.

On January 27, 1967, three American astronauts were killed in their Apollo space capsule while it was still on the ground.

On April 19, 1967, an unmanned lunar probe, *Surveyor 3*, made a soft landing on the moon, sent back photographs of the moon's surface, and dug a trench with a mechanical shovel in order to obtain and analyze small rock and dust samples: an essential prerequisite for a manned lunar landing.

The U.S. launched the *Saturn 5*. At more than three thousand tons weighed as much as a navy destroyer. On top of it sat the first unmanned Apollo space capsule to be sent into orbit.

On April 24, 1967, Soviet cosmonaut Colonel Vladimir Komarov, who had been successfully launched into space, was killed when the parachute on his spacecraft failed, and he was hurled to earth.

Missiles and Bombs

On September 18, 1967, U.S. Secretary of Defense Robert McNamara announced the creation of an anti-ballistic missile (ABM) defense system.

On November 3, 1967, McNamara gave details of a new Soviet nuclear bomb, which could be delivered through a low-orbiting space rocket (the Fractional Orbital Bombardment System—FOBS), and of the American ability to shoot down this bomb at any point during its first orbit.

On November 8, 1967, the U.S. began building intercontinental ballistic missiles (ICBMs), of which 656 were Polaris missiles in submarines.

Collins Radio

At Collins Radio, I went to work on the LORAN (long-range radio navigation) system.

Collins had built this system for the U.S. Navy. The LORAN system was housed in a cabinet that was six feet tall and nineteen inches wide, and that weighed a couple hundred pounds. It was an analog operation, with a series of synchronous motors that kept the time differential from three shore stations. The system was so slow it couldn't track over one hundred nauts. Collins wanted to replace the analog system with new digital technology to do the same thing, only much faster.

I didn't know much about computers, so I was sent to C-8500 computer school in Cedar Rapids, Iowa.

Believe it or not, I traveled back and forth to computer school each day in a Gulfstream airplane.

Collins had a couple of Gulfstream II turbo-prop planes that were used to test out their aircraft systems. One was housed at the airport in Addison, outside Dallas, and the other was in Cedar Rapids, Iowa. Each morning before 8:00 AM, and each afternoon after 5:00 PM, each would fly to the opposite airport to transport executives, employees, and equipment between plants. Occasionally, they'd fly to Newport Beach, California, if an executive needed to get there in a hurry. I'd drive out to Addison every morning, fly to Cedar Rapids, and be there by 9:00 AM for class. At 5:00 PM, I'd catch the return flight and be home in time for dinner.

One evening as we were returning, the aircraft took a nose dive. The pilot acted fast, turning off the autopilot and regaining control of the aircraft. As I was breathing a sigh of relief, I saw the co-pilot come out of the cockpit, open an access panel, and bang on the autopilot unit. Then the pilot turned the autopilot back on, and everyone relaxed.

Fifteen minutes later, the plane took another nose dive.

Again the pilot switched off the autopilot, and again, he regained control. Again, the co-pilot came back, took off the access panel, and kicked the autopilot unit. But the pilot didn't go back into autopilot mode; he flew the plane the rest of the way home. Rather than risk another problem, he'd let the ground crew handle the bad unit.

At C-8500 school, I learned about digital logic, binary and hexadecimal numbering systems, minimization and minterms, Karnaugh mapping, and logic design. I learned that Boolean algebra has two values, zero and one, and three basic functions: and, or, not. Any Boolean function can be constructed from these three. I learned the algebraic law of classification: identity, dominance, commutativity, associativity, distributivity, and DeMorgan's Laws.

This was great, state-of-the-art stuff!

Our task was to integrate a device-control air transport racking (ATR) unit with a 100 mHz LORAN receiver. The device-control unit had a small tape cartridge on the front to load the stored program automatically when power was applied. Tapes could be quickly interchanged to give the LORAN system new parameters or to automatically bench test the LR-104 system. The system was to take the incoming signal, measure its amplitude, and send back a digital number to the emitters of the eight 100 mHz receiver's amplifiers. It was to be a digital servo loop or automatic gain control (AGC), to control the signal and keep it from overdriving the stages and distorting the signal. Once the receiver's signal was under control, the computer would lock onto the master's transmitted signal. Then, it would lock onto A's transmitted signal and then B's signal. It would keep track of the time difference between the master and A and the master and B, and send these signals to another computer, which would calculate the longitude and latitude of the craft. The combined unit dimensions were those of an Aeronautical Radio, Incorporated (ARINC) unit that weighed 35.5 pounds.

We tested the system and found it would track up to 2,000 nauts. I understand it worked very well on F-4 Phantoms in Viet Nam. While we were developing our LR-104 LORAN C/D navigation receiver system, we were the first unit to use the new Collins equipment compiler system (ECS) on the C-8500. We'd submit our logic design to the mechanical engineering department, which was in charge of the ECS testing. A while later, we'd get a set of glass plates with the circuits laid out on them. Then we'd have to follow every trace to ensure it went to the right place and not somewhere else. It was about ninety percent accurate, with ten percent of the traces going somewhere they weren't supposed to go. We'd mark our corrections and resubmit our project. A little later, we'd receive another set of glass-wiring plates and check them out. The ten percent we'd corrected would be right, but now another ten percent, that was previously okay, would be messed up. One day we received the following memo:

MEMORANDUM

TO: LR-104 Engineers

FROM: W. C. Sheard, Jr.

DATE: November 7, 1967

SUBJECT: Kit Lists and ALWUs for LR-104 Circuit Cards.

REFERENCE: Warning! Warning! The computer has done it again!

Those of you who are astute and conscientious engineers have probably wondered how the factory can possibly create a product from the never-ending deluge of paper disgorged by the system. You have most likely discovered by now that the secret is in the kit list and assembly line write ups (ALWUs).

However, *be ye hereby advised* that compared to the current computer output of subject documents, Hogan's goat was a paragon of virtue!

These documents are strictly computer generated, and something (a faulty diode, perhaps?) has given the beast the idea that it can design on its own. Its favorite stunt is to seize upon some specific item of the L/M that particularly warms the cockles of its processor and call out 101 of them. As a suitable alternative, it will also form leads on bulkhead-mounted, glass-trimmer capacitors.

Being one of those aforementioned engineer types, I have spent a couple of weeks trying to locate the faulty diode. To date the results have been about as successful as trying to shove toothpaste back into the tube.

It is, therefore, the opinion of this office that since creation of the kit lists and ALWUs is not an engineering responsibility, and the possibility of getting the output squared away in time to be of any assistance in the upcoming build is remote and nebulous indeed, all engineering efforts along those lines should cease and desist.

We should, however, out of human compassion, be prepared to offer manufacturing a generous portion of help when they begin their build.

The ATR-type-case device control unit we used was about three inches wide, eight inches tall, and fifteen inches deep. It had four kilobytes of magnetic core memory, wound by some little old Japanese or Chinese lady using a microscope. It had a discrete arithmetic logic unit, an accumulator, and some shift registers. At the time, Intel was developing the 4004 microprocessor, which would later do the same thing on a 100-millimeter chip. The same functions today can be done easily with an 8080/85 or Z-80 microprocessor. In fact, I've seen hand-held units the size of cell phones that can do the same thing. However, they were all rendered obsolete with the global positioning system (GPS).

Watching the Cowboys

In 1967 our home team, the Dallas Cowboys, played the Green Bay Packers for the NFL championship. Because the game wasn't sold out, the TV broadcast in Dallas was blocked. But the game was being televised in Sherman, Texas, ninety miles away, on channel twelve.

A few days before the game, our vice president, Bill Roodhouse, came down to engineering to requisition a channel twelve amplifier. He wanted to watch the game.

I started building the channel twelve amplifier. No sooner had I begun than there were orders for ten more. I built a total of twelve systems. We installed a TV antenna on top of our new corporate headquarters (called the Ivory Tower) and ran cables down to Mr. Roodhouse's office.

On the day of the game we got great reception. According to a later newspaper article, a lot of other people in Dallas did not—they'd attached bent coat hangers to their TVs to serve as antennas. The reason for the great reception on Saturday was the weather. It was overcast on Saturday, the day before the game, but the skies were crystal clear on Sunday so all the signals went straight through the atmosphere. However, there was a storm—a storm of protest from a lot of unhappy people who claimed the FCC had turned down the power so they couldn't receive the game. But this wasn't true.

Anyway, because our antenna was a thousand feet in the air on top of the Ivory Tower, Bill Roodhouse enjoyed the game.

The Biggest Surprise of My Married Life

After I'd finally managed to edge out my competition and eliminate the other suitors from Joan's life, I proposed and she accepted. We were married on Christmas Eve, 1965.

I never expected the surprise that was in store for me.

When I'd moved into my efficiency bachelor apartment, I'd bought a four-roll pack of toilet paper. But I wasn't home much—after all, I was working eighty-eight hours per week. So any time I needed to go to the bathroom, I was at work. When I got married, I still had three of the four rolls left.

Soon after we were married, Joan and I went grocery shopping together. She put a twelve-roll pack of toilet paper into the shopping basket, and I thought to myself, "One-time set-up charge."

But the next week, when we went shopping again, Joan put another twelve-roll pack of toilet paper in the shopping basket.

I said, "Hey, wait! We just bought a twelve-roll pack last week."

Joan replied, "We're nearly out, so we need another twelve-roll pack."

That was the biggest surprise of being married—needing eight rolls of toilet paper per week.

Mexico

Our honeymoon to Acapulco, Mexico, was a result of Joan's job.

Joan worked for Hershel Hancock at Hancock Insurance. Mr. Hancock was an independent insurance agent, specializing in business fire insurance. He could write policies from several insurance companies, choosing whichever one fit his customer's needs the best. He would go to the customer's place of business, review the existing insurance, and do a survey of available fire equipment already installed. If he could reduce the customer's premiums by having an automatic sprinkler system or other fire equipment installed, he'd work with the customer to get the necessary improvement implemented. Also, he could write all the other insurance the customer might need, such as employee insurance or specialized insurance for management, and he could take care of the owner's personal insurance needs at a reduced premium in a package deal.

He had several multi-national corporations as customers. Several had operations in Mexico. One problem customers had was that no Mexican insurance was good in the U.S., and no American insurance was good in Mexico. His Mexican clients would always have to make a special stop at his place to get local insurance when they came to town. He was working on a deal with a Mexican insurance company where he had a receptacle licensing agreement. He wanted to write insurance from the Mexican insurance company, so his American clients would not have to buy short-term, high-priced insurance when they went to Mexico. Also, the Mexican insurance company could write policies from an American company for their business clients coming to the states.

When Joan asked for a vacation to go to Acapulco, Mexico, for our honeymoon, Hershel asked if she would be his agent and go see Vice President Señor Medina Moray Becerra, of Agencia De Seguros y Fianzas in Mexico City. Although the round-trip from Mexico City to Acapulco would not be deductible, we could deduct our trip from Dallas to Mexico City and back. Well, that sounded good to me. Hershel had Joan type up a letter of introduction, and he called Señor Medina Moray Becerra and told him we were coming.

We packed our bags into a new Oldsmobile and took off for Laredo, Texas. When we crossed the border, we had to buy travelers' insurance at the high rate. After we cleared customs, we drove on to Monterey, Mexico and spent the night in a hotel in the middle of town. The next morning we were awakened by roosters crowing. We looked out our hotel window on the ninth floor and saw all kinds of farm animals on top of the building next door and in the yards.

We had breakfast and took off. The highways were pretty good but had only two lanes. Northern Mexico is Chihuahua desert and pretty flat with nothing but cactus and yucca plants. There was very little traffic on the highways. We went to Saltillo next, and then to Queretaro, where we stopped for the night at a Camino Real Motel. It was a pretty nice motel—white stucco in the pueblo style with roof timbers sticking out of the walls and high arched gateways. We freshened up and went down to get dinner.

Boy, were we surprised. The waiters all had tuxedos and white gloves, and each had a towel over his arm. It was like a five-star restaurant in the States, but the prices were really reasonable. I think our meal was twenty dollars.

The next morning we decided to do some sightseeing. We saw in the magazine in the room that there was a silver factory close by, so we thought we'd visit it. Boy, was that interesting. It was a long metal building about 150' x 50'. On the back wall were several furnaces where they melted the silver. There was one big electric motor at one end of the building, with a wide belt that went up to a pipe in the rafters. This pipe ran the length of the building and had belts running down to the silversmiths' workstations about every ten feet. The silversmiths would take the molten silver out of the furnace and pour it into a mold. When the silver hardened and cooled, they would break open the mold with a bang, take the silver item out, and put it on the workbench where the pulley ran brushes for polishing and grinding wheels for shaping. There was an anvil on each bench for hammering cups and pitchers. The tour guide said the silversmiths received twenty-five cents a day working in this hot, dusty, dirty place.

We continued on to Santa Rosa where we visited a sixteenth-century Spanish hacienda. It was gorgeous, with wide, arched passageways, terracotta tile-covered walkways, and sculptured plants and hedges. The ceilings must have been fifteen feet high, with hand-hewn timbers and huge black wrought iron chandeliers in ornate designs. The candles had been replaced with light bulbs, but they were in candleholders so it was hard to tell the lighting was electric. The walls were decorated with beautiful multicolored tapestries. There were huge oil paintings of conquistadors, matadors, big black bulls, señoritas, gauchos on horseback, and other Spanish images, and there was lots of gold and silver everywhere. There were beautiful hand-woven rugs on the ornate stone floors, and big high-backed oak chairs with huge oak tables. It was just magnificent.

We continued on into Mexico City and found a hotel. As we climbed out of the car, I could smell gas. After we checked in, I asked the concierge if he knew a good auto shop with a reliable mechanic. He said the hotel had an auto mechanic on duty and asked if I wanted him to take a look.

He called the garage, and this guy showed up who could barely speak English. He understood enough, though, to know what I was talking about. We went out and raised the hood. He looked around and couldn't see anything, so he asked me to start the engine. Sure enough, gas poured out of the fuel pump. I turned off the engine and asked if he could fix it. Probably. He took off the fuel pump and asked if I wanted to go with him to get a new one.

I figured this would be a good way to see parts of Mexico City that tourist never see. We went to five different auto-parts places, but none had Oldsmobile parts. They had Ford and Chevrolet parts, but no Oldsmobile parts.

So he came back, took the fuel pump apart, and pulled the butterfly valve out. We went around the loop of parts houses again, this time comparing the butterfly valve in each of the available units. None was exactly like the Oldsmobile's.

However, he did find one that had the same rubber part with the exact hole patterns, but with a different type of shaft. He explained that we should buy it. He said he could take both parts to a friend of his who would take the bad unit's shaft off and put it on the new rubber part. Okay, I said, let's try it. I bought the part and paid $3.49 for it. His friend replaced the shaft and charged us a dollar. We took the parts back to the hotel, he installed the fuel pump, and it worked perfectly. I asked how I owed him. We'd spent about eight hours on this project. He said $5.00. I paid him and gave him a dollar tip. I couldn't believe how inexpensive it was.

The next day Joan called Señor Raul Becerra Torres and Vice Presidente Señor Medina Moray Becerra, and they invited us to their office. They discussed the terms of the deal, and then Señor Medina Moray Becerra invited us to his home for dinner.

He picked us up at our hotel at 7:00 PM and took us to his house where we met his family. Everyone was really very nice. During dinner, they asked what we were going to do next. We said we were on our honeymoon and were headed to Acapulco. They wanted to know where we'd be staying. We said we didn't have a reservation, that we'd see what was available when we got there. Señor Becerra hit his forehead with the palm of his hand and said, "Oweee, you are not going to Acapulco in the height of the tourist season without reservations! Here, let me see what I can do." He picked up the phone and made a couple of calls. Then, he took out his business card and wrote a name and phone number on the back. He said to go see this guy at the tour-boat office, and he'd take care of us.

The next day, we drove to Acapulco and went to the tour-boat office and gave the man in charge the business card. He was delighted to meet us and wanted to know where we'd like to stay. We said we wanted some place off of the main beach road, because everything there was so expensive—we wanted an inexpensive place a block or two away. He said he knew the perfect place for us, so he took us one block off the beach to the Motel Caribe. No one there spoke any English, but we got along just fine. We got a room with a bath and three meals for eleven dollars a day. Amazing—Acapulco for eleven bucks a day!

But Acapulco was a tourist trap. Everywhere we went, someone had his hand out for something. If you drove around, boys wanted money to watch your car to keep it from being stolen or broken into. If we hadn't paid them, I'm sure they'd've broken a window. We went into several of the hotels on the beach, but they were all higher than a cat's back. Most wanted over a dollar for a cup of coffee and you couldn't even sit in one of their chairs on the beach without paying a ridiculous fee. We stayed only five days instead of the seven we had planned, and decided to head back.

The folks in the middle of Mexico are the greatest people you'll ever meet. It's the ones in the border towns and tourist towns that have been spoiled by the dollar.

We drove through Mexico City and headed north. After Santa Rosa, it was about 250 miles to the next large town. We'd filled up in Santa Rosa and were flying at eighty-five miles per hour, trying to get to the next town before dark. About 120 miles past Santa Rosa, the car started sputtering, and the engine stopped. We coasted to a stop.

We got out the map to figure out where we were. We'd been given this emergency package by AAA for the Green Patrol in Mexico. Thirty percent of Mexico's economy is tourism, so they take care of their tourists. Every couple of hours, the Green Patrol drives along the highway to look for stranded tourists. The AAA package contained a card to fill out, saying who you were and where you were. It said to give the card to a passing motorist who'd deliver it to the police in the next town. The police, in cooperation with the Green Patrol, would dispatch a repair truck.

We filled out the card and waited for another car to come by. It took half an hour before someone came by, and when they did, I flagged them down. The driver didn't speak English but the passenger did. I explained our problem and that I needed him to drop the card off at the police station in the next town. The guy said the driver was the mechanic on his ranch and asked if I'd mind if he took a look under the hood.

I raised the hood, and the mechanic asked me to start the engine. I tried, but it just sputtered. They said they thought the problem was in the carburetor float. The mechanic went to their car and came back with a toolbox. He took the float apart and a minute later showed me a piece of lint. The rancher said, "Here's your problem." They put the float back together and said to start it. It started and ran like a top.

I asked him how much I owed him, and he said, "Nothing. Just have fun in Mexico and return soon."

As I said before, the people in the middle of Mexico are the greatest, nicest people you will ever meet.

We gave him one of Joan's business cards and said for him to call if he was ever in Dallas.

The next day, we stopped in another little town and went shopping in the plaza. Joan found a beautiful topaz ring for twelve dollars and bought it for Martha.

We spent the night in Monterey, and the next day we drove back to Dallas.

Bonza Bottler Day Again

Six months later, on June 6, 1966, it was Bonza Bottler Day again. The day, the month, and the year were numerically the same: 6/6/66.

I recalled the exact moment—eleven years, one month, and one day earlier—when our teacher at South Oak Cliff High School had pointed out the date: 5/5/55.

I could picture in my memory the room at school, and how the sun had been shining in the window that day, and how the numbers had looked, written on the blackboard. I remembered the discussion we'd had in class, about when Bonza Bottler Day would occur again and what we'd be doing on this day.

Now I thought about the next Bonza Bottler Day, 7/7/77, and wondered what I'd be doing then.

Learning—Really Learning—Math

By now I was taking calculus at Southern Methodist University. It was the most exciting course I've ever taken. For all those years in high school, my math teachers would always say, "Just use the formulas. Don't ask questions. Don't worry about where they come from."

Now, in college calculus, I learned how to prove the theorems and how to create the formulas. I finally found out where the formulas came from.

I wished I'd been shown this stuff in the second grade. That would've inspired me to want to learn more. I firmly believe that the fundamentals of college-level calculus can be taught to second-graders.

The principles are very simple. The first derivative with respect to time is velocity. The second derivative is acceleration, or the rate of change of velocity. The third derivative is the rate of change of acceleration.

I demonstrated this to the children of my friend Paul Ketchum by having Paul drive down the freeway. We traveled at sixty miles per hour. I told them that was the velocity. I explained that this was also one mile per minute, or eighty-eight feet per second. We came to a complete stop by the side of the freeway, and then accelerated to sixty miles per hour. As we increased speed, I told the kids to watch the speedometer register the velocity change, or acceleration. I explained that this was why the gas pedal was called an accelerator. I had Paul speed up and slow down to demonstrate the third derivative, the rate of change of acceleration.

To me, it was very simple, and I think the kids understood the concept—even if they couldn't do the math.

Pink and Green

Joan and I bought our first house. It was at 413 Vernet, in Richardson, Texas, and we lived there until we moved to California. Our purchase price was $13,500. I was scared to death that with my $3.75 per hour wages, I wouldn't be able to make the $102 monthly mortgage payment.

Meanwhile, Joan attended a Mary Kay demonstration. Mary Kay was a cosmetics company run by a native Texan named Mary Kay Ash. Ash launched her company in Dallas in 1963, but her cosmetics weren't sold in stores, they were sold by independent sales consultants, who sold products to friends at demonstrations or "shows," usually in their homes. The company eventually became extremely successful, and Mary Kay gained fame for living in a pink mansion, packaging her products in pink cases, and rewarding her top sales reps with pink Cadillac. Mary Kay once said, "Those who get ahead in life, get out and find what they need to succeed. If they can't find it, they create it."

The show that Joan attended was hosted by her friend Abbie Orevec, and Abbie immediately recruited Joan to become a consultant.

The idea behind the company was that each consultant enlisted other consultants, who then worked under her. For this privilege the enlistee gave the recruiting consultant a percentage of her earnings. That consultant kicked up a percentage to the consultant above her, and so on, right up to Mary Kay herself.

I'm told the Mafia also works this way.

Abbie introduced Joan to Dalene White. Dalene was making considerable money, because she had a number of consultants working for her and was getting a percentage of all of their sales.

It was a classic pyramid sales scheme. Pyramid or Ponzi schemes were illegal, but Mary Kay was successful in skirting the laws because she pretended that each consultant was in business for herself. Mary Kay, however, always had the final word in all matters—unless it was a lawsuit, in which case the consultant was on her own. I didn't know it at the time, but later when I learned about independent contractors, I found out this set-up wasn't completely kosher.

The consultants on the higher steps of the pyramid had fancy-sounding titles—some even had more than one. Dalene White was Abbie Orevec's "senior director" and was one of the first "area directors" with Helen McVoy. Pope McDonald was "vice president of sales."

So now Joan was a Mary Kay consultant. But there were problems. She could attract people's interest in the products, but her religious background restrained her from using good sales techniques to set up shows and close sales. Dalene held consultant sales training sessions to try to teach her. After I attended all the training with Joan, I decided to try selling Mary Kay products myself.

I joined as a consultant and set about lining up my first five demonstrations. It was harder than I'd expected, but it was good sales training. Actually I got my five shows booked before my wife did, and I became a full-fledged consultant before she did.

Another problem Joan had was taking money from people. Now you'd think that having helped establish three churches, she wouldn't have a problem asking for and receiving people's money. But she did.

Dalene taught us to go around to each customer at the end of the show and kneel down on one knee so your head would be below hers. (Note that this is the same position a man is supposed to use when he asks his girl to marry him.) Once in that position, we were to ask the customer if she needed help filling out her order. If she hesitated, we were supposed to take the order card and fill it out for her.

Again, this went against Joan's religious convictions. But I had no such scruples. I did exactly as Dalene taught, and it worked!

Next we went to work recruiting consultants.

A Waitress and an Oil Baron

We knew a woman named Blaze Morrow. Blaze's husband Gordon was finishing his master's degree at SMU, and she was working as a waitress at the Hilton Inn on Mockingbird Lane and Central Expressway. We often had brunch there on Sundays after church. Joan recruited her into Mary Kay.

Blaze had an old car with no muffler, and it burned oil like crazy. Blaze and Joan would go downtown to Neiman Marcus's Zodiac Room for lunch. They were quite a sight coming out of Neiman's: two beautiful women dressed in their finest, with hats and gloves like proper ladies wore in those days. Once, as they were leaving, the parking attendant brought the old car around. It was smoking badly and when he stopped, it backfired. The two beautiful ladies got in. Blaze said to the attendant, "She's a bomb, ain't she?" and they drove off in a cloud of smoke.

Meanwhile, Joan was working at Hershel Hancock Insurance, Suite 407, Stemsons Tower West. She found it hard to set up shows because she was stuck in an office all day. She'd done demonstrations for all the secretaries in her building and needed new customers. She decided to change jobs and go to work for Olsten's, a temporary-help agency. That way, she'd be going to different companies and meeting new people, and could book shows at each new place.

One of her assignments was with Ruddman Oil in Providence Towers in downtown Dallas. She was executive secretary to Bill Ruddman, whose company had leased five floors in the Providence Towers office building. She asked if she could use his board-of-directors conference table to have a Mary Kay Cosmetics demonstration for the secretaries in the building. He said okay. She invited thirty secretaries to lunch in the boardroom and gave each one a facial.

She made over $500 on her lunch hour—a lot of money in those days!

Bill was really enthusiastic. He said she made more money in an hour than he paid her for a week of work. He was so impressed that he invited us to come out to his place that Saturday to tell him more about this Mary Kay business.

Perhaps you've heard the story of the Texan millionaire who, when asked the size of his spread, replies, "One hundred acres." The joke is that a hundred-acre spread is no big deal at all in Texas—it's a huge state, and some ranches here have well over 100,000 acres—until the millionaire gives the second part of his answer, "In downtown Dallas."

That Saturday, we went to Bill Ruddman's ranch. It was in the swanky North Dallas area, just off Preston Road above Beltline Road. I don't think it was a hundred acres, but it was huge for north Dallas.

The place was a lot like South Fork on the TV show *Dallas*. The house looked the same. He had a big swimming pool out back, next to his tennis courts, and a large bathhouse with an exercise room full of equipment. The bathhouse/exercise room had sliding glass doors and mirrors and a full fifteen-foot wet bar conveniently located close to the pool.

Just outside the exercise room was an enormous leather massage table. Joan had him lie down on this while she gave him his facial. As she was finishing and patting on the special "Mr. K" lotion for men, she said, "This will make you feel like a million dollars."

At this, Bill Ruddman—a bona fide Texas oil baron who was worth well over $100 million—sat up and looked at me and said, "Make me feel like a million dollars? Is that a put-down?!"

We all had a big laugh over that one.

While we had drinks, Bill brought up the real reason for inviting us out. He saw what a great salesperson Joan was, and he wanted to put her in business for herself, or at least as a subsidiary of Ruddman Oil. He said he could take the Mary Kay products to his chemists and have them analyze the contents. Then he could manufacture a similar product and put her in charge of sales and marketing.

This went against Joan's religious convictions. She turned him down.

The Lady Herself

As Mary Kay consultants working in Dallas, Joan and I often crossed paths with Mary Kay Ash.

We'd attend sales meetings with motivational speakers. Mary Kay always had great motivational speakers. One that I recall was Zig Ziglar, of the Success Motivation Institute in Waco. He gave a famous presentation known as his "Biscuits, Fleas, and Pump Handles" speech.

Another speaker I remember demonstrated how to manipulate people's feelings. He came on stage and pretended to be a Russian ambassador. He opened by telling a funny story. This got everyone laughing. Then he told a sad story. A lot of the ladies started crying. Then he started to run down the United States. This made a lot of people angry; some were so mad they got up and walked out, while others started to boo him. Mary Kay's son, Richard Rogers, had to come up on stage and stop the speech and explain what the guy was doing.

Boy, was that speaker good!

Mary Kay always seemed a little jealous of or afraid of my wife. Joan, with her perfect dress, hat, and gloves, was a terrific model for Mary Kay cosmetics, and I think Mary Kay could see the same potential that Bill Ruddman had seen. Joan had real charisma—she nearly always stole the show, no matter where she was or what she did. People just naturally wanted to associate with her.

Poor Mary Kay was just a little short dumpy lady with a lot of teased, bleached-blonde hair—and that was probably a wig.

Dallas Theatre Center

We became volunteers at the Dallas Theater Center and gave tours of the theater on weekends.

The building was stunning. It was the only theatre that Frank Lloyd Wright ever designed, and its shape was unique in that it had no 90-degree angles. Plays were performed on two stages, the huge main stage and a smaller downstairs stage. I really liked the intimacy of the downstairs stage. It seated no more than thirty patrons in seven rows—this meant you were always right near the stage and could easily see the expressions on the faces of the actors.

I'll never forget seeing the play *Rain*. There was a metal bucket on the side of the stage, and during the performance water dripped into this bucket to simulate rain. One of the female characters came down a staircase after finding another one had committed suicide. The actress was standing not more than five feet away from my seat. When she began to cry, her tears nearly fell in my lap!

There were a lot of *great* plays in both theatres.

Trinity University—which had degree programs in acting, directing, costume design, and stage production (space planning, sound systems, lighting systems, acoustics, and rigging systems) managed Dallas Theater Center's directing and training functions, offering creative workshops in nearly all aspects of theatre. Regular professional actors performed in a repertory of plays, but there were also productions for new actors and teenagers, plus a children's program. It was, and still is, a well-rounded theater.

In addition to the Dallas Theater Center, Joan and I attended several performances at the Dallas Music Hall. The one I remember the best was *My Fair Lady* with Rex Harrison.

The Gopher of the Stars

As one of her temporary assignments, Joan became secretary to Al Mayo, the advance planning chairman for the American Astronautical Society. He was manager of advanced studies at LTV's astronautics division and was organizing the society's thirteenth annual convention, to be held March 30, 1967, in Dallas. The planning committee also included Austin N. Stanton, president and CEO of Varo Corporation, Garland, Texas (he died in November 1995), and Dr. Ross Pevy, vice president of the Graduate Research Center of the Southwest. One of Joan's duties was to call all the top scientists, engineers, and managers in the society to confirm that they were coming to the convention, and to set up any special arrangements, such as hotel reservations, plane tickets, or rental cars.

Joan asked Al Mayo if I could attend the planning meetings. He said yes, because there was nothing secret about the convention. I attended all of the night meetings.

After one of them, someone said to me, "What do you think of the way we're handling the planning?"

I said, "I think you have too many chiefs and not enough Indians."

I was asked if I wanted to help. Of course, I wanted to help! They put me on the committee and gave me a big white committee badge. I became the planning committee gopher—as in "go-fer this, or go-fer that."

Over a hundred top scientists and managers in the space program attended the meeting—some of them were the greatest scientists in America. When conference time came, I was the one who got to go to the airport and pick up many of the attendees. I didn't get business cards from everyone, but the list of those I met included:

Dr. C. Stark Draper. He was brilliant. He designed the Lunar Lander computer and control system used in the Apollo space program. He was also the biggest, most pompous asshole I've ever met. Draper Labs at MIT is named after him.

Dr. Michael Murphy. An expert in space psychology, he founded the Esalen Institute.

Dan Rather. Well known as a correspondent for *CBS News*, he initiated me into the Turtle Club.

John B. Campbell, William J. Coughlin, and Irwin Hersey. Respectively, they were editor of *Space/Aeronautics*, editor of *Technology Week*, and contributing editor for the American Institute of Aeronautics and Astronautics.

Thomas L. Burkett. He was director for the southern region of the Martin Company, which later merged with Lockheed to become Lockheed Martin Aeronautics Company.

Wesley V. Hurley. He was engineer of research and development for the Port Authority of New York.

Randall O. Eakers. At General Electric, he served as manager of aerospace and defense for GE's defense programs.

Of course, Dr. Werner Von Braun was at the meeting. Unfortunately I didn't get to meet him.

I did, however, manage to get introduced to astronauts Charles "Pete" Conrad, Jr. and Richard "Dick" Gordon, Jr. They were on the next-to-last of the Gemini flights, *Gemini 11*, and flew faster (at 17,500 miles per hour) and farther (850 nautical miles) than any of the previous flights. They were the honored guests at this convention.

I also met Kirk Debus, head of the Kennedy Space Center.

I learned from Al Mayo that the range safety officer for the U.S. space program was from the State Department. His orders were to immediately destroy rockets if they looked like they were going into orbit. The State Department didn't want our rockets flying over the Soviet Union and causing another political scandal like the U2 overflight had caused. Dr. Werner Von Braun had already built a couple of rockets (*Atlas* and *Jupiter C*) that could orbit the earth, but the State Department wouldn't let them fly over Russia for fear of upsetting the Kremlin.

During the conference I had a conversation with Dr. Ross Pevy. He explained that he wanted to interconnect all of the colleges and universities in the Dallas metroplex area via a local TV network. This would enable any class at any institution to be attended by students at other institutions. He said Boston, Chicago, Los Angeles, and San Francisco already had such systems and they were great.

Martha's Chevrolet in the Middle of Stemmons Freeway

Martha had received her driver's license, and we were letting her drive around town. There had been several accidents reported in the news; the drivers of cars that had broken down and were on the side of the road had been hit by speeding autos, and some of the people were killed. We told Martha that if she ever had a car problem to leave the car and go to the nearest phone and call one of us; we would come to assist her.

Well, it hadn't been a month after that when Martha was traveling north on Stemmons, just past Woodall Roger's freeway, in front of the Cabaña hotel when she ran out of gas. She was in the middle lane of this five-lane freeway. When the car stopped rolling at the bottom of the little hill, she jumped out and ran like hell, leaving the car in the middle of the freeway, where it could not be seen by cars coming over the hill. An eighteen-wheeler just happened to be following her and had slowed down as she slowed down. When she jumped and ran, the truck driver stopped his truck and turned on his emergency blinkers. Cars could see the eighteen-wheeler stopped in the middle of the freeway and nearly killed each other swerving to avoid it. Several cars stopped and the drivers came over to assist the truck driver in pushing our car off of the freeway. Fortunately, no one was hurt, but it scared the life out of a number of people.

Joan Tries Skiing

Joan and I decided to go skiing with the ski club where I worked. That Friday night, Joan met me at work and we all boarded a bus to Ruidoso, New Mexico's Sierra Blanca Ski Resort. We slept on the bus and arrived just in time to start skiing the next morning. I'd learned to ski in Germany, but Joan had never skied. She took the beginners' one-hour ski class and learned how to snowplow, but not much else. I was going to let her ski on the kiddies' slope while I rode the lift to the intermediate slopes. Nothing doing—she insisted on going up to the intermediate slopes with me. If I wanted to ski, I had to take her with me.

Up we went. She had a little problem getting off the lift as the slope at the end was steep, but she managed to snowplow down the ten feet. The next slope was about a mile long, but it would take some traversing and some good intermediate skiing. When she saw that slope, she knew better than to try to ski on it, so she took her skis off to walk down. Immediately she sank in the snow up to her hips. The ski patrol had to come and dig her out and help her down the mountain. I skied down and waited for her in the snack bar. She didn't ski on Sunday, and I had a few good runs before boarding the bus home at 4:00 PM. We arrived back home in time to go to work. No more was said about skiing.

Looking for Greener Pastures in California

After a few months, Joan decided the Dallas metroplex area was saturated with too many Mary Kay consultants, so she wanted to find greener pastures in California.

I sent Joan and Martha to California to find a place to live. Joan decided to check out the state and make of listed of potential relocation spots. My only requirement was that it had to be near a University of California campus so I could continue my education. I really wanted to go to California Institute of Technology, but because of English, my grades weren't good enough to get in. I was thinking UCLA would be nice.

They flew out to Los Angeles, rented a car, and started up the coast highway. But they never even stopped at UCLA. Their first stop was Santa Barbara. There's a great University of California campus there with a lot of research. They checked out Santa Barbara, put it on their list, and headed up the coast highway (Highway 1) to Monterey.

First they had to go through Big Sur. Joan saw the sign for the Esalen Institute before they reached Big Sur and remembered speaking with Dr. Michael Murphy, the director of Esalen, when she'd invited him to the American Astromical Society meeting. She decided to drop in and say hi. They were totally unprepared for what they found. Parked outside the gate were lots of hippie buses, painted with psychedelic colors and peace symbols. A bunch of woolly looking hippies guarded the gate. The hippies wanted to know why a couple of pretty, straight-looking ladies wanted into Esalen. When Joan told them she was a friend of Dr. Michael Murphy they snapped to, opened the gate, and escorted them to the main office building.

Dr. Murphy was not in, but his assistant took them around to look the place over. There were several large buildings where meetings were held and a lunch-room. In the back were the baths where everyone was nude, getting massages and rubdowns. There were no conferences going on at the time—just a bunch of hippies high on marijuana cleaning up the place—so they left.

They drove on up to the Big Sur Inn, where they secured a room for the night. Then they met Mary Fleenor (Margie) Post, owner, operator, and waitress of the roadside restaurant Post Rancho Sierra Mar. They asked about places to stay, and Margie told them about the Riverside Campgrounds a mile north that rented cabins for thirty-five dollars per month in the winter. They spoke with Mr. Clemons, operator of the camp, and gave him a deposit to hold a cabin until we could get out there in September.

They stayed at the South Coast Motel, just south of Big Sur, then went on up to Carmel. Joan said we'd live there someday, but first she had to find a place for me to work and go to school. Over in Monterey there was a good University of California campus, so that made the list.

They headed up the coast highway to Santa Cruz and then on to San Francisco. San Francisco was out—it was too big a city. That left Santa Barbara, Monterey, and maybe Santa Cruz on the list. They drove to Napa and Sanoma, but did not find any possibilities for work or school, so they returned to San Francisco. They turned in the rental car and flew back to Dallas.

As soon as they returned, we started to getting our stuff together to move.

We rented our house on 413 Vernet in Richardson and moved in with a couple of engineers I worked with at Collins radio.

Advances in Technology

On December 3, 1967, Dr. Christiaan Barnard carried out the first human heart transplant. His patient, Louis Washkansky, survived for eighteen days. Dr. René

Favaloro developed the coronary bypass operation. A new technique, mammography was introduced for the detedtion of breast cancer.

On December 14, 1967, the U.S. announced the multiple independently targetable re-entry vehicle (MIRV).

Chateaubriand Restaurant

Louise Bennett, the lady Joan worked for at Olsten's, wanted to take us out to the Chateaubriand Restaurant for a farewell dinner. It was a very impressive and expensive five-star restaurant on McKinney Avenue, where all the waiters had tuxedos and towels over their arms. We sat down for drinks before dinner, and Joan got ready to order her favorite drink, "between the sheets." Hershel Hancock had introduced this drink to her. The question came up of what was in the drink. No one at the table knew. I said the waiters at these expensive restaurants go to school to learn which wines go best with the food, so they should know what was in the drink.

I motioned for our waiter to come over. I asked him, "What is in between the sheets?"

With his head held high and looking straight ahead he said, "Discriminating couples, I presume," then he turned and walked away.

Later, Louise told us the story of her daughter Sue's cat dying. She called Louise and asked if she could bury her furry companion of nine years in Louise's backyard, as Sue did not have a burial place in the apartment complex where she lived. Louise said yes, but she needed to do some shopping for some new shoes. Louise asked Sue to meet her at the North Park Mall to give her the body, and Louise would take the precious feline home and give it a proper burial. Sue found a Neiman Marcus shopping bag with some tissue paper in it. She lovingly wrapped her furry friend up in the white tissue paper and gently put it in the Neiman Marcus shopping bag, with the beautiful furry tail laid over the tissue paper. At the mall, the two went into a shoe store, and Sue put the shopping bag on the floor while they were trying on shoes. When they were ready to leave, they couldn't find the shopping bag. They paid for their shoes and started down the mall. A large crowd was gathering in the fountain area, and paramedics were rushing in with a gurney. As the crowd parted to allow the paramedics in, there on the mall floor was a very large black woman, out cold. Beside her, the Neiman Marcus shopping bag lay open with the dead cat half way out, still wrapped in the tissue paper all except its beautiful furry tail.

Louise and Sue looked at each other, walked away, and did not claim the cat, so the poor thing never did get buried in the backyard.

Tammy

Tammy was our Pomeranian. Joan wanted puppies, so we had Tammy bred. When she was about ready to have her babies, we decided to name them Neiman and Marcus. Well, she only had one, so we named it Neiman.

Tammy went into shock with the birth and would not take care of the new baby, so Joan kept it on her neck during the night. When the puppy was hungry, Joan would put it down on Tammy and let it nurse. When it went back to sleep, she'd put it back on her neck. Needless to say, Neiman thought Joan was its mother and spent the rest of its life in bed with us.

Entertainment and Preparing to Move

Doris Day and some financial investors built the beautiful Cabaña Hotel on Stemmons Expressway at Continental Avenue. It had a Mediterranean motif with a lot of Roman statues. The bar was named Nero's Nook, and the waitresses wore Goddess outfits—tunics with very short white pleated skirts and tops with a single shoulder strap. It was really the "in" place to go in Dallas.

David and Lisa was one of my favorite movies. Later *The Heart is a Lonely Hunter* starring Alan Arkin was another one.

I quit my job at Collins Radio. We called a moving company and had our furniture picked up. We loaded up Joan's Chevrolet Nova convertible with our three dogs and Siamese cat, and off we headed to California.

Chapter Nine:

Living The American Dream
(1968 to 1975)

Go West, Young Man

Like the Forty Niners before us, we set out for California to seek our fortunes. Carmel hadn't made Joan's original list, but after a lot of discussion, we decided to settle there anyway. We'd set up Joan's Mary Kay office in Carmel, and I'd find electronic work.

Big Sur

When we arrived in Big Sur in January, 1968, we stopped at the Post Rancho Sierra Mar diner just past Nepenthe, owned and operated by Margie Post, whom Joan had met on her earlier trip. Margie's great-grandfather had settled in Big Sur in the 1800s, and a huge parcel of land had been handed down through the years to his descendents. We had a bite to eat, and Joan called Mr. Clemons of the Riverside Campground, where she'd reserved a cabin for us. He said, "Go ahead and move in—I'll meet with you later."

When we got to the Riverside Campground, I found that we had to use a rope bridge to cross the Big Sur River. This was really getting away from everything! It was magnificent, a place of beauty and serenity. We found the one-room cabin, and growing right beside it was a big redwood tree. I mean big: it was eight feet across at the base. The tree was huge, but the cabin was tiny, and here we were—Joan, myself, three dogs, and a pregnant Siamese cat—in just one room. The bathroom and showers were down the trail in the bathhouse.

The rest of the camp was inhabited by pot-smoking, loud-music-playing, naked hippies (some of them worked at Esalen, where they took workshops in things like yoga and physics). Our first priority was to find and set up a Mary Kay office for Joan, so early every morning we'd get up about the time the hippies were going to bed. We'd dress in our finest: Joan in her business suit, with hat, long gloves, and high heels, and me in my business suit with white shirt and tie. Then we'd step carefully across the swaying rope bridge. One morning we met this hippie girl who was coming back across the bridge from a night of doing God-knows-what in the forest. When we said, "Hello," she just looked at us, like we were some kind of weird hallucination, and said, "Far out." The rest of the camp could not believe what they saw. I think some even stopped smoking pot after seeing us.

We'd head into Carmel and try to find a place for Joan's Mary Kay office, and I'd look around for electronic work. After a couple of weeks, Joan had a conversation with Margie Post and wound up doing a Mary Kay demonstration for some ladies in Big Sur. After the show, Joan packed up her stuff and left. The next morning, when we were at the diner having breakfast, Margie came over and sat next to Joan. She wanted to know where she could get some of the cosmetics Joan had shown them the night before. Joan said she had a whole trunk full of it. Margie filled about eight orders of items for the ladies.

One of the first things I wanted to do was to go to Cannery Row. I'd enjoyed John Steinbeck's book about the place so much I just had to see it for myself. First we went to Doc's Western Biological Laboratory, where Doc ran the mail-order company that took orders for various sea creatures that high school and college biology labs need. Then we went across the street to see Lee Chong's grocery store. Next we walked down to the shack at the end of the row to see the Palace Flophouse and Grill (owned by Lee but inhabited by "Mack and the Boys"). Then we walked back to Dora Flood's Bear Flag Restaurant (Dora Flood was a madam, and the Bear Flag was a "respectable" whorehouse). By this time it was a pizza place with a Dixieland band and the entrance through an old telephone booth.

After visiting Cannery Row we drove down Lighthouse Avenue and saw the LaPort Auction house. Later we visited LaPort's Auction once a month when they auctioned off antiques for estate sales in Pebble Beach.

To meet more people in Big Sur, we joined the local Grange association. For the Grange's upcoming fall festival, Margie's cook and a guy named Buzz Brown (more about him later) were working on a play, *Paint Your Wagon*. Joan and I were invited to participate. We were given the role of lovers. It was easy—there were no lines to memorize, and all we had to do was walk across the stage hand in hand. It was fun working with and getting to know the locals.

Buying Property in Big Sur

Sometime before this, I'd read a book review in *Reader's Digest* of *The Spaces In Between: An Architect's Journey* by Nathaniel Alexander Owings. I'd looked for the book at several bookstores, but had never managed to find it. (About fifteen years later, when I lived in New Hampshire, I went down to Cambridge, Massachusetts, wandered into an out-of-print bookstore, and asked the fellow behind the counter if he had a copy of *The Spaces In Between.* He said, "I just saw it a couple of days ago. Let me see if I can find it." Sure enough, he came up with the book, which I bought and which made for fascinating reading.)

Anyway, back to 1968. That book's author, Nat Owings, was quite well known at the time—he'd been on the cover of *Time* magazine, and was a cofounder of Skidmore, Owings and Merrill (SOM), one of the largest architectural firms in the US. South of Big Sur, on a cliff on the ocean side of Route 1, Mr. Owings had a gorgeous A-frame cabin called "Wild Bird."

By now I was fed up with hippies and the tiny one-room cabin, and we wanted a place of our own. We met Reginald Dewar of Strought Realty in Big Sur. He showed us twenty-five acres just south of "Wild Bird," a place that had five level building sites. It was owned by a lady in LA. We thought we could invest in the land, then sell off four of the five building sites and make enough to cover our site. We had a $500 deposit and were ready to make an offer on the property. But the lady in LA wanted $100,000! Where were we going to get that kind of money? Joan said she could get it from Austin Stanton, her friend who was CEO of Varo, a manufacturing company in Garland, Texas. So we put down the deposit.

Joan had met Austin when she was ten years old as he came through southern Illinois with a geological sounding rig prospecting for oil. Austin had taken Joan and her brother Albert to town and had bought them ice cream. Years later, when she came to Dallas, she looked him up and they renewed their friendship. Eager to buy the land, we headed back to Dallas to meet with Austin Stanton. I presented him with our proposition, but he said he didn't have that much capital to loan us. So we lost our deposit—which we could ill afford.

Nepenthe

Nepenthe was near to the camp and was a famous restaurant that had been featured in the movie *The Sandpiper* starring Elizabeth Taylor and Richard Burton. Several scenes were filmed by the restaurant's magnificent round fireplace.

Buzz Brown was a sculptor who lived in Margie Post's old barn, on top of a small hill behind the diner, and he used the tool shed as his sculptor's workshop. The barn wasn't air conditioned, so sometimes in the summer he'd take his bed out on top of the hill and sleep under the stars.

He found an old redwood stump left over from a lightning strike. He carved it into a sculpture he called *The Dark Angel* and sold it to the Phoenix Gift Shop in Nepenthe. Then he sold another sculpture to some corporation back east and invited everyone to come to a celebration party at the barn.

He had a lot of linen-covered tables filled with gourmet hors d'œuvres, pate de foie gras, and tongue. That was the first time I'd ever tasted tongue, and it was really good on crackers. He had several bottles of very good wine, both red and white. There were two tables filled with all kinds of exotic fruits. It was a great party.

Carmel

We continued to look for a place to set up Joan's Mary Kay office, and we finally found a two-room space in Carmel, on the corner of Dolores Street and Seventh Avenue, above a German restaurant. There was a dentist's lab in the other half of the upstairs. We rented the two rooms and started to fix the place up. In one room there was a telephone trunk line coming in that must've had a hundred circuits.

The only two businesses I could think of that used that many lines were an answering service or a bookie joint.

Anyway, we spent the nights in one room while we painted and carpeted the other room. When we finished the first room, we moved the bed and clothes into it and painted and fixed up the other. Then we had it carpeted. We brought in a desk and some storage shelves for Joan's office and for her cosmetics. In the other room, we set up folding tables for giving facials, holding shows, and holding sales meetings.

We opened P.O. Box 5072, Carmel, California as our mailing address. There's no mail delivery in Carmel, and the daily trip to the post office is a sacred ritual for Carmelites.

Edelweiss

The German restaurant was later renamed Edelweiss restaurant, with Sergeant John Vance as chief cook, and his German-born wife Lilo (Lil) as waitress. We became very good friends because we ate or drank coffee four or five times a day down there. They wanted to know if we'd ever eaten artichokes. Well, being from Texas, no. The first time we ate artichokes, John and Lil stood behind the kitchen door and watched as I tried to chew the whole leaf, sharp spine and all. Finally, they came out laughing and showed us how you're supposed to scrape the pulp off the leaf and leave the ribs and sharp spine, and how to dip the leaf in lemon mayonnaise sauce before you eat it.

More Scientific Innovations

On September 11, 1968, the anesthetic epidural technique to ease childbirth was announced. Dr. Christiaan Barnard performed his second heart-transplant operation. The patient lived for seventy-four days: fifty-six days longer than in the first operation.

On October 11, 1968, the *Apollo 7* spacecraft was launched from Cape Kennedy (as Cape Canaveral had been renamed) for an eleven-day mission during which it made 164 orbits around the earth.

Our First Apartment in California

By now, we'd lived in the tiny cabin among the hippies, lost $500 on the failed land deal, and spent nights braving paint fumes in the new office while we were fixing it up. In May we rented an apartment under a house next to a garage on San Antonio Avenue, several doors north of Ocean Avenue. It was only one room with a bath, but it was better than sleeping on the floor in the office.

Church of the Wayfarer

We joined the Church of the Wayfarer and started Bible study classes with Dr. Freeman, a professor of ancient languages at the University of the Pacific. He spoke and read Latin, Greek, and Hebrew. He could read the Bible in these languages, and in Bible study, he would read the original passages to us aloud, then translate and discuss the translation. One of the words he discussed was "plumatah," which was used to describe a vision of God as experienced by a prophet. Dr. Freeman explained that "plume," the root word, could describe a plume of feathers or a plume of smoke. He would point out archaic words that were not found anywhere except in the Bible; thus, there were no known cross-references through which the exact meanings of certain words could be determined. He left it up to each of us to decide what the writer was trying to get across.

He also pointed out that the order of words in the ancient languages had special meaning to the people of the time. He said that for centuries mankind did not have any numbers or numerical digits in their language. Instead, early civilizations used the letters of their alphabets to express numbers. This idea or principle was especially true for the languages of the Bible. Both the Hebrew and Greek alphabets had numerical values for each letter in their alphabets. Now not only does each letter in the Hebrew and Greek alphabets have a standard and fixed numeric value attached to it, but this same principle can be applied to words as well. And not only to words, but to complete thoughts and sentences. For example, there's the well known number 666 from the book of Revelations, and the number seven that is mentioned "fifty four times."

So I purchased the book *Theomatics* by Jerry Lucas and Harry Lorayne. It was interesting reading, but I wasn't impressed. I'd seen numbers manipulated to prove points in a lot of sales pitches.

Tuck Box

We liked to eat at a place called the Tuck Box on Dolores Street. It was an old-fashioned-looking place, with an old English-style thatched roof and Hansel-and-Gretel gingerbread trim. It was also very small, with only about ten tables inside and outside. Because the seating was limited, they'd ask if you minded sharing a table with strangers. This made it a great place to meet people. We met a woman named Violet Hess, and a man who, as it turned out, was the inventor of the original No-Doz tablets.

Around this time we also got to know Joan and Neal Akers, who lived in LA. Neal liked to ride Harleys.

Jack and Mary Heidemann

At the Tuck Box, we met Jack and Mary Heidemann of Carmel. Jack was originally from Holland, and he'd met Mary, a librarian, while attending the University of Michigan during the depression. After graduation in the thirties, Jack and Mary moved to California. The depression was still going on, and jobs were scarce. When he applied for employment, the only position available was working for the state of California in Sacramento, taking care of the capitol grounds. Jack had once worked for the King of Holland as a gardener, and the state of California was apparently impressed—he got the job. As a result, Jack and Mary had a good income all during and after the depression.

They were avid readers and had a library collection that was second-to-none in classical and antique books. When they learned that I liked Shakespeare, Joan and I were invited for dinner and a discussion on good books. They had read them all, had copies of most in their library, and relished discussing them with anyone who would listen. We enjoyed many great evenings at their place.

Jack always started the evening with a salute of a shot of vodka and a "skoal" toast. I learned that Vikings would use an enemy's "skoal" or skull as a chalice or drinking cup. Making a toast and drinking mead or ale, or sometime even blood, from the skull of a fallen enemy was the ultimate celebration of victory. Thus the word "skoal" came into general use.

Barbara West

Barbara and Roger West were another couple we met at the Tuck Box. Barbara was born in London and was the daughter of a famous English architect named P. Morley Horder. Her interest in theatre began at an early age when her father transformed his garage into "London's Littlest Theatre." She studied at the Central School of Speech and Drama at the Albert Hall. For some years she was on the London stage and played in Shaw and Shakespeare with the Sybil Thorndike Company. She was part of an experimental group called The Garden Theatre, which was sponsored by G. K. Chesterton who wrote an original play for the small stage, as did John Van Durten, Sir John Squire, and many others. Actors of note, among them Sir John Gielgud, often participated in this experimental theatre group. After her marriage to engineer Roger Rolleston West, they came to Canada, where Barbara directed at the Vancouver Little Theatre and studied painting with Frederick Varley, of the famous Canadian Group of Seven. She worked in speech and drama for the Canadian Musical Festivals and the Dominion Drama Festival at Ottawa. Then she went to Hollywood, where she joined the Lawrence Olivier Company, touring with Olivier and Vivien Leigh in their production of *Romeo and Juliet*. Later, in San Francisco, she directed plays for the Theatre Arts Colony and the San Francisco Municipal Theatre.

When the Wests came to Carmel, Barbara, always interested in painting, studied with John and Patricia Cunningham at the Carmel Art Institute. Soon painting began to replace theatre as a full-time job for her. She became a life member of the Carmel Art Association and exhibited in San Francisco and around the Monterey Peninsula. She also had one-person shows in San Francisco, Vancouver, and Carmel and won awards at the Madonna Festival in Los Angeles and other religious art exhibitions. During this period, she maintained her own gallery studio at Sunset Center.

Her husband, Roger Rolleston West, graduated from Cambridge University. He joined the DeHavilland Company, became their chief engineer during World War I, and was credited with developing one of the airplanes that beat the German Fokker and the famous Red Baron. After that, whenever he toured the DeHavilland factory to review the new work they were doing, the employees would stand and cheer for him.

After we met Barbara West, Joan decided to combine her Mary Kay office and training room, and use the other room as a religious art gallery to display Barbara's religious works. We hung her paintings and displayed a sign above the portal to the upstairs: "The Upper Room Gallery—Barbara West's Religious Paintings."

After the death of her husband in 1975, Barbara returned to England and took up residence in Clifton, Bristol, where she continued to paint and exhibit in a number of small shows. She died on August 29, 1986, in England.

Hog's Breath

We liked to eat at the Hog's Breath cafe because they had a *great* bowl of French onion soup. The Hog's Breath cafe was owned by Clint Eastwood, and often he would be in the cafe having lunch, or dinner, or just coffee. We knew him from the local business and professional meetings we attended, and he would sometimes invite us to sit at his table.

Mary Kay in Carmel

Another couple we met were Maryanne and Tom Koppand. Maryanne was one of Joan's first Mary Kay recruits in California. But Tom was a pharmaceutical salesman, and his territory was changed to Montana. So they moved to Missoula, Montana, after only a short period of time. Phyllis and Howard Boyd were two other early recruits.

There's a hierarchy in the Mary Kay organization that involves having each sales associate recruit other new sales associates, who then recruit their own sales associates in turn, and so on. Once a sales associate has done enough successful recruiting, that associate is spun-off by her superior (called a director or a senior director) and becomes a director in her own right. Joan quickly became a senior director and spun off Gaile Letchworth as her first director. (Gaile's husband, Major Rod Letchworth was a Top Gun instructor at El Toro Marine base. He taught dog-fighting tactics he'd learned during two tours of duty in Viet Nam.) Then Joan spun off Pam Rowe in Oregon, Harriet Metcalf, Phyllis Boyd, Isabel McGee of Sunnyvale (wife of Max), Betty Sheldrep of Carmel (wife of Al), and others.

Two more recruits were Remmy and Sy Umpinco of Guam, and when they had a Mary Kay party, they made it a luau. The luau approach was a great success—at least thirty people showed up for the facial. Joan seated about twenty women around a central table, while ten more sat in chairs on the outside and in the living room. I think this was one of her biggest shows ever.

When Joan became a director, she received a stock option for five hundred shares of Mary Kay stock at thirty dollars per share, and at that time the stock was going up like crazy. Joan also won the use of a new pink Cadillac for a year. I enjoyed driving the new Cadillac and listing to the demo tape that came with it and the *Concerto Symphonique* by Henry Litolff.

Circular Slide-Rule Calculator

One of the problems the Mary Kay consultants had was dealing with the requirements for becoming a director. The system was set up for a three-month period, to ensure that the consultant really had a unit to direct. The first month, the consultant had to announce her intent to become a director and "sell" (which meant purchase) $1,000 in merchandise and recruit ten new consultants. The second month she had to sell (purchase) another $1,000 in merchandise and recruit another ten consultants, and her twenty recruits had to sell a total of $10,000 in merchandise. The third month she had to sell (purchase) another $1,000 and recruit ten more consultants, and her thirty consultants had to sell a total of $15,000 in merchandise.

A consultant on the director-track had to maintain a minimum of twenty-one consultants doing a total of at least $15,000 per month. Most new consultants wanted to debut at the September shindig in Dallas to experience all the glamour that Mary Kay herself put into the event. Either that, or they wanted to debut at the January seminar in Dallas.

I created a circular cardboard calculator to make it easy for Joan to explain the process. It had a fixed back plate divided into twelve slices, and the months were printed on the outer rim. Then I created a movable disk slightly smaller so the months would show on the fixed disk. I created a large arrow with "Debut" printed on it. I marked off the prior three months and divided the wedges into four concentric circles. The first circle was for the sales the consultant had to sell for each month. The middle circle was for the number of recruits for each month and the total needed. The innermost circle was for the sales the recruits had to maintain for each month. All the consultant had to do was put the big "Debut" arrow on the month in which she wanted to debut and read the previous three months' requirements. It was a simple little circular slide-rule calculator, but it demystified the process and made it easy for a consultant to see what she needed to do to become a director.

When Pope McDonald, of the core Mary Kay team in Dallas, came out to see how we were doing, I gave him one of my director calculators. He was amazed. He took it back to Dallas, and by the time of the next seminar, he had thousands of professionally generated director calculators, and he was giving them out to all units.

Catering Woes

We'd go up to the Bay Area on weekends and hold Mary Kay workshops. In advance, we'd talk to catering managers at large hotels and negotiate a price-per-place for a sit-down lunch. Then we'd would book a room at the hotel and charge the consultants the agreed-upon lunch price. In San Mateo, a couple of times we booked the Radisson Hotel, which looked like a castle; it is a Dunfey Hotel now. But we had some problems with it.

After we finished the day-long seminar, when we paid our bill it was ten percent higher than what we'd calculated, given the number of consultants and guests. I figured it was the hotel's way of making sure the catering department made a profit. This happened more than once. We'd talk to the catering manager, get a price, estimate the number of participants, and multiply by the agreed price—but the caterer always came up with a head count that was about ten percent higher than ours. And after the seminar was over, there was no way to verify the count.

I decided that next time, I'd pay the bill at the time of the serving. We booked a room at the Holiday Inn in Chinatown in downtown San Francisco, and when the food was being served I cornered the catering manager and told him I wanted to pay for the service right then and there. He was really upset, stammered and stuttered, said he was in the middle of serving, came up with every excuse to dissuade me. I said okay, stop serving if you have to, because I want an accurate count *now*. Finally he agreed, and the bill came out to be what we'd calculated.

Glad I got a handle on that rip-off!

Perception

When Joan went to Dallas for her week of director's training, Mary Kay put all the prospective directors through a number of tests to see if the prospect would make a good director. I think it was more nosiness on Mary Kay's part than anything else. However, there was one test that Joan just blew away.

Each test was supposed to have a bell curve for the intended group of testees, with the majority falling in the middle and no one finishing the test. If anyone finished the test, then it was considered not to have accurately tested that person's abilities.

Joan had been an executive secretary for several years and was used to typing full pages of numbers for monthly profit-and-loss statements, income-and-expense statements, and all kinds of statistical financial information. These had to be correct or someone could get fired. The perception test was a full page of numbers, in which the second column of numbers was supposed to match the first column. There were four sets of columns, with fifty numbers per column. The theory was that the reader would read the first number and compare it with the number in the second column. If they were not identical, the reader was to circle the wrong number.

This was a timed test, twenty minutes I think, and no one had ever finished. Most would barely get halfway through before time ran out. They told Joan's group to start, and Joan turned the paper over and began. She marked all twenty-five of the wrong numbers in about thirty seconds and tried to turn in her paper. The tester told her to check her work. Joan went back and sat down and looked at the paper again. The same twenty-five numbers came up bad. She turned in her paper after a total of ten minutes and made a 100 on the test. The tester couldn't figure out how she did it. It was supposed to be impossible. No one could read all those numbers and compare them in twenty minutes, much less do it twice in ten minutes.

The way I see it, Joan had this incredible ability of perception. She could walk into a room of fifty people and immediately tell which ones had their noses out of joint. She could analyze the whole scene in a glance. On the test, the wrong numbers just jumped out at her as not being right. The numbers to her was like looking at a Moiré Interference pattern—the wrong numbers jump out at her as not being right and kind of quiver on the paper or creates large-scale changes in the moiré pattern where the wrong numbers are.

Basic Business Problems

Mary Kay consultants had two basic problems to overcome.

The first was not having any business sense or money sense. The Mary Kay business was fairly simple. You invested $1,000 in product and sold it for $2,000. Of the $2,000, there was no more than $250 profit after subtracting $1,000 for the next set of product, plus $750 for gas, lunches, clothes, cleaning, and miscellaneous expenses. Success was possible only if you kept to a tight budget. The $250 could evaporate very quickly if you weren't extremely frugal. If you reinvested every penny ten times and held to your $250 clear profit, you'd have $2,500, from which you could repay the original $1,000 (usually borrowed from a husband or a kid's piggy bank) and still have $1,500 to keep the business going.

However, most (but not all) consultants would borrow $1,000 from the husbands, sell $1,500 worth of product, take the $1,500 cash-in-hand, and make a down payment on a new $15,000 automobile. Now they were almost out of product, out of cash, and $13,500 in debt—not to mention the debt to their husbands.

The second biggest problem consultants faced was husbands who typically did not want their wives to have their own money. No kidding—these guys really wanted to keep their women barefoot and pregnant! It took us a couple of years to figure this out.

Joan would recruit a consultant, and the new lady would go like a house on fire, selling product like crazy. Then she'd start missing sales meetings, and she'd quit booking shows. When we'd talk to her about her sales, she'd say her husband wanted her to be home more. Soon she'd drop out of MK entirely. Admittedly, some of the problems were due to the expenses the consultant was running up, but most were a result of the male chauvinist pig mentality. Once we became aware of the root of the problem, whenever Joan found someone who wanted to become a consultant, we'd go over to her place and have a discussion with the husband.

One question we'd ask was, "How would you feel if your wife made more money than you?" That was the real test. Eighty percent said *no*! That really floored me.

I guess I'm the dumb one. I thought your wife making more money was a better way of life. Then you both could do the things you wanted to do without bankrupting the marriage.

Highland Inn

As we became integrated into the community, we developed a long list of pre-ferred places to shop and eat and hang out. The Mediterranean Market, with smoked meat, cheese, and bread, was one of my favorites. And I enjoyed going to hear Nancy Ballard, a harpist with the Monterey Symphony, who played at a number of hotels and restaurants in the area.

Mr. and Mrs. Ramsey, owners of the Highland Inn, always had a great Christ-mas festival. Mrs. Ramsey would recite the tale of the piping of the Yule log, and you could hear the piper outside and the eight men carrying the Yule log into the house and placing it in the fireplace. Then Mrs. Ramsey would put the ashes of last year's log on top of it and pour some wine on the log and light it. It was a very nice annual Yule party. After the lighting of the Yule log, they served punch and cakes and Christmas candies.

For entertainment, there'd be a piano player in the Sunset Room of the inn. His name was Chuck Row, and he looked like Colonel Sanders.

There was also a palm reader in the Sunset Room, and her name was Dede McBride. We saw her there for several years before we became acquainted. One year Joan decided to have her palm read, so we invited Dede over. She did a won-derful reading. She was so good that I asked her to read my palm, too. It was amazing; she was right on about my temperament. From then on we became good friends.

Finding a Job

When we'd arrived in Big Sur, one of the first things I did in Monterey was to start my job search by filling out an application at a local employment office. The counselor looked at my application and chuckled. He said, "Monterey is for the rich and famous, and most of the jobs around here are service-related, except for those at Fort Ord. There are no electronic jobs in our files. Good bye, and good luck."

Well, that was not what I wanted to hear. I decided to see what I could find on my own.

Monterey had been built by railroad tycoons in the 1800s. They'd built their own railroad down from San Francisco and it ended at Pebble Beach, where they built their mansions. And yes, most of the jobs there were service-related. Car-mel, an artist colony, had been started just over the hill from Pebble Beach by a bunch of starving artists, and it, too, didn't have any electronics jobs.

But there must be some in the area, I told myself.

I found Comsat's satellite ground station in Carmel Valley, and I put my resume in there. Then I learned that Litton Industries had an office at Fort Ord. I went there and filled out an application. They had a lot of electronic work supporting the Army's Combat Development Experimental Command, but there were no openings. I found three other places also, but again, no openings.

I continued to help Joan set up her Mary Kay business in Carmel. After a month, we were getting low on cash, so it was definitely time for me to find a job. I bought the *San Francisco Chronicle* and the *Los Angeles Times* Sunday papers, the ones with all the employment ads. In the *Chronicle*, I found an ad for the Stanford University Linear Accelerator (SLAC) and in the *Times* there was an opening at California Institute of Technology (Caltech) in the physics department. I'd always dreamed of attending MIT, Caltech, or Stanford and getting a physics degree, but my English grades kept me from even trying. Now, here was a chance to work for the smartest people in the business.

SLAC

I went up to Stanford and applied for the opening. They interviewed me and thought I was a good candidate, so they took me on a tour of SLAC. Boy, this was *great*. They drove me to one of the target sites first, and I noticed these little radiation stickers posted in various places, such as on telephone poles and walls. They took me inside to the target area and showed me where I'd be working. It was fabulous.

There was a big control room with consoles and IBM mainframes, and the lead-lined target with all the coax cables running from the photo detectors and calorimeters to the mainframes. There were other target areas similar to this one. There was one IBM mainframe computer dedicated to controlling the beam and switching the beam between target areas, depending on the scientists' needs. I'd be installing and maintaining all of this electronic equipment. I nearly dropped dead.

I had died and gone to heaven.

I kept hearing this sound—"ding," and then "ding ding," and again later, "ding." I asked what the dinging was all about. My guide explained that it indicated when the high-energy beam got out of containment. He pointed out the radiation stickers all around. He said, "That shows where the beam hit." They had a radiation monitor who went around all day with a Geiger counter, and when he found a hot spot where the beam had hit, he'd put one of those radiation stickers on it. (The half-life of the radiation was fairly short, a matter of days, and would decay pretty quickly.)

My guide drove me down the long two-mile accelerator tunnel in a golf cart. It was the world's longest and straightest pipe, made perfectly straight by lasers. It was filled with some inert gas. Every few feet, there were accelerator magnetic coils to boost the particle's speed and move it along. The accelerator would move a 1.66×10^{-28} g particle to 99.999 percent of the speed of light, and it would weigh over a ton when it arrived at the target. At the beam-generation end, he showed me the klystron that generated the particles. He explained how they selected the different types of particles they wanted, such as electrons or protons or neutrons, for whatever experiment the scientist was doing.

This was *great stuff.*

We went back to the office, and they offered me the job! I told them I would have to think about it and get back to them on Monday. Talk about nerve. But I wanted to go to Caltech and see what they had to offer.

I flew down to Los Angeles, rented a car, drove out to Caltech, and interviewed. They were building spacecraft to monitor x-ray and gamma-ray radiation in space to get more data on the Big Bang theory. This was my kind of stuff, the stuff my dreams were made of. Besides going to MIT, Cal Tech, or Stanford University, I'd always dreamed of someday building and maybe even riding in a spacecraft. When they offered me the job, I accepted it immediately. I'd have to move to LA and get an apartment, but I would have had to move to Palo Alto if I'd accepted the job for SLAC at Stanford. I flew home and told Joan: the good news was that I had a job, but the bad news was that it was in LA. Well, she was both happy and disappointed. Happy I had a job, but sad it was so far away.

California Institute of Technology

I packed up my little VW with just enough stuff to get by, loaded in our two Pomeranians and pregnant Siamese cat, and drove to Pasadena. Joan and I spent a few days with Bill Allen and his wife Leslie, who were living in San Pedro, while we looked for a place for me to live. I'd already started work and was lucky to get a shower, change suits, and grab a few hours' sleep on their couch. Joan and her two hysterical little yaps hung out with Bill and Leslie, though. Joan would boil frozen turkey parts for the dogs, who would then scatter the parts around the apartment. The smell of decaying turkey parts in the morning was horrendous.

I finally rented an apartment at 1124 East Del Mar Boulevard in Pasadena, at the rear of the Caltech campus, and moved there in March 1968. Joan drove the VW and dogs back to Carmel.

I could walk to the physics building and work as long as I felt like it. I worked until midnight a lot of nights. When I had some free time in the middle of the week, I'd go to the Millikan Library and read. It had 100,000 books that interested me. If questions came up, all I needed was to write them down and ask someone the next day.

I really had died and gone to heaven.

The library was open all night so the students could study. Several nights it was 3:00 AM before my eyes started slamming shut and the words became so blurry it was difficult to read.

Dr. Vogt was head of the Caltech physics department, and Dr. Garmire was the principal investigator for the "experiment to measure the differential energy spectra of electrons from 0.15 to 2.8 MeV and hydrogen and helium isotopes from 0.5 to 40 MeV/Nucleon using the interplanetary monitoring probe (IMP-H and –J)." The experiment was designed to provide information on energetic particles in the interplanetary medium and trapped in the tail of the magnetosphere. The measurements of the differential energy spectra of electrons, proton, deuterons, tritons, He^3, and He^4 and their variations with time, from solar emissions and from galaxies, would be used to support studies in the following major areas:

Energetic particles of galactic origin. Studies on their origin, interstellar propagation, and solar modulation in the interplanetary medium.

Energetic particles originating in solar events. Studies of solar particles' acceleration processes, and the propagation and storage of these particles in the interplanetary medium.

Energetic particles trapped and originating in the magnetosphere tail.

Although the scientific objectives were broken into groups they were, in fact, all interrelated. The investigation was part of Caltech's continuing studies of the astrophysical aspect of cosmic radiation, which were being supported by observation with satellites and a balloon-borne detector system.

I was hired to build the detectors.

My background at TI and Collins Radio, along with my NASA soldering certificate, made me ideal for the project. Dr. Garmire and an electrical engineer designed the detectors, but they needed a skilled assembler to put them together. I was right at home working in the lab, which was in the third basement of the physics building.

The IMP-H spacecraft was to carry a very sensitive magnetometer to measure the very low magnetic fields in space. The flux monitor for the space craft was so sensitive it was picking up the magnetic fields of the cyclotron on the first floor, so it could not be used in the physics building. It just pointed up to the cyclotron all the time, which was much more powerful locally than the earth's magnetic field here. The physics department had a lead-lined safe where they kept the calibrated radioactive sources such as uranium, caesium iodide, sodium iodide, calcium fluoride, and other x-ray and gamma-ray sources for testing and calibrating the detector.

On my lunch hour, or when I had to go to the chemistry building to get cleaning fluids for the soldering such as trichloroethylene ($CHClCCl_2$), carbon tetrachloride (CCL_4) and grain alcohol (C_2H_5OH), I'd stop in at the other labs and ask what they were working on. They were usually glad to talk about their research. One lab I stopped in was doing eye/brain research. They had subjects fitted with contact lenses on one eye and the lens had a mirror attached to the side. A laser beam was bounced off the mirror and onto a photo detector. The subject's head was fixed, and there was a TV in front on which the researcher displayed various patterns. The laser would draw the pattern that the eye took to analyze the image. For example, they'd display a girl in a bikini and watch how a guy analyzed her. It was really interesting.

On Friday after work, I'd take a bus from Pasadena to Los Angeles Airport (LAX) and catch a commuter flight to Monterey. Joan would pick me up at the airport. On Monday I'd return to LA. One weekend I arrived in Monterey and called home for Joan to pick me up. There was no answer. So I took a taxi home and waited. About 10:00 PM the phone rang. It was Joan. She had driven to Pasadena without telling me she was coming. Now she was in Pasadena, and I was in Carmel. Boy, was she mad!

The Tower of Power

On several occasions when Joan came to Pasadena for the weekend (and found me there!), we went to the Tower of Power, which was the name for Dr. Robert Schuler's drive-in church in Anaheim. He was building the Crystal Cathedral at the time. I never did get to go to a service in the Crystal Cathedral, but we did attend services where he had the entire church wall slide back, so the members in their cars could have a clear view of the podium.

Earthquake

One night about 8:00 PM I called Bill in San Pedro to talk about some of the things I was doing at work. While I was on the phone with Bill, an earthquake hit. We both hooted and remarked that it was our first earthquake. We were jostled around by the quake but didn't sustain any damage. At the time, Bill was living on a cliff overlooking Catalina; years later, he returned for a visit and found that a half-mile of the cliff had fallen into the Pacific, but his old apartment was still intact.

Mom's Visit

My mom flew out to visit me, and she stayed in our apartment with our pregnant Siamese cat. While she was there, the cat gave birth to six kittens. The poor cat didn't know what to do with these new things. They were all bunched up together, and their umbilical cards got tangled up and dried before I came home. I took them to the vet, but he couldn't save them. They all died. So much for fathering.

The Child is Father of The Man

These are some of the big questions the younger me would dream about, but had no one to ask for answers:

Can we know the universe, or are our sensors too limited?
The theory of the Big Bang is that one marble-sized object blew up and created all the stuff in the universe. Why doesn't it seem logical to me?
Did the initial Big Bang create living matter, or did living matter develop later out of basic chemicals?
Is the external world (the one we see) the real world?
What is consciousness?
Could there be multiple universes?

Even as I write these memoirs, I still haven't completely answered those questions. My brain needs parameters and context in order to reason and calculate. Am I part of the problem? Does my way of thinking imply that the external world is totally outside my power of investigation? I feel that I'm intellectually a prisoner within myself. But after years of thinking and reading and working with great thinkers at places like Caltech, I found that I knew more than when I began—at least I'd made a start. Here's what I've come up with over the years to answer the kid that, in many ways, I still am:

We may never "know" the universe, certainly not completely. But just within the past six hundred years, humankind has made progress. Or rather a handful of geniuses has learned things for the rest of us. A scientific genius of the highest order—a Newton or Einstein, for instance—is a pioneer with a mind almost beyond comprehension. The difference between one of those giants and, say, any U.S. President is like the difference between a man and a dog. When they have an idea and prove it, they can change the way the whole world thinks. When our species was still evolving 200,000 years ago, we didn't know anything. Then 10,000 years ago, when the first human history was being recorded, we still didn't know anything relating to the big questions. The sky was of great interest and being studied; patterns of celestial movement was being recorded by most cultures but we were only projecting our ignorance onto the night sky, seeing animals and gods in the stars. Later, 6,000 years ago, we still didn't know anything, but thanks to the ancient Greeks around 300 B.C., especially geniuses like Aristotle, we were looking with a more scientific eye. But we still hadn't made any real progress; the work of the Greeks was still being taught in school.

Then single-handedly in 1512, Copernicus figured out that the earth and planets move around the sun—a brilliant step that would send us leaping forward in the study of astronomy and cosmology. The big limitation in Europe at the time was that the Catholic Church with its Inquisition killed or tortured or imprisoned people who didn't believe that the earth was the center of the universe. (The church also still thought the earth was flat.) Galileo, armed with those first primitive telescopes called "spyglasses," improved on Copernicus, as did Kepler with his laws of planetary motion. Then Newton, along with G. W. von Leibniz, made the biggest leap yet by inventing calculus, which allowed people to understand the mechanics of how gravity controlled the laws of motion.

For the next two hundred years, scientists refined Newton's work to the point where it would enable us to go to the moon. Then came Einstein's general theory of relativity, which overshadowed Newton's view of the universe. We now understood a new physics that explained everything from atoms to stars. Because of Einstein, whose theory split the atom, the world was changed forever with the explosion of the atomic bomb.

Then also in the 1900s came quantum mechanics, developed by Bohr, Summerfield, and Heidelberg. Quantum mechanics has enabled us to discover a whole new subatomic world so contradictory to what had come before that almost no one comprehends it. It shows a multi-dimensional universe that doesn't behave the way we thought it did and isn't predictable. This was all accomplished by mathematics and can't be proven by the scientific method. But many physicists intuitively know it has merit because of the uncannily predictive nature of math.

Meanwhile, technology has developed instruments of all descriptions that can study the universe on many levels. We have reluctantly accepted the Big Bang—a theory that the universe began with the explosion of one tiny object that sent matter flying throughout empty space.

As I mentioned in a previous chapter, perhaps the most conclusive, and certainly among the most carefully examined, piece of evidence for the Big Bang is the existence of an isotropic radiation bath, known as the "cosmic microwave background" (CMB), that permeates the entirety of the universe. In 1964, two young astronomers, Arno Penzias and Robert Wilson, accidentally discovered the CMB using a well-calibrated horn antenna.

Their findings and other instruments tell us that we are not close to seeing the outer edge of that expanding matter because it continues to expand. Also we may never witness the Big Bang itself. Scientists now think that the universe is 15.6 +/ - 4.6 billion years old, and that we are currently able to see back to 14 billion years with the Hubble space telescope. However, this could be a trick of perspective like an Ames room (a distorted room used to create an optical illusion), and the true shape of the universe is not known. Some of the illusion comes from our math. Pi and prime numbers are a product of our decimal numbering system. Pi and prime numbers do not exist in the binary numbering system. This could cause our perspective to be wrong.

Maybe we need a revolution in math equal to the addition of the zero to the numbering system. The significance of the zero addition to the numbering system is that it has no value except as a place holder. If mathematicians can replace time with just a place holder, it would eliminate time but make the calculation more relative. Something like removing time from our math and replacing it with a type I duality function of bosonic or fermionic function, depending on the application. Instead of spacetime, replace with spacebosonic or spacefermionic, depending on whether it is about an infinite brane or a particle.

Reading about modern particle physics, quantum mechanics, and string theory in Gary Zukav's *The Dancing Wu Li Masters* and *Warped Passages* by Lisa Randall is like reading Louis Carroll's *Alice In Wonderland* and *Through the Looking Glass* when I was a child. Particle physics is very interesting and created by brilliant people, but is not connected to reality—just mathematical constructs not blessed by nature with existence.

The story *Alice In Wonderland* is universal. Its theme of relativity belongs to all time and all ages. Since it was first published, it has never gone out of fashion and never will as long as children love wonder-stories and grown-ups have young hearts. Those who read the book when it was first published saw a book different from the one a child of today reads. Not in text, that has not changed, but the outer circumstances have—the world has changed. At the time when the book was first published, there were certain poems that appeared in every reader and were read over and over again until they were unconsciously learnt by heart. All the poems in the story are parodies upon once familiar rhymes.

So back to the rest of my original youthful questions and some partial theories.

An eccentric White Knight shows Alice several worthless inventions.

I have a theory, whether you like it or not, that what we call the universe (i.e., all the bosons, fermions, photon, gluon, quark, hadrons, atoms, neutrinos, molecules, earth, planets, moons, suns, stars, galaxies, nebula, and clusters) is a solid—not a very dense one and with lots of imperfections (voids). Some of the space is filled with only one hydrogen atom in a liter or more of space (about 10^{-30} grams per cubic centimeters) of space. This solid is a klein bottle that does not have an inside, outside or edges. This space is also filled with gravitational waves or branes, electromagnetic like fields, i.e., lines of force, cosmic rays, x-rays, gamma rays and other stuff, bosonic and type I strings, that we know nothing about. Each atom is the resultant of the intersection of three or more of these lines of force out of phase. Like the leading edge of a square wave the shell has an infinite number of lines. Each line of force has an associated vector, which I will call "μ", "$đ$", "ε" and "υ," corresponding to what physicists call an electron, neutron, or proton with both an equal matching between bosons (particles that transmit forces) and fermions (particles that make up matter). Supersymmetry relates the particles that transmit forces to the particles that make up matter. The atomic weight of each of these vectors is the amplitude of the resultant vector (i.e., "particle?"), which has all the properties of strings. An electron weighs 1.66×10^{-28} g when at rest. A proton weighs 1,836 times as much as the electron. Valence bands are harmonics of the resultant vector. Changing from one valence band, first harmonic to a higher level, second harmonic, absorbs energy and conversely changing from a higher level valence band to a lower band emits or gives up some of its energy. Odd harmonics are unbalanced and can produce a covalence bond. Even harmonics are balanced and can not accept covalence bonding.

If the new harmonic is not balanced it will create a hole or positive charge enabling it to synchronize to other odd harmonics. If the highest harmonic is balanced it will be negative and repel or not be able to synchronize with other bands.

When a particle accelerator separates out an electron, proton, or neutron, it is selecting the dominant vector, which is to be used and amplified. The other vectors are reduced to a minimum. As the particle accelerates the amplitude of this single vector is increased. Increasing this one vector and reducing the others spreads the particle out like a balloon being flattened by air pressure. When it reaches the target and encounters other resultant vector's fields (atoms) the resultant collision and creation of new particles are just harmonics and new vectors created by the violent addition and subtractions of the fields. The leading edge of the spike creates a number of noise spikes as well as harmonics and new resultant vectors (particles) are like fractals. The universe solid is like a beaker of water or a crystal but with different properties with T-duality. We cannot see all of its properties, but we can measure some of them like the water and crystal.

Time

Time is like the white rabbit that is always late.

Scientists seem obsessed with time, like the white rabbit, saying, "I'm Late! I'm Late! Oh dear! I shall be late!" Then the Rabbit actually took a watch out of its waistcoat pocket, and looked at it. Scientists seem to have fallen down the rabbit hole into the topsy-turvy alternate world of Wonderland (particle physics, quantum mechanics, and string theory). Time is a human construct used in math to separate events. The division of time into past, present, and future is the product of our language and is a useful description in explaining many physical phenomena, and is now the accepted view of the physical world. Some languages do not have time.

In the universe it does not exist. Einstein once expressed this point in a letter to a friend regarding the subject. "To us who are committed physicist," he wrote, "the past, present and future are only illusions, however persistent." The reason for this is that, according to relativity theory, time does not happen bit by bit, or moment by moment: it is like space, in its entirety. Time is simply there.

To understand why this is so, one must first appreciate that your now and my now are not necessarily the same. This is because the simultaneity of two spatially separated events is entirely relative. What one observer regards as happening at the same moment but at another place, a second observer, located elsewhere, may regard as happening before, or after, that moment. On an astronomical scale, the effect is enormous. An event on a distant galaxy, which is judged to be simultaneous with noon today in a laboratory on earth can be shifted by centuries, from one's point of view. Like time, "Reality is merely an illusion, albeit a persistent one," says Einstein.

So Alice sat on, with closed eyes, and half believed herself in Wonderland, though she knew she had but to open them again, and all would change to dull reality in the after-time.

Black Holes and the Big Bang

Black holes are like the rabbit's hole.

Black holes are generally thought to be created by supernova. Such a violent event is sufficient to reduce the lines of force in the immediate vicinity to zero or maybe even break them. The resultant effect is a null or void in the solid. The pressures created by all the stuff in the solid tends to try and fill the void or repair the break. This is viewed as mass being sucked into the black hole by the enormous gravity created by a neutron star or singularity. Matter spirals around a black hole like water spirals when it goes down a drain thus creating spiral galaxies like the Milky Way.

"Curiouser and curiouser!" cried Alice. "Now I'm opening out like the largest telescope that ever was! Good-bye, feet!" (For when she looked down at her feet, they seemed to be almost out of sight, they were getting so far off). "Oh, my poor little feet."

Matter is pulled apart by the enormous gravity. But I think the pressure of the solid is pushing the matter into the void trying to fill the hole or repair the break.

Some scientists believe that when one gets beyond the event horizon of a black hole there will be a brilliant white light because of all the light that is unable to escape. This must be the bright light at the end of the near death experience (NDE) tunnel where all the souls supposedly go.

The radiation and other emissions emitted are created by the reduction of this matter to zero. The reason no light can escape is there is no light in there to escape. Given enough space, the hole or void will be filled, and the black hole will cease to exist. In the meantime, to answer the question of "where does the matter go?"—it goes to zero so it is not being spewed out in another alternate universe somewhere. A fish cannot imagine or create fire, man cannot image empty space, a null or a void, or create it—he can only change the state of matter; that is one reason why man created a heaven. The other reason is to explain away the fact that we are going to die. Scientists must put something in there even if it is just a ball of strings the size of a marble and give it super powers, super density and no Pauli repulsion. Dumb as fish.

The Big Bang theory isn't logical to me because, from a human point of view, there is never just one of anything. So maybe there are many marbles, many expanding universes, maybe connected like the soap bubbles that clean my laundry.

We still don't know whether the Big Bang expelled organic matter along with inorganic. If it did, that might make a reasonable case for some sort of higher or supreme being. Joan's influence aside, I've always been what I call a doubtful deist.

If some higher being is playing with marbles and creating universes, I sort of doubt that it is keeping an eye on whether life is evolving from his marbles. Also, if there is such a being, my thinking is that there are more of them, maybe playing marbles together. And maybe they are looking over their shoulder to try to see what higher being created them.

Speed of Light

A tsunami wave travels through sea water at between 600 and 700 miles per hour. Sound travels through air about 700 miles per hour at sea level. A shock wave travels through a crystal or a steel bar at 1,000 mile per hour. Light travels through the universe at 300,000 kilometers per second. The instantaneous transfer of energy through a crystal or a steel bar is the property of the lines of force but the shock wave will be much slower.

A photon (a particle of light) visits all the points along its path at the same moment: to a photon, it is no distance at all to cross the universe because the lines of force that create it are connected to all the points. Which brings us to the classic question: "Is it a particle or a wave?" It is both, by duality.

So T-duality obscures the difference between large and small distances. What looks like a very large distance to a momentum mode of a string seems to a winding mode of a string like a very small distance. This is counter to how physics has always worked since the days of Kepler and Newton. The photon cannot itself travel across the universe because, like the shock wave through the steel or the tsunami through seawater, it has its speed limit.

The white queen comes running wildly through the wood and talks to Alice, only to morph into a sheep before Alice's eyes. Sounds like superstring theories that are related by duality transformations known as T-duality and S-duality. These dualities link the quantities of large and small distance, and strong and weak coupling, limits that have always been identified as distinct limits of a physical system in both classical and quantum physics.

There are several ways theorists can build string theories. Starting with the elementary ingredient: a wiggling tiny string. Next decide: should it be an open string or a closed string? Then ask: will I settle for only bosons (particles that transmit forces) or will I ask for fermions (particles that make up matter), too? (Remember that in string theory, a particle is like a note played on the string.)

If the answer to the last question is "Bosons only, please!" then one gets bosonic string theory. If the answer is "No, I demand that matter exist!" then we wind up needing supersymmetry, which means an equal matching between bosons (particles that transmit forces) and fermions (particles that make up matter). A supersymmetric string theory is called a superstring theory. There are five kinds of superstring theories. The final question for making a string theory should be: can I do quantum mechanics sensibly? For bosonic strings, this question is only answered in the affirmative if the space-time dimensions number is twenty-six. For superstrings we can whittle it down to ten. How we get down to the four space-time dimensions we observe in our world is another story.

If we ask how to get from ten spacetime dimensions to four spacetime dimensions, then the number of string theories grows, because there are so many possible ways to make six dimensions much much smaller than the other four in string theory. This process of compactification of unwanted spacetime dimensions yields interesting physics on its own.

Types IIA and IIB, and heterotic SO(32) and $E_8 X E_8$ string, are so small the average size should be somewhere near the length scale of quantum gravity, called the Planck length, which is about 10^{-33} centimeters, or about a millionth of a billionth of a billionth of a billionth of a centimeter. They cannot be seen with any of our instruments. It is like taking half the distance to something. One can never get there because there will always be half the remaining distance.

The question as to what is consciousness now seems self-evident to me. I wonder why I ever questioned it. I am conscious because I am aware of myself and my environment. When I think and then act, I see cause and effect. I know that I exist because I am conscious of it; when I do not exist I will no longer be conscious.

I am constantly astonished by how much we've learned so fast, as least by our own standards. I think that the universe is full of life, and who knows how far along they are in their understanding? Humans are inferior to most other animals in virtually every way, except that our big brains can out think anything on earth. We now seem to be unstoppable in our quest for knowledge. What will we know in just another 600 years—or 6,000? A lot depends on whether we survive. If we don't, all our efforts will be no more than the single blink of a firefly in a dark, lonely night.

Bill gave me the book *Labyrinths* by Jorge Luis Borges, and I immediately recognized the story of *The Library of Babel* as Einstein's general theory of relativity. We had several great discussions about the other stories in the book.

In conclusion, this older me really can see the external universe, albeit indirectly. I now see the real world, as I call it, at least from a partly informed human perspective, through a glass darkly. There are infinite levels that we cannot see either directly or indirectly. But what we do see is not illusion. I now feel able to form some opinions that go way beyond the known universe, something most scientists are hesitant to do without more information. In the words of singer James Taylor, "I don't need no details, just give me a clue."

Giving Up My Dream

Joan telephoned. She said that Litton Industries at Fort Ord, one of the places where I'd submitted my resume, had called. They had a job opening and wanted to hire me. Fort Ord was only five miles north of Monterey, so it would be much closer to her. She said I should take the job. When I went home that weekend, she was still mad about going to Pasadena and me being in Carmel. It was decided, for the sake of my marriage, that I would call Litton on Monday and take the position.

Now, I have some idea of what King Edward VIII felt when he abdicated the throne of England to marry Wallis Simpson. King Edward VIII only had to give up the British Empire. I had to give up knowing the universe as well as I wanted to.

Joan didn't understand. She had no concept of the general significance of what was going on at Caltech or how important it was to me. She was a creationist and believed that questions of the universe should be left to God. She had a good religious intellectual vacuum (in other words, a closed mind). She was afraid that new knowledge might make her rethink her beliefs.

After I gave my notice, I loaded up my VW and drove down to San Pedro to see Bill and Leslie. I spent the night before driving back to Carmel. Bill had just returned from a Los Angeles Newspaper Guild banquet, where he won the Cal-State Journalist of the Year Award for his weekly column at Long Beach State. We talked and he gave me a poem he had written about nurturing children, which he thought was relevant to my situation.

Stillborn

Little Mozart!
I saw you
I saw you today
Do not deny it
There was something about you
(A glow?
A spark?)
that you were not able to conceal
It could have been no one else
Do not deny it
It wasn't until your screaming mother
snatched you from me
and threw you in the shack
out of harm's way
that the glow
the spark
grew dull and died
telling me you were gone

"Bill," I said, "do you remember a painting called *El Hombre*, by the Mexican artist Rufino Tamayo, that was in the Dallas Museum of Art near the Planetarium? It was of a man reaching for the stars while his dog was going for a bone."

I told him that *el perro's* bone now seemed like a more obtainable goal for me. Bill thought he heard me growl when I left.

The Graduate

The film *The Graduate* had just come out, and the song "Mrs. Robinson" by Simon and Garfunkle was very popular. I always sang it when I drove my VW from Pasadena to Carmel—along the road Dustin Hoffman took to stop the wedding in the movie. The scene of him by the two tunnels on Highway 101 still brings back many memories of my days in California.

Renting Our First House

We rented our first house, on the south side of Camino Del Monte, just off Highway 1 at the Serra Avenue exit. That's when we found out about the Universal Commercial Code. The Universal Commercial Code was a deal that moving companies and storage companies had worked out with the U.S. congress and state governments. The law ensured that they were to be paid in full before releasing any part of people's goods. So when we went to get our household goods out of storage, we found there was a lock on everything until we paid our bill. We slept on the floor for a couple of months in an empty house until we saved enough to pay our storage bill. Finally we were able to get our stuff out of storage and start living like decent folk again.

Choking

One Saturday Joan cooked a pork roast for dinner and put the leftovers in the fridge. Sunday night, after we came home from church, Joan stopped in the kitchen to grab a bite of the roast pork while I went into the bathroom. When I came out, I heard strange sounds coming from the kitchen. It was Joan.

She was literally choking to death.

When I reached her, she was trying to drink some water to wash down the meat lodged in her throat. But it was not working. She was ready to pass out.

I spun her around, locked my hands around her middle, and bent her over at the waist as I squeezed and tried to pick her up.

It worked. The stuck piece of pork roast spurted out on the floor, and Joan took a long, grateful breath.

Later I learned that what I'd instinctively done was called the Heimlich maneuver.

Apollo 8

On December 21, 1968, *Apollo 8* was launched. Three days later, after a sixty-six-hour flight from earth orbit to moon orbit, the three astronauts on board began circling the moon, each orbit taking two hours. Close-up pictures of the moon's surface were sent back to earth, as well as the view of planet earth as seen from the moon.

We spent Christmas of 1968 at Blaze and Gordon Morrow's place in Palo Alto and watched the crew of *Apollo 8*—Commander Frank Borman, Command Module Pilot Jim Lovell, and Lunar Module Pilot William Anders—circle the moon and do a live television broadcast from lunar orbit, in which they showed pictures of the earthrise. They ended the broadcast with the crew taking turns reading from the book of Genesis. After *Apollo 8* had completed ten orbits of the moon, it returned through space to the earth's atmosphere. It re-entered on

December 27 at 25,000 miles an hour, before decelerating and splashing down in the Pacific Ocean.

Soon after that, on March 3, 1969, the lunar module, the landing vehicle that would in due course have to land on the moon, was given its first test in space.

Litton Scientific Support Lab

Litton Scientific Support Lab (LSSL) at Fort Ord was a think tank designed to work with the Army's Combat Development Experiential Command. Military-scientific field experimentation was a relatively new technique for the enhancement of combat effectiveness in the 1960s. It permitted practical examination of military concepts well in advance of possible need. The general purpose of this joint military and scientific endeavor was to integrate general experimental scientific methodologies into simulated combat and related areas. Experimentation for the U.S. Army was conducted by the Combat Development Experimentation Command with special scientific support from LSSL. One study that I remember hearing about was called "suppression": the Army wanted to know how combat solders would react to enemy suppression fire.

Fort Hunter-Liggett

Most of the really hard experimentation was conducted ninety-three miles south at Fort Hunter-Liggett, in the Gabilan Valley near Jolon, California. It was a huge valley where the Army could play war games.

Getting there and back was the long part. I'd leave Carmel at 6:00 AM, take Highway 1 to Monterey, and Highway 17 to Salinas. Just before I entered Salinas I'd turn onto El Blanco Road, pass the Spreckles Sugar Company factory to Sanborn Road, and hit Highway 101 south of Salinas. Then I'd head past Gonzales, Soledad, and Greenfield to King City. Just before King City I'd turn off on Highway 14 to Jolon, go over the inland mountain range to Sulphur Springs Road, and turn right into Fort Hunter-Liggett. Then past Mission San Antonio de Padua, the third mission founded by Father Junipero Serra on July 14, 1771.

After I finished work at 4:30 PM, I'd take the same route home. By this time in the afternoon, the Salinas Valley had heated up to over 100 degrees, creating a low-pressure area. The cold wind from Monterey would come in at about forty miles per hour. My little Volkswagen could barely make fifty miles per hour against that head wind, so it was slow going. It took an extra twenty to thirty minutes. Instead of an hour and a half, which was what it took driving down, it would be a two-hour drive back up.

I lived the story of Steinbeck's *East of Eden*. I had been to all of the places talked about in the novel: Salinas, Gonzales, Soledad, Greenfield, King City, Paso Robles, Santa Cruz, Watsonville, Castroville, Moss Landing, San Jose, Stanford, and San Francisco. I could see the mountains on both sides of the Salinas valley as I drove Highway 101 almost every day for three years. I could see and smell the eucalyptus trees planted in long rows across the valley for windbreaks. I worked and climbed over the poor rock-covered rounded foothills west of King City where Samuel Hamilton's farm was. I could see and smell the wild flowers and grasses that grew on them. I followed the Salinas River and could feel the heat in the summer and the cold in the winter. I could feel the hot wind that blew south down the Salinas valley in the afternoon. I went by the Spreckles Sugar Factory, and I rode the train from Paso Robles to King City to Salinas and got off of the train at the Salinas station. I knew the streets on both sides of the tracks. I knew people like the ones in the story.

The Army had five CDC-6600 mainframe computers housed in portable buildings atop the mountains. There was a direct range measuring system (DRMS) at each mainframe to measure the time to each player in the valley and what he indicated he was doing. Through triangulation from the five DRMS sites, the position of each player could be determined. Each had a DRMS back-pack that was connected to a round counter on his weapon, and a wrist band that had a keypad to give information to the computer as to what he was doing. If he was just advancing up a road he would press "A." If he had spotted the target he would press "B." When he attacked the target he would press "C." And so on.

Each scenario had its own list of events. Each of the targets was equipped with a gas discharge gun simulator and a pop-up cardboard target, each was controlled by the computer program. The cardboard target had aluminum foil on both sides. When a round went through the target it would short out front to back and indicate a hit. There were microphones mounted next to the target that would record the shockwaves of nearby rounds. The computer then knew how many rounds had been fired, how many had hit the target, and how many near misses there were, and it could calculate the effective fire power of each type of weapon and/or tactic used.

DRMS Test Set

One day the Army complained they were losing data from troops in the field. They wanted LSSL to look into it. Litton proposed a DRMS test fixture to check out the system. The Army agreed and gave LSSL a contract to develop it.

Another engineer and I were assigned to build it. The other engineer had been working with the DRMS system for a couple of years and was designing a test set based on the design of the system itself. When I saw his plans I told my manager, Roger Tardiff, that his plan was all wrong for a test set. It needed to be very stable, at least ten times more accurate than what it was testing, and not built just like the system it was testing. Roger wanted to know if I could design and build it myself.

"Sure. Piece of cake."

"Then do it!"

He put the other engineer back in the field and let me design and build it alone. He'd already bought a nineteen-inch rack and a digital chassis, so all I needed to do was define the cards to be put in the chassis. I made a preliminary design based on a high-speed-clock (ten times faster than the DRMS clock frequency) system and a lot of digital gates. I explained my design to management, and they approved the purchase of the clock and the digital circuits boards.

I ordered these and went to work wire-wrapping the chassis. It took three months to wire up the chassis. I did it in stages. First, I hooked up the power supply, clock, and count-down circuit. When I finished with the very stable clock, I started coding the pulse identification timing for all of the master transmitters. After that was complete and working, I did the pulse identification and timing for all ninety-nine remote units. I was surprised how good it really was. Then I needed to get the RF signal from the radios into the digital system. I ordered a pulse detector and some attenuators from a local electronics supply house in San Jose and put them all together. It really came out nicely and worked better than I had imagined.

Then came the critical test.

I hooked up one of the remote units and there it was, a perfect match. The coding and timing were right on; they locked onto the remote DRMS signal the first time. I checked out all remotes, and they all responded perfectly. I located one of the masters, and it tested out exactly as the remotes.

So what was causing the problem?

I scratched my head a little, then realized the remote systems were all being aligned in the air-conditioned maintenance shop, then taken out and operated in

Hunter-Liggett's 100-degree-plus temperatures. I took my heat gun and started to heat the remote DRMS system.

Bingo!

Sure enough, the pulse timing began to slide out of alignment, and the test set began to lose data from the remote DRMS units. Eureka! Here was the problem. I called Roger in and showed it to him. Boy, was he surprised—and elated—that it worked so well and demonstrated the problem so accurately.

Now, what to do about the problem? The solution was simple: the DRMS remote systems needed to be built like my test set. But we knew the Army would never go for that. The only useful suggestion was to align the remote units in the temperature they were to operate in. That was a reasonable conclusion. I wrote up the analysis and conclusion and presented it to management and the Army. Everyone agreed it was an excellent job. The next step was to deliver it to Hunter-Liggett, but first there had to be an operations and maintenance manual for it. I started writing the operating instructions and gathering my design notes and timing diagrams for the maintenance manual. It took me several months to describe everything it did and how to use it.

Finally, it was all done.

I'd always known that all that childhood tinkering in my ham radio shack would pay off. Now I'd completed my first totally independent engineering project. I was truly a hardware design engineer!

M16

The Army was having a problem with a its M16 rifles. Calcium deposits were building up in the gas tube and in the bolt carrier. The small gas rings on the bolt would not allow the bolt to turn, causing the rifle to seize. During light firing with limited automatic use, such as for target practice, and with frequent cleaning, the weapon functioned normally. But without frequent cleaning, and with excessive automatic firing, such as in a firefight, the M16 could jam.

In Viet Nam, the army and marines were having real problems with the weapon. When a firebase was overrun and then retaken, they'd find GIs dead in their foxholes or bunkers with their M16s disassembled. Investigation showed that the M16s were jamming. This meant death to a combat solder in a firefight. The Army contracted LSSL to try and figure what exactly was going on, and why.

I was assigned the job of analyzing the problem. First I had to instrument the M16 to measure the timing of the trigger, the hammer, and the bolt. This was not easy. The M16 was anodized, so it would not conduct an electrical signal. I had to glue strips of aluminum foil on the trigger, the sear, the hammer, and the bolt. Then I had to mount the M16 in a vise so I could accurately aim it at the bullet catcher. After that, I had to set up an oscilloscope and camera so I could catch the timing from the moment the trigger was pulled until the sear released the hammer, and know the amount of time it took for the hammer to hit the bolt.

Once I had the single-shot timing, I needed to get the same timing for automatic firing. I had to move my operations out to a firing range at Fort Ord for the automatic testing. Finally, I had all the timing information and presented my findings to management and the Army. It appeared that the weapon was cycling at a rate faster than that at which it was designed to operate. We requested that some special ammunition be made up with different amounts of powder loads. I tested several, and a slightly lower powder load seemed to work a lot better. The Army immediately ordered that all M16 ammunition would have less powder, and the problem lessened.

The conclusion is that basically the ammo didn't match the weapon. Also the humidity in Viet Nam caused the barrel/chamber and bolt to corrode very quickly if not taken care of constantly. The jamming wound down towards the end of the war because of this discovery, but there was still hated of the M16 by the troops.

Apollo 11 Lunar Landing

In May 1969 a Soviet satellite landed on the surface of Venus.

On May 18, 1969, three astronauts were sent into space to test the docking procedure between the service module and the lunar module. On May 22, 1969, these three astronauts and their craft approached to within eight miles of the lunar surface in the lunar module before returning to the service module, and then back to earth.

On July 16, 1969, *Apollo 11* was launched from Cape Kennedy for the actual moon landing. Five days after blast-off, with the spacecraft in orbit around the moon, the lunar module, codenamed *Eagle*, was detached from the command module and descended to the surface of the moon.

On July 20, 1969, I watched the *Apollo 11* lunar landing at the home of Dick Reeves. He was the manager of advanced planning and analysis at LSSL and lived on the next street over from us in Carmel. *Apollo 11* Commander Neil A. Armstrong and Lunar Module Pilot Edwin E. "Buzz" Aldrin Jr. landed *Eagle* in the Sea of Tranquility on the moon, while Command and Service Module Columbia

Pilot Michael Collins continued in lunar orbit. After the landing and the lander were secured, a ladder was slowly lowered from the lunar module until it nearly touched the surface of the moon. An estimated 600 million television viewers, as many as 80 million of them in the U.S., were watching throughout the world. "I am going to step off the LM now," said Neil Armstrong, and as he stepped on to the lunar surface, he said, "That's one small step for man, one giant leap for mankind."

During their stay on the moon, the astronauts set up scientific experiments, one of which involved a laser rangefinder reflector. They also took photographs and collected lunar samples. The module took off from the moon on July 21, and the astronauts returned to earth on July 24.

McDonald Observatory in Texas made the first laser range measurements of the distance to the moon with an accuracy of 1 centimeter. It was later determined that the moon's orbit was increasing away from earth by 3.8 centimeters per year.

Litton in Sunnyvale

I was sent to the Litton Sunnyvale office to help out on a tape-to-tape converter project. Bob Heart was the project manager. IBM had sold thousands of typewriters with magnetic tape recorders in them to store the documents that secretaries typed. The problem was that the format of the typewriter tapes did not match any standard computer tape protocol. Litton Sunnyvale was developing a system that would read the typewriter tapes and convert the data to standard nineteen-inch high-speed computer tapes, so the data could be read into a computer.

It was a neat project, but the best part was listening to Bob Heart tell stories.

One of his stories was about what happened when he worked on a project that had this room full of equipment that took computer data and wrote it on a magnetic drum. He said that if you ran a sheet of paper over the drum it would put a static charge on the paper where the data was imprinted. If the paper was subsequently passed over a vat of powdered ink sensitive to the magnetic charge, the ink would stick to the paper, creating the same image that was on the magnetic drum. Then, if you passed the paper over a heating element, the image would become permanent, like a typewritten page. One day management brought in a group of potential investors and demonstrated how the process worked. After the demonstration, management decided to form a separate company to manufacture and produce a smaller version of the process. They asked Bob to be part of this new start-up company. "No," he said, "I'm very happy with the job I have now."

A few years later, he learned the name of that start-up company.

It was Xerox.

He could have retired a very young millionaire.

Another project he worked on at IBM was the drum random access memory system. This used strips of magnetic tape about a yard long, which hung in a drum configuration. When the computer requested data from one of the tapes, the drum would rotate and drop the selected tape in a chute at the bottom for it to be drawn across a set of magnetic read- or write-heads. Remember this was very advanced from the IBM punch card, before the days of the computer hard disk.

Bob's favorite word was "smegma." Whenever he had a problem he'd say, "Smegma." He said it so often that most people he worked with also used the term, even though they didn't know what it meant. One day they were making a presentation to a group of doctors as potential investors. Bob was manning the slide projectors, while the salesman up front explained how the system worked. Someone asked the salesman what the tapes were made of. He didn't know exactly, so shrugged and said, "Oh, smegma and iron filings."

The doctors looked at each other and laughed.

Bob said he nearly fell off his chair.

I watched the *Apollo 12* launch on November 14, 1969, and the landing on the moon on November 19 at the Sea of Storms. The Command Module was named *Yankee Clipper* and the Lunar Module was named *Intrepid.* The crew included Charles Conrad, Jr. as the commander and Richard F. Gordon, Jr. as command module pilot—both of whom I had met at the American Astronomical Society meeting in Dallas—and Alan L. Bean as lunar module pilot. They returned to earth on November 24, 1969.

PG&E

Each month with our Pacific Gas and Electric (PG&E) bill, we received a little newsletter. I saved the following from a newsletter received in 1969:

The average age of the world's great civilizations has been 200 years. These nations progressed through this sequence:

From Bondage to Spiritual Faith

From Spiritual Faith to Great Courage
From Great Courage to Liberty
From Liberty to Abundance
From Abundance to Selfishness
From Selfishness to Complacency
From Complacency to Apathy
From Apathy to Dependence
From Dependence back again to Bondage
In seven years the U.S. will be two hundred years old!

ATS

Ralph Neal, the Litton director, yearned to become a vice president of Litton, and Lou Erdle, the maintenance supervisor, wanted to be director. The two decided that the way to get promoted was to bring in more revenue. So they brought Richard King and I back to the Fort Ord office to write proposals in hopes of gaining more contracts. We wrote ten proposals in a year but got no new contracts. One day, as we were watching a group of GIs paint the barracks across the street, I said to Richard, "It sure would be nice to point to that at the end of the day and say, 'See what I've done.'" I pointed to the painted barracks across the street. Richard agreed. We'd been working our tails off writing proposals and had nothing to show for it.

Then one day the word came in from Litton's Mellonics Division in Sunnyvale that we'd won the Frankford Arsenal Automatic Target System (ATS) contract. There was great joy that night, but the next day, panic set in. Now we had to build it!

We couldn't use Fort Ord property due to a perceived conflict of interest. So they sent us to Litton's Defense Sciences Laboratories (DSL) on Abrego Street. We went to work on the electronics. We ordered a PDP-8/i mini computer from Digital Equipment Corporation (DEC) to do our calculations and hired a programmer to program it. It had an ASR-35 teletype input-output device with a paper tape reader and punch. It came with a three-pass assembler on punched paper tape.

When Richard needed to assemble his program, it was an all-day affair. First he'd read in the paper tape assembler, which would take about twenty minutes of chug-chug-chugging per pass of the assembler program. Next he'd read in his program, which he'd punched out onto paper tape the previous day; this took another twenty minutes. Then the assembler would punch out a pseudo-code that only it understood onto paper tape—consuming another twenty minutes.

Now the programmer would re-input the pseudo-code, and the assembler would grind on that for another thirty minutes. Richard drank a lot of coffee watching the thing chug along.

Neither Richard nor I had any experience with optics, so we hired Dr. Gene Crittenden, chairman of the physics department at the Naval Postgraduate School, to develop the optics. The system was to score ammunition rounds from 5.56 millimeters up to 20 millimeters. It had to score these rounds at automatic-weapon speeds of up to 6,000-rounds-per-minute to test the mini-guns.

The current method at Frankford Arsenal was used in the underground ranges. They had a roll of three-feet-wide brown wrapping paper in the floor at the target area. The paper was stretched from the ground to a take-up reel in the top of the range. There was a 2,000-pound block of concrete and steel on the firing line where the weapon barrels were mounted in vises, to prevent them from moving. The barrels were classified as new, medium, and old, depending on the number of rounds of ammunition that had been fired through them. The Army quality inspector would randomly select one hundred rounds from a lot of one million. Frankford Arsenal's job was to accept or reject the million-round lot, based on this random selection. Frankford Arsenal would take a new barrel and shoot ten rounds at the paper. Then they'd roll the paper onto the take-up reel, fire another ten rounds, roll the paper, fire a final ten rounds, and roll the paper. They'd change to a medium barrel and repeat the thirty rounds, then go through the same process with an old barrel. That would leave an extra ten rounds out of the hundred, in case of misfires or other problems.

Next the roll of paper was taken into the office and each set of ten rounds was cut into a separate sheet. The analyst would sit down with a ruler and a Friden or Madas mechanical calculator. He would measure the distance between the rounds; get the maximum horizontal and vertical spread; calculate the mean horizontal and vertical distance; calculate the standard deviation, mean radius, and center of impact; and come up with a what was called a "figure of merit." It would take the better part of a week to make the decision.

This was the process they wanted to automate.

We proposed putting a light beam across and up with photo detectors spaced one-tenth of an inch apart. When the round went through the light beam, it would blank out the detector and we'd get an X and a Y position. Also, we spaced the X and Y light planes a foot apart. We rigged it so that when the round broke the first plane, it started a counter, and when it broke the second plane, it stopped the counter. The computer could calculate the velocity of the round at the target.

The target area was one meter square. To protect the target from stray rounds, we ordered a bullet-protector made of hardened steel tank armor.

Naval Postgraduate School Test

Dr. Crittenden wanted us to come over and watch a test he'd set up in the basement of the Navy Postgraduate School. The test used a long, plastic magnifying glass, like the ones for reading fine print in books. It was half round and about two inches long to cover two one-inch photocell circuits. Behind this was a small light bulb, and the glass would collimate the light into a flat beam. On the other side was a photoelectric cell with sensors spaced a tenth of an inch apart, and a transistor amplifier. Dr. Crittenden intended to fire a 30.06 hunting rifle into a steel bullet-catcher, and capture the photo-detector's response on an oscilloscope.

We were ready to go, with our ear protectors on, when he fired the first round. All hell broke lose!

Fire alarms went off. The building was evacuated.

Dr. Crittenden had forgotten to notify anyone about the test. A couple of years earlier, the Navy Postgraduate School had hired a contractor to install new green chalkboards to replace the old blackboards. The contractor used a nail gun to attach the new boards to the cinder-block walls. Unfortunately, the first nail he fired went through three walls and three classrooms before it stopped. The classrooms were full of students and instructors at the time, and now, years later, they were still gun-shy.

After everyone settled down, we went back to work. The tests went great. The photo-detectors responded very well to the high-speed projectile. We went to work to get the boards manufactured and the system assembled, and the programmer toiled away programming the PDP-8/I.

The Territorial Imperative, Dom DeLuise, Arnold Schwarzenegger, and Don Knotts

I was reading *The Territorial Imperative: A Personal Inquiry into the Animal Origins of Property and Nations* by Robert Ardrey and *On Aggression* by Dr. Konrad Lorenz. The authors described how lions, who had rather poor eyesight, could detect the slightest movement and knock prey down with one swat of their big paws, while primates, on the other hand, had excellent eyesight for detection of colors and shapes, because their food supply consisted of colorful berries and uniquely shaped fruit.

I filed that useless tidbit of information somewhere in the back in my brain. The following Friday I went to the local grocery store to cash my paycheck. While I was waiting in line, in front of me was a lady with a stroller and a half-asleep kid less than two years old. As she approached the cashier, the kid jumped up, suddenly wide-eyed, and stretched out his arms to the gumball machine by the cashier's window. I thought, "Aha! There's that principle at work!" The kid wasn't old enough to know the sweet taste of the colorful gumballs. I'm sure his parents had never given him any, because of the danger of choking. But he'd obviously been born with the knowledge that colorful round things were good to eat!

I read Dr. William H. Sheldon's theory of somatypes in *The Varieties of Human Physique* and *The Varieties of Human Temperament*. He proposed three basic body types and associated them with personality characteristics. The first, the endomorph, is characterized by a preponderance of body fat, typically has a soft-skinned pear-shaped body, with wider hips than their shoulders. This type gains weight easily and tends to be curvier than the other body types. An example would be Dom DeLuise. The second type, the mesomorph, has wide shoulders and narrow hips. This type is marked by well-developed muscles, packs on muscle easily, can lose weight easily, and is rarely either fat or skinny. An example would be Arnold Schwarzenegger. Finally, the ectomorph, generally small-framed and thin, is distinguished by a lack of excess fat and muscle. This type has narrow shoulders and hips, and has trouble gaining weight and building muscle. An example would be Don Knotts.

Apollo 13, Apollo 14, *and* Apollo 15

I watched the *Apollo 13* launch on April 11, 1970. On their way to the moon, a malfunction forced cancellation of lunar landing. The command module was *Odyssey* with John L. Swigert, Jr. as command module pilot, and James A. Lovell as the mission commander. The lunar module was named *Aquarius* with Fred W. Haise, Jr. as lunar module pilot. The lunar module *Aquarius* was used as a lifeboat, and it saved the crew from the lack of oxygen. I and several million others watched as they returned to earth April 17, 1970. It was the world's greatest rescue mission.

I later watched *Apollo 14* when it was launched on January 31, 1971, and the landing on the moon on February 5, 1971. The mission commander was Alan B.

Shepard, Jr., and the command module pilot was Stuart A. Roosa commanding *Kitty Hawk*. Edgar D. Mitchell was the lunar module pilot, and the module was named *Antares*. They returned to earth on February 9, 1971.

I also watched the *Apollo 15* launch on July 26, 1971, and the landing on the moon on July 30. David R. Scott was the mission commander and Alfred M. Worden was the command module pilot. The command module was named *Endeavor*. The lunar module was named *Falcon* and was piloted by James B. Irwin. They returned to earth on August 7, 1971.

General Freeman and the Anti-War Protestors

At Litton's Defense Sciences Laboratories on Abrego Street, we all worked in cubicles. The cube next to me was occupied by General Freeman, a retired four-star Army general. Litton had hired him after he retired for his insight and Army expertise. He'd show up at work around 10:00 AM and call the secretary to his cube to take dictation. I'd hear him dictating, saying he'd had breakfast with General So-and-So or General Such-and-Such at Fort Ord that morning, and that they'd discussed several upcoming projects for which the Army would like some help. The secretary would type up the document and send it to Litton's main office. Then General Freeman would get on the phone and call Caracas, Venezuela, and ask to speak with El Presidente. Eavesdropping from next door, I would hear him on the phone, talking to El Presidente as if they were old friends. This would go on all day. He knew people all over the world!

On those rare days when he didn't have a lunch appointment with some general or have to telephone some foreign big-shot, he'd lean over the cube wall and say, "Hey, Charlie, you want to go to lunch?" I went to lunch with him a lot that year.

When Litton won the Singapore defense contract, they called General Freeman to ask if he knew anyone in the navy who'd been a port commander. They needed that expertise. Sure enough, he immediately called this admiral friend of his who'd been in charge of Pearl Harbor a few years back.

When we were ready to do our preliminary live-fire test on the target, we went to General Freeman. "Can you get us a couple of automatic weapons and some ammunition and a firing range, so we can test our target?" we asked him.

"Sure," he said. "No problem! Just come with me out to Fort Ord."

I went with him to Fort Ord, and boy, talk about an impressive experience! He had a plate with four stars on it attached to his front bumper. As we drove through the front gate at Fort Ord the MPs snapped to attention as if it were the

Commander-in-Chief himself. Everywhere we went, people jumped through hoops for him. He had four stars on his briefcase, too, so that got quite a lot of attention. We went to the armory, and he requisitioned an M16, an M1 carbine, and a thousand rounds of ammunition without question. We put them in the trunk of his car and drove off.

On the way back to Abrego Street, we noticed several California highway patrol cars were stopped on the road. A highway patrol officer was directing traffic, as were a number of Monterey police officers. They were there because of an anti-war demonstration against "military-industrial complex" companies working with the military in the area. Fortunately the officers didn't stop us and search the car—although I'm sure General Freeman would have gotten us off the hook.

Back at the Abrego Street office, we secured the weapons and ammunition in a steel locker. By now the demonstrators were gathering outside our front door to protest Litton's involvement with Fort Ord. As a joke, I said to my manager, "You want me to get rid of the protesters? I've got two automatic weapons and a thousand rounds of ammo in that locker."

He freaked out. "Don't *ever* say that again. Don't even think it! We could all be canned for such thoughts."

He decided it would be best if we took the rest of the day off. He closed the office, and we all slipped out the back door and went home.

The test at the Fort Ord range went perfectly. We finished up the system, shipped it to Philadelphia, and I went with General Freeman back to Fort Ord to return the weapons.

"What weapons?" the authorities at Fort Ord asked us. "We're not missing any weapons."

When we turned them in, the Army guys looked up the serial numbers and said that according to their records, the weapons we'd just returned had been sent to Germany. I could've kept the two automatic weapons and no one would've been the wiser!

Final Test and Acceptance

Our target was installed in an underground three-hundred-yard range. There was a small problem getting the bullet protector down there because it was so heavy, but we managed to assemble it and get it working. We made a few test firings to check the alignment. It was amazing. Most of the guys working there were master gunsmiths. In fact, they made all their own barrels, and had some of the best machining tools I've ever seen. They were also master marksmen. If we asked them to put a round at a special location to test a photocell, any one of them could do it within a tenth of an inch at three hundred yards. They loaded their own ammunition, balancing the rounds and measuring the exact number of grains of powder.

Finally we were ready for acceptance testing. Richard King had been called back to Monterey and then sent to Singapore to work on the new contract, so it was now mostly my project. We started the test, and as fast as they would fire the weapon, the computer would print out the X and Y locations and velocities. At the end of ten rounds, it would print out the standard deviation, mean radius, and figure of merit. Everyone was completely amazed. What had previously taken a week of calculations could now be done before they could get the barrel unclamped from its mount. And that was only the small stuff.

Now came the big twenty-millimeter test. The guy mounted the first barrel, but as he was getting ready to shoot, I stopped him and said I thought it looked a little low. But he was skeptical—he was in charge and had been doing this for years. He yanked the lanyard and *boom*. Then *bang*! The round hit the bullet protector about two inches off the ground. We sure were glad it was there. The round was just a lead projectile, and it shattered into a thousand pieces when it hit the armor plating and went under the target. Fortunately, it wasn't an armor-piercing round, like the ones I'd seen in another range where they were testing a new high-velocity, spent uranium thirty-millimeter gun. I later read a *Newsweek* article that said that in the Seven-Day War between Israel and Egypt in which this gun was used by the Israelis, the rounds went through the Egyptian tanks like they were butter.

Flechette

One day, as I was running down the range to get the paper roll, I stepped on a flechette fin, which is a steel missile or dart. It stuck in my right shoe heel and the point of the flechette stuck in my left ankle. It was a rusty one, too. I pulled it out, put it in my pocket, went back to what I was doing, and finished the day. Then I went to the dispensary to get a tetanus shot.

The nurse, a sweet, young Florence Nightingale-type, asked what happened. I showed her the rusty flechette.

"What does that do?" she asked.

I said, "It kills and maims. When it hits you it tumbles and tears up your insides."

She cringed. "My goodness, that's a terrible thing!"

I said, "This is Frankford Arsenal! What do you think they do here? They make things that kill and maim."

She was still cringing as she gave me a tetanus shot, cleaned the wound, and applied a bandage.

Teaching Certificate

While Richard King and I were designing the ATS, Richard had a job at Monterey Peninsula College teaching a course called "Introduction to Digital Logic." When he was sent to Singapore for the new contract, the college asked him to recommend someone else to finish the semester. He recommended me. I finished the semester. However, the college needed me to have a teaching certificate, so I signed up for a teaching certification class that summer. It was a very informative course. We learned how to create a lesson plan, do research, and present to a class.

There's more to teaching than meets the eye. A lot of research goes into creating a lesson plan. Then, after the lesson plan is created, presenting the information is not an easy task either. Trying to write on the chalkboard and talk to the class at the same time is an art. Write, point, turn, and talk.

After several weeks of learning how to teach, we were told to create a lesson and deliver a typical class. This teaching certificate class was for all teachers in the college, so there were all kinds of professionals getting their certificates, such as artists, businessmen, and teachers. The two I remember were the artist and the businessman. The businessman was starting a scented soap company. The scented soap guy was one of the first to teach his class. He was great because he was a salesman, so his presentation was smooth and informative.

Then I gave my presentation on the computer numbering system. I thought the topic was pretty straightforward, but I was unable to get the ideas across to some of the artists.

Next the artist gave her presentation, and I couldn't understand the concepts of hue, value, and other stuff she talked about.

Fortunately I received my teaching certificate for adult education in the junior college school system of California.

Dr. Crittenden and the Limitations of a Ph.D.

One day at lunch I asked Dr. Crittenden about his doctorate in physics.

He said he'd received it in the late thirties. After he graduated, he taught physics at a small college in Oklahoma. In the early forties, he heard that a friend, Dr. Lawrence, was working on a very important project in California. He wrote to Dr. Lawrence to see if he could work with him. Dr. Lawrence said, "Come on out."

It turned out Dr. Lawrence was participating in the Manhattan Project, extracting enriched uranium for the first atomic bomb. Dr. Crittenden worked with Dr. Lawrence for several years before coming to the Naval Postgraduate School.

His secretary worked for my wife in Mary Kay Cosmetics. One day I asked her how much she thought Dr. Crittenden made. She said about $30,000 per year. I said to myself, "He has a Ph.D. and that's all he earns? My wife makes over $40,000 a year and drives a new Cadillac."

Here was a well-educated Ph.D. with thirty years' experience, and he was making less than my own wife!

That's when I realized education was not the road to riches. I began planning to start my own business.

Conflict of Interest

I talked to the other engineers I worked with and learned that several had similar ideas about starting a business. We began getting together, meeting when we could, to put together a business plan and investigate our options.

A few guys wanted to bid on a contract Litton was bidding on. But I said no way—we needed to find opportunities elsewhere. After a couple more meetings, they were still talking about bidding on the same contract. It was easy money, after all, and it was what they all knew best. They'd been doing similar work at Litton for years.

I pulled out of the group and looked for other things to do. I began investigating burglar alarm systems that would alert the police in the event of a break-in. But I abandoned that idea because I figured the police would stop responding after a few false alarms.

Meanwhile, the other guys at Litton formed a company and submitted a bid in competition with Litton. Two weeks later, they were all fired. I'd seen that one coming. Litton had eyes and ears all over the place, and when they got wind of what was going on, they acted promptly to eliminate any conflict of interest.

Lou Erdle, Fort Irwin, and the Evaluation of the Sheraton Tank

The engineering manager was named Lou Erdle, and he had his own Cessna 182 airplane. When he needed to go to Hunter Liggett, he'd just fly down and back. One fall, he flew to Colorado to go deer hunting, and while making an approach to a small airport on top of a mountain, his landing gear caught a power line and flipped the plane over on its top. Lou was killed.

Meanwhile, one day at Litton we received orders to go down to Fort Irwin in the Mojave Desert. We'd endured the scorching heat at Hunter Liggett, and we were really in for it now. We packed up a couple of trucks with our instrumentation and took off, heading down to Bakersfield, over to Barksdale, and out into the Mojave Desert near Death Valley. There we made a left turn in the middle of nowhere. After fifteen miles, we passed the Goldstone Deep Space Satellite Tracking Station. The government situates such things away from civilization, because civilization produces noise that interferes with the weak signals coming from satellites million of miles away in space.

We were already in the boonies, but now we went another ten miles until we saw a sign saying, "Welcome to Fort Irwin." It was all you could see for miles in any direction. We kept going and finally came to some buildings. It was the Army's tank testing range. We were given a tour and shown several Army tanks. The new Sheraton was the one of greatest interest—it was the latest and most modern. It had been proposed to shoot normal projectiles and launch wire-guided missiles from the same barrel. This had never been done, and some were saying it couldn't be done. But while the debate was in progress someone ordered 1,000 to be built, and now they were coming off the assembly line.

We were asked to evaluate whether it would work. We had to have a reference, so we instrumented one of their regular tanks and a couple of Sheratons. Farther out in the desert, next to some mountains, the Army had laid down railroad tracks and set up a remote-controlled train to pull targets for the tanks to shoot at. They also had some of those new thirty-millimeter spent uranium rounds; the tank commanders were ordered to never raise their gun barrels over 15 degrees above the horizontal for fear the rounds would go over the mountains and all the way into Barksdale.

We spent a month out there in the hellish heat evaluating those tanks. It was amazing how they could be going thirty miles an hour and hit another target two miles away going thirty miles an hour in a different direction. Boy, were these guys good!

LHA Threat Evaluation

When we returned from that project, I was sent back to Los Angeles to work in Litton's Woodland Hills office on the landing helicopter assault (LHA) threat evaluation project. Litton had won a contract to build the navy's new LHA ship, the *Saipan*, LHA-2. The military was evaluating defensive electronic systems proposed by General Dynamics and Raytheon to defend the ship against enemy (Russian) aircraft and missile attacks. What they needed was someone with top-secret clearance to program Russian equipment signatures into the computer. I was the only one with top-secret clearance. Everyone else only had a "secret" clearance designation. Top-secret background investigations were too expensive for commercial companies unless absolutely necessary. But mine just involved getting an update.

I was given the top-secret list of all the known aircraft and missile signatures compiled by the CIA and other intelligence organizations. I was locked in a windowless room to do my programming. I worked on the coding for nearly a week. I was almost finished—I lacked only about twenty more threats to complete the list—but they needed to get the programming sheets over to keypunch immediately so they could punch the IBM cards for a 4:00 PM computer run. It was complicated: they had to take all the magnetic material out of the computer room and go over everything with a demagnetizer before and after the top-secret computer run. I figured I could program the last twenty threats while the keypunch operators punched the first twenty-five pages of my programming sheets. I called the security courier to come over and pack up the information and transport it over to the computer building. He put everything into a secure courier pouch, and we went to the computer building. We couldn't ride in the same vehicle, and the top-secret pouch had to go by itself.

I went over in a company car. When I got there, the courier was going into the computer building and checking in. I tried to follow him but was stopped. They wouldn't let me in because I wasn't cleared for that particular building. So much for the last twenty threats! They ran the program, and I was told I could go home.

Albert Ketchum

Joan's father was Albert Ketchum. He liked nice cars, and when he came to Carmel to visit us, he owned a Studebaker Avante. Back in the 1930s when her parents had first moved to California, he'd had a Lincoln Zephyr with air horns that could be heard ten miles away. But Grandma Ketchum didn't like California, so she baked pies and sold them house-to-house to save up money to buy a forty-acre spread in Illinois.

Buying a House

Speaking of buying property, we'd been trying to buy a house in Carmel for some time. It was a highly competitive seller's market. Each Sunday morning I'd get up early, get the *Monterey Peninsula Herald,* and see what houses had come on the market. One Sunday I was awake at 2:00 AM, so I went and picked up the first paper off the press. We found a couple of places for sale in Carmel, but by the time we went to look at them they were already sold. Talk about a hot real estate market!

So we went to Carmel Realty and asked Betty Setchel what we had to do to find a house. Betty said Carmel was enormously popular. A lot of people with money wanted a place there. Folks would set up $10,000 escrow accounts with realtors, so that even before a place came on the market, the realtor could make an offer and place a down payment on the property. Then the customers could check the place out, and if they liked it, they would exercise the offer; if not, they could always find a way to weasel out of the deal.

We asked if she had anything coming on the market. She said the Carmel Library was going to be selling a small two-bedroom, one-bath old gingerbread place on Camino Real. An elderly man had died and left it to the Carmel Library; the library needed the revenue and was ready to sell.

We asked if we could see it. She didn't have a key but said we could walk down and peer in the windows. We did, and it was exactly what we were looking for—one block from the ocean with an ocean view! The house had two bedrooms separated by a bath on the south side, a kitchen and breakfast nook across the back, and a living room on the north side. It lacked a garage and driveway, but it did have a beautiful old oak tree in the middle of the patio.

We asked Betty how much to offer. She said small places like that were going for about $30,000. This was a lot of money—more than twice the price of our place in Richardson, Texas. But we had her put together the offer and present it to the library. The library took a couple of weeks to decide; they had to have a board meeting and vote on the offer, but finally it was accepted.

Now all we needed was a mortgage.

We applied, and the mortgage company said it would be a couple of weeks before they'd approve anything. Well, a month went by with no word, so we asked Betty what the holdup was. In another couple of weeks, she got back to us.

There was a problem with the title. In California, property can only be owned by one of three entities: a person, a corporation, or a trust. The Carmel Library was none of these, so there was no clear title, and the mortgage company wouldn't make a loan without one.

I went to the city attorney and explained the problem. He put together a quick deed in the name of the chairman of the board at the library.

We waited yet another month and still hadn't heard anything, so we went back to Betty to ask what the problem was. The title company had surveyed the property, but it was not a legal lot. The fence of the owners on the south side was two feet inside the property line. I explained the new problem to the city attorney and asked him to write letters to the property owners on the south to get them to move their fence. It was a very sticky situation, because there's an old law that says if someone uses your land for more than twenty years and you don't formally protest, then they own or have a right to your land. I was glad the city attorney was working with us.

A couple *more* months went by and again we didn't hear anything, so again we stopped in to see Betty. The property owners on the south side did not claim the property, but the doctor on the corner refused to pay to have the fence moved. That Saturday night, some "vandals" tore down the fence on the south side and stacked the wood in a ditch by the doctor's house. On Monday, I called the mortgage company and asked them to survey the property again. They did, and it was now a legal lot.

The sale could go through with one stipulation—when we sold the property, we'd have to go through the same mortgage company, as they knew the history of the title and would be willing to grant another clear title.

But now there was *another* problem! The house didn't meet city building codes. The wood was full of termites, the wiring had to be replaced (it had the old 1920s-style knob wiring), and the hot water heater was located in the bathroom, which was forbidden. The house—our house!—was condemned. It was going to be torn down.

I went to the city building inspector and asked for a list of the violations and for suggestions about exactly what I needed to do the remedy them. I said I didn't want any problems when he inspected the corrective work. He said he didn't want any problems either. I later found out that California building inspectors were a little nervous at the time, because a building inspector who'd gone to a house in a hippie-infested marijuana-farming region of the Santa Cruz mountains had disappeared and was never heard from again!

I went right to work. First, I had to re-wire the house. I installed a new fuse box with circuit breakers and ran oversized wires to the outlets. I had extra lines and low-amperage circuit breakers, so no one could overload the circuits with all the new electrical appliances people plug into the outlets, such as radiant heaters, hair dryers, and stereos. (I didn't want the wires to heat up, because these old houses were tinderboxes and could go up in flames in thirty seconds.) The wires were all run parallel, one inch apart, and were supported every two feet (per NASA spacecraft specifications!). When I was done, the building inspector said he'd never seen a house wired with such caution and precision.

The next issue was to move the hot water heater from the bathroom to a pantry closet off the kitchen. I had to do a complete plumbing job, running both hot and cold lines from the bathroom to the back of the house. When that was done, we had the termite people put a tent over the whole house and fill it with malathion. The house passed the building inspection with flying colors.

Finally, we had a guy come out and lay down a thick red carpet. We were ready to move in. We had a place in Carmel!

A Salmon Fisher and a Carpetbagger

Two people we met at around this time were Jack Flores and Gordon Parmeter.

Jack Flores, our carpet layer, worked on a salmon boat in Alaska during the fishing season. His dad had a fishing boat in the Monterey harbor, and one day he brought us a large bucket of squid. He taught us how to clean and cook squid. After his next trip to Alaska, he brought us a twelve-pound side of fresh salmon preserved only in salt. All we had to do was slice off a piece, and we had the best fish in the world.

Gordon Parmeter was president of the Pilgrim Products marketing division in Saratoga, California. He was supposed to be a preacher. But he was really a carpetbagger. He had a mail-order business that would send out thousands of pieces of junk mail each month. He calculated that 1 percent of the recipients would buy the junk he advertised.

The End of My Job

Litton eventually lost their big government contract when it came up for bid and another company underbid them. The new company offered most of us at Litton jobs at 10 percent less than what we'd been making. I guess that 10 percent was the reason they'd been able to underbid Litton.

I started looking for another job.

Philco Ford, Aeronutronic Ford, Ford Aerospace

When Litton lost the government contract, I began seeking work in the Bay Area. One weekend, our friends Blaze Morrow and her husband Gordon came to one of Joan's Mary Kay meetings. During lunch, I mentioned I was looking for a job, and Gordon said they needed a digital engineer to help develop digital multiplexer at the company where he worked. It was a division of Ford Motor Company. I went in on Monday and interviewed with several engineers on the project. They thought I could handle the digital part of the multiplexer. The next week, they made me an offer and I accepted it.

Originally, the facility had been named Western Development Laboratories (WDL). When Ford bought it, they combined it with their Philco Ford division. Later, Ford purchased a Los Angeles company called Aeronutronics. They combined this with their Palo Alto company, and the name became Aeronutronic Ford. Later, in the seventies, the company was bidding on an international communication satellite contract (*Intelsat V*) and Ford wanted its name up front, so they renamed it Ford Aerospace.

At around that time they made a thirty-minute video to tell the world about Ford. The video discussed Ford's multiple business units, including Ford Motor Cars, Ford Trucks, and Ford Systems. Ford Systems consisted of the Steel Division, the Chemical Division, and the catch-all "other" division. We fell into the "other" category, so our director didn't even get a speaking part in the video.

We felt rather slighted, especially considering the overall financial picture. Ford Systems' net income for the year was about $3.5 billion dollars. The Steel Division's net income was about $1.5 billion, and the Chemical Division had a net income of about $1 billion. Ford Aerospace—jokingly referred to as Ford's "Buck Rogers Division"—was getting a lot of attention at the time, but their annual *gross* income at that time was only $580 million dollars. The net income for the "other" division was $1 billion, much more than the gross income of Aerospace, and certainly not on a par with the other two divisions.

Our director deserved that speaking part!

SMS

The development of the synchronous meteorological satellite (SMS) for National Oceanic and Atmospheric Administration (NOAA) was a quantum leap in global weather analysis. It provided the meteorologist with the overall picture of large weather systems every twenty minutes, which had been unavailable until now. Large system analysis had been performed by pasting together local hourly weather faxes that never quite covered the whole event.

SMS was later renamed geostationary operational environmental satellite (GOES) by NOAA when it was launched. A GOES satellite can provide frequent visible (VIS) and infrared (IR) images to monitor both routine and cataclysmic weather events such as hurricanes. They can relay meteorological observation data from surface collection points to a processing center and broadcast processed graphic and image weather data.

The GOES satellites are maintained at earth-synchronous altitude, 35,800 kilometers above the earth's surface. These spacecraft orbit in the earth's equatorial plane. At this altitude, their west-to-east motion equals that of the earth beneath; therefore, they appear stationary at a desired longitude. Radiation from a very small area on earth is focused on detectors that convert radiation energy to electrical signals. Imaging is achieved by scanning an optically focused spot over the earth. Radiation enters the telescope at right angles to its optical axis via the optically flat scan mirror, which is equipped with an angularly positioned stepping mechanism to facilitate scanning of the terrestrial scene. The spacecraft maintains a spin rate of approximately one hundred rpm and aligns the spin axis perpendicular to the earth's equatorial plane. The scanned pattern, as processed for use, resembles a conventional television image. The VIS image has 14,565 lines and 22,000 pixels per line, with a resolution of nearly 0.5 miles. The IR image has 1821 lines, with five-mile resolution. The visible infrared spin scan radiometer (VISSR) output is eight analog VIS and two IR analog signals. The digital multiplexer receives the eight analog VIS and two analog IR signals, converts them to digital signals, and multiplexes them into one twenty-eight-megabit data stream.

Digital Multiplexer

I found myself on the design team of the most advanced technology for weather observations and communications. Gordon Morrow was project manager for the VISSR digital multiplexer. Laird Kirby was the 6-bit-analog-to-digital-converter design engineer, and Jules Vile designed the 8-bit-analog-to-digital converter. I took the digital output of the 6-bit and 8-bit converters and multiplexed them into one 28-megabit data stream, along with some other telemetry data and status information and also built in redundant cross-switching to work around failed components.

Toastmasters International

Gordon Morrow was a busy man, so he promoted me to assistant project manager. He asked me to attend the weekly staff meetings in his place and update management on the progress of the project.

I was really scared the first couple of times I went to those meetings. Management always had lots and lots of questions about the project. I figured I'd better learn how to best present the information. One day I saw a bulletin-board notice that a Toastmasters International club was forming at Ford and all were invited to attend. I went and heard all these great speakers from other companies do fantastic presentations. They were excellent at explaining things. This was exactly what I needed to learn, so I signed up.

The program involved making speeches in front of an audience. After each one, the speaker received constructive analysis from Toastmasters' monitors. I wasn't bad at the spontaneous speeches because I was a good bullshitter. But the first ten prepared speeches were really hard for me. The analysis from the monitors was really valuable, however, and by the last one, because of the practice and the help, things became a little easier for me.

When the time for me to present my designs to the design review board and defend them against NASA scientist and engineers, I felt very confident.

Toastmasters International really helped me in my presentations. I stayed in the club for four years, and during the last year I was the vice president of our unit. I highly recommend Toastmasters International to anyone who feels uneasy giving speeches.

Zucksworth's Zoo

I worked directly for project manager Gordon Morrow, but Sam Zucksworth was our department manager. He had a closed-in office, not a cube, with a secretary at a desk just outside his door. Anyone wanting to see Sam would speak to his secretary first, and she'd phone him and announce the visitor.

Every week, we were required to turn in time cards for NASA. We were supposed to fill them out daily, but usually the cards just sat on our desks until Friday when we had to turn them in. Then we'd hastily fill in the numbers, creatively estimating the time we'd spent on each of the projects.

One day the NASA auditor came to inspect our time cards. The auditor looked like George Gobel, with a blond crew-cut and blue eyes. He had a military-style bearing—perfect posture, at stiff attention—and he carried a clipboard under his right arm as if it were a swagger stick. An assistant secretary, who had a clipboard under her arm also, followed him everywhere he went, one pace behind and to the right.

These two went up to Sam's secretary and asked to see him. Sam's office door was directly to the auditors' left. The secretary phoned Sam and announced that the NASA auditors were there to check the time cards. No sooner had she hung up than Sam yanked open his office door, right in front of the auditors, and screamed as loud as he could, "*Everybody fill out your time cards!*" There was a mass stampede as we all raced to our desks to fill out the cards.

I was fortunate to work with a lot of people who were much smarter than I was. Everyone was well-educated in very diverse subjects. We'd have discussions on all sorts of topics, and I learned a lot from them.

For example, Laird Kirby was a follower of Ayn Rand. We had a lot of great discussions about her book *The Fountainhead* and "Who is John Gault?" (Read the book and you'll learn the answer.) One week, we'd been discussing the philosophy of Jean-Paul Sartre, Georg Wilhelm Friedrich Hegel, and Arthur Schopenhauer; the psychology of Carl Jung and Sigmund Freud; and the theology of Friedrich Nietzsche, Soren Kierkegaard, and Martin Buber.

When I came into work the next day, I found the following scrawled on my desk pad:

Do be	Sartre
Be do	Buber
Do be do be do	Sinatra

Deficit Manufacturing and Dummy R&D

Every so often, the boss would ask us to estimate the cost of building a new system.

There would be three main scenarios.

First, there were always parts that have already been built and need only a few little changes to meet the new specs. For these, we could get the cost figures from the finance people.

Second, there were the never-been-done-before parts. For these, there'd be a lot of head scratching. But from past experience, we knew what was involved in creating something similar, so we could extrapolate and come up with a figure we thought might be feasible.

Third, there was the totally unknown part. In all new developments, there are unforeseen problems that can throw a monkey wrench in the works. So we'd always include a "fudge factor" in the final estimate that we'd submit to management.

Typically when we won a contract and the budget figures came in, they'd be only half of what we'd estimated! We'd complain we couldn't possibly complete the project in that amount of time for so little money. Marketing's response would be, "We knew the customer (NASA and NOAA) wouldn't go for the figures you gave us, so we cut them in half."

To make the budget even tighter, there was another problem. Whenever we'd submit an estimate, management would issue us a set of dummy research and development (R&D) numbers to which we'd charge our time and material as we did the advance work on the proposed project. (For example, we'd buy material and parts, and test out our design theories.) When we won the contract, finance would dump all the dummy R&D charges into the new contract. This meant that half the budget would be used up before we even started on the project for real!

So we'd go to work, and in a few months we'd run completely out of project money. We'd continue working, of course, and often when NASA or NOAA would see our test data, they'd request changes. So we'd charge for "change of scope" time, since we'd already run out of money, and this would make up the difference so the project would be profitable. But if NASA or NOAA didn't have any changes, management would simply tell them they'd have to pay extra if they wanted the project completed. This worked well because when we were halfway through a project, they wouldn't want to start over with another contractor. NASA or NOAA would go to Congress and request and get the extra money.

This practice was so common that we coined a phrase for it: "deficit manufacturing." Congress was operating on deficit spending, and the military-industrial complex of companies was using their same tactics. I think the consumer is doing the same with credit-card debt. When you die or the economy collapses you want to be a million dollars in debt, so you won't lose anything and you'll know had a great time while it lasted.

In our space-craft business, things had to be made perfectly because there was no way to go up into space and repair a problem. In the commercial sector of the electronics business, however, there's a saying, "There's never enough time or money to do things right the first time, but there's always enough time and money, if the product sells, to come back and fix it later."

I'll elaborate on this in the Intel section.

Bill's Big Break

When Bill Allen graduated with an MFA from Iowa State, he'd already received an invitation from Professor Robert Canzoneri of Ohio State University to come and teach writing and start a literary magazine. He had the whole summer of 1972 to kill before reporting to work. Then out of the blue, he was urged to join Eddie Greding and his wife, Marcia, in El Salvador, where Eddie was teaching biology and collecting rare frogs. Bill jumped at the chance, envisioning a paradise of volcanoes and rain forests to explore while he helped catch frogs with Eddie. But things were to turn out quite differently.

While he was there, there was a revolution in the capitol city of San Salvador. The military eventually took control of the whole country. The revolution started on the university campus, so the fighting was hitting close to home. Eddie came home in a panic and told Bill to grab as many of his one-of-a kind frog specimens as possible; they had to get the hell out of there. They drove non-stop to the border, and Eddie had to bribe the border guards to exit. They continued non-stop day and night, only stopping for gas and a sandwich, going through Guatemala and Mexico without a break. Eddie slept in the back seat most of the way, only waking up wanting to go to the bathroom, which Bill said he could do at the next gas station, of which there were very few.

After a month or two back in the States, Eddie received a request by the new revolutionary government to return and finish his teaching at the university. Marcia was against it, but they went back anyway. Meanwhile Bill reported to Ohio State and started his new job. Caught up in the "publish or perish" demands of university teaching, Bill started writing travel stories and selling them to *The New York Times*. The first was called "How Eddie, Marcia and I Smuggled Pickled Frogs out of El Salvador and Lived to Tell All."

The article turned out to be Bill's big break. In fact, it caused an international incident.

Bill used black humor to get back at El Salvador, but nobody south of the border thought it was funny.

Guatemala canceled three years of advertising in *The New York Times*. Over fifty articles in Spanish were written denouncing the "Gringo Allen." One article called him "Hyena meat a la *New York Times*." One Central American president said, "First the Pentagon Papers and now this!" Bill received a bag full of fan and hate mail, now part of Ohio State's literary archives. In no time, he was writing for *Saturday Review* and *Esquire* and landed a two-book contract.

Needing new ideas to write about, he sent all the Spanish articles to Eddie for him to translate, saying, "I'll do a follow-up piece and really get the bastards."

Big mistake.

The package never reached Eddie—it was confiscated by the government, which arrested Eddie and Marcia and threw them in jail. Marcia, put in with hookers, spent her time in jail sobbing that she was just a schoolteacher and had done nothing to deserve this kind of treatment. Eddie thrashed around in a cell full of drunks, saying he was going to die without his medication. After a couple of days of incarceration they were deported out of the country.

Bill had written me when the article originally appeared in *The New York Times*, suggesting I buy a copy, but it was sometime before I learned about the repercussions. Too bad about Eddie and Marcia, but Bill got early tenure because of the exposure.

"Material Is Immaterial"

We built the SMS satellite for NASA using the purest and most expensive parts and material in the world. All the parts had to be "flight-qualified for space." That meant they were of the best quality and radiation-hardened. The aluminum chassis that housed the electronics had to be an excellent thermal conductor. There's no air in space, so the temperature generated by the electronics had to be conducted to the outer shell of the chassis to be radiated into space. The aluminum had to be homogenous and free of impurities or imperfections that might create hot spots and impede thermal conductivity (temperature flow) away from the electronics. We had a special aluminum ingot made to exact material specifications, poured into a mold, and slowly cooled over a week to prevent hot spots from developing that would reduce heat dissipation.

This ingot was cut to the approximate size of the chassis, put on a numerical control milling machine, and bored to exact specifications to house the electronics. Then the chassis was gold plated. The cover was installed with thirty-two titanium screws. Later, the titanium was replaced with beryllium to save a couple of ounces of weight. NASA was giving $10,000 per pound of weight savings because the rocket and payload were overweight.

All this special handling cost a lot of money. For example, if we used a ninety nine cent transistor, by the time we burned it in and power-cycled it to eliminate the infant mortality rate and did everything else we had to do, the transistor would cost over $100. But it was necessary.

Anyway, the total cost of the hardware was less than 10 percent of the cost of the project. We coined another phrase: "Material is immaterial."

By now it was 1971, and my favorite hit record was "When You're Hot, You're Hot, and When You're Not, You're Not," by Jerry Reed.

In 1971 Intel introduced the microprocessor, a minute device on a single "chip" for processing information within a computer

In 1971 surgeons developed a fiber-optic endoscope, for looking inside the human body, enabling innumerable probing and diagnoses to be made without the need for incisions. The Microsurgical Instrumentation Research Association introduced diamond cutters; titanium technologists and microscope makers introduced the diamond-bladed scalpel.

I watched *Apollo 16* launched on April 16, 1972, and the landing on the moon on April 21, 1972. The mission was commanded by John W. Young, and the command module, named *Casper*, was commanded by Thomas K. Mattingly II. Charles M. Duke, Jr., commanded the lunar module, named *Orion*. They returned to earth on April 27.

Apollo 17 was launched on December 7, 1972, and it landed on moon December 11, 1972. The mission was commanded by Eugene A. Cernan, and Ronald E. Evans was the command module pilot of *America*. Harrison H. Schmitt was the lunar module pilot of *Challenger*. They returned to earth on December 19.

Brainstorm

One day in 1973, while I was designing a microprocessor-based test set, I was running out of RAM—random-access memory. At the time I was using 8 kilo-byte (KB) memory chips. There were 16 KB memory chips available, but they were about three times the cost of an 8 KB chip. While I was working with this limited memory, I realized that the quantum property of the electron, known as spin, could be used as a binary-storage element. Doing this could enable billion and billions of units of memory to be contained in a very small space. The main problem would be detecting and changing the spin. Spin had just recently been identified—or at least, I'd just learned about it, so thought I'd had a really important brainstorm.

I raced to the technical library and asked the librarian to do a search to see if anyone else was having the same thought. When I got the print-out of the search results, I learned that the search on spin, bubble, and memory had yielded about twenty pages of brief descriptions of people working on exactly that problem. Several Ph.D. programs and a number of aerospace companies were hard at work on the issue and were publishing papers on their results.

Many years later I read a June 2002 *Scientific American*, and the cover story was called "Spintronics, A New Twist in Computing." It said, "Microelectronic devices that function by using the spin of the electron are a nascent multibillion-dollar industry—and may lead to quantum microchips. Devices that rely on an electron's spin to perform their functions form the foundation of Spintronics (short for spin-based electronics), also known as magneto electronics. Information-processing technology has thus far relied on purely charge-based devices—ranging from the now quaint vacuum tube to today's million-transistor microchips. Those conventional electronic devices move electric charges around ignoring the spin that tags along for the ride on each electron."

Our Shoot-Out with Hewlett Packard

Every now and then, NASA would issue two identical contracts to two different companies and see which came up with the best design. While we were working on the VISSR Digital Multiplexer for SMS, which was a 28-megabit system, NASA issued a contract to both Ford Aerospace and Hewlett Packard to build a 100-megabit system.

As we started to design the new system, we dubbed the project our "shootout with Hewlett Packard."

The new system required substantial gigahertz wave-guides to pipe the high-speed data around. We laid this all out on a 4' x 8' plywood board and got it working. Then came the day for NASA to test both systems. They took our plywood board over to Hewlett Packard and ran the test. Our system worked as advertised but HP's did not. So we won the shootout and the next satellite contract!

For all of this high-tech research and development, we coined yet another phrase: "pushing back the foreskin of science."

The USN Pueblo

On January 23, 1974, the U.S. Navy's *Pueblo* spy ship was captured by the North Koreans. Apparently, the guys aboard the *Pueblo* had the same problem getting rid of classified documents that we'd had back when I was in the Air Force and our plane got into trouble. After that, we'd switched to water-soluble paper for all classified material. But the *Pueblo* crew didn't have water-soluble paper and their paper shredder wasn't fast enough to get rid all classified material. The North Koreans found a gold mind of intelligence information.

Playing the Stock Market

I had a friend named Manny, and he and I were always talking business. Manny was a retired chief from the navy. He owned some apartments and a deli, and always liked to talk money. He would quote Jimmy Dean's sausage advertisement—"There is a very good reason for that"—when a business venture looked a little expensive. Manny and I both invested in the stock market.

I'd meet him at the Merrill Lynch office in Palo Alto at 7:00 AM. That's when the New York Stock Exchange opened. We'd meet in the morning and "place our bets." Then we'd come back at lunch and see how we did.

We'd research a stock and if it looked good, we'd buy a few shares. Neither of us had much free cash, so we mostly dealt in penny stocks. One day I was researching a stock and read that Merrill Lynch, the brokerage firm I dealt with, "made a market" in the stock. What did that mean? I asked my broker. He said it meant that if the market started down, Merrill Lynch would buy the stock to prop up its price.

What the hell! If they'd prop up the price, how could I lose? I bought a hundred shares at $30 per share. This was in early 1974. Well, in 1974, the stock market crashed. I put in my sell order when the stock Merrill Lynch supported dropped to $25 per share. I figured a 16 percent loss, or $500, was not too bad. But it was three days before Merrill Lynch bought the stock, and by then it was at $9 per share. I asked the broker, "*What the hell happened?!*" He explained that because so many big investors were selling large blocks of the stock, it took Merrill Lynch a couple of days to buy it all. Boy, that really sucked. I lost over $2,000 of my $3,000 capital listening to Merrill Lynch.

Years later, when the stock market crashed in 1988 and a guy in Florida went to the Merrill Lynch office with a machine gun and shot his broker and the broker's boss, I must admit I practically cheered, thinking, "If more people did that, maybe those bastards would play fair."

Of course, the way to make money in the stock market was obvious—buy low and sell high. But it seemed every stock I was interested in skyrocketed, and when I wanted to sell, it crashed. Too many times I wound up buying high and selling low.

You want to know how to become a millionaire in the stock market? Start with two million.

The Origin of the Moon, Chlorofluorocarbons, and World Population

In 1974 a theory was proposed that the moon was part of earth and was knocked off by an impact of another body, a comet or large asteroid. It came from analysis of the moon rocks, returned by Apollo astronauts that showed their composition was similar to earth's.

Also in 1974 man's impact on the global ecological balance was alarmingly brought to the forefront when M. Molina and F.S. Rowland warned that chlorofluorocarbons (CFCs), as used in domestic fridges and as propellants in domestic aerosols (such as hair sprays and deodorants) were likely to be damaging the atmosphere's ozone layer which filters out ultraviolet radiation from the sun.

In 1974 the population of the world reached, and passed, four billion.

Outrunning the California Highway Patrol

By now I was commuting from Carmel to Palo Alto, 90 miles each way. I'd leave Carmel before 6:00 AM and arrive at Ford at 7:30 AM. If I left home after 6:00 AM I'd get tied up in traffic in San Jose and wouldn't reach work until after 9:00 AM. I'd go down Highway 1 through Moss Landing to Santa Cruz and take Highway 17 over the mountains to San Jose, then Highway 101 to the San Antonio Avenue exit where Ford Aerospace was located. In the afternoon, I could leave before 4:00 PM and make it home to Carmel by 6:00 PM.

I'd made this trip quite a few times in my little 1958 Volkswagen bug and knew the winding road over the mountains like the back of my hand. This stretch of Highway 17 was known as "Dead Man's Alley" because of all the auto deaths every year. It was slow going uphill, but I could fly down the backside.

One time I was going so fast, almost ninety, that I floated a valve and sucked it into one of the cylinders. I had to have the engine rebuilt by a friend.

Another time I topped the mountain going home and shifted into high gear for the run down the backside of the hill to Santa Cruz. I'd typically take all the right-hand turns in the outside lane, and all the left turns on the inside lane. On this day, it seems there were cars in the inside lane on right turns and in the outside lane on left turns, so I could easily pass on the best side. Boy, was I flying!

What I didn't know at the time was that a California highway patrol car had sneaked on to my tail at the top of the mountain. As I passed each car, there was another car in the other lane and they'd get together just after I passed and block the CHP car. This happened about eight times going down the hill. When we flew onto the flat section at the bottom of the hill, I was doing 70. The speed limit was 60.

Here came the CHP, and boy was he mad! When I stopped, he came stomping up to the car and was furious. He wrote me up for doing over 70 in a 50-mile zone, which was the speed limit coming down the hill. He hadn't been able to clock me, but he said he knew I was going over 70. I tried to explain that there were other cars on the road and I couldn't have been going that fast.

He said, "Tell it to the judge."

So I went to the court to tell the judge.

During the hearing, I asked for a display board so I could draw a picture. I drew the winding road, which everyone was very familiar with in San Jose, and the position of all the cars as I passed. I showed that they were blocking the CHP, and that I wasn't going that fast in third gear—only 50—while the cars I passed were only going 35 or 40. The fact that they were going so much slower just made it look like I was doing 70, I innocently explained.

Well, the judge thought about it for a minute, then told me to go get my speedometer checked and send him the certified results. I thought I'd won my case.

So I got my speedometer checked and sent the certified results, just as the judge had requested. And a week later I received a response saying I was *guilty*. I had to pay an $80 fine.

So much for telling it to the judge.

Robert Crown Law Library

It appeared that many things we wanted to do to promote Mary Kay were against various company rules. So I started to read up on the laws affecting independent contractors.

Ford had access to Stanford's facilities, so I went to the Robert Crown Law Library research desk to get started in the right direction. They suggested I read the *Fundamentals of Business Law* by Robert N. Corley and William J. Roberts. I thought it would be better to have my own copy, so I bought a used one at a bookstore. They also suggested *Principles of Business Law* by Essel R. Dillavou and Charles G. Howard, so I bought that one, too. Then I bought a book on contracts by John D. Calamari and Joseph M. Perillo.

Once I'd read all these, I tackled *Business Organization (Cases and Materials)* by Alfred F. Conrad and Robert L. Knauss. This was an eye-opener; it really made me question some of the things the Mary Kay company was doing to unsuspecting consultants.

I was getting good legal training in the School of Hard Knocks.

After that, whenever the Mary Kay people said we couldn't do certain things, I'd challenge the company regarding their position on independent contractors. Boy, did they start to cover their asses. All kinds of written legal stuff started coming our way.

My Finest Hour

At Ford, we'd just won the TIROS-N low orbiting (800 miles) polar satellite contract when Gordon asked me to take a look at the cylindrical electrostatic analyzer (CEA) spec. He knew I'd worked on the x-ray and gamma-ray radiation detectors at Caltech. The scientist was proposing an obsolete method. He wanted an analog piecewise linear exponential curve and had put in a 10 percent error budget.

I stuck my neck out and I proposed a digital stair step, where each step would be flat at a given voltage for the full time period, with no need for a special smoothing algorithm, and each segment would be an exponent of the previous section. My digital design suggestion would eliminate the problems in the scientist's proposal. By doing things digitally, I could reduce the error rate considerably. Gordon and Sam both liked my proposal, and I presented it to NASA and the chief scientist, who was impressed that I understood the details of the problem so well.

They said, "Build it."

I did and it came out a thousand times better than I'd dared to hope. When we had the final design review, everyone was very impressed with my design and with the final test results. They'd never seen a system that came as close to the design specifications as this one did. I'd now made my mark in hardware design and was as good as the best in the world.

Meanwhile, I'd decided to get out of hardware design. Computer-aided design and computer-aided manufacturing (CAD/CAM) were just becoming popular in that industry, so it made sense for me to go into computer programming.

Once again, I was looking for a new job, this time in computer programming or sales.

Ocean and Casanova

Meanwhile housing prices in California were rising at a compounded rate of 33 percent a year. So the next time our Mary Kay stock reached $60 and split two for one, we sold off half of the shares and bought a couple of houses on the corner of Ocean and Casanova.

This was prime Carmel real estate. The property would surely go up, and we'd make a bundle. We rented our little place on Ocean and Camino Real and moved into one of the new places. We left the other one rented, while I remodeled the one we lived in. I tore out the old kitchen and replaced it with all new cabinets and appliances. Each night when I came home from work, we'd paint and fix up another room. After dinner, we'd go to work on yet another room, often staying up until 1:00 AM. Then, after a quick nap, I'd go to work, come home, and go at it again. This went on for a year, but in the end, we had both places completely remodeled.

The problem was that all of my commuting and working on the houses at night didn't leave much time for Joan and I to spend together as a couple. So we began to drift apart. We'd been working our tails off ever since we'd hit California, and our marriage just could not stand the strain.

Joan arranged for a divorce. I gave her all our houses in Carmel, because they were bought with her Mary Kay money. Because I'd made the down payment on the first house we'd bought, she gave me the house in Richardson, Texas.

As happened to many of the Forty-Niners, my California gold rush went bust.

Part Four:
With New Beginnings

Chapter Ten:

New Beginnings
(1975 to 1980)

The Chris and Dick Show

I moved to Chris Holombo's house at 347 Aldean Avenue in Mountain View, California. Chris was an avid steam train buff. He would travel all over the world just to ride a special steam train. Chris also built model railroads. He'd converted his dining room into a den and had all his models on shelves over the dining room table, which he used as a workbench.

Dick Magney lived there also. Our friend Paul Ketchum referred to the two of them as "The Chris and Dick Show." Dick had a brilliant older sister who had a Ph.D. in chemistry, and Dick's parents had always held it against him that he was not as smart as his sister. When Dick was in the navy, he'd worked on marine diesel engines and loved it. All he wanted to do was work on diesel engines. And he was good at it. But his parents would never let him live down the fact that he hadn't gone to college. So he left home, and when I met him, he was living with Chris and working as a janitor at the local church. He also did odd jobs around the neighborhood.

When I moved there, I stayed home most evenings because I was paying off all the credit-card debt built up during our marriage, and I couldn't afford to go out and party. Dick and I would listen to Antonin Dvorak's *Symphony Number 9*, and I enjoyed it very much.

Recombinant DNA

Researchers had just discovered how to cut and splice together the DNA of disparate species and were beginning to contemplate the cornucopia of experiments this opened up. "Recombinant DNA was the most monumental power ever handed to us," said California Institute of Technology president David Baltimore. "The moment you heard you could do this, the imagination went wild."

But a number of scientists at the time raised concerns about whether such experiments might create dangerous new organisms, microscopic monsters that could sneak out of the lab undetected on the sole of a shoe or down a drain and threaten public health.

The Asilomar conference on recombinant DNA was convened to discuss the regulation of biotechnology in February, 1975, at the conference center at Asilomar State Beach in Pacific Grove, California. The conference was a turning point in molecular biology: a defining moment for a generation, a milestone in the history of science and society. A group of around 140 professionals (primarily biologists, but also lawyers and physicians) participated in the conference to draw up voluntary guidelines to ensure the safety of recombinant DNA technology. It is generally considered the main event in the history of biotechnology and the regulation of science and technology.

After much haggling, the group settled on a set of safety guidelines that involved working with disabled e-coli bacteria that could not survive outside the lab.

Psychology of Self Evaluation

I bought a book called *Tales of Power* by Carlos Castaneda and learned about astral projection and being in two places at the same time. Then I read *The Phenomena of Astral Projection* by Sylvan Muldoon and Hereward Carrington, and *The Astral Journey* by Herbert B. Greenhouse. Every night I'd try to astral project, but I never succeeded. The best I could do was have some flying dreams. I read *Seven States of Consciousness: A Vision of Possibilities Suggested by the Teaching of Maharishi Mahesh Yogi* by Anthony Campbell. I also read *The Probability of The Impossible: Scientific Discoveries and Explorations in the Psychic World* by Dr. Thelma Moss. It was about Kirlian photography, bioenergy, psychokinesis, and acupuncture.

Meanwhile, I was never going out but was itching to be around people. One day I saw a flyer from Foothill Junior College advertising a course called Psychology of Self-Evaluation. I thought it might be a good way to review my personal assets and set some goals for the future, plus a way to get out of the house, so I enrolled.

It was a great course. In fact, I took it again two years later with a lady friend who was at loose ends. They gave us a standard personality and career preference test and sent it off for evaluation and analysis.

In the meantime, they had a list of good books to read and discuss, such as *I'm OK, You're OK* and *What Color is Your Parachute?* The teacher was an intelligent single woman with a Ph.D. in psychology. We had a lot of terrific discussions on motivation. Most of my knowledge of motivation came from my wife's days in Mary Kay Cosmetics and the sales meetings with Success Motivation Institute (SMI) seminars. Two of the Mary Kay classes had been about self-confidence. I was full of that, and had a lot to say about it.

Meanwhile, our tests were back. Mine showed exactly what I thought it would. My scores for jobs such as nursing and healthcare provider were about as low as they could get; my empathy was low, and I had no time for sick people. However, my mathematics, science, engineering, and military aptitudes were as high as they could be. That figured. There wasn't much in-between for me. It was either black or white. Either I liked it and was good at it, or I didn't like it and couldn't care less

Chinese New Year in San Francisco

"I Left My Heart In San Francisco" by Tony Bennett was one of my favorite songs.

The instructor of the Psychology of Self Evaluation course had given everyone her home number in case any of us needed to talk with her. It was Chinese New Year, and I was watching a news broadcast about the big parade that would take place that night in San Francisco. I wanted to demonstrate to the teacher how confident I was. So I called her up and asked if she'd like to go to San Francisco that evening and see the parade. I suggested that afterward we could go to the Hyatt Regency near the Embarcadero and have a couple of drinks in the bar that rotated around on the top of the tower while we gazed down on the city below.

She said yes, so I jumped in my Lincoln Continental, picked her up, and headed toward downtown San Francisco. We arrived, parked at the Hyatt, walked through the lobby, and came out on Front Street and Clay Street where they were staging the parade. The parade dragon was over a block long, and I didn't want to take time to go two blocks out of our way, so I ducked under the dragon as they were working to get it ready. I stopped for a second and told the teacher, "You can't get any closer to the parade than under the dragon!"

We walked ten short blocks and came out right where the reviewing stand was. We had the best viewing place of the whole parade. She could not believe it.

We arrived just as the parade started. After two hours, as the parade was nearing the end, I grabbed her and said, "Let's go." We returned to the Hyatt and took the elevator to the rotating bar and restaurant. We had a couple of fantastic drinks and made a couple of rounds to see the city. Then we left.

On the way home, she just kept saying, "I can't believe it. That was one of the most fantastic evenings I've ever spent."

Disco Chuck

After I'd paid off the credit-card debt, I looked around to see what was happening. It was the disco era. I let my hair grow down to shoulder length and had it styled. I bought three disco-style leisure suits, one navy, one white, and one maroon. I also bought three silk shirts, one navy, one black, and one Chinese red. With this new look I needed some shoes, so I bought five pairs of patent-leather shoes: navy, white, red and white, black, and brown. I could mix and match and be coordinated. The guys at work called me Disco Chuck.

One of my favorite places was the Jumping Frog Disco. It was just like the one in *Saturday Night Fever* with the lighted dance floor. So eat your heart out, John Travolta. I was ready for action.

HMO

My healthcare at that time was through an HMO. These companies make their money receiving payments for enrolled patients, and then not providing any services to them. One of the big problems I ran into was never being able to get an appointment. The phone lines were always busy. After trying to get someone on the phone for three or four days, when I did get someone, the appointment would be for six to eight months later.

Vasectomy

That summer, Bill Hartman and his wife went to Italy for their vacation, and I was invited to housesit for them. I decided to have a vasectomy while I was staying at their place, because it was near the Kaiser Hospital. I knew I didn't want kids, so this looked like as good a time as any. But the hospital was a little afraid to do it. I had to bring in my divorce papers and sign several consent forms and promise not to sue them. I went in and had the procedure done. The doctor snipped the vas deferens on one side of the scrotum, pulled it out, and tied a knot in the upper end. Then he did the same thing to the other vas deferens, and I was ready to go. I had to have a friend along to drive me home so he took me over to Bill Hartman's house, where I sat on an ice bag for two weeks to keep the swelling down.

I thought that if any woman every tried to trick me into marriage by claiming she was pregnant, I'd tell her, "Name it 'Houdini,' if it got past those knots."

When Bill Hartman and his wife returned, they told me all about Rome. One of the places they visited was the Vatican. Across the street from the Vatican was the Banco di Santo Spirito or Bank of the Holy Ghost. I thought it was an interesting idea—letting a ghost handle your money. I don't think it is so much different from a normal bank. If anything happens to your money you will never be able to find anyone responsible.

My Brush with Death

One night I woke up at 3:00 AM. There was a searing pain in my abdomen. I knew I had appendicitis—which in those days meant certain death if your appendix ruptured.

I managed to get dressed and crawl out to the car. Somehow I drove the three blocks to Kaiser Permanente Hospital. In the busy emergency room, the nurse took a quick look and put me on a gurney in a curtained-off area and said the doctor would be right with me. This was at 3:25 AM.

I lay there on the gurney for over two hours, the whole time thinking my appendix had ruptured. I knew I was going to die. The nurse had probably just put me there behind the curtain, out of the way, to die alone while she tended to patients who could be helped. I kept waiting for my life to flash before my eyes, or to see that long tunnel with the blinding white light at the end of it.

Finally, at nearly 6:00 AM, the day-shift doctor came in. He listened to my abdomen with a stethoscope. He prodded and poked around. Then said, "Go home. Eat crackers. Drink 7-Up. You've got that flu that's going around."

IRAN Team

Now that I was single, there was nothing to tie me down, and I wanted to see more of the world. I'd reached the pinnacle of hardware design and noticed that everything was going the way of computers—CAD/CAM—so I'd have to reinvent myself to stay on top. I considered getting into software design, with a long-term goal of going into management. I'd read that most managers came from sales, and so would keep that in mind.

In the meantime, I thought I might do a little traveling. I kept my eyes open for opportunities. I looked in the company listings of job openings and found one for what was called the "IRAN Team." These were depot-maintenance people who flew all around the world installing, upgrading, and maintaining the Air Force's satellite tracking stations. That sounded like great fun—traveling around the world at the company's expense. I applied and was selected for the job of maintaining the Univac 1200 computer system at all ground stations.

In the new department, I was introduced to the team: John Paletiere, Bill Akins, and Scott Robinson. Bill Walsh was the antenna man. And there was another team member named Jimmy Pitts.

After a short training session, I was off to Manchester, New Hampshire, for my first look at one of the ground stations. We stayed in the Queen City Inn in Manchester—on the company's dollar, of course. The team had been regular customers, staying there off and on for years, so when we booked our reservations, they were ready for us. On the marquee in front of the inn was the slogan: "Georgia has the peaches, but we have the Pitts"—referring to our team member, Jimmy Pitts.

After clearance into a secure area at the ground station, we were led to the operations room where most of the electronics were located. There were several nineteen-inch racks full of equipment, mostly receiving and transmitting radios and analyzers. Over to one side was the computer. It had a ten-platter hard-disk system that carried the computer program. Each disk platter was programmed to tell the satellite what its next photo mission was to be at the Blue Cube at Moffitt Field in Sunnyvale, California. There was a special "page overlay keyboard" that matched the computer program on the platter. Both items were required for data input to the satellite, and both were classified top secret and carried by separate couriers to the ground-station sites. At no time were they to be both in the same place at the same time, except when being created or when operating at the ground station.

The antenna was a one-hundred-foot, one-hundred-ton monster that moved like a fine-jeweled watch. It had to move very quickly to track the satellite as it flew overhead at 17,000 miles per hour. The information downloaded from the satellite was not stored at the ground stations, but was immediately sent back up to a communications satellite and relayed to the Blue Cube in Sunnyvale.

The ground station would be taken off-line for only a few hours, for our IRAN Team to do maintenance and upgrades. During downtime, whether in the middle of the night or whenever, was when we had to do our thing. We'd be standing by when downtime came, because we had to get everything finished in the allocated period. I was fortunate that the computer system had a self-diagnostic program disk—all I had to do was get it started and the disk did all the work. If it found any problems, it would tell me which card to replace. After we finished with our stuff we could go home. All our work was at the ground stations, and there wasn't much to do when we got back to Palo Alto except fill out our travel forms. Then we'd have a big post-project party over at a team member's place.

One day we were notified that the bearings, on which the New Hampshire antenna rotated, had developed problems. There were flat spots or cracks in some of the ball bearings, and they needed to be replaced at the next downtime. This meant raising the hundred-ton antenna about two feet, pulling out the old one-ton bearing, and replacing it with a new one.

Obviously, this was no small task.

Fortunately our man Bill Walsh, who'd erected the antennas, had worked in steel construction his whole life and knew exactly what to do. We'd need to schedule a longer downtime, because the antenna dome would have to be partly removed so that heavy lifting cranes could get cables on the antenna. While we were working on the electronics, Bill was getting the lifting cranes in place. As we finished with the electronic part of the job, Ford management and the Air Force suddenly called a meeting to discuss the overall process and plan the best time to lift the antenna. While they were having their discussion, Bill, oblivious to the meeting, lifted the antenna, replaced the bearing, and put the antenna back down on the new bearing.

He was buttoning it all up when someone invited him to the meeting. When he came in, the managers and Air Force guys began telling him when he could raise the antenna. He said, "Wait a minute—it's already done!" and turned and left the meeting. Ford management and the Air Force guys just looked at each other.

That's the kind of guy Bill was. He had two daughters, whom he took with him on a lot of his jobs. The girls made good gophers and could climb all over the antenna without bending anything. He'd let them sit out on the antenna horn and operate the theodolite surveying instrument, which was used to define the parabolic curve of the antenna.

At one of the "post-project" parties in Palo Alto, I met Bill's daughter Leslie, whom I later married.

Not-So-Greenland

Our next trip was to the American military base of Thule, in northern Greenland. It was a grueling journey.

We flew five and a half hours from San Francisco to Kennedy International Airport in New York.

We caught an hour and a half bus ride to McGuire AFB in New Jersey.

We boarded an Air Force C-141 Starlifter airplane, sat facing the rear of the plane, and flew for six more hours to St. John's, Newfoundland, Canada, where the plane refueled.

Then we flew *another* six hours to Thule.

The Air Force guys had a saying: "There's a girl behind every tree in Thule."

There are no trees in Thule.

And Greenland is not green. It's literally a *great big ice cube*, about 5,000 miles long. The Greenland glacier dates from before the Ice Age, 65 million years ago.

Not surprisingly, Greenland is always cold.

The U.S. Air Force has maintained a base there for many years. During World War II, our B-29 bombers were stationed there. Later B-36 and then B-52s were added. But now there wasn't much there except the Ballistic Missile Early Warning System (BMEWS), a large radar site, and the satellite tracking station. All the airplanes and support personnel were gone. There were only a few thousand military personnel stationed there on eighteen-month rotation schedules. The Air Force supposedly had about two hundred women airmen stationed in Thule, but they were as hard to find as trees.

So there wasn't much to do except get drunk.

There *was* a library at the base, but it housed only a few shelves of materials. I went through most of the books and magazines the first two weeks I was there. The bowling alley was in terrible shape. Other than drinking at the club, the most activity that went on was in the workshops. Some guys worked on cars; some made things out of wood or metal. But if you were only there for a couple of weeks like we were, those kinds of projects were out of question.

Everyone seemed to be drunk and/or bored. One night at the club, I watched a guy dance with the post in the middle of the dance floor. No one except me even seemed to notice. He must have done it a lot.

While I was there, I learned about this weather problem in Greenland, called "phases." What happens is the sun shines on the top of the glacier and makes the ice crack into fine splinters. Then when a weather front comes in, because of pressure gradient forces and something called "the Coriolis effect," the wind kicks up the ice slices, and it blows horizontally like crazy.

Phase Three was considered the mild one, with winds up to about twenty knots and visibility of less than a mile. During a Phase Three, you were allowed to travel around the base in trucks but could not go off base. Phase Two involved winds between twenty and thirty knots, visibility of less than fifty feet, and temperatures below zero. During a Phase Two, only extreme emergency personnel were allowed outside. Phase One was the worst: winds were above fifty knots, visibility was less than ten feet, and temperatures were less than 30 degrees below zero. You couldn't even see the barracks across the street. Yet if you looked straight up, you could see blue skies.

One time when I was there I experienced a Phase One, with the wind blowing at seventy knots and the temperature at 65 below zero. During a Phase One there were absolutely no outside activities. Who cared? I sure didn't want to get out in that!

There were stories about some airman who disobeyed orders and tried to go to the next barracks during a Phase One and was never heard from again. But it was so cold that even when there were no phases at all, I didn't want to spend much time outside. While I was up there, a female airman went jogging off base on a nice summer day with no wind and was never seen again. They suspected she might have fallen into the fjords.

Blue Cube, Moffitt Field

My next trip was to the Blue Cube at Moffitt Field in Sunnyvale, California. The name accurately describes the building. It looked like a large blue cube, with only one entrance. Several huge satellite-tracking antennas loomed outside of the building. These were mostly used for communicating with other tracking stations around the world—and with the Pentagon and the CIA. Because this was a main operations facility for the Air Force, every floor and every room was guarded. You were only allowed to enter the specific area you needed to be in at a given time.

The stuff that came through there was really of futuristic quality—the resolution in the photos was outstanding. For example, I saw a spy photo taken of Soviet officers in Moscow, and you could practically tell which of the Russian guys needed a shave. That's a little exaggerated, of course, but you could definitely see the epaulets on the officers' uniforms and could make out what rank each one was. In one picture of an airplane, you could count the rivets in the airplane's wings.

Hubble Problem

Years later, I ran into a friend I'd worked with at Ford and I heard the following story. When the Hubble space telescope was first launched in April 1990, reports came back almost immediately that it was out of alignment and was sending back out-of-focus images of space. When Bill Allen and I had studied telescopes, one of the first things we learned was the importance of the mirror grinding. Even ninth graders know about mirror grinding, so how could the Hubble telescope be out of focus? Astronomers determined that the conic constant of the mirror was −1.0139, instead of the intended −1.00229. Bill and I really believed that the mirror grinding could not have been screwed up. In reality, my friend said, the telescope's primary mirror had been ground to the wrong shape at the facility where U.S. spy satellite mirrors were manufactured. The Hubble's mirror had been intentionally calculated for a focus of about eight hundred miles—which was the distance above the earth where the Hubble orbited. It was not manufactured to focus out into deep space, but to look at the ground!

After the Hubble space telescope was launched and the problem was identified, the CIA and NSA had the use of the telescope for over three years—to spy on earth—until NASA came up with a fix and modified the telescope in December 1993. Whatever happened to the identical back-up mirror? Do you think it was used in a CIA telescope?

Vandenberg AFB

Another trip we took was to Vandenberg AFB on the central California coast. The Air Force was building a facility just like the one at Kennedy Space Center in Florida, so they could launch their own shuttle flights. They already had Titan and Minutemen silos buried in the coastal plain: it was where they tested the missiles and trained the launch teams.

The Air Force was also trying to develop a ballistic missile defense system they'd built in the South Pacific on Kwajalein Island. They held what the media called "the Great Turkey Shoot," in which they launched a ballistic missile from Vandenberg AFB and, as it entered the atmosphere near Kwajalein Island, tried to shoot it down. I suspect that they missed on purpose, so the Soviet trawlers would not get good information about our defense system.

Alaskan King Crabs

We also went to Alaska, to the Aleutian Islands, and worked on the satellite tracking station there.

Before we came back, Jimmy Pitts bought some Alaskan king crabs at the local fish market and had them frozen so he could bring them home. When we boarded the commercial plane in Anchorage, Jimmy asked the stewardess to take care of the frozen crabs for him. He expected her to put them in the refrigerator, but the refrigerator was full, so she put them in the hanging baggage locker, on the floor.

About halfway home, they thawed out some—and started to leak on the floor and smell. The stewardess came on the PA system and said, "Will the man with the crabs please come forward."

Jimmy jumped up and yelled as loud as he could, "*Quick! Say Alaska King!*"

The whole plane just broke up laughing. The stewardess would not show her face again until we started to land.

AFOS

I was summoned to help another installation team with the Automatic Forecast and Observation System (AFOS) that the National Oceanic and Atmosphere Administration (NOAA) had ordered and that Ford had built. It was to be installed at three hundred sites across the U.S.

The first unit was installed at NOAA's main office in Silver Springs, Maryland. That sure was an interesting place to work. All kinds of scientists and engineers there were working on some fascinating weather problems. It was really intriguing to talk to them.

The AFOS system consisted of a Data General Eclipse computer system, a lot of telephone connections to special weather-data nets, and the display console. It was all painted blue. From the display console, the operator or analyst or meteorologist could call up information on any of the weather-data nets or satellite images and either display the data, or combine it with other information in overlays. It was quite an improvement over the old paper facsimile machines the meteorologists had previously worked with.

After that system was installed and working, we moved to Sutland, Maryland. There, in the largest building I ever saw, except the Pentagon, was just about everything the U.S. government had. The main building was five blocks long and had five enormous buildings running off perpendicular to it. We were installing the AFOS system on the third floor, at the end of the middle building, next to the satellite ground station and the big IBM mainframe that assembled the satellite pictures. There must've been four or five IBM mainframe computers in this part of the building.

At lunch, I'd wander around that huge building and investigate what was in it. On the first floor, I found the Small Business Administration (SBA) bookstore, which had a copy of every book the SBA had ever printed.

About a week later, I'd worked my way down into the basement in my wanderings. There, I found the "Land Sat" library. "Land Sat" is a polar orbiting satellite that takes photographs of a mile-wide corridor of the earth as it orbits. They had photos of every square inch of the globe in there. I looked and found the image of Carmel, California, and sure enough I could identify my house in Carmel. It was amazing!

AVAWOS

Next I was assigned to work on the Aviation Automatic Weather Observation System (AVAWOS) at Patrick Henry Field, Newport News, Virginia. This system had a Data General 1200 computer connected to all the weather instruments at the airport—a digital thermometer, a digital anemometer, a digital wind-direction indicator, and three rotating-beam ceilometers placed around the airport.

The rotating-beam ceilometers were a recent innovation. Each contained a tube with a slot in it. A small motor caused the tube to spin around a light bulb. The tube had an infrared filter over the slot. A series of infrared light pulses would be sent into the sky, and when the pulses were reflected off of clouds, a message about the angle of the reflection would be transmitted to the computer. The computer calculated the tangent of the angle to get the distance between the cloud and the earth.

There was also a visibility detector, which was simply a strobe light and a light-intensity detector. The strobe would flash at regular intervals and if no reflected light was detected, visibility was clear. If a reflected flash blinded the detector, it was probably snowing. There was a gradient range between clear and snowing that included various intensities of rain and fog.

An NOAA scientist had run experiments, set up some data points, and fed the information into the computer. Once the computer had all the figures from these various contraptions, it crunched the data and came up with the "present weather at the airport." Then it transmitted the report on one of the weather data-nets to the entire AFOS system. The data was also sent to a voice synthesizer that would convert the data to a voice message (which sounded like Robbie Robot) and transmit it over the local vertical omni range (VOR) station. Any pilot interested in landing at Patrick Henry Field would key in the frequency of the VOR to get his directions to the airport, and would receive the present weather information generated by the AVAWOS.

I understand from my pilot friends that there are now a lot of these systems installed around the country. NOAA picked Patrick Henry Field to test the prototype system because of the diverse and severe weather conditions that constantly went through the area. For example, one day during a storm, the temperature was 29 degrees and falling. It was raining cats and dogs, plus snowing and sleeting—all at the same time. The wind was blowing at thirty knots when the light went out on one of the rotating-beam ceilometers. I was elected to go fix it.

Boy, was it cold! When I opened the ceilometer's box and bent over to replace the light bulb, the wind blew my raincoat over my head. I was immediately soaked with freezing, bone-chilling sleet and snow. It was the coldest I've ever been!

The next week I had to go back to Thule, Greenland, where it was 35 degrees below zero, and that didn't feel nearly as cold as the day I got soaked at Patrick Henry Field.

Yorktown

While in the area I decided to do some sightseeing, so I went over to Yorktown, where General Cornwallis surrendered to George Washington at the end of the American Revolution.

All my life I'd heard about the Revolutionary War and what a huge deal it was. George Washington! Thomas Paine! Valley Forge! Crossing the Delaware! Our Army had a tough time of it, yet our weary but determined soldiers fought bravely on, in spite of cold, hardship, and numerous defeats, to win our independence from England.

What a disappointment. As battlefields go, it was a tiny place. It didn't take me long to walk over the entire grounds. There could not have been more than 10,000 troops there on both sides. The separation between the lines wasn't over three hundred yards. I don't think the cannons of that day could shoot much farther.

Maybe it's because I'm from Texas, but I expected it to be bigger.

He Who Needs the Most, Loses

Late one afternoon while I was still in the Newport News area, on a gray and overcast November day I took a drive to see more of the historic sights that had helped shape our nation.

I turned off the James River Road at the Newport News Shipbuilders Dry Dock Company. And there I came face-to-face with the blinding facts of the present reality, the life that lay open to me, the almost unalterable future like a tunnel from which I could not escape.

I saw, as if in a vision, two workers, picketing there with their steel hard hats and their fat beer bellies, while a policeman wrote a parking ticket for a car at the curb. There was a large, closed parking lot with signs demanding four dollars a day from the poor workers, who, judging from the way they were dressed, were already straining to live amid inflation on their meager incomes. Farther down the dirty, lonely street, half of the buildings were closed and boarded up. Of the ones open, three were theaters featuring XXX-rated movies and topless go-go girls, and most of the others were pool halls and beer joints. A few black males lounged outside or leaned against shiny new cars. The only other businesses were loan companies and pawnshops.

I drove a bit farther down the street, to where the banks and savings and loans outnumbered gas stations for corner locations. The newest, brightest, and cleanest buildings were the city, state, and federal buildings that occupied the area near the new neat park overlooking the bay. Farther still, down at the mouth of the river, Fort Henry stood guard against a MAD nuclear attack like a dinosaur facing the incoming asteroid of doom.

Here, where the Ironclads first met—and all over our great country—the fighting continues. Nation against nation, US Steel versus Japan's steel, North against South, black against white, business against unions, management against workers, the overweight beer-belly crowd versus affluent American society. Just as a falling barometer can forecast foul weather, so these were—and still are!—the omens of a bleak future to me.

He who needs the most, loses.

New Concerns

In 1975 many new areas of concern were being opened up by scientific research. These included the danger of cancer among asbestos workers and their families; the effect of excessive intakes of lead on mental retardation; and the increase in mercury poisoning among fisherman in the Mediterranean as levels of mercury, discharged as industrial waste, were continually increasing.

Also in 1975 a whole-body X-ray scanner came into use. Skin-grafting techniques were improved. Methods of combating tooth decay were enhanced.

Lincoln Continental

While at Ford in California: as a perk, the Ford executives got to drive new cars every year, turning them in after twelve months for new ones. The company had a program where employees could purchase these executive car turn-ins. The returned vehicles were advertised in the company paper, and you could go to the transportation department and buy them at about 20 percent under their Kelly Blue Book prices. To eliminate profiteering, the company made you agree to keep the car for a year before you sold it.

By now, I'd paid off my credit card debts and had saved a few thousand dollars, besides. The stock market had taken a nose dive, so instead of investing, I decided to buy one of these car turn-ins. I purchased a beautiful 1974 dove gray Lincoln Continental Mark IV with a maroon velour interior. Boy, was it nice! I still drove my little VW bug most of the time, and I kept the Lincoln parked in the executive's lot in front of the main lobby so the security guards on duty could watch over it. When I'd get a lunch date, we'd head out the lobby door, get into my fine big car, and drive to lunch. On Sundays, I'd pull up to the church in my shiny Lincoln.

I kept it for a year and then sold it for a $1,000 profit. Not a bad deal.

The next year I bought a snow-white 1975 Lincoln Continental Mark V. It was even nicer than the Mark IV.

My Quest to Start My Own Business

I decided to investigate starting my own computer business. To research the idea, I went to the J. Hugh Jackson Library, Stanford's business library, which was named in honor of a retired dean. The librarians there helped me find some great books. Almost immediately, I learned that I didn't want to be part of any business organization except a corporation. Sole proprietorships and partnerships seemed like just asking for trouble, as they appeared to involve taking on someone else's problems and debts.

I came across a book called *Guide to Venture Capital Sources,* by Stanley M. Rubel. Boy, was that an eye opener. I liked it so much I went out and bought the fourth, or 1977, edition. I'd figured I'd need to borrow between $250,000 and $500,000 for my new enterprise, but the book said very few venture capital companies would be interested in loaning such a small amount. Small amount?! To them, it was. Most wanted to lend between $10 million and $100 million. Wow! I never knew there was so much money available. But there was a catch. They only wanted to lend to companies with proven management teams. Very few would get involved in a high-risk start-up business—and if you were a new company and your staff lacked proven management experience, you were considered high-risk. Most venture capitalists wanted you to have at least five successful years in business before they'd even talk to you.

My Invention

Meanwhile, because data-entry seemed to me to be a major stumbling block of small-business and personal-finance programs, I'd decided my business would involve marketing a product to address this challenge. My product idea involved a ten-line by twelve-column keypad that would fit on a desk in place of a desk pad and could replace the standard PS-2 keyboard. My keypad would be covered with a Mylar overlay, similar to the page-overlay keypad that I knew the Air Force had adopted for programming and operating its satellite ground stations.

My overlays would be customized to each customer's applications, and there'd be a set of standard overlays if the customer needed them. The edge of the overlay would fit under a holder on each side; the holders had sensors to detect the application of the overlay and configure the software for input. Once the overlay was in place, the user would simply press the key under an icon showing the desired type of input, and then enter the amount on the ten-key keypad. There'd be one overlay each for "income," "expenses," "assets," "liabilities," and any other specialized application a customer might have. The software would do the rest.

So all the user had to do was slip in the correct overlay. The sensors would immediately recognize it and configure the software. Then the user would press the correct icon-button and input the data. There was no need to know anything about accounting—the software would insert each bit of data in the correct place. Every thirty days, the system would output monthly statements for such things as income, profits, losses, and net worth or value of the business. It would also do a ratio analysis and compare the results with both the previous month and the national average for that small-business inventory control (SBIC) number.

I personally thought it was a great invention, and I couldn't wait to get my company started. But I needed money.

Bonza Bottler Day Again

On July 7, 1977, it was Bonza Bottler Day again. The day, the month, and the year were numerically the same: 7/7/77.

Now I thought about the next Bonza Bottler Day, 8/8/88, and wondered what I'd be doing then.

Voyager I and Voyager II

Voyager I and *Voyager II* were launched about two weeks apart on their historic flights through the universe.

On August 20,1977, *Voyager II* left the launch pad at Cape Kennedy. On board the space probe were photographs of earth, verbal greetings, and a medley of musical compositions. *Voyager II* was our invitation to other intelligent species in the universe to please come and visit our planet earth.

On September 5, *Voyager I* was launched, also from Cape Canaveral aboard a Titan-Centaur rocket.

Richard White

In November 1977 I found another wonderful book, *The Entrepreneur's Manual* by Richard M. White, Jr. He was a senior consultant and general manager of Business Solutions, a consortium of consultants who specialized in supporting new business start-ups and spin-offs. I tracked down Mr. White and, after getting him to sign a non-disclosure agreement to protect my ideas, I actually had a meeting with him. I outlined my business plan, and showed him a document I'd entitled "Application of the Page Overlay Keyboard (POK)." It included the following points:

> Product
> Configuration
> Price or Profit/Costs
> Market Segment/Direction
> Models or Types
> Outputs—Programs to Generate
> Inputs—Programs to Generate
> Trademarks
> Expansion and Diversification Plans

Mr. White thought I had a good business plan. And he liked my product idea. But unfortunately, he said the whole project was just too small for his interest.

Business License, Business Banking

I went down to San Jose, to the state tax office, and got a business license. I was now more than just plain old humble "Charles Willingham." I was now proud and delighted to be "Charles Willingham doing business as (or dba) Small Business Accounting and Analysis, also known as (aka) SBA^2, or SBA2."

Then I went to the Bank of America in Palo Alto and opened a business checking account. There I found some very helpful people who gave me a lot of *Small Business Reporter* pamphlets on topics such as "How to Read a Financial Statement and Ratio Analysis of Business." I also began to order books and management aids from the Small Business Administration in Washington, D.C.

Business Lessons

I received a copy of J. K. Lasser's book, *How to Run a Small Business.* One of the things he talked about was buying an existing business. I went to a couple of seminars about buying existing businesses. These seminars were interesting. I learned that, no matter what the product was, a business was in business to make money. Everything in the business had to be reduced to its lowest common dominator—money—and analyzed to see if it was worth doing.

Another thing I learned was that the big boys (large corporations) did not like to get into a bloody fight of survival over a niche product. They preferred to sit back and let the struggling young businesses slug it out in the marketplace. If the product was something they were interested in, the big boys would watch the market and see which firms survived their first five years. If a firm survived, and if the product was an asset the big boys wanted, they'd move in after five years and make the small business an offer. The offer usually included the owner and management signing a five-year contract to work for the big boys. This way, they could get the market niche plus a proven management team. It was no skin off their backs, just some stock options.

A Tempting Opportunity

One of the going businesses I researched was a printed circuit board assembly operation in Scotts Valley. It had an owner, a variable number of assembly employees, a wave solder machine, about ten folding tables for assembly work, plus sufficient soldering tools and miscellaneous parts for the operation. The owner drove in from Silicon Valley, picked up orders to be assembled, and dropped off finished boards. The asking price was $50,000. I had just enough for the purchase but would be left with no working capital.

Meanwhile, in the quality control (QC) department where I worked, there was a black lady I really liked. One day I was discussing the pros and cons of buying the business with her. She suggested that I buy the company and hire her as the president. She could also do the QC work and hire her mother—who was not only half black, but was half American Indian, *and* blind!—as vice president of human services. I would be the vice president of manufacturing.

Wow! Black, Indian, and handicapped. With all those minority points, I knew we could easily get a lot of money from the U.S. government—probably much more than we needed.

I was really tempted to take her up on the idea.

OPM and Other Great Business Tips

While I was studying to start my business, I came across some information about financing. It said that rather than using your own financial resources, whenever possible, you wanted to use OPM. This stood for "other people's money." That way, if the business didn't make it, and most don't, you hadn't lost anything of your own.

Another tip was: "A good way to keep money in the corporation is through pension plans." The corporation could use the pension money at a very low rate of interest, a rate set by the corporation's board of directors. The corporation could turn this cash over and over again, making money with it.

A few years ago, when there were discussion in Congress on TV about the liability of pension plans, I was floored to hear a couple of CEOs say they'd have to rethink their pension plan policies, as the plans were never intended to pay out money to the pensioners. The plans were just designed to keep cash in the corporation. And in Dallas, when LTV filed for bankruptcy, one of the things that came out in court was that they didn't want to pay retirement benefits for all the people retiring from the company. It was too much of a liability and could bankrupt LTV, so management tried to have the bankruptcy court throw out all of their claims. Well, the court saw it differently and ordered the company to sell off its assets to pay the retirement benefits. Boy, was this a blow to the business! It is different now. Too many large corporations are filing for bankruptcy, and they have found sympathetic judges who allow them to dump their pension plan and let the Pension Benefit Guaranty Corporation (or PBGC) pay some but not all that is owed.

A couple of more tips involved market surveys, market analyses, and test marketing. A market survey, no matter how badly written, was stressed as being absolutely essential before starting a business. On this subject, one book used the example of the Great Atlantic Lobster Company. I was familiar with the name: when I'd drive up through Fremont, Hayward, and San Leandro to Oakland to go windsurfing, I could see this big warehouse on the bay with "Great Atlantic Lobster Company" emblazoned on the side.

The example in the book told of a young man who was getting his degree in business administration from one of the Bay Area universities. As part of his studies, he was to write a market survey. He created his survey to find out if there was a market for Atlantic lobsters in the Bay Area. He mailed a thousand letters to local restaurants asking if they were interested, and if so, how many lobsters per month would they need and how much would they pay for them, wholesale. Well, his questions were not real specific or well-written. So in response, he received twenty-five letters with checks to cover advance orders for over a thousand lobsters! He received over three hundred more letters with interest in purchasing his Atlantic lobsters as soon as a price was negotiated. He also received some three hundred additional letters with positive interest in Atlantic lobsters. He immediately dropped out of school, flew to New England, purchased two thousand lobsters, flew them back to Oakland, and sold them at a huge profit. He was in business within a month, and it has been growing ever since.

Another example in the book involved a young engineer who developed a sort of chastity belt for dogs and cats. It had a small battery and could pulse the battery voltage to a transformer and build the voltage spike to 500 or 600 volts. This was not enough to hurt a dog or cat, but when the male mounted the female he would get a discouraging jolt. The engineer ran a test-market program. Within a month of the announcement of his new product, articles began circulating in newspapers, and in all of the pet publications, about the new "animal torture belt." Some of the articles falsely claimed that animals had been burned to death, while others were crippled for life. Additional articles said the inventor was a convict who'd gotten his idea from the electric chair while in the prison. None of this was true, but the engineer had stepped into the $250 billion-per-year "pet product market"—and no one making money in that market wanted any limits to the pet population.

Carolyn Johns and Pheromones

Carolyn Johns, of 1941 Edgewood in Palo Alto, was a Mary Kay consultant and later became a director. I met her at a Mary Kay seminar, and she was a lot like my ex-wife: beautiful, smart, and aggressive. She was also a Coldwell Banker real estate agent. After I was divorced, I looked her up because she lived in Palo Alto just down the street from Ford Aerospace where I worked. We dated for over a year. Her full-time employment was as secretary to a Ph.D. chemistry professor at Stanford University. His department was doing research on synthesizing pheromones of a number of agriculture pests in an attempt to control infestation.

The researchers had successfully isolated several of the pheromones, and they were running a field test to determine if the synthesized chemicals they'd developed would attract male bugs. They set up a large test area in the Sacramento valley and had petri dishes positioned on posts in grid arrangements over several acres. After a few weeks, they found they'd caught a disappointingly low number of male bugs. They were baffled by the low capture rate. They were sure they'd synthesized the pheromones correctly, and they painstakingly reviewed their research and their processes. While they were going over their notes, a storm blew through the valley and washed everything down.

The next day, they returned to the test area to put more synthesized pheromones in the petri dishes. Lo and behold! They found hundreds of male bugs in the traps. It turned out that the synthesized pheromones they were using were ten-thousand-times too strong. The researchers concluded that the poor male bugs were beating their wings off miles away, down the valley, looking for the females they were sure were there.

The Church of Scientology

I went to a Business and Professional Women's dance and met a pair of very nice-looking ladies about my age. After a bit of small talk, they said they were with the Church of Scientology. I'd heard of Scientology but hadn't known it was a, quote, church, unquote. They gave me their business cards and invited me to come in and see their "church." So one Tuesday night after work I stopped by.

Man, there were a lot of good-looking women in there!

I was introduced all around and then taken into a briefing room—by a man—with some other people—also mostly men—interested in "joining" this "church." They introduced us to a book called *Dianetics,* by a writer named L. Ron Hubbard. Hubbard was the founder of the "church." The Scientology guys told us that Dianetics was "The Modern Science of Mental Health." I didn't much care about this—I wanted to get back to where the beautiful women were—but I sat there and listened politely. They explained the levels you have to go through to get "cleared," as they called it—in other words, to get rid of all the bad and negative programming in your mind. They showed me a chart with the various "clear" levels on it. It sure looked like a pyramid, and I wondered if this was another version of Mary Kay, which also had a pyramid structure. They said they offered a basic Scientology course at a $1.00 per lesson.

When I found out the course was taught by some of the stunning women I'd seen earlier, I couldn't sign up fast enough.

At first, the course was great fun. I got to sit about two feet away from and directly across from a different fine-looking woman at each class. It was wonderful, gazing into their beautiful eyes, smelling their fragrant perfume, seeing their gorgeous skin and hair. It was sure cheaper than taking them out to dinner. All you had to do was repeat a series of dumb phrases for an hour. "Do fish fly? Do birds swim?" It reminded me of those "Dick and Jane" books you read in first grade. The theory was that the silly phrases were supposed to find your hot buttons and grind them down to remove your sensitivities.

That was the theory, but in practice, it didn't seem to go anywhere. After a while, I got tired of just staring and saying stupid things. I wanted some action. Like a date. Or a kiss. Repeating dumb phrases over and over was boring if it didn't lead to something more.

I guess they'd found one of my hot buttons!

At this one meeting, they introduced a new manager. He'd bought a "franchise" for $25,000 and had gone completely through the entire training in only a week. He was pronounced "cleared."

I figured it would take me 25,000 weeks to get cleared at a $1.00 per lesson. That seemed like too much effort for me, especially if I couldn't even get a kiss.

I also figured the man had been cleared, all right—cleared of $25,000 he could have invested elsewhere.

Mormons

I gave up on Scientology but remained open to other possibilities. I heard about the Church of Jesus Christ of the Latter Day Saints—also called the Mormons. I got my hands on their basic religious text, the *Book Of Mormon*. I read it, but was not impressed.

Buddhist Temple

Next came Buddhism.

To learn what that religion was all about, I started to going to a Buddhist Temple in Mountain View, California. They talked about how everything moved in cycles—birth, life, death, then rebirth. I liked a lot of their philosophy because they seemed to have a more realistic interpretation of life than I'd found in Christianity. Buddhism appeared to encompass all of life. Instead of focusing only on man, it took into account the animal, vegetable, and mineral elements of life: all the animals (humans and other species), vegetables (plants), and minerals (earth, sun, moon, and stars).

What I didn't understand was the idea of worshiping the Buddha. I could see that he'd been a great man and all, but it seemed there was a lot of mindless repetition of prayers. It wasn't that much different than the dumb phrases of Scientology.

And there weren't as many beautiful women.

The Seth Tapes

Next came channeling, and Seth.

I'd heard about a book called *Seth Speaks* by Jane Roberts. I read it. "Seth" was the earthly name for a "personality" from another world who spoke to us humans through the "borrowed" body of the spirit-medium-channel Jane Roberts. She "channeled" Seth so that he could communicate with humans. I learned there were some Seth tapes, but you couldn't just buy them, you had to go listen to them at special sessions.

So I went to one of those sessions. The people there reminded me of typical charlatans, like mediums that talk to dead relatives, palm readers, tarot card readers, and healing preachers. That was the end of my interest in Seth.

Los Altos Methodist Church—And Another Beautiful Woman!

So it was back to traditional Christianity. I joined the Los Altos Methodist Church. The couple who were my key contact invited me to dinner so we could get to know each other and they could answer any questions I might have about the church. That night I found out that the wife had a sister who was a cocktail waitress—and she was single. I didn't have many questions about the church, but

I had a lot about the sister. By skillfully inquiring while I feigned interest in the church, and without seeming too eager, I was able to learn from the woman that her sister and I had a lot in common. The sisters were Air Force brats, had been in Germany at Wiesbaden, and had also been at Incirlik, Turkey. I managed to get almost her total life history before the evening was over.

Next chance I got, I went to the bar where the sister worked. I sat in her section. She was a knock-out! It was a slow night, and I invited her over to my table. She had no idea who I was, so I said I was with the Secret Service and had been assigned to monitor her activities for the years when she was overseas. I began to tell her about all her past and how I'd been following her. Needless to say, she was floored. Here was a total stranger who knew her life's history. From the church couple, I'd managed to find out a lot of personal things about her, so I had information that that only someone very close could have known. It was too good of a setup. The poor girl didn't have a chance!

Bible Study—How I Became the Devil

Los Altos Methodist Church was very active, with a lot of intelligent members in its congregation: professors, teachers, retired military officers, doctors, lawyers, and businesspeople. They had weekly Bible study classes at night, and these were enlightening, interesting, and enjoyable. We really had some very intellectual debates and discussions. In one class, for example, we discussed *Siddhartha* and *Narcissus and Goldmund* by Hermann Hesse.

We also discussed Christian existentialism stressing the subjective aspects of the human person considered as a creature of God; especially the theory emphasizing the natural desire of the creature to seek his creator (as in the philosophers and thinkers Augustine, Pascal, Nikolai Berdyaev, and Gabriel Marcel) or the distance between guilty man and omnipotent God (as in Soren Kierkegaard, *Either/ Or*, and the dialectical or crisis theology of Karl Barth and Emil Brunner). Nietzsche proclaimed the "death of God" and went on to reject the entire Judeo-Christian moral tradition in favor of a heroic pagan ideal. Part of the discussions were about the work of the French writer Albert Camus, *The Stranger* and *The Fall*, which are usually associated with existentialism because of the prominence in these works of such themes as the apparent absurdity and futility of life, the indifference of the universe, and the necessity of engagement in a just cause.

After a couple of years, the preacher decided to change the regular class format for our study of the book of Job. He assigned parts in the story to several class members. We were to read our parts and play that role in the discussion. I was assigned the part of the Devil—which some said was type-casting. I thought I did a very good job of defending the Devil. My position was that the Devil did nothing to Job on his own, but talked God into bringing all the problems to Job.

One week we were unable to use the church library for our Bible study class. Another group needed it for their meeting, so we all agreed to meet nearby at a class member's house. We formed a car caravan, and I was the second car in line. I'd just turned the ignition on when someone tapped on my window. I rolled the window down and there was a another member of the class who said, "I should have known the Devil would be driving a white Lincoln Continental and would have a blond on his arm."

I had been dating a lot, and one day I noticed that I started to drip and had a little pain when I urinated. I went to Kaiser Permanente Hospital for a check up. Sure enough I had gonorrhea, so the doctor said to drop my pants. He gave me the biggest shot of penicillin I had ever seen and gave me a note. Then he told me to go sit in the emergency room for an hour. If I did not have a reaction to the penicillin I could go home. If I did have a reaction, the doctor in the emergency room would know what to do. I sat in the emergency room for over an hour and did not have any adverse effect. A week later it was cleared up. We had gone to see the movie *The Poseidon Adventure* with the song "There's got to be a morning after" by Maureen McGovern. Now, every time I hear that song I think of what happened and that big syringe with the square needle. Some thought it was providence; God was getting even for defending the Devil.

The Real Story Behind the Personal Computer

Because I was working with computers so early on, and viewed, first-hand, the genesis of the personal computer industry, I want to set the record straight about the development of the personal computer.

In 1973, in return for some programming he'd done for the Intel company, called the "Intel system-install system" (ISIS) operating system program for their microprocessor development system (MDS). He was a professor of computer sciences at the Naval Postgraduate School in Monterey and was playing around with a small, 8080-based computer given him by Intel. His name was Gary Kildall. In working with that small computer, he wrote what was later called CP/M. The small machine Intel had given Kildall was equipped with a monitor and paper-tape reader and was certainly advanced for its time, but Kildall became convinced that magnetic-disk storage could make the machine even more efficient.

Meanwhile, at IBM a team of engineers led by Alan Shugart had invented a data-storing system that used a disk drive. Trading some programming time for a disk drive from Shugart, Kildall first attempted to build a drive controller on his own. Because he felt he lacked the necessary engineering ability, he contacted a friend, John Torde, who agreed to handle the hardware aspect of interfacing the computer and the disk drive. At the same time, Kildall worked on the software portion, refining the ISIS operating system he'd written earlier. The result was "control program/monitor" (CP/M).

The prototype of CP/M developed by Kildall in 1973 underwent several refinements. Kildall enhanced the CP/M debugger and assembler, added a BASIC interpreter, and did some work on an editor. CP/M was widely used by many Intel-based computer systems until the appearance of the IBM personal computer.

Intel released an 8-bit 8080 chip in 1974. The 8080 was designed not to make computing a part of everyday life, but to make industrial machines and home appliances more intelligent.

In the meantime, in Albuquerque, New Mexico, a former Air Force officer and engineer named Edward Roberts had formed a company called Micro Instrumentation Telemetry Systems (MITS). Roberts was working to design what would eventually become the Altair 8800—the world's first microcomputer that used Intel's new 8080 processor. He enlisted the help of William Yates and Jim Bybee to help design the computer.

In January 1975, the Altair 8800 appeared. It was, quite simply, a metal box with a panel of switches and lights for input and output, a power supply, a motherboard with eighteen slots, and two additional boards. One of these board was the central processing unit (CPU), with the 8-bit Intel 8080 microprocessor at its heart; the other board provided 256 bytes of RAM. This miniature computer had no keyboard, no monitor, and no device for permanent storage, but it did possess one great advantage: a price of $397 in kit form. MITS planned to sell 4K memory boards at $150 as add-ons; these could be installed into extra slots in the back.

Programming initially meant manually setting eight switches for each byte—or character—of the program and its data. Each switch represented the binary equivalent of one byte being entered. If you've ever seen a computer program, or tried to open a file in the wrong program and had your screen fill up with gibberish—nonsensical letters and symbols—you can guess how many characters are involved in even the simplest program. The task of manually setting eight switches for each byte was obviously tedious and error prone. A single program required throwing hundreds, even thousands, of switches.

Programming the Altair 8800 was further complicated by the fact that you had to enter every byte in sequential order. A single mistake meant you had to start over, beginning the entire series from scratch. Imagine throwing 1600 switches and finding you'd made a mistake with the third byte! In addition, to enter even a simple command you had to throw still more switches that were incorporated on the front panel of the unit. So for the Altair to be a real success, someone had to come up with a computer language, a way to program it more easily to perform worthwhile tasks. With its limited memory and its lack of an efficient programming technique, the Altair was merely a high-priced toy for technophiles.

Computer languages that circumvented the switch-throwing system were already known in the mainframe world. A U.S. Navy officer named Grace Murray Hopper (known as the mother of the computer) was really the first to solve the problem when at the Pentagon she helped invent a computer language—English words that the computer itself translate into binary code. A language meant that rather than just flipping switches, users could simply use a keyboard and type lists of instructions into a computer. But Hopper's language—COBOL—didn't work on the Altair.

To help promote the machine, Roberts convinced two men from *Popular Electronics,* Arthur Salberg and Leslie Solomon, to run a feature on it. So the front cover of the January 1975 issue displayed a color picture of the Altair 8800. The article mentioned that this fabulous device was available in kit form from MITS for only $375.

When the article came out, a young computer programmer named Paul Allen was visiting a friend at Harvard.

Oh, and you've probably heard of this friend—his name is Bill Gates.

Anyway, Paul Allen saw the article and realized that the two of them could create the language needed for the computer. After discussing it excitedly for three days, Bill Gates called MITS to talk to Edward Roberts. He exaggerated their position, claiming they already had a working version of a language for the Altair. But Roberts had heard this story before—since the article had come out, he'd gotten dozens of calls from people with similar claims. So he told Gates that the first company to deliver the language had the contract.

Allen and Gates now had to turn their boast into reality by creating a language for the Altair. The task before them was immense. They had to produce a language that could operate in only 4K of memory and still have room for its resulting program to run. For the next seven weeks they worked night and day at the Harvard computer center, using a PDP-10 minicomputer to simulate the Intel 8080. But as impossible as the task seemed, Gates, Allen, and another student named Monte Davidoff, forged ahead with the work. Finally they'd done it—they'd created a prototype. But they didn't have an Altair to test it on. For the test, they'd have to take a copy of the language to MITS in Albuquerque. Paul Allen was elected to go.

When Allen got to Albuquerque, Roberts met him at the airport in an old pickup truck and drove him to the MITS "corporate headquarters." Allen was shocked—the "corporate headquarters" was located in a run-down strip mall, between a massage parlor and a Laundromat. To make the day even more unnerving, Allen hadn't brought enough cash to cover his hotel expenses and had to borrow from Edward Roberts.

Next morning, Roberts picked up Allen and took him back to headquarters to test the language. Allen had recorded the language on paper tape so he could load it into the Altair without flipping toggle switches 30,000 times. The paper tape meant he only had to use the switches to boot the computer. Allen booted the computer and gave the command "Print 2 + 2." The Altair quietly printed out "4"!

Everyone was stunned; it had worked.

The Apple

At about this time, a twenty-six-year-old named Steve Wozniak had also been dabbling in computers. In 1976, he designed what would become the Apple I, based on the Motorola MC-6502 microprocessor. His twenty-one-year-old friend, Steve Jobs, had an eye for the future and insisted that he and Wozniak try to sell the machine. On April 1, 1976, the Apple Computer company was born in the Jobs' family garage.

Home Brew Computer Club

I was a member of a group at Stanford University called the Home Brew Computer Club. We were just a bunch of guys who were mostly trying to improve on what now seems a primitive video game, called Pong, that was all the rage in the mid-seventies. At this time, the only direct exposure most Americans had to computers was through games like this—and they were quickly growing in popularity. So inventing the next hot computer game seemed the only really profitable use of computer technology for the masses. Many of us had hopes of making a fortune by inventing the next great game.

All tolled, there were probably about 20,000 of us trying to catch the golden ring by creating a best-selling game.

And none of us took the Apple I very seriously.

Apple didn't take off until 1977, when the Apple II debuted at a computer trade show. The Apple II was the first personal computer to come in a plastic case and include color graphics; it was an impressive machine, and it sold for less than $5,000. Orders for these computers multiplied quickly after the trade show. Then, with the introduction in 1978 of the Apple Disk II, the most inexpensive and easy-to-use floppy drive yet to appear, Apple sales further increased. With just 4K of memory and the flexibility that customers could use their own TV sets as monitors, the Apple II became the first mass-marketed personal computer.

By now, a lot of the hobbyists at our Home Brew Computer Club were taking the Apple more seriously. We'd seen Apple II sales take off, and we wanted to jump on the bandwagon. So we quit writing games for the Intel-based 8080/85 machines, such as the Altair, and began writing programs for the Apple II.

Apple II soon became the system of choice for anyone interested in microcomputers, because it had useful applications that went beyond games. And it had the most games, too. It was fairly easy for us hobbyists to port our 8080/85 game programs to Apple's MC-6502 processor. Soon, there were word processors, accounting programs, database programs, interpreters, compliers, assemblers, debuggers, and all kinds of programs for the Apple II.

Apple Grows Up

With the increase in Apple sales, however, came an increase in company size. By 1980, when the Apple III was released, the company had several thousand employees and was beginning to sell overseas. In response to this growth, Apple had taken on a number of experienced mid-level managers and, more importantly, several new investors who now staffed the board of directors. Before their arrival, Apple had been a place where a bunch of twenty-somethings would stay up all night creating the next Apple invention, or would sit around in bean-bag chairs and eat pizza from the box during company meetings. Now, the older, more conservative men turned the previously fun Apple organization into a "real company," much to the dismay of its original employees.

In early 1983, Jobs began to court John Sculley, then president of Pepsi-Cola. In April, he was successful, and Sculley became president and CEO of Apple.

Jobs believed Sculley would help Apple "grow up," but had no idea how right he would turn out to be. Eventually, the decision to hire Sculley cost him his own job at the company he'd co-founded.

Meanwhile, Back at the PC

So Apple had the most games, and the most useful programs. And Intel-based 8080 systems were lagging behind because there was no "killer application" for the Intel systems. True, WordPerfect was becoming the standard word processor, and the publishing community selected it from among a number of competitors as their program of choice. But there wasn't a really great application for Intel systems until VisiCalc.

VisiCalc was a spreadsheet application developed by a couple of guys at Harvard. They created it to do the "what-if" homework in their business accounting class. But they failed to patent VisiCalc, and so other companies wrote their own versions of the spreadsheet program.

In 1978, Intel introduced 16-bit technology with the 16-bit 8086. The personal computer (PC) was now truly coming into its own. This new chip represented a major step ahead in performance and memory. The 8086 was faster and more powerful than its predecessors, and it had more memory. Moreover, the old 8080/85 code could be compiled to run on the 8086.

When the 8086 arrived on the scene, Microsoft, like other developers, was confronted with two choices: continue working in the familiar 8-bit world or turn to the broader horizons offered by the new 16-bit technology. For a time, Microsoft did both.

Acting on Paul Allen's suggestion, the company developed the SoftCard for the popular Apple II, which was based on the 8-bit 6502 microprocessor. The SoftCard included a Z80 microprocessor and a copy of CP/M-80 licensed from Digital Research. With the SoftCard, Apple II users could run any program or language designed to run on a CP/M machine.

At the same time and, coincidentally, a few miles south in Tukwila, Washington, a major contribution to MS-DOS was taking place. Tim Paterson, working at Seattle Computer Products (SCP), a company that built memory boards, was developing an 8086 CPU card for use in an S-100 bus machine.

Paterson was introduced to the 8086 chip at a seminar held by Intel in June, 1978. He had attended the seminar at the suggestion of his employer, Rod Brock of SCP. The new chip interested him because its instructions all worked on both

8 and 16 bit, and there was no need to do everything through the accumulator. It was also lightning-fast and could do a 16-bit ADD in three clocks.

After the seminar, Paterson began work with the 8086. He finished the design of his first 8086 CPU board in January, 1979. By late spring he'd developed a working CPU, as well as an assembler and an 8086 monitor. In June, Paterson took his system to Microsoft to try it with stand-alone BASIC, and soon after, Microsoft BASIC was running on Seattle Computer's new board. Microsoft only had a couple of programs and could not support a complete computer system.

During this period, Paterson also received a call from Digital Research asking whether they could borrow the new board for developing CP/M-86. Though Seattle Computer did not have a board to loan, Paterson asked when CP/M-86 would be ready. Digital's representative said December 1979—which meant Paterson would have to use stand-alone BASIC for a few months longer before switching to another operating system.

While Paterson was developing 86-DOS, the third major element leading to the creation of MS-DOS was gaining force at the opposite end of the county. IBM—which had seemed oblivious to most developments in the microcomputer world—now turned its attention to the possibility of developing a low-end work-station for a market it knew well: businesses and business people.

IBM

IBM representatives from Boca Raton, headed by a man named Jack Sams, went to Intel to buy hardware, and Intel had a deal waiting for them. For their introduction into the personal computing market, IBM could have Intel's 8088 chip—the one that no one else wanted—and for a low price. Such a deal! Intel had developed the 8086 16 bit CPU first but then discovered that CPU development out paced the peripheral chip development. So in their infinite wisdom they decided to limit the I/O (input/output) of a new chip (8088) to take advantage of all their 8-bit peripheral chips. This made no sense to independent hardware designers like me. Why design a CPU board with only an 8-bit bus because when the 16-bit peripheral chips came out the whole board would have to be scrapped. The 16-bit bus could use all of the 8-bit peripheral chips just by connecting to the lower 8 bits of the 16-bit bus. Using an 8086 chip and a 16-bit bus, only a small part of the board would have to be redesigned when the 16-bit peripherals came out. The study group was not told that the software would run at half speed because of the 8-bit bus. The use of the 8088 chip in the IBM PC set back personal computing performance ten years.

When I worked at Intel, I learned that IBM and Intel had been partners for some time. Indeed, IBM owned about 25 percent of Intel, and Intel supplied memory chips to IBM at a discount. Intel had also built add-on memory systems for IBM's 360 and 370 mainframes.

The IBM group also went to Microsoft and shared their interest in developing a computer based on a microprocessor. They acknowledged that IBM was unsure of microcomputing technology and the microcomputing market. They said that IBM wanted a complete set-up, with an operating system and applications, which Microsoft did not have. As mentioned above, there was still no "killer app" for an Intel-based system. The spreadsheet was a move in this direction—close, but no cigar. Microsoft had only a few applications, including a basic interpreter, a spreadsheet, and a database. And Microsoft did not have an operating system. IBM was accustomed to long development cycles—typically four or five years—but the company was aware that lengthy design periods did not fit the rapidly evolving microcomputer marketplace.

Needing an operating system, Bill Gates's Microsoft partner, Paul Allen, contacted Rod Brock at Seattle Computer Products. Allen told Brock that Microsoft wanted to develop and market SCP's operating system, and that the company had an original equipment manufacturer customer for it. SCP, which was not in the business of marketing software, agreed, and it licensed its 86-DOS operating system to Microsoft. Eventually, SCP sold the operating system outright to Microsoft for $50,000, plus a favorable language license and a license back from Microsoft to use 86-DOS on its own machines.

The 86-DOS system, of course, became MS-DOS. The prototype ran for the first time in February 1981. In August 1981 MS-DOS debuted.

Apple Again, and Lisa

Back at Apple, Jobs and several other engineers began to develop the powerful Lisa computer. The Lisa was based on the Motorola MC-68000 microprocessor, an innovation that would eventually redefine personal computing. The Lisa had two Motorola 16-bit microprocessors: one for the CPU and the other for display controller. This design made the Lisa one hundred times more powerful than anything comparable at that time. The system off-loaded the enormous CPU-intensive display process to the second MC-68000, leaving the main CPU free to perform computing and other processes. It was an extremely efficient design.

By now John Sculley, formerly president of Pepsi-Cola, was firmly ensconced at Apple. Sculley was a successful beverage businessman, but it soon became clear that he didn't know computers. Just as Pepsi and Coke had for decades hoarded their secret formulas, Apple now hoarded its own information. Apple would not give out the critical data that third-party software designers needed to use to write programs for Lisa. Instead, Apple intended to write *all* its own application software. The goal was: "Sell software bundled with the computer, thus boosting profits." The reality was: "Kiss of death."

This stupid sales strategy eliminated tens of thousands of potential software developers for Lisa. The strategy might have worked if Apple had developed its own killer applications for the Lisa—but it hadn't, and it didn't. Sculley failed to understand the important role third-party software developers could play when a new system was introduced. At this point, most independent software developers jumped ship from Apple and started writing programs for the IBM PC. As if this weren't bad enough, Apple made a second major mistake in pricing its new computer: Lisa cost more than $10,000—about twice what PC buyers expected to pay.

Over the next few months, Apple laid off a fifth of its work force, some 1,200 employees. The company also posted its first quarterly loss.

Fortunately, Steve Jobs was able to save Apple with his pared-down (no pun intended!) Macintosh and killer-app desktop publishing. But Lisa, with her two Motorola 16-bit microprocessors, remained the most powerful thing the industry had seen, even for ten years after her demise. It wasn't until 1995 that the IBM PC could match the performance of Lisa.

My Job at Intel

I still hadn't started my own business, but I left Ford when I was hired as a senior quality engineer for Intel's microcomputer development systems (MDS).

Intel was having problems with the MDS's reliability. The company needed a troubleshooter—someone who could come in from the outside to analyze problems and recommend changes. Well, as a lot of people have said, "Trouble" is my middle name, so I joined the team at Intel.

Intel had been developing and delivering the MDS for several years to promote the use of its microprocessors. The vice president (VP) in charge of MDS development was very proud of the system. He took every available opportunity to tell anyone who would listen how great this system was. Just before I was hired, he'd attended a national sales meeting in San Diego. At this meeting, when he tried to praise the MDS system's reliability, the salesmen literally laughed him off the stage. What had happened was that middle management had been underplaying the reliability issue, so the magnitude of the problem had not reached the VP level. This was why they needed an outside engineer—me—to do the evaluation.

Having worked within the super-reliable NASA environment, I had very high standards and so was the ideal candidate for the job.

I'd decided to change careers from hardware design to software design because most hardware development was going toward microprocessors and CAD/CAM. This meant that, to get ahead in my field, I needed additional training on either microprocessors or CAD/CAM. The job at Intel was a great opportunity for both. At Ford Aerospace, during a lunch-class TV-training program, I'd had some microprocessor design and development training. And I could analyze the Intel systems with the eye for reliability that I'd learned from NASA. I could also take courses on the latest microprocessor systems and learn high-level software development.

My first task was to take fifty MDS systems from the warehouse, set them up in a room, run them for thirty days, and troubleshoot any failures. So I did. Well, the salesmen were right. Every single system failed within thirty days! And about a third failed more than once. Some had as many as five failures in thirty days.

I calculated the overall MDS system failure rate at 154 percent in thirty days. This was amazing to me—154 percent! Coming from a NASA environment where the satellite had a guaranteed minimum of five years' operation, this failure rate was unbelievable.

This is when I first heard the phrase I mentioned earlier: "There's never enough time or money to do things right the first time, but there's always enough time and money, if the product sells, to come back and fix it later."

Intel's manufacturing cost for the MDS was about $2,400, and they were selling it for a minimum of $16,000. So they would just fix any failure brought to their attention. A lot of the failures never even came to their attention—they were repaired by Intel's customers for the MDS, all of whom were microprocessor design engineers who were usually using another MDS system to fix the failed ones.

I set about analyzing the failures and writing my report. Typically a report for this kind of project is short. But the problem was so severe, my report ran to thirty-two pages. The problems were grouped into two major categories: bad parts, and electrostatic discharge.

Bad Parts

The basic concept of integrated circuit design and manufacturing grew out of transistor design and manufacturing, which I'd learned early on back at Texas Instruments (TI). The core issue is yield. What percentage of "ideal" items will the manufacturing process yield? In integrated circuit design and manufacturing, hundreds of identical circuits are etched on each silicon substrate. Given millions of variable parameters in the process, the number of ideal chips that are developed is about 25 percent—and that's a high estimate. The real numbers are highly classified industry secrets.

This 25 percent figure means that only 25 percent of the chips manufactured will work exactly as they were designed to work. The other 75 percent will exhibit varying degrees of failure. But perhaps 20 percent of the total will function in all but one or two areas—and depending on which areas are effected, the chips may be sold at a reduced price. For example, if they are just a little slow (running at, say, 100 mHz instead of the design of 200 mHz), then they can be sold at a percentage of the price of the ideal ones. If the math co-processor is defective it can be sold for an application that uses slower software math routines. With 25 percent being ideal and 20 percent being slow but useable and saleable, the yield rises to 45 percent.

Now comes the real challenge: what do you do with the dogs, the other 55 percent? Well, Intel had a system of what was called "S specs." Each of its microprocessor chips were tested by a computer, and if a chip didn't meet the first few parameters, it was given an "S spec" designation. This meant the chip could be reviewed and possibly used in some application that didn't require the deficient parameter. The cause of the high MDS system failure rate was this: some idiot manager in the semiconductor group had a bunch of "S spec" microprocessor chips that he needed to get rid of. So he dumped them on the MDS development group.

Intel had different quality designations for its parts. The number one or highest-quality parts sold for around $350 each. The number two parts, the slower versions, would sell for around $250. And on down the scale, to a price of $100 for the slowest chips. To keep costs down, the MDS group was not allowed to use the highest-quality parts, as these were needed to generate profits for Intel's semiconductor division. So the whole design of the MDS system was flawed from the beginning.

But it gets worse.

There was no incoming-parts inspection process—not for parts from internal sources, and not for parts from outside vendors. It was obvious to me that if a company wants to produce a reliable product, this is one of the first reliability checks it needs to perform. Every vendor has low-quality—or at best, unknown-quality—parts that it needs to get rid of. Even reputable warehouses will ship unknown—quality parts, if they know a receiving company doesn't do inspections: "Hey, let the receiving company find the bad apples." But if you monitor and inspect incoming items and return them, the burden falls on the vendor to make the order right with acceptable parts. So after you've returned a number of their shipments, at their expense, because they didn't meet your incoming inspection quality standards, then that vendor becomes more vigilant about the quality of the parts they send you.

I said "even reputable warehouses," because everyone knew there were enterprising groups of individuals who went around to the semiconductor manufacturers in Silicon Valley and collected their trash. Instead of taking it to a landfill, they'd go through it to find the discarded semiconductors. They'd sort out these discarded and obviously defective parts and run them through a bath to take all the markings off. Then they'd re-mark the chips with military-quality marking (which sold for the highest prices) and sell them to local warehouses. At the warehouses, these "bad and dead" chips, or unknown-quality chips, would get mixed in with good chips and could get shipped to unsuspecting customers. Because some of these chips had found their way into military hardware, the FBI had gotten involved and was trying to track down these crooks.

Back to Intel. There was no incoming inspection program, and "S spec" microprocessor chips had been dumped on the MDS development group. At least these chips were identified as defective by the "S spec" designation. This meant that a board designer could work around a known problem. But the real issue was what's called, in the computer industry, "infant mortality rate."

All semiconductors have an infant mortality rate, which means that a chip can burn out quickly when it first comes into use. The infant mortality rate can be identified by a "burn-in process" of applying a good operating voltage, heating the device in an oven, and letting it run for a month. When this is done, all the high-infant-mortality-rate chips will die off, and the rest will probably function for a hundred years. If you really wanted to get tough with the chips, as we'd done at Ford, you could cycle the power on and off about every hour, because most semiconductors fail when power is applied. (Another test involves cycling the temperature down to the required lower temp and quickly bringing it back up to the upper temp while you cycle the power simultaneously, but this was way too much to expect of a commercial company like Intel.)

I suggested that an incoming inspection be implemented, to catch chips with parameters that did not meet our specifications. I recommended an eight-hour heat test as part of an incoming inspection process. This test would catch about 85 percent of the infant-mortality chips.

To digress a bit, we had a number of floppy disk drives from Shugart, one of the biggest and best drive manufacturers in Silicon Valley. The drives either did not work or had high infant mortality. I met with management and suggested doing an incoming inspection process on the Shugart drives. But they declined. They didn't want to upset Shugart, because Intel had some kind of sweetheart deal with them. Probably for half-price units.

Electrostatic Discharge

Electrostatic discharge (ESD) was the other major area that caused failures in the MDS system. The fundamental circuit design worked, but zero attention was paid to the grounding, power arrangements, and ESD. When I was at TI in Dallas, the group next to the ASR 5 production line was building the first all-integrated-circuit computers. The system consisted of one-foot square boards, with about a hundred semiconductor chips per board. These boards were packed fairly densely in the prototype computer in three rows of fifty boards each. Semiconductors draw a little more current than just a transistor, so when the power was first turned on, the ground rose to the minimum voltage of the gate (2.1v), even though the designers had put in 1" x 2" copper bars for power and ground busses. All the boards had to be modified, and each had to have its own voltage-regulator circuit built on the board. Even then, when the circuits switched, large voltage spikes could occur on both the ground and power busses, causing other circuits to switch when they weren't supposed to. The designers had to put capacitors all over the boards to supply the needed voltage and current spike when some circuits switched.

The Intel MDS system had none of this. The power and ground busses were thin-etched traces running all over the boards. I recommended that the boards be redesigned so that each would have a power and ground plane, and so that there'd be capacitors on each circuit chip's power and ground pins.

A major problems was the 20,000-volt static discharge caused when someone simply walked across a carpet and touched the keyboard. The static discharge would kill the microprocessor and other chips in the keyboard, then would travel down the cable to the back plane and kill the CRT microprocessor and some more chips. Sometimes it would even kill chips on the motherboard before it finally found ground in the power supply.

The Failure of the Incremental Approach

When I presented my report and explained all the issues, management was stunned! We had some long discussions about the problems. Their first thought was that redesigning all the boards and putting in power and ground planes and capacitors would cost over $1 million. That was out. So they held meetings about how to make minor changes in each production run to reduce the failures.

I tried to explain that this "incremental approach" was useless, because the non-modified boards would exhibit the same failure rate; the system failure rate would remain the same 154 percent.

Well, they had my report, so assigned me to the board and system test area. I was to sit with the test engineers, monitor the testing, and conceive a way to improve throughput. The whole concept of the MDS was to be able to see inside a microprocessor chip and get a handle on what was going on in the hardware. The MDS system had an in-circuit emulator (ICE), which took the place of the microprocessor chip on a board and which made all of the internal microprocessor registers, RAM, and input and output gates available to the test engineer. If the test engineer was having problems with a specific register or gate, a short subroutine could be written to exercise just that part of the board and observe the output. This way the test engineer could troubleshoot any board problem.

All this flexibility caused more problems. The design engineers had created pre-set parameters that the boards were supposed to meet before they could pass the test. If too many boards were failing a specific test (mostly because of "S spec" parts), the pre-set parameters could be adjusted so more boards would pass the test. The parameters changes were supposed to be discussed with the design engineer before any changes were made. But this slowed production. And production seemed to be everyone's primary goal. So test-system management would let the changes be made, saying, "I'll discuss it with the design engineer later"—which, of course, they never did.

After sitting with the test engineers for over a month of production, I wrote another report. This one suggested that the QC group keep a set of standard test programs. Every Monday they should pick up all of the test programs in the testing area and issue new quality-certified test programs.

This didn't go over very well because it would have slowed production.

I was assigned to troubleshoot twenty-five CRT boards and follow them through board test until they were all fixed, to see how many times through the system it took to get good boards. The stated policy was "If a board does not pass the third time through the test, scrap it." I proceeded to troubleshoot the twenty-five boards. Twenty were repaired by the second pass, leaving five for their third pass.

Unfortunately, on the day of the third pass, I was off sick. But I'd left behind a huge sign saying, "These boards are bad! Don't touch them without first contacting Charles Willingham or QC management!" When I returned after my sick day, the box and the bad boards were nowhere to be found. After two days of searching, I learned from a test engineer that because it was the end of the month, the bad boards had been put into systems and shipped out to meet monthly quota requirements.

I couldn't believe it.

The next day, I happened to be sitting in a cube next to the board test manager. The system test manager had stopped by to talk to him. I eavesdropped. I heard them discussing the big bonuses they'd receive for making that month's production quota. They were actually getting paid *extra* to ship bad systems!

So much for quality.

A couple of months later, the vice president got called on the carpet by some major customers for the low quality of the MDS systems. Boy, did the stuff hit the fan! He had a meeting with MDS management and with the QC manager. This was the genesis of Intel's reliability improvement program (RIP). They brought out my thirty-two-page report and went through it item by item.

They decided to try the incremental approach first—the one I'd said was useless. They redesigned one of the boards a little and started a new production. Just as I'd predicted, the failure rate stayed the same. The next month they redesigned another board and tried again. Again the failure rate was the same.

The VP halted production for an entire month and ordered that each of my recommendations be implemented. All the boards and systems in production were scrapped.

My friend Ed Grady and I went to the scrap sale. I bought ten each of all the system components and hauled them over to Ed's garage. The boards were selling for $1 apiece. We stayed up nights for three months and made six of the systems completely operational. We cannibalized good parts from the worst boards to make the systems work. We set them up in his garage and let them run for several months to eliminate infant mortality. Then we soldered heavy-duty buss wire on the power and ground traces, and soldered as many capacitors on the back of the boards as we could find places for. Ed still has two of these systems in his garage, and they are still running.

After the complete redesign and after production restarted, the failure rate plummeted to 21 percent. Not bad, since it had been 154 percent when I'd joined the company. But now they demanded a 5 percent failure rate, so more tweaking was in order. They implemented the incoming inspection program that I'd recommended, and the failure rate dropped to 14 percent. We were getting closer. To get to a 5 percent failure rate would mean abandoning "S spec" parts and doing temperature and power cycling, and I didn't think they would ever implement those.

Education at Intel

While I was doing my regular job, I took a lot of classes at Intel. They offered their microprocessor design courses on the 8085/86, on the device control micro-processor 8741, and on all of the peripherals chips Intel was making. I took all of these. I also took assembly language programming for the 8085/86, C and C++ and Pascal high-level programming. I used a lot of what I learned in writing test programs for the board and system test quality control procedures. In the 8741 course, I designed, built, and programmed a single-chip microcomputer to create five letter code groups of International Morse Code for code practice. The code speed and spacing were variable, from one word per minute to one-hundred words per minute, and it had a volume control. I played with it for a couple of months, but I was too busy to really sit down and get my code speed up. I stored it in the garage, thinking I might want to get my code speed up sometime in the future.

Land of Opportunity

One day when I was sitting with the test engineers at Intel I had a conversation with Tom Nguyen. Tom, who'd been a professional civil engineer in Viet Nam, was working as a test technician because the U.S. did not recognize his engineering degree. We wound up going to lunch together quite often, and I found out he was one of the "boat people" rescued off the coast of Viet Nam. I could remember seeing photos of these survivors who'd been hauled up on an American war ship wearing nothing but rags and loincloths.

Tom explained that he and some other survivors had been brought to Mountain View, California, and put up in an apartment. There were fourteen people in his family. All the men immediately went to work, taking any job they could find, regardless of their education. Each man brought home his paycheck and gave it to "Papa San," the head of the family. Papa San bought all the living necessaries—and did so frugally. For example, he'd buy a hundred-pound bag of rice and some vegetables, and that was what they'd all eat for a month. Whatever money was left over, Papa San invested.

One day at lunch with Tom, I watched an old Asian lady pick milkweed from the vacant lot next door to Intel. I asked Tom what she was doing. He said the milkweed was one of the staple foods people ate in Viet Nam. She would take it home, wash it, mix it with rice, and make a meal.

If the "boat people" kids were old enough to attend school, their families demanded that they spend twelve hours a day either in class or studying. A lot of the ladies and girls worked also, and of course they, too, gave their money to Papa San.

Well, after about five years, Papa San's frugality paid off and he had enough money to buy a restaurant. The ladies and girls in his family worked in the restaurant, with their pay being only the left-over food that they got to eat for free. Because the workers were paid this way, the restaurant had no labor costs. The restaurant made a good profit, and after another five years, the family bought another restaurant, which was run by Papa San's number-one son.

Their life reminded me of my early childhood years, and I respected their frugality and their work ethic.

I suspect that if American's don't shape up and learn to stay out of debt and stop squandering our wealth, in a few years we'll all be working for these boat people and others like them!

Golden Dragon

But maybe not. Because some of the kids of these people from Asia have adopted our bad habits.

For example, there was a Chinese restaurant called the Golden Dragon, up on the corner of El Camino Real and San Antonio Avenue. They had a terrific "Happy Hour Buffet" on Fridays, so my friends and I would often meet there after work. As a result, I had several conversations with the owner and his wife.

They had a son who was born and raised in the states. Unfortunately for them, he'd became Americanized. He didn't want to work hard. He didn't want to be an entrepreneur. He didn't want to take over the family's restaurant business.

Instead, he was out playing around, goofing off, and burning the tires off of his American-made muscle car—just like the American kids.

Meanwhile, Back at Intel

The 4004 was named for the approximate number of old-fashioned transistors it replaced. The 8008 addressed 16 kilobytes (Kb) of memory; this was the chip used in the Traf-O-Data tape-reader built by Paul Gilbert. Its paper-tape reader could manage the 16-channel, 4-digit binary-coded decimal (BCD) tapes generated by traffic-monitoring recorders. The 8080, a faster 8-bit chip, could address 64 kilobytes (Kb) of memory.

Intel was selling 16-Kb memory chips like crazy and had been for almost two years. They had sampled 32-Kb memory chips and were tooling up to go into full production as soon as they got as much money as possible from the 16-Kb chip market. They'd also developed 64-Kb memory chips and were selectively sampling them out to large customers such as IBM, GE, and some aerospace manufacturers. But they were holding off on full production until they sucked all the money they could from the 32-Kb memory market.

Suddenly, the Japanese dumped 64-Kb memory chips on the U.S. market at two-thirds the price that Intel had planned to sell its chips for.

"Foul! Did you see that rabbit punch the Japanese gave us?" Intel cried to Congress. Intel demanded that Congress ban the import of 64-Kb memory chips, or at least put a huge tariff on the chips so they'd be more expensive than Intel's. Congress refused, and so everyone in America bought the cheaper, Japanese-made 64-Kb memory chips.

By the time Intel got into full production of 64-Kb memory, Japan had dropped its price to about one-half of Intel's and had started shipping 128-Kb memory chips. Intel had the technology for 128-Kb chips but was not yet ready for production. So Intel was caught with its pants down. By the time Intel started shipping 128-Kb chips, Japan had dropped its price on these and had begun shipping 256-Kb chips.

The cheap availability of high-memory Japanese chips was great for us hardware design engineers, but it was killing America's semiconductor companies.

Electron Microscope Micrograph

One day a couple of engineers from Intel's semiconductor unit came over to brief our part of the company about their new products. To show us the new level of deposition they'd achieved, they brought several electron-microscope micrographs.

The electron-microscope micrograph product showed a one-atom-thick layer deposited on the substrate. To my amazement, they looked like steel ball bearings all arranged together! I don't know what I'd expected atoms to look like, but this was not it.

From the descriptions I'd read about electrons circling around a nucleus, I guess I thought they were scaled-down version of the everyday world and would look like a small solar system. From Newton's masterpiece *Principia,* and from Copernicus, Kepler, and Galileo, I'd gotten the concept of a mechanical universe. Then, at the turn of the nineteenth century, the atom was thought to be a sphere with a spring sticking out. But that had all changed.

The use of the word "atom" is just an informal way of talking about a mathematical algorithm. It's a helpful means of encapsulating the abstract concept in physical language, but it doesn't explain the atom's own set of physical attributes, such as a definite location in space, and a definite velocity through space. The term "atom" itself is merely a code word for a mathematical model.

Anyway, seeing the atoms was an enlightening experience that challenged my own conceptions—or misconceptions—of what an atom looked like.

"Umphrese"

I sold my place in Richardson, Texas, and bought a house at 1018 Evelyn Avenue in Sunnyvale, California, on October 28, 1976.

I had to go back to Dallas to sign the sales contract for the Richardson house. The morning of the signing, I went to the real estate office on Preston Road and Allen in North Dallas. A new shopping center was being built there, and as I sat waiting for the real estate office to open, I watched the workmen put up this new fancy sign on the side of a new and expensive-looking dress shop. First they put up a U, then an M. I wondered what the new store's name would be. Then they put up a P—but by now, the real estate office was open, and I had to go inside and sign the papers. When I came out, the workmen had completed the sign on the new dress store: "Umphrese."

I recognized the name as that of my old acquaintance from years before in the T-N-T singles group at the church Joan and I had attended: Joe Umphrese Willman. So I went in and asked this very proper saleslady if I could please speak with the owner. "Just a minute," she said, "I'll get Mr. Umphrese." Sure enough, here came Joe Willman! So we went across the street to a coffee shop and caught up on old times. He and his sweetheart Pattie had been married for several years and were very happy together. And their business was booming.

Leslie Lee Ann Walsh

I'd been dating Leslie Lee Ann Walsh, daughter of Bill Walsh, the antenna man I'd worked with on the Iran team at Ford. She was working two jobs: at Frankie, Johnnie and Luigi Too, and at the 94th Aero Squadron. The *San Jose Mercury* ran a survey of the longest lines in the Bay Area. Star Wars was first and Frankie and Johnnie was second. Frankie and Johnnie was the local hangout and pick-up joint/pizza parlor for the eleven colleges and universities in the area. They only served wine with the pizza, so the crowd would bring their own bottles in a cooler and stand in line for over an hour to get a seat. One night a couple of graduate students brought their chimpanzee with them. He was their research project and they were teaching him to sign. His name was Zippy. Some people complained that monkeys were not allowed in restaurants. Leslie said, "He is a seeing-eye monkey."

He was much better behaved than a lot of the students.

94th Aero Squadron

Leslie was a bartender at a bar and restaurant called the 94th Aero Squadron. Located on the San Jose airport property, it was designed like a World War I aerodrome. They had old bi-planes parked outside, guarded by World War I machine guns in sandbagged foxholes. The entrances were all sandbagged, and they played World War I songs on the PA.

Leslie worked downstairs in the bar. She'd learned her bartending skills from George Andrew at the Quincy Bay Inn. More about George later. Margaritas were the bar's most popular drinks. When Leslie made her Margaritas, she made the ones for the ladies without any tequila. The women just loved her Margaritas because they could drink all night and not have a hangover the next day.

Then the business was sold, and new management took over. The first thing they did was raise the prices and put in automatic drink machines. The ladies didn't like the new Margaritas, so they stopped coming. Then the men stopped coming because there were no women.

The place went out of business within two years.

Leslie and I were married in Los Altos Methodist Church on December 18, 1975. We went to the Jumping Frog Disco for our reception, where we danced to "Evergreen" by Barbara Streisand, which was our song. Scott Robinson was my best man, and Kathy Robinson was Leslie's maid of honor. We flew to Boston for our honeymoon and visited her parents there at Christmas. We went to Christmas Midnight Mass at the family's Catholic Church. I noticed that for their communion, they served the best wine! It was nothing at all like that old Welch's grape juice the Baptists served.

After we returned home to California, one night we sat down at the dining-room table, and I asked Leslie what she wanted to do with her life. She hadn't thought much about it. (I now know why. She just wanted to stay home and raise babies, like her mother had.) Anyway, I told her that schools in California were only charging $3 per semester hour, and I suggested she take a few classes to see if she liked them. If she didn't like a class after giving it a chance, I said she could just drop it. At that time the newspapers were all packed full of job ads for programmers, so I suggested she take a beginning computer programming course.

She enrolled in a basic programming class, a beginning law class, and ground school for a private pilot's license.

She was a whiz at basic programming—but she hated it with a passion. However, she loved the class in beginning law. She said it had a lot of logic to it, like the programming course, but was in English and not hieroglyphics.

Meanwhile, the thought of working a job where she'd be stuck behind a desk just scared the hell out of her. She didn't mind hard work—after all, she'd helped her dad erect hundred-ton antennas outdoors and she'd enjoyed that—but she cringed at sitting behind a desk. So she loved the idea of getting a private pilot's license.

We had several discussions as the year progressed. Programming was out of the question because she hated it so much, so that left piloting and law. Women had broken into the "all-male-commercial-pilot" stronghold, and they were now being promoted to captains and first officers at the major airlines. So becoming a commercial pilot was, by now, a very realistic long-range goal. As for the possibility of her getting a law degree, I said it would be great for her to have one, no matter what field she went into. But if the law degree wasn't from a prestigious school such as Harvard, Stanford, or Yale, then the country's major law firms would probably be out of her reach. Still, she could make a good living chasing ambulances, or doing divorces.

She decided to work on her commercial pilot's license.

She invited her ground-school instructor, Betty Hicks, over for dinner one evening. That evening, Mrs. Hicks mentioned that she was a consultant to a golf ball manufacturer and that she often flew from one golf tournament to another. I knew nothing about golf at the time, much less women's golf. But two years later, when I saw a PBS special on Babe Zaharias, I learned that Babe Zaharias and Betty Hicks were the number one and number two players for over seven years in women's professional golf.

I'd had a star athlete right there as a guest in my home, and I hadn't even known it.

Leslie started working hard on getting her commercial pilot's license. On Sundays we'd fly from the San Jose airport to the Nut Tree airport, halfway between Oakland and Sacramento. It was a favorite place for fly-ins—gatherings of people who flew small planes. We had fun flying all around. For example, one time Leslie flew us up to Tahoe, so she could get some high altitude, air density experience.

Every birthday, anniversary, and every time Leslie passed a pilot's test we would celebrate with an intimate dinner for two at The Carnelian Room. The Carnelian room is located on the top, 52nd, floor of the Bank of America Building providing a fabulous panoramic view of the city skyline and the Bay. It is literally the highest restaurant in San Francisco. The romantic Tamalpais Room has an elegant European style décor, reminiscent of an English manor. The interior is accented by warm tones, enhanced by antiques and flowers at the tables, and the lighting is via chandelier. Leslie was always impressed by the fact that her menu did not have any prices in it but mine did.

Adventures in the Air

Leslie's grandfather, Adam Walsh, had been a member of Notre Dame's famous "Four Horsemen and Seven Mules" football team and had played under the legendary coach Knute Rockne in the 1930s. Adam was a large man who'd been the center on the team. By the time Leslie and I were married, he was living down in LA in a retirement community, so we went to see him, and he flew up to visit us. After his visit, when Leslie flew him down to LA, I went with them.

When we started to make our return flight back home, we found that one of the magnetos (a type of alternator) did not work. Leslie's flight training had taught her not to fly with only one magneto, so she called the flight school where she'd rented the aircraft, and they said they'd be down to fix it during the week. Because I had to be back at work Monday morning, I took a taxi to LAX and caught a commuter flight to the San Francisco airport, SFO. The flight school came down and fixed the airplane as promised, and Leslie flew back home late on Monday.

Sky Diving

One of the reasons Leslie was comfortable with flying an airplane was that she'd had a lot of experience doing sky diving.

When I met Leslie, she'd already racked up about twenty jumps from when she'd lived back in New England. We hadn't been dating long when she decided to see what sky diving was like in California. (Sky diving!? What kind of woman was I hooking up with?)

She asked me to go with her.

(Jump out of an airplane? Me? Was she nuts?)

Wanting to maintain my reputation as a "macho man," I couldn't turn down her invitation. When my friends heard about it, they began betting each other that I wouldn't actually do it. So I asked them to come along and verify that I, "Macho Chuck," had actually jumped out of an airplane. Ed Grady (an engineer I worked with) and a couple of others who thought they were tough guys came with us, so we could see who had the biggest cohones.

We all had to get up early on Saturday to drive out to Antioch, east of Oakland. The ground school started at 8:00 AM and lasted for eight hours.

They taught us all the dynamics of falling through the air: how fast you fall, how long it takes to get to the ground without a parachute (twelve seconds from 12,000 ft), and the instability of the human body as it falls. This was in just the first couple of hours, and if it didn't scare the hell out of you, the rest of the course was a piece of cake.

They marched us outside and showed us how to arch our backs and heads, stick our stomachs out, and bend our knees. They told us to hold our arms out to stabilize the unruly aerodynamics of the human body, in hopes of getting some control of our fall—before it killed us. Then, back into the classroom for more education: the procedure of getting in the airplane and out of the airplane, and what to do during the jump.

Next we were herded back outside to practice a posture called the "frog position." They introduced the parachute landing fall (PLF), and explained how to keep your legs together, bend your knees so the leg muscles would act like a spring and absorb the energy of the fall, and put your arms up over your face to protect your head. (Protect your head? Yikes!) They said that when you first touched the ground, you should turn slightly to the left or right so you wouldn't slam your face into the ground. (Slam your face?) They said to roll up in a ball on impact and roll over either shoulder—this would further absorb the energy of the fall, so you could complete the landing without breaking any bones. (Breaking any bones! What had I gotten myself into?)

They let us practice all this new information and training by jumping off a four-foot-high platform, and by doing forward and backward falls for about an hour. Then it was time for lunch. At lunch, we sat outside. We watched the other sky divers preparing to go up, and we observed the jumpmasters checking them all out before they boarded the plane. Most of the experienced jumpers had a ritual for anyone jumping with their first square parachute. They'd all line up along the runway and "moon" the airplane as it took off!

After lunch, the training focused on emergency procedures. (Emergency procedures?) We were told what to look for and how to correct problems, such as the chute not deploying, the shroud lines twisting, or a shroud line coming over the canape. They also taught us how to cut away the main parachute and deploy the emergency chute. Then it was outside yet again for an hour of practice of all the things we'd learned, and a lot more practice PLFs.

And we weren't done yet. Back inside for still more emergency procedures. They hooked each of us up in a harness and hoisted us off the floor. Hanging there in the harness, you had to go through all the procedures, including cutting away and dropping out of the main chute. Then back outside for another hour of PLFs.

During all of this training, I never let on that I was the slightest bit worried. "Macho Chuck" strutted around like a bantam rooster. I'd even managed to gain confidence and to convince myself there was nothing to worry about. After all, these people were experts. They must have trained hundreds, even thousands of jumpers. They did this every weekend. Nothing to worry about. It would be a piece of cake.

By 4:30 PM, ground school was over.

It was time to jump.

I was still full of confidence as the first group went up. When it came my time to go, getting into the airplane was easy enough, but as it started down the runway, I knew there was no turning back—I'd have to jump.

I prayed all the way up.

It seemed like it took forever to get to 12,000 feet. I had plenty of time to convince myself that I'd lived a pretty nice life. It had been a good forty years. I'd had some great friends, experienced some fun times, worked some interesting jobs. I was satisfied. I was ready to meet my maker. I told myself that God would forgive me for committing suicide in this way, since I hadn't really intended to take my own life.

We made the first pass. The first jumper made it out the door. Then another. Next, it was my turn.

I moved gingerly forward. I eased myself into position, sitting in the door. I let my feet dangle out in mid-air, but I didn't dare look down.

The jumpmaster checked out my chute and my harness. He connected the static line to the aircraft. We approached the jump zone, and I still was not fully prepared. A million ideas were racing through my mind. I thought maybe he should check my chute again. I thought, is my harness really on all right? I thought, what if God won't forgive me for committing suicide this way? I thought—

"*Go!*"

The jumpmaster had slapped me on the back and yelled, "Go."

It scared me so much, I jumped out of the plane and screamed.

Before I knew it, I was looking up at the plane flying away. It was a moment of sheer terror. I forgot everything from the training. In the frog position, you're supposed to be looking down, not up. But I could only stare frantically up at the plane as it soared away. My heart was racing. My whole body was trembling. There was an enormous lump in my throat. I felt like I was going to cry or throw up—or do both at the same time.

After a fraction of a second, the chute opened. The opening shocked me out of my terror. I was overcome with gratitude, because—glory be to God and Jesus!—my parachute had worked. I mentally started to review the procedures: I checked to see if the shroud lines were twisted, or if there was a line over. But there were no problems; I was in a perfect chute. I began the turning procedure, first to the right, then to the left. Everything worked perfectly.

By now, I could hear people on the ground hollering. I looked around and saw a beautiful setting sun. It was one of the most gorgeous sights I'd ever seen—and I thanked the Lord for letting me live to see it. I prepared for my PLF, legs together, knees bent, arms over the face.

Bang! I was on the ground.

The PLF went smoothly. Just as we'd practiced.

It was *great*. I was alive. I was exhilarated.

I couldn't wait to do it again!

Second Jump

I must be a masochist, because the next weekend I was back at that sky-diving school in Antioch, ready for another jump.

My second jump was the most exciting thing I have ever done. Not all students in the class had gotten the chance to jump that first Saturday, so we all had to come back the next weekend and finish jumping, while a new class was going through ground school.

Having survived my first jump, I was no longer just "Macho Chuck." Now I was "Macho, Macho, Macho Chuck!"

I went through the procedures with my new jumpmaster, and passed inspection with flying colors. Then we went up in the plane.

This time, because we were in a smaller aircraft, I didn't have to sit in the door before the jump. Instead, I was instructed to step outside on the aircraft's landing gear and hold onto the wing brace. This left me with one foot hanging in the open air, and it put the full force of the wind in my face. The first time I'd been scared I was about to die—but now I had a really magnificent feeling of freedom. When the jumpmaster said, "*Go!*" I released my grip on the wing brace.

Suddenly I was in the frog position, falling free. It was fantastic!

My chute was opened automatically and perfectly. On this jump, because I wasn't frozen in terror, I had plenty of time to look around and turn the chute completely in a circle in both directions. I had a stunning, panoramic view of the gorgeous rural landscape of northern California. I loved it all—the view, the feeling of freedom, the adrenaline rush! Then I prepared for my landing, and it went flawlessly.

In the meantime, Leslie was taking a couple of sky dives herself. She was doing more advanced jumps: five-second freefalls. My friends and I had all finished our dives, so we sat in the shade of the building, watching her. Her first jump went exactly as planned—nothing special, just a routine freefall until her chute opened. It was almost boring to watch.

Her second jump was very different.

We sat and gazed up at her, there in the sky, as she plummeted down. And down. And down. I kept waiting for her chute to open.

It didn't open.

I watched as her form fell through the air, faster and faster, until she went behind the little hill next to the airport.

No chute opened.

I sat there, frozen.

Meanwhile, the paramedic on staff and a couple of jumpmasters charged out to their Jeep and climbed in. They raced toward that hill with their sirens wailing. My friends hustled me into our car, and we sped after them.

Everyone feared the worst. I was shaking like a leaf as our car rushed to the scene behind the paramedic's Jeep. It was awful. I couldn't even bear to think about it.

Then, just as the Jeep up ahead crested the hill, there was Leslie! She came walking casually up over the hill with her parachute in her arms. Everyone was yelling and screaming. They were all so happy to see her—especially me. She waved to us, then climbed into the Jeep and rode back to jump school.

As soon as they arrived, the jumpmasters marched her into the office and chewed her out royally. They also grounded her.

On the way back home that evening, my friends and I asked her what had happened.

"Nothing really," she chirped. "I was just having a great time!"

"A great time?" I said, raising one eyebrow.

"Yes! I was doing some turning maneuvers and would count, 'One-thousand-one,' and say to myself, 'Turn right, Leslie,' and 'Oh, isn't that a beautiful sight?'"

"You were watching the scenery?"

"Right," she continued. "So then I'd count, 'One-thousand-two,' and think, 'Turn left, Leslie. Oh, isn't that just gorgeous down there!? One-thousand-three. What a wonderful, beautiful day!'"

It turned out that her thoughts between counting meant that she took eight seconds to count to five, for her five-second freefall. When she noticed she was getting really close to the ground, she pulled the cord and her chute opened just in time to break her fall.

Green Shit Box

Leslie's best friend was named Kathy. One day Kathy asked me for financial advice. She'd saved $5,000 and had the opportunity to buy some land in Redmond, California, with her father putting up an additional $5,000.

Or she could use the $5,000 to buy a fancy green Datsun 280Z sports car.

I explained that whatever she did, she should *not* buy a car as an investment. A car always depreciates, at least for the first fifty years, after which it might, I said *might*, go up in value—but only if some antique collector wants it, and only if you've stored it on blocks in a temperature-controlled garage for fifty years.

I also told her I normally would not recommend buying land as an investment. Land just costs you more money in taxes, maintenance, and improvements, unless you're going to farm it. However, since her dad was putting up half of the cost of the land, she'd double her investment right away, so it wouldn't be such a bad idea. Even if land values decreased a little, she could still sell it at a profit.

I also told her to consider buying gold. It had gone from $35 per ounce to $65 per ounce in the past year. I thought it would continue to increase in price for a while longer.

Well, she bought the green 280Z.

After a couple of months, she decided to drive to Tahoe for the weekend. Halfway up the mountain, the car stopped, and she had to have it towed. It needed a $1,500 engine overhaul to get it running again.

Then, about six months later, the engine died a second time. She had to spend another $2,000 for a new engine. That, plus some other work she'd had done on it, meant that her investment had cost her $5,000. So she was now up to $10,000 for her "green shit box," as she called it.

In the meantime, gold had gone from $65 to over $800 per ounce.

All tolled, her "green shit box" had cost her roughly $70,800. And of course, I never let her forget it!

Gilroy and Watsonville

Gilroy, California, was the onion and garlic capital of the world. It was just south of San Jose on Highway 101. Every day I drove through Gilroy, on the way to and from work in Palo Alto. It was a pleasant twenty minutes, if you loved the smell of onions and garlic. When I'd first arrived in California in 1968, a ten-story IBM building stood on the south edge of San Jose. From there to Gilroy, approximately twenty miles away, there were only orchards on both sides of Highway 101, and a lot of roadside fruit stands. You could buy fresh apricots, oranges, lemons, cherries, and peaches just picked from the orchards. Boy, were they delicious! In Gilroy, you could pick up bunches of green onions and bags of fresh garlic.

The last time I drove down Highway 101, in 2002, I saw house after house after house. All the orchards had been bulldozed. A lone garlic store stood for-lornly beside the road in Gilroy. And going south of Salinas, all the farms on the way to Paso Robles were now wineries and vineyards.

Watsonville, California, the artichoke capital of the world, was just north of coastal Highway 1, south of Santa Cruz. There was not much of a smell from the artichokes, but just knowing they were there made my mouth water.

Another Adventure in the Air—A Balloon Flight

For my 39th birthday, Leslie gave me, Paul Ketchum, and her best friend Kathy a balloon ride in San Jose.

Most balloon rides are early in the morning or late in the afternoon, when the winds are low. That way, the pilots can better control their balloons. Our ride was in the morning, and it was very quiet.

Balloon rides are the exact opposite of sky diving. There's none of the excitement or terror or exhilaration—it's just very quiet and peaceful. You can hear voices on the ground a thousand feet below.

We flew up and down the valley south of San Jose over Gilroy. The winds blow in different directions at different altitudes, and that's how the pilot flies the balloon, by changing altitude until he finds the winds blowing in the direction he wants to go. We flew south down the valley, then changed altitude and flew back north. So we weren't far from our launch site when we landed.

Afterward the flight, we made a champagne toast to the gods for letting us have a safe trip. We each had to get down on our knees with our hands behind our backs and drink the champagne from a paper cup we picked up with our teeth while the balloon captain poured champagne over our heads. It was great fun!

Disneyland

For the next birthday, my 40th, Leslie gave me a four-day, three-night visit to Disneyland. We flew down on Friday, took the Disneyland limo to the Disneyland Hotel, and checked in. We rode the monorail over to the park. I got to ride all the rides twice that day. However, on Saturday and Sunday the place was crawling with too damn many urchins, rug rats, curtain climbers, and cookie crunchers. I could only get in about half of the rides. But Monday, I was able to get in all the rides again before we had to come home.

As another gift, Leslie bought me a gold pocket watch. It had a music box in it and a little animated scene. She'd paid a bundle for it in a shop in Ghirardelli Square in San Francisco. Later, when we had a robbery, they stole the gold pocket watch and some coins Leslie was saving.

Symphony and Ballet

Leslie and I went to the Ticketron to get tickets to the San Francisco Symphony. While we were standing in line, we noticed that the Blues Brothers were having a concert in the Oakland outdoor theatre so we purchased tickets to it also. We went to the San Francisco Symphony and enjoyed the concert very much but were disappointed by the audience. They kept milling around and talking all during the concert. They had no class at all.

The next week we went to the Blues Brothers concert in Oakland. We expected the crowd to be a little rowdy, and they were, but overall they behaved a lot better than the SF Symphony patrons. Probably because there were over a hundred huge, uniformed policemen standing in the aisles and all round the area. There was a lot of pot smoking, which the police ignored. I didn't inhale; I held my breath for an hour and a half.

One couple close to us had brought a large shopping bag of homemade Alice B. Toklas brownies, which they shared. The brownies were great, and I couldn't get enough of them.

Next we went to the Oakland Ballet to see Rudolf Nureyev. First he performed classical ballet. I couldn't believe how high he could soar in a grand *tours en l'air*. It appeared he was hanging suspended in mid-air. After the classical ballet he performed several modern dance routines, which was something he'd always wanted to do. During one of the routines his male partner was supposed to support him during a cartwheel but slipped and dropped him on his head. Boy, was he pissed. Besides that one mishap it was a memorial performance.

We also went to a performance with Mikhail Baryshnikov and he was *great* also.

Motorcycle

Meanwhile, Leslie had always had a motorcycle since she was sixteen. One of the first things I did after we were married was buy her a motorcycle. It was the year that Honda came out with their CX-500, which was water-cooled and shaft driven. Also, it was low enough that Leslie could stand up and straddle it to hold it up. It was really a nice bike.

Voyager 1 and Voyager 2

The closest approach to Jupiter occurred on March 5, 1979, for *Voyager 1*, and on July 9, 1979, for *Voyager 2*. *Voyager 1* used the gravity of Saturn to slingshot up and out of the plane of the ecliptic. Then *Voyager 2* went on to explore Uranus and Neptune, along with dozens of their moons. Their extended Interstellar Mission was intended to explore the outermost edge of the sun's domain, in an area called the heliosheath, the zone where the sun's influence wanes, and to beam back new information about the final frontier of our solar system.

Decision to Move

Leslie's astral thread was anchored in New England. She wanted to return home, and she kept saying how boring the weather was in California. She missed the four seasons in New England and the great skiing on Mount Washington in the White Mountains. She also missed the beautiful fall colors along the Kancamagus Highway.

It began to sound appealing. I would be near MIT and could visit it as often as I liked. Besides, all of her family was there. Nothing would do except that we move back there. I knew that if I left California, I'd never be able to return because everything was increasing in cost at an unbelievable rate.

We sold the house on 1054 Evelyn and moved in with a friend of Leslie's on 2273 Zoria Circle in San Jose. This was before the 94th Aero Squadron went out of business, and the friend was a part-time bartender there. He had a degree in computer science and worked full-time for the post office, delivering mail door-to-door. He said the work was easy and took only three or four hours. So he'd spend the rest of the day at the pool hall, shooting pool or bowling. He said the job was pretty secure—unlike employment in the Silicon Valley rat race of the time. Also, it was stress-free, and the benefits were great. His salary was only a little below what I was making. But considering the amount of time he put in, he was actually making twice what I was earning, plus he had a guaranteed pension.

$10,000

I'd never seen $10,000 in my life and wanted to know how it felt to handle that much money.

With the sale of our house we made $40,000 profit, and I put it all in the bank. But I told the teller I wanted $10,000 cash, and I wanted it in hundred-dollar bills with sequential serial numbers from 00 to 99.

The teller's jaw nearly hit the counter.

Her face turned white, and then she blushed red. She said she'd have to speak with the manager. She disappeared into the back and came out a little later. To get that much money with sequential serial numbers, they'd have to order it from the Federal Reserve Bank. And it would take six months. I didn't have six months, as we were preparing to move to New England, so I said I'd take it in any kind of hundred-dollar bills.

She said that would take at least two weeks, as they didn't keep that much extra cash at that bank.

So I put in my order in for $10,000 cash and said I'd be back in two weeks to pick it up.

Thirteen days later, I found a great deal on a couple of floppy drives I needed for my computer system. I phoned Leslie and asked her to go to the bank and get $1,000 in cash and bring it to work so I could buy the floppy drives. She went to the bank, cashed a check, and asked the teller for hundred-dollar bills. The teller said, "Oh, no. That's not possible. A man is coming in tomorrow to get all our hundred-dollar bills."

Leslie laughed and said, "I know. He's my husband."

So I bought the floppies, and the next day at noon, I went to the bank. The teller took me into a back room and carefully counted out the $10,000. She said, "That is the most money I have ever seen."

I was surprised. I thought that, working in a bank, they'd come across that much every day.

She had me sign a couple of forms. One was an IRS form. Banks are required to notify the IRS when anyone withdraws more than $9,999.

I put the money in my brief case and took it home.

At home, I took it into the bedroom and spread it all out on the bed. Then I jumped on top of it and wiggled and rolled and wallowed around for thirty minutes. I just wanted to see how it felt to be in that much money.

I played with it for a while, and let Leslie play with it, but she was too scared and embarrassed to get on the bed with it.

Finally I gathered it all up, put it back in its binder, and put it in the brief case. Next day, I took it back to the bank and deposited it.

"I have met the enemy, and I am he."—quote by Charles Willingham.

Gathering of the Clan

Leslie's sister, Taylor, had moved to Hawaii a couple of years earlier but now she, too, was ready to come home. She didn't have enough money to get herself and her belongings back to Massachusetts, so we decided to help her.

She could ship her stuff to San Francisco, and we'd load it onto a U-Haul with our stuff and take it across the country. She could help us drive the U-Haul. She flew into SFO and we picked her up, but it was another week before her stuff arrived.

Soon we'd be in New England.

Chapter Eleven:

The New England Years (1980 to 1985)

Driving East

Paul Ketchum did the best packing job I've ever seen. We loaded up the big U-Haul truck and headed east with the three of us in the cab. We arranged a system where the person sitting next to the passenger-side door was in charge of the cooler. If anyone wanted a cold drink, the "cooler meister" would pass it over. If the driver wanted a sandwich, the person in the middle would hold the bread while the "cooler meister" layered on the meat, cheese, and dressing, then passed it to the driver.

We drove south, through Bakersfield, Barstow, and Needles, to Kingman, Arizona. Next day, we passed through Flagstaff and Gallup, New Mexico, spending the night in Albuquerque. Early the next morning, we took off for Amarillo and Wichita Falls, then headed down to Fort Worth where we picked up Highway 183 north and turned toward Dallas. I wanted to go past Amon Carter Field and show my wife the old B-36 parked out there. But when we got there, it was gone, as was Amon Carter Terminal. Meanwhile the old U-Haul truck was spitting, coughing, and backfiring, so I nursed it along to Mom's house. Once there, I dropped off Leslie and Taylor and sputtered over to the U-Haul shop on Skillman Avenue just before they closed at 5:00 PM.

I insisted that they fix the truck immediately, telling the U-Haul guys I was not about to unpack and then reload all our stuff into another truck. They took a quick look and discovered the points were completely burned up. It took only fifteen minutes to replace the points and the condenser. The truck now ran like new. This was November 12, 1980.

Kenneth, my nephew, lived in the duplex next door to my mom's house, and he came by and said he wanted to ride with us to Massachusetts. We said he was welcome to come along. Now we had four people in the cab. It was a little crowded, but the cab was big, so it wasn't too bad. There was still plenty of room for our cooler-meister system and assembly-line sandwich making. The only change was that the person sitting next to the driver now had to do the gear shifting.

We rotated drivers every three or four hours and took potty breaks and stretched our legs. We made it to Forest City, Arkansas, the first night, and to Louisville, Kentucky, the following day. On the third day, we reached Columbus, Ohio, and I contacted Bill Allen, who was now divorced and was still teaching English there at Ohio State. We rented a Motel 6 room for two nights and spent Thanksgiving with Bill and his girlfriend.

It snowed most of Thanksgiving Day. Being from Texas and having lived for years in California, I was unused to that much snow. Nevertheless, we took off early the next day and followed the snowstorm east to Buffalo, New York. There we picked up a little "lake effect" snow, but we continued along Highway 90 to Syracuse and Albany. We crossed the state line into Massachusetts, passed through Springfield and Worchester, and arrived in Boston at around 8:00 PM. We made straight for Braintree, where we rented a room for the night. Then we went to Scituate, and Leslie and Taylor's parents' house. Bill Walsh had already rented a storage room for us, and he and Leslie's brother helped unload all our stuff and put it in storage before returning the truck to U-Haul. The next day we took Kenneth to the bus station for his trip back to Dallas.

Meanwhile, on December 6, 1980, a new American communications satellite, Intelsat 5, was launched into orbit around the earth. It was able to relay 12,000 telephone calls and two color-television channels at any one time. I had a special interest in this endeavor as it was the project I'd helped bid on while I was working at Ford Aerospace. This was the program that had prompted Ford to change the name of our facility, putting the Ford name up front, and which had caused the company to create a special video promoting Ford Motor Company.

Now that we were in Massachusetts, I started to look for a job. The only transportation we had was a Honda CX-500 motorcycle, so on December 16, 1980, I bought a cheap little new 1980 Chevette from Best Chevrolet, 128 Derby Street, Hingham, Massachusetts.

Advant Corporation

I went to the offices of Advant Corporation at 440 Amherst Street in Nashua, New Hampshire (Nashua was more or less a suburb of Boston). Robert Holzner, the vice president of marketing, interviewed me. He was impressed by my resume, and the interview went well, so he offered me a job as a sales rep at that first meeting.

Advant had a line of in-circuit emulator devices, or ICEs. I'd acquired five Intel MDS and ICE units when I was starting my own business in California. When Bob learned I had these, he wanted me to bring one into the office and train him on it. Now I knew he didn't really want to be trained—what he wanted was to have the Intel MDS and ICE systems in the office so he could compare the two for his customers, demonstrating the cost saving of Advant's ICE. The Intel MDS started at $16,000, and when you added the ICE component, the cost soared to $21,000. The comparable Advant ICE, at $4,995, was quite a bargain.

Most of what we did as sales reps was to call customers who'd filled out and mailed in "bingo" cards from electronics publications requesting information about our products. Once we'd received a card, we'd immediately send out literature and make a follow-up phone call to ensure the prospect had the literature. After we talked with customers and found out what their interests were, we tried to set up demonstrations. A lot of our prospective customers were engineers. I always enjoyed going into an engineer's laboratory, plugging in our Advant ICE unit, and writing a small program to troubleshoot his system. Several times I found problems in an engineer's system that he'd been troubleshooting for days and had been unable to solve. Whenever I did this, it almost always resulted in a sale.

Advant also sold an MS 4100/MS 4110, and an MS 2800 system. The MS 4100 was a large-capacity data-storage subsystem designed for use with the full line of Intel development systems, such as the Intellec 800, Series II or Series III. The Advant controller had been designed to emulate many of the Intel disk subsystems.

The Royal Crest Apartments, and the Art of the Dance

As soon as I got the job at Advant, we rented a place at Royal Crest Apartments on Spitbrook Lane in Nashua. Spitbrook Lane was the first street in New Hampshire on Route 3 north, and Royal Crest Apartments were right across the street from Sterling Electronics.

It was a storybook-type location. The apartments consisted of a large group of four-plexes designed in the gingerbread Black Forest architectural style. They were built on winding roads over forest-covered hills and were near a pond and a stream. The small pond would freeze over in the winter, and ice-skating was permitted when the ice was thick enough. Nearby was a Sheraton Hotel built in the form of a castle.

Dances

There were some dance clubs not far from our apartment on the north side of Nashua. One large complex had several clubs connected together, including a country-and-western club and a place called MG's with an old MG car over the bar. We liked to go to MG's because it played disco dance music.

Speaking of dancing, Leslie loved attending ballet performances. She cherished a painting she had of a pair of worn-out ballet slippers. (To me they just looked like a couple of old shoes!) Growing up in Massachusetts, Leslie had always loved the Boston Ballet's Christmas presentation of *The Nutcracker*. So when the Christmas season arrived, it was mandatory that we see the performance not just that December, but every single December after that.

New York Trade Show

Twice a year, some folks from Advant and I had to attend an electronics trade show on the south corner of Central Park in New York. Here we demonstrated our ICE system. But this was all we were allowed to do. The powerful New York unions blocked us from doing almost anything else. You couldn't carry your own system into the convention center. You couldn't set up your own tables. And heaven forbid you'd infringe on the electrician union's territory by daring to stick an electric plug into a power strip. Someone from each of the different unions had to do every little thing. And it took forever even to get the jobs started, let alone completed. Sometimes it seemed like by the time the union guys arrived, it was time to start breaking down the systems to go home.

Anyway, during the shows, a large corporation would often buy out an entire club for a night and invite customers out for the evening. One time, Bob managed to get tickets to the Copacabana. I really enjoyed dancing the samba to the tune *At The Copa*. It was something I'd always wanted to do since learning to dance.

Flight to Toronto

One day Bob asked me to go to Toronto, Canada, to attend a sales meeting and distribute some new brochures. Leslie needed some cross-country flight time, so I asked if I could have her fly me up there. He thought it was a great idea, so we arranged to rent a plane and took off from Hanscom Field. Shortly after we crossed into New York State, we ran into a storm front heading straight toward us.

I said to Leslie, "It's not worth it! This sales meeting isn't important enough to risk flying a light plane in bad weather. Find the nearest airport, and let's just land."

We tied the airplane down at a small airport in upstate New York and went into the coffee shop to have lunch while the storm passed. After two hours of rain, the skies cleared, and the weather briefing predicted good flying all the way to Toronto. We took off again, and a little later we flew over Niagara Falls and into Canada.

We landed at Toronto International Airport, taxied to the quarantine ramp, and waited for the inspector so we could clear customs. After customs, we secured the aircraft and took a taxi to the hotel where the sales meeting was being held. By now the meeting was over, but I found the sales manager and gave him the paperwork. We had dinner with him and several salesmen.

After dinner, we went sightseeing. Toronto seemed totally British to me. Most of the men were wearing three-piece suits and bowler hats and were carrying umbrellas. The ladies were all impeccably dressed. The town itself was spic-and-span. There was no graffiti anywhere, and no trash on the streets. We went down into the subway: it was squeaky clean, was well lit, and had all kinds of shops and boutiques. It was just like being in a mall.

The next day, we flew to Montreal. The difference between the cities was like night and day. Montreal was typically French. It had beautiful churches and buildings that reminded me of France. The people always spoke to you in French first, and sometimes, but not always, would switch to English if you spoke English. The style of dress was casual, with big sweaters and berets. We had a delicious dinner with a wine that was wonderful. We liked it so much, we bought a bottle of the same wine to take home with us. The next day, we flew to Burlington, Vermont, checked through U.S. customs, and returned to Hanscom Field.

IPO

We went merrily along like this for a year. My marriage was strong, and I enjoyed my job. Bob let me file my own travel expense reports, but he never filed his. The company was in the process of issuing an initial public offering (IPO) and going public on the over-the-counter stock market. After the company was finally listed on the stock market, Bob took a couple of weeks and did nothing but fill out three years' worth of travel expense reports, which he submitted for reimbursement. Those expenses sucked a lot of new investment capital out of the company.

The Ninety-Nines

Meanwhile, Leslie completed the requirements for her commercial pilot's license and her instrument license. Then she received her certified flight instructor's (CFI) license, and she started to work for Daniel Webster College at the New England Aeronautical Institute.

I encouraged her to join a group called The Ninety-Nines. It was the world's largest and oldest organization of licensed women pilots, and had been founded in 1929 by ninety-nine female pilots. She did join, and we attended several meetings. She became very comfortable with the group. She was surprised that there were so many lady pilots, but she felt secure because she had more certifications than most, although she had fewer flying hours.

That would soon change.

My Times among the Mafia

During this period I met George Andrews and his girlfriend, Margie. George was a bartender at the Quincy Bay Inn, where Leslie had worked as a cocktail waitress before going to California with her father. George had taught Leslie how to bartend, which meant Leslie would always be able to find a job wherever she went.

Leslie said the Quincy Bay Inn was the local Mafia local hang out. Half of the guys there were Mafia, she said, and the other half were vice cops. Leslie had a lot of stories about the Quincy Bay Inn. A typical cocktail waitress's trick is to tap a guy on the back while passing behind him carrying a tray of drinks through the crowd; this would let him know she was there so he wouldn't back up and spill her drinks. Leslie said every time she tapped someone on the back at the Quincy Bay Inn, she could feel his gun. She said the guy would just readjust its position and never miss a word in a conversation.

According to Leslie, the Mafia guys were real gentlemen. They were much better tippers than the cops, and they'd always walk her and the other ladies to their cars at night. Leslie's friend George always talked about being friends with the local Mafia capos and said he was one of the few people who could joke with the local Don.

One of the big rackets around New England involved stolen credit cards, and one night, without knowing it, we spent a night on the town on someone else's American Express card. The Mafia men would get someone's credit card, through pickpocketing or otherwise, and would sell it for a hundred quick bucks. Then the person who'd bought it could have a spending spree. But only for a day.

One time George invited us out for dinner and drinks. Later, when he went to pay the bill, we found out he had a stolen credit card. He told us if anyone from the restaurant ever asked us, we were to say his name was Luke something—the name on the card.

Our friend George ran a catering business and had a bar and restaurant in Mansfield called the Road House. For several years, George had dated his girlfriend Margie, but George was married and couldn't get a divorce. He said this was "for legal reasons." Because he knew so many Mafia guys, I suspect these "legal reasons" may have had to do with the fact that a wife cannot testify against her husband. Eventually, however, George did get divorced, and he and Margie got married.

George asked me to be his best man, and Margie asked Leslie to be her maid of honor. He said we were the only "legitimate" persons he knew! Leslie and I decided to fly them to Nantucket to get married, and Leslie rented a twin-engine Beechcraft from the flight school where she taught. After the wedding, George had a sit-down dinner for about three hundred guests at the local union hall as his wedding reception. His catering business did the food service.

The receiving line stood in front of the head table. Under the table was a huge box for the gifts. As the guests came through the line and introduced themselves, George would whisper to me, identifying some as his Mafia friends. Each one of these guys gave an envelope to George with money in it. A few gave the envelope to me, and you can be sure I put mine in the box under the table. George later told me he made $30,000 on that dinner. These guys take care of one another.

Wiscasset, Maine and the Music Box Museum

Leslie adored music boxes. When she learned there was a Music Box Museum in Wiscasset, Maine, we just had to drive up to see it.

Talk about a kid in a candy store! It was as if Leslie had died and gone to heaven.

The museum was located in an old Victorian house, and it was full of wonderful music boxes of all makes and sizes. There were some of the most elaborate musical contraptions I'd ever seen, complete with drums, bells, and all sorts of mechanical instruments. There were a lot of very expensive antique and one-of-a-kind music boxes.

We took a two-hour tour of the place, during which some of the more famous units were played while we learned their histories. At the very end of the tour, we were introduced to the principal pianist of the Dallas symphony orchestra. We were told there'd be a candlelight concert in a couple of weeks, at which more of the music boxes would be played and which would end with the principal pianist accompanying some of the music boxes in a duet. Naturally we signed up.

The show was an evening event. We were seated in a magnificent room in the Victorian mansion, a room with hardwood floors and that was lit by candlelight. There was a lovely grand piano right there in front of us. I'd been to a number of terrific piano performances, but I'd never been seated on stage right there within an arm's length of the piano as the maestro worked his magic. You could feel the vibrations from the grand piano on your face, and could feel them through the floor with your feet. It still sends goose bumps up and down my back and makes my hair itch when I think about it. If you like classical piano music, you can imagine what a fabulous experience this was!

Our Texas-Style New England Home

We met a woman named Dar Rowe who worked for Dube Realty, and we asked her to find a house for us to buy. On November 4, 1982, after a long search, we purchased a home at 2 Hancock Lane in Merrimack, New Hampshire. We financed it through Indian Head National Bank in Nashua. We had a thirty-year mortgage for $63,000 at 13.5 percent interest.

Leslie never really liked that house. It was a ranch-style house typical of Texas, rather than the two-hundred-year-old New England saltboxes or Victorian homes she preferred. But it had been well built by the owner/contractor. It was only five years old, and had a full basement and an attic that was well insulated and sealed off from the rest of the house. This meant it was relatively cheap to heat in the winter and didn't require air conditioning for the summer. It had a full two-car garage under the house that took up part of the basement. It also had three bedrooms, two full baths, a living room, a kitchen, and a den. The house also had a formal dining room in which I hung our Italian Florentine ceramic pink rose chandelier. It lit the room, including the twelve-foot dining table we purchased along with eight matching high-back formal dining chairs.

We knew we'd spend most of our time in the den and kitchen. So we left the living room empty, as a sort of ballroom for dancing on the hardwood floors. We put our couch, coffee table, TV, and stereo in the den next to the kitchen.

Cats

Leslie's job teaching flying at the New England Aeronautical Institute at Daniel Webster College in Nashua allowed her to often throw parties for her students. When the students arrived, before they were admitted they had to surrender their car keys. That way, if they had too much to drink they'd have to spend the night, take a cab, or walk home.

One night at a party, after we'd just acquired a kitten, I brought up the subject of naming our new pet. I suggested Smegma, which got me knocked in the head by Leslie. One of the students really laughed because he knew what it meant.

We already had a female feline, which we'd named Bella. Bella took the new male kitten under her wing and helped us raise him. He was the king of everything he saw and pranced around like he owned the world. He even tried to push Bella around. She'd take it for a while and then swat him a good one when she was tired of playing. Undeterred, he'd march off as if he had just won a prizefight, and so we named him Bully.

After several months had gone by, once he'd shown us that he knew where his home was, we let him go outside more and more often with less and less supervision. One day when he was outside and Bella was inside, we heard a terrible cat fight going on out back. We ran to the door to see who was killing our dear little kitten. Here came Bully shooting across the backyard making a beeline toward the back door with the local Tom Cat hot on his heels. We opened the door to let Bully in, and Bella shot out the door straight for Tom Cat. Talk about turning tail and running, old Tom took off for the hedges with Bella hot on his tail. I don't know what happened after that, but we never saw old Tom Cat again. Meanwhile, Bully just strutted around as if to say, "You took good care of that one, so I'll go get you another one."

Wheeler-Dealers

We had some great times at that house. Leslie also gave parties for our adult friends. These included George Andrews and Margie, Bob Holzner and his wife, Dar Rowe and her husband, Leslie's mother and father, and some couples from the college.

My boss, Bob Holzner, was what we called "a legend in his own mind." He thought he could outtalk, outsell, and outdeal anyone. But George Andrews was the real "carpetbagger." Because George ran numbers for the Mafia, he always carried several thousand dollars in his pocket to make payments to the Mafia bagmen. George could do a deal in such a way that his "victims" knew they'd been taken but enjoyed being taken. That's why he got along with the Mafia so well. They enjoyed a slick scam as long as they got their money.

Naturally at one of our parties George and Bob got into doing a deal. Bob was wearing a leather jacket that he'd bought for cheap at a flea market in California. He began bragging about what a great buy he'd made and how a coat like his was worth $150. George said he'd give him $100 right there for the coat, then pulled out a roll of hundred-dollar bills that would choke a horse. Bob jumped at the deal because he'd probably only paid $60 for the coat. George claimed he knew a guy who'd give him $250 for it tomorrow.

While all the wheeling and dealing was going on, I took off my patent leather shoes and tried to sell them. But not being a wheeler-dealer like Bob or George, I didn't sell my shoes. However, I did misplace them. About an hour later, I was still without my shoes and my feet were getting cold. I went to my closet to get another pair of shoes and when I bent over, I was too tipsy to maintain my balance and fell over. As I lay there on my shoes it didn't seem too uncomfortable, so I decided to take a nap. Later Leslie came to look for me and found me asleep in the closet. Several of the partygoers wanted to wake me up, but Leslie said to let me sleep it off. I slept in the closet that night.

The Ram in the Thicket

Leslie received her multi-engine license and then her multi-engine instructor's license. On October 3, 1982, she had her airplane single- and multi-engine instrument flight instructor's certification. I was really proud of her.

One day we were looking for someplace to go that afternoon and night. We looked in the paper, and I noticed the movie *2001: A Space Odyssey* was showing in Peterborough. Leslie had never seen the movie, so I decided we'd go see it.

Heading out Route 101 west toward Keene, we passed a beautifully carved sign for a place called The Ram in the Thicket, a bed-and-breakfast set back from the highway. We decided that after the movie we'd have dinner there.

The movie theater in Peterborough was a hoot. I think it was in an old school auditorium, with wooden floors, folding chairs, and a pull-down home-movie-type screen on the stage. Seeing the movie was great fun for me, but I don't think Leslie enjoyed it.

The film started early, so by 8:00 PM. we headed back to The Ram in the Thicket. It was an old Victorian house, with only eight tables in three rooms for dinner. The owner was the cook, and his wife was the waitress. They had only one meal choice that was served to everyone, but the cuisine was outstanding. After the delightful dinner, we went into a parlor, which had a bar where the cook/owner was also the bartender. We sat around and had drinks until midnight, talking to the owner.

We learned he'd been a priest and had left the priesthood to start this bed-and-breakfast. We had a great time discussing a lot of topics, and when they were ready to close up for the night, he invited us to come out to the barn, which he was remodeling.

It was fabulous. He'd put in an enormous swimming pool, which was half inside and half outside. The master bedroom was on the first floor, off to the side of the pool, and it had a huge bathroom with a shower next to the pool. You could jump out of bed and go right into the pool or the shower through sliding glass doors. There was a nice living room, kitchen, and dining room on the first floor. The loft had been converted into a den and library with thousands of books. It had a balcony overlooking the living room and pool. We spent several more hours up in the library discussing books. We stayed until nearly 3:30 AM and headed home after a most enjoyable evening.

Reorganization

My boss Bob went back to Advant's home office in San Jose, and I received a memo from one of Advant's head honchos, a guy named Dick Lolewecke. The memo said:

In an effort to become more efficient, Advant has initiated a new Sales/Marketing Program aimed at increasing sales and cutting cost.

The reorganization consists of:

Consolidating the five territories into three regions (Eastern, Central, and Western).

Converting from Direct Sales to marketing through Technical Sales Representatives.

Strengthening our Regional Technical Sales Support to help the Sales Reps and maintain our excellent Field Service support.

As a result, Advant is closing the office at 440 Amherst Street, Nashua, NH and you will be reporting directly to me, at the new Eastern Regional office in Landing, NJ, as my Technical Sales Manager (Eastern Region).

Because of your remote location, you will be required to maintain an office in your home or apartment to support your added responsibilities.

The writing was on the wall, so I started to look for another job.

Alden Electronics

I applied to Alden Electronics for a programmer's job I'd seen advertised in the *Boston Globe*. At the interview, when I started to talk about my background—designing part of the geostationary operational environmental satellite (GOES), working on and installing AFOS (automation of field operations and services) and AV-AWOS (aviation automated weather observing system), and working at the National Oceanic and Atmospheric Administration (NOAA) offices in Maryland—the interviewer's eyes lit up like Christmas trees. I didn't realize until later that Alden was strongly connected to NOAA and the National Weather Service. The interviewer thought he'd hit the jackpot, and I was offered the engineering manager position at Alden.

The History of Alden

I learned the background of the company while I worked for Alden. John and Priscilla Alden had disembarked from the Mayflower in 1620, and there'd been an Alden in business in Massachusetts ever since.

Around 1900, the family had an office furniture company in Brockton, Massachusetts, where they built metal desks, filing cabinets, and storage closets. It sounded to me like what we Texans call "a hammer and bang shop," but it was one of the largest office furniture businesses on the east coast.

In about 1910, when Marconi and Atwater Kent started to manufacture their tuned radio frequency (TRF) radios, Alden manufactured the bakelite knobs, ceramic standoffs, and tie-wrap posts onto which the components were soldered. The company spun off a division called Alden Components.

During World War II, Alden won a contract to build paper weather facsimile machines to print the National Weather Service's hourly forecast. Soon Alden became the world's leading manufacturer of these machines. Interestingly, the machines were all owned by Alden and were merely leased to the National Weather Service; that way Alden could charge to maintain them. The company had built up an excellent traveling service-and-repair department to maintain its systems. After the systems were in the field for about ten years, Alden would swap them out for new ones. Then they'd bring the old ones back, refurbish them, and sell or rent them to third-world countries. For example, Chili and Peru in South America had one of the largest weather networks in the world after the U.S. and Russia, and both of these South American countries used Alden's machines.

Alden spun off its Alden Electronics division to manufacture and maintain its paper weather facsimile machines. The paper in the machines had to be soaked in special chemicals to keep it from catching fire when the spark burned the weather-pattern image onto the paper. Alden also owned the company that made these chemicals.

In 1980, after I'd installed the AFOS system, the National Weather Service had decided to go all-digital and get rid of the old paper facsimile machines. This was bad for Alden, as it could potentially have put the company out of the National Weather Service. So when I applied for a job, they saw me as their ticket into the digital weather business.

Working with the Weather

My first task was to develop a digital weather graphic system for Alden. A couple of Alden engineers had microprocessor design and programming experience, but little else. They were all great electrical and mechanical engineers and knew the National Weather Service's data links, but they hadn't had the chance to develop programming or a graphics system. Now was their chance.

I worked out a plan to buy a mini VAX (virtual address extension) system from Digital Equipment Corporation (DEC), so everyone would have a terminal to work on and so there'd be a central place that housed all our information. Most of the National Weather Service's computer programs were developed by colleges and universities on DEC systems under federal contracts. This meant most of the software was available for free, or at a minimum copying fee, from the federal government. The mini VAX system I proposed could run any of these programs. All we needed to do was search through the government catalogs, pick out the programs we needed, order them from the federal government on nineteen-channel magnetic tape, and download them into our VAX.

I presented my proposal, and Alden management decided to go for it.

We ordered and installed the mini VAX and set about developing the first digital weather system. In a year, we had the first system up and running. It was really tough to get everyone to work on the computer—a lot of folks had what I call "terminal fear." Remember that this was still in the early days of computers, and even though the digital weather terminal was very straightforward, a lot of the Alden people were intimidated by it.

We'd take the data off of the National Weather Service digital data lines and just display it. Then we started developing the C2000R color weather display system. Now we had some expertise, so next, we developed the APT-4A color weather satellite receiving system; this system digitally processed, displayed, and stored picture transmission from NOAA and the Russian METEOR orbiting satellites, as well as WEFAX (weather facsimile) transmissions from GOES, GMS (geostationary meteorological satellite), and European METEOSAT geostationary satellites.

Meanwhile, on March 23, 1983, I watched President Reagan on TV as he delivered his speech about the future defense of the U.S. He proposed putting laser weaponry in space. This became known as his "Star Wars Address."

Santiago, Chili

Alden management wanted me to design a satellite ground station. Now this was a tough one! But I completed a proposal for an Alden satellite ground station, and management liked it so well they decided to send me to their biggest customers to see if someone would buy it. They scheduled me to go to Chili and Peru to pitch the new station.

One of the engineers regularly went down to South America to work on Alden systems and to train the customers' technicians. He came to me and said he'd handle all my traveling arrangements as he was going down the week before I was. He booked my flights and hotels for the whole trip. I flew to Miami, where I changed planes and cleared customs. For the flight from Miami to Santiago, Chili, the engineer had booked me in first class. The flight to Chili was a long one, and first class is really the only way to fly that distance.

When we landed in Santiago, I was surprised to see the terminal surrounded by federal troops. As the plane stopped at the gate, a voice came on the airplane's

PA system and said, "Charles Willingham, please come forward." I went up to the flight attendant, and she turned me over to a couple of the soldiers, or "federales." They escorted me to a waiting black limousine and drove me to the end of the terminal. All I could think of was that *Midnight Express* movie. But it turned out to be exactly the opposite experience.

I was escorted into the terminal and taken to a private room. After a few minutes, a soldier came in and politely asked for my passport and plane ticket. I surrendered these and sat waiting. After another fifteen minutes they brought my bags in, stamped my passport, and welcomed me to Chili. Then they put me in the limousine and drove me to my hotel. My engineer friend had apparently contacted the head man at Chili's federal weather service, which had arranged for my VIP treatment.

When I arrived at the hotel, I met the engineer and we went down to the bar where I bought him a couple of drinks. When the waitress came over, I noticed she was wearing a little pin on her dress from the U.S. Navy for the P-3 Orion subchaser aircraft. I asked her where she got it. It turns out that one of the boys from Moffett Field in Mountain View, California, had given it to her. They'd made regular trips to Santiago while searching for Russian subs.

The next day I gave the presentation about my Alden satellite ground station to several generals and top executives of the Chilean Weather Service. They were interested. But they thought the cost—over a million dollars—was too high.

I was really excited about seeing the Southern Cross constellation while I was so far south of the equator, but it was not to be. The weather was overcast the whole time I was there.

Lima, Peru

Next I flew to Lima, Peru, and made the same presentation to some more top executives, most of whom were generals. This time, the guys were really interested—in spite of the cost! But they were concerned about the system's one-meter resolution, which gave an accurate picture but was not highly precise. They told me they needed higher resolution. This was because they flew Cessna jets and Lear jets over the Andes every day at a cost of $1,500 per hour, and they were always really disappointed when the weather was overcast and the rainforest was obscured by clouds.

The generals said they wanted the system to predict the weather, but they insisted they needed higher resolution. They wouldn't elaborate on why they needed better resolution for flying over the mountains. I think what they really wanted was a spy system that could assist as they searched the jungle for rebels.

After the presentation, a colonel took us to dinner at a restaurant on a pier over the Pacific Ocean. During our dinner conversation I learned he'd been at Lackland AFB in flight training when I was there in basic training. It's really a small world!

Madrid, Spain

Because I'd been so willing to make the trip to South America, management scheduled me to go to Europe and make the same presentation to the Spanish and German Weather Services. This time Alden's president, Joe Girouard, would come with me to do some politicking and renewing acquaintances with his biggest European customers.

We flew to Madrid, Spain on October 18, 1983. My lasting impression of Madrid was the air pollution from the diesel fumes. Anyway, I gave my presentation, and again the prospects said the cost was too high.

Afterward, the local Spanish Alden sales rep took us north of Madrid to the underground church and tomb topped with a 500-foot stone cross in the Valley of the Fallen, or *El Valle de los Caidos*. It had been built by order of General Francisco Franco to commemorate soldiers on both sides who'd died in the Spanish Civil War.

As we approached, from miles away you could see the church on the mountainside and the cross on top of the mountain. It looked like a nice little church. But as you got closer, it seemed to get bigger and bigger. We parked and began the walk up to the church. It was carved out of a granite mountain. The steps in front were twelve to fifteen feet high, and the church doors were over twenty-five feet high and made of solid oak. Each door had a small, normal-sized door through which people could enter.

The church interior was one long tunnel. After passing through the first set of arches, you came to a huge colonnade a couple of hundred feet wide, thirty feet deep, and over fifty feet high. Then came a set of enormous doors. Through these doors there was another arcade of about the same size, but this one had huge arches allowing entrance to pews. The pews spread out in a cross fashion. At the center of the cross, the ceiling was open to the top of the mountain. The pews led down to General Franco's tomb, which took center stage. There was a preacher's podium and a choir loft. Behind the choir loft, a wrought-iron grate separated the nuns or priests from the church. On the back side of the mountain was a Roman Catholic monastery, which had tunnels into the church.

Frankfurt, Germany

After the Madrid presentation, we went to Frankfurt, Germany on October 20, 1983. Following our arrival at Rhein-Main Flughafen—Frankfurt International Airport—we took a taxi to our hotel in Frankfurt. The taxi driver was a speed demon. As we were rocketing down the autobahn, I noted the speedometer read 160 kilometers per hour, and I pointed out to Joe that this converted to 110 miles per hour. He turned white.

As we checked in at the hotel, Joe said he was tired and wanted to get some rest for the presentation the next day. I told Joe I was going out to the American Air Force base.

I went to the hauptbahnhoff, which now had a subway station. When I'd left Frankfurt in 1963, the city had just begun building its subway system, and now it was complete. It was beautiful. Under the street in front of the station was a fabulous underground shopping center. The subway itself was very well lit, and there was no trace of graffiti.

I took the bus out to the air base, just like in the old days. When I arrived at the main gate, I just walked in just as I always had. The guards didn't even look at me. I passed the terminal where I'd arrived and departed, and the big hotel for in-transit families. I walked past the old operations building, but it was now being used for something else. The chain link fence and guardhouse were gone and the "6916 Radio Squadron Mobile" sign had been replaced with another outfit's name. I went by the old library, the service club, and the rocket club, and they were pretty much the same.

I headed down to the barracks where I'd been billeted. The mess hall out front was gone, and the barracks building was now a nursery—I guess you could say it was a nursery when I was there, taking care of all of us young airmen! Next, I crossed the antenna field, the same way I always had when I'd been stationed there. I found the shop, but I didn't think it was still a secured area. Aircraft were still worked on in the hangar, but they didn't look like the ones I'd worked on twenty years before. I think the Air Force has retired all the RC-130s and replaced them with newer RC-135s.

The next day I gave my presentation to the German Weather Service. Same story: they were interested, but the system was too expensive. We returned to the

U.S. on October 21, 1983—twenty years and ten months after I'd left Germany the first time.

Corporate Pilot

Meanwhile, Leslie was still teaching flying, as she'd been doing for a couple of years by now. One of the things a pilot must do to keep his or her license current is to periodically have a checkout ride with a certified flight instructor. The president of Rule Industries had previously had Leslie do his checkout ride, so they knew each other. Being president of this conglomerate, he had board meetings all over the U.S., Canada, and Europe. His company had purchased a Cessna 210 with extra fuel tanks, and had completely outfitted it with the latest of everything, even radar. He used the Cessna to fly to each of his meetings, but actually this was really dangerous. Attending a board meeting all day left him exhausted, and then he had to fly at night to his next meeting. And so when Leslie was doing his checkout ride he asked her to come work for him, to take care of the Cessna and handle the flying duties.

She'd always said she wanted to do corporate flying. Now, here was her chance! So she went to work for Rule Industries. While they were at the home office, she didn't have a lot to do, so he put her in charge of the computer system because she'd had some programming experience.

So that BASIC programming class she'd taken in California turned out to be useful to her after all.

One day she had to fly the Rule president to Portland, Maine, so she decided to go shopping there while she waited for him to finish his meeting. She bought an expensive red velvet hat—and I mean very expensive, much more than we could afford. She knew she'd have to tell me about it, but how would she keep me from being upset over this extravagant purchase?

Well, when I came home from work that night, there she was in bed wearing only the red velvet hat.

How could I be upset? I always was a sucker for a beautiful lady wearing a hat!

Flight to England

Several months after she went to work at Rule, the president needed to go to England for a meeting and wanted Leslie to fly him there. Of course, they would take turns flying and would have to make several stops. She did all the flight planning, and he approved the route. They left Hanscom Field and flew to St. John's,

Newfoundland. Then they hopped over to Greenland and on to Reykjavik, Iceland. The big jump was from Iceland to Scotland. They returned via the same route.

Stealing Identity

One of the engineers who worked for me bought a new pickup truck and applied for credit through a local bank. A couple of weeks later, American Express called to thank him for applying for a credit card. But he hadn't applied for a credit card, so he called the local police. The investigation revealed that someone at the auto dealership had copied his credit application and made one of their own to American Express.

There were a lot of creative people in Massachusetts.

Little Girl Lost

When I'd first met Leslie, she'd been a little girl lost, a fish out of water. She was a good, hard-working Catholic New England girl out in the swinging, free-loving, pot-smoking, anything-goes atmosphere of 1970s California. She'd gone out there to be with her dad, who'd worked with me at Ford Aerospace. Our work for the Air Force satellite tracking station took us around the world, so when we were back in Palo Alto, all we had to do was fill out our travel reports, get some updated training, get ready for the next trip—and party, party, party! I'd met Leslie at one of these parties. She was so quiet, I'd hardly noticed her except that she was sitting at the table with her dad, who was my friend and colleague. Just to be sociable, I'd asked her to dance. She hadn't had much dancing experience and was very shy and hesitant, but her dad said to go ahead and dance with me. On the dance floor, I told her I'd been a dance instructor and would teach her. This immediately killed any confidence she had. But she was really a good follower and picked up the steps quickly. After several dances, she began to relax and enjoy the evening, knowing I wouldn't embarrass her.

Meanwhile, here she was in this godforsaken land, trying to establish a life for herself. She'd worked for her dad on a number of jobs and was really a hard worker. In California she had those two jobs mentioned earlier: one as a waitress at the Frankie, Johnnie, and Luigi Pizza Parlor and the other as a bartender at 94th Aero Squadron.

When we got married, she wanted to move back to New England. At the time, I knew that if I left California, I might never be able to go back.

When we moved to New England, and bought our house, her life was almost complete. When she landed the corporate flying job at Rule Industries, all her objectives had been met.

Except one.

Kids.

The wicked "baby wants" sat in.

Like her mother, all she'd really ever wanted to do was stay home and raise children. By now she'd passed her thirtieth birthday, and her biological time bomb was ticking louder everyday.

But babies were not in my future. In fact, before we were married, we had several counseling sessions with our minister, during which the subject was examined from every angle.

In those session, it was made very clear that I did not want children, I did not like children, and I'd taken measures to ensure that I'd never father any children.

I didn't want to raise anyone else's kids, either.

But here we were in New England, and her biological clock was pounding away.

We went to some more counseling sessions, this time with a Catholic priest, and I think the purpose was to try to change my mind. But that didn't happen.

And so there was only one solution—divorce.

Leslie moved out and got an apartment. I told her to take the furniture because she would need it. Besides, I didn't want to haul it to wherever I was going. I was left in the house with just a twin bed and a chest of drawers with my jewelry case on top. The jewelry case contained my broken high-school ring, a couple of tie tacks, some collar stays, my medals and ribbons from high-school ROTC, my Air Force ribbons, and my military dog tags. I had my computers in boxes in the basement, along with my over fifty cartons of books I'd collected over the years, plus some tools and the motorcycle.

The rest of the house was completely empty.

One day when I came home from work, I noticed my jewelry box was open on the kitchen counter. The front and back doors were open. The burglar had obviously picked the wrong house. Intending to steal something of value, he'd wound up taking only some wine bottles from the fridge. I told the police, but they didn't bother to fill out a report, because nothing of value had been taken. However, they did say someone was working the area, because there'd been several break-ins reported. They said they'd mark my place on their crime map.

I bet that burglar was very disappointed to have broken into an empty house!

Frozen Pipes

Having been raised in Texas, where I'd slept as a kid on a screened-in back porch, I was used to having fresh air at night for sleeping. So I always liked to keep the bedroom window open, even when it was cold.

One night in New England it was 10 degrees outside, and I was so bundled up with all the covers in the house that I could hardly turn over. But I was cozy and warm, and was dreaming about a South Pacific island with a steamy tropical rain forest and a beautiful waterfall. With all that water in my dream, I woke up and had to pee. So I threw back the covers and headed for the bathroom.

My feet hit the floor with a big splash of hot water.

The forced-hot-water heating pipes under the bed had frozen and burst. Hot water was streaming all over the bedroom.

I ran downstairs and shut off the water into the house until I could figure out what to do. I went to the bathroom and then downstairs to examine the heating pipes. The set-up was a two-zone hot-water system. One zone was for the living room, den, dining room, and kitchen. The other zone was for the bedrooms and bathrooms. I turned off the valve to the bedrooms and bathrooms, but turned on the water to the rest of the house so the other half would stay warm and no more pipes would freeze.

I headed back to the bedroom and began mopping up. But the damage to the hardwood floors had already been done. Plus some of the water had leaked down into the basement and ruined about twenty cartons of my books.

The problem turned out to have been caused by the placement of the bedroom thermostat. It was in the hall near the kitchen, which meant that the heat from the living room/kitchen zone kept the temperature in the hall up, but the heat never came on in the bedroom zone.

I called a plumber. The pipes in the other bedrooms had frozen also, so the plumbing bill was over $600. Later I called a floor sander to refinish the hardwood floors. That was another $1,500.

The Count

While the floors were being refinishing, I moved my twin bed and chest of drawers downstairs to the basement. The first night with this new sleeping arrangement, I felt something on my face. I thought it might be a mouse running around the basement ledge and knocking some dust onto me. The cold water pipe entered the basement through a large hole, so I figured a mouse was coming through the hole. I decided that the next night I'd wait until he came in, and then block up the hole so he couldn't get out. The next night when I went to bed, I left the light on so I could see the little varmint. I'd been waiting there for about thirty minutes when I saw this *bat* come flying around the basement!

I named my bat "The Count."

There wasn't much for him to eat in the basement, so I thought he must want to go out. The next night was much warmer, so I opened the cellar window. A couple of nights passed, and he just flew right past the open window. Then he flew up and landed on the ledge and looked out, but flew back into the basement. The next night he flew up to the open window, misjudged his landing, and fell out the open window. Immediately, he flew back inside. The following night he flew out the window, and I never saw him again.

House for Sale

After I had the floors redone, I put the house up for sale. In about a month I received and accepted an offer. I quit my job and packed up what little was left, including the books that had survived the water damage. Then a month went by and I still hadn't heard from the real estate people. So I called.

The deal had fallen through.

The loan company had learned that the buyer had filed for bankruptcy and had a bad credit history. The real estate salesman hadn't done his homework and prequalified the buyer. So there I was without a job and running out of cash. I had about $3,000, which I could stretch for two months, but I needed $1,000 to rent a U-Haul truck to get back to Texas. While waiting for the house to sell, I took long walks around the neighborhood and enjoyed the changing colors of the fall. I also spent a lot of time in the local Burger King, drinking coffee and reading the free newspaper.

And I prayed to God every day that He'd send me a qualified buyer.

At the very last minute, the real estate people came up with a young guy from the Midwest who had sufficient cash for a down payment and a short but acceptable credit history. His loan went through with flying colors.

So you see, prayer does help!

I took the profits from the sale, subtracted my down payment, and met Leslie at her bank to deposit half into her account. She drove with me to the U-Haul office, where I had them put the little Chevette on an auto-dolly to hook it up to the truck.

We said our tearful good-bye, and I was off.

U-Haul and Moving—Again

I steered the big truck down Highway 495, followed Route 90 west to Sturbridge, and took highway 86 to Hartford, Connecticut. As I drove, memories of Leslie kept running through my mind. I passed the Pratt & Whitney jet engine factory and airport—where Leslie had once almost landed. I passed the Hartford Airport—where I'd once had to pick her up.

Then I traveled along Highway 91, to Highway 95 toward New York City, where I crossed the George Washington Bridge and headed for the home of our friends Steve and Barbara Lenny. Leslie had made arrangements for me to spend the first night there, in the hope that they'd talk me into coming back to Boston.

They didn't, and I continued my trip.

Visiting Bill Allen in Columbus

I took Interstate 80 to Cleveland and then Interstate 71 to Columbus, where I planned to stop and visit Bill Allen. As I traveled along 71, just before the turnoff to Bill's place, a passing motorist signaled that something was wrong with my trailer. When I looked in my rearview mirror, I could see smoke coming from the driver's-side tire on the towing dolly!

I pulled over immediately and jumped out of the truck. Sure enough, the dolly tire had gone flat and was burned up. I disconnected the auto-dolly with the Chevette, left it on the side of the road, and drove on to Bill's place, where I arrived at about 10:00 PM. I didn't want to leave the car on the side of the road all night, so I phoned U-Haul's twenty-four-hour road service and explained the problem. They said they'd be right out. Bill and I drove back to Highway 71 in the U-Haul and waited. Sure enough, a wrecker showed up, lifted the auto-dolly off of the ground, and replaced the whole tire and rim in less than fifteen minutes. I hooked it onto the truck again, and by midnight we were back at Bill's place having drinks.

I spent several days at Bill's, revitalizing my spirits after the trials of the previous months: the divorce, the flood in the bedrooms, the stress of worrying about selling the house, etc. We relaxed on Bill's deck, charcoaling steaks, and Bill introduced me to boiled corn-on-the-cob with Tabasco hot sauce. It was a fabulous and glorious visit. There was a huge maple tree in front of his house, and it was radiant with Ohio's gorgeous fall colors of orange and red. Chipmunks were scurrying everywhere, gathering and storing nuts for winter. We had a great time, drinking and laughing and joking. At one point, we hooted and hollered so loud that Bill's neighbor John could hear us.

Later we settled down, got more serious, and talked about what we were going to do with the rest of our lives. Bill said he thought I should take my $20,000 nest egg, go to Costa Rica, and buy acreage near that country's rainforest. He knew people in Costa Rica who could help me. A lot of English-speaking environmentalists were there, trying to save the rainforest. Bill said that once I got established there and built a house and barn, he'd come down during the summers and hold writing workshops for the environmentalists; they needed to learn how to write about the problems they encountered, so they could gain support for their causes. Bill said we could take trips into the rainforest and see all the colorful birds and howler monkeys.

I probably should have followed his advice, but I didn't. I was still heartbroken over the failure of my marriage. I just wanted to get back home to Texas.

Poetry and Music

I left Bill's place in Ohio and headed straight to the Lone Star State.

Once I was back in Texas, in October of 1984 I wrote the following poem, based on Edgar Allan Poe's "Annabelle Lee":

Leslie Lee

It was many and many a year ago,
    In a kingdom by the sea,
That a maiden there lived whom you may know
    By the name of Leslie Lee,
And this maiden she lived with no other thought
    Than to love and be loved,
    For that's what it meant: to be.

She loved with a love that was more than love—
    This fair maiden Leslie Lee,
All that she wanted was a child,
    And to live in a kingdom by the sea;
She was a child and had her child,
    In this kingdom by the sea:

But her high-born kinsmen came
    And bore her child away, you see,
To sequester and protect her,

This is the way it had to be.

But her love was stronger by far than the love
    Of those who were older than she—
    Of those "far wiser" than she—

And neither the angels in heaven above,
    Nor the daemons down under the sea,
Can ever dissever the soul of the child
    From the soul of Leslie Lee.

And this was the reason that, long ago,
    In this kingdom by the sea,
That I broke the heart of this maiden
    The beautiful Leslie Lee.
For she couldn't have children
    And be married to me.

But the moon never beams, without bringing me dreams
    Of the beautiful Leslie Lee,
And the stars never rise, but I feel the sad eyes
    Of the beautiful Leslie Lee.

I also made a CD of music that reminded me of Leslie. It contained the following songs:
Moments to Remember—The Four Lads
Graduation Day—The Four Freshmen
Moonlight in Vermont—Margaret Whiting
Old Cape Cod—Patti Page
Canadian Sunset—Andy Williams
Autumn in New York—Frank Sinatra
MTA—The Kingston Trio
Love Walked In—The Hill Toppers
When I Fall in Love—The Lettermen
The Way You Look Tonight—The Lettermen
No, Not Much—The Four Lads
Music Box Dancer—Frank Mills
My Special Angel—Bobby Helms
You Are So Beautiful—Joe Cocker
Evergreen—Barbra Streisand
After the Lovin'—Engelbert Humperdinck
You Light up My Life—Debbie Boone
Autumn Leaves—Roger Williams
Perhaps Love—John Denver and Placido Domingo
The Way We Were—Barbra Streisand

Somewhere My Love—Robert Goulet
I'll Never Love This Way Again—Dionne Warwick
I'll Be Seeing You—Frank Sinatra
Till Then—The Hill Toppers
I Wish You Love—Frank Sinatra

Maybe I should have included *How Can You Mend a Broken Heart?* by the Bee Gees, because at this point, that's what I needed to know.

Chapter Twelve:

The Texas Years (1985 to 1999)

Back in Texas

When I arrived back in Texas, my brother Rocky invited me to me move in with him and his wife. He said I could store all my stuff in his garage and attic, but I didn't want to take up all his space. So I bought a small metal building and erected it in his backyard to hold some of my things and leave room for his stuff in the garage. I stayed with my brother until I found a job, which took about six months. While I was there, I decided to sell my Intel microcomputer development systems, so I ran an ad in thc Sunday paper. A young engineer came over and I sold him the whole lot for $1,000. I was now out of the computer hardware business.

Motorola Four-Phase

I found a job at Motorola Four-Phase on LBJ Freeway and Josey Lane. Now that I had an income, I moved into the Green Tree Apartments, on the Dallas North Tollway (DNT), just before Belt Line Road. It was a beautiful upstairs studio apartment, with a high ceiling over the living room and an upstairs balcony bedroom. From the bedroom I could see through the floor-to-ceiling windows facing north, out over the nearby buildings. I watched the weather blow in by day, and gazed at the stars and lights at night. The tollway was being rebuilt at the LBJ and DNT interchange, and this created major traffic problems. I was only five miles from my job, but it took an hour and a half to get there and another hour and a half to get home. So to avoid the rush hour, I'd work until 6:00 PM, then go across the street to the Summit Hotel, have a drink in the bar, and fill up on the free hors d'oeuvres as dinner.

Motorola had bought Four-Phase Systems and included it under the Motorola corporate umbrella. Four-Phase System's claim to fame was its data-input system for IBM mainframes. In general, one of the major challenges of data entry is the sheer number of operators inputting data. This causes what is known in the computer industry as being "I/O (input/output) bound": the amount of data being entered overloads the mainframe's input-output system, so the system needs hours to catch up, consuming expensive computer time. Four-Phase developed a process via which all terminals were connected by a coax cable, with a signal circulating through this long electrical loop. This loop, in effect, acted like a large memory-storage area. As the data-entry operators keyed in data, the information went into the memory on the cable immediately, so there was no I/O backlog. This process increased the throughput tremendously. Then the system could batch-load data into a mainframe at machine speeds, and so expensive mainframe computer time wouldn't be tied up waiting on I/O data input. It was a very effective and efficient system.

Newcomers like me were sent to school for two months learning to support and program the system. Most of the programming involved learning various data-entry applications. Every application had its own way of presenting information, so each application had to be programmed to display the presentation format on a screen and convert it to an IBM batch format that could be sent to the mainframe.

We went to school for another month learning to support Motorola/Teledyne's 68000-microprocessor system. It was state-of-the-art technology, with a scaled-down Unix operating system.

Once we'd been trained, we went to work supporting all this stuff.

Windsurfing

I'd windsurfed in California, and one of the first things I wanted to do when I arrived in Texas was go windsurfing. So I started visiting all the lakes and scouting potential windsurfing areas. Most windsurfers tended to gather all in the same area, Windsurfer's Point, just as hobby cat sailors hung out at Hobby Point.

The main problem with sailing in Texas is the wind. There are no land features to create any differences in weather. In this rolling-hill country, during summer the winds are usually light and variable, except when there's a thunderstorm blowing in from the northwest at a hundred miles per hour, which is no good for sailing or windsurfing.

I found a place on Lake Ray Hubbard that rented windsurfers. They had a regular school for training, and I visited it several times and watched the "newbies" falling in the water. Trying to stand and hold that sail up without any wind is not easy. In California, after we'd learned how to windsurf we'd wait until the afternoon, when the steady winds of twenty to thirty miles per hour would come. It was easy to windsurf in those conditions, because the sail would hold you up.

Now, in Texas, I was waiting for a steady wind of about thirty, without a thunderstorm, before I rented a windsurfer. It seemed to take forever. Finally one day there was a good thirty-miles-per-hour wind and no thunderstorm. Excited, I left work early, jumped on my motorcycle, and raced to Lake Ray Hubbard and the windsurfing school.

When I ran in and tried to rent a windsurfer, the guy told me sullenly, "We don't rent out windsurfers when the wind gets over twenty."

"What!?" was my incredulous reply. "That's when the sailing is the best!"

"Our insurance won't let us," was his answer.

I drove straight to the windsurfing shop and bought my own windsurfer.

Meat Market

Sometimes after work, some of the single guys would go over to a place nearby (D&B's) to play pool, drink beer, and chase women. The place was a real meat market. The bar was in the middle of the room, with pool tables on one side and some dinner tables on the other. The ladies would sit at the bar. A guy would circle until one of the girls caught his eye, then he'd strike up a conservation. If things worked out they'd leave together.

Lay Off

After I'd been at Motorola about three months, my first performance review was coming up. My boss, Ron Sands, wanted to go with me over to Motorola's Communications Division on Beltline Road and see how well I could troubleshoot an SNA (systems network architecture) communications link between the Dallas mainframe and a mainframe in Schellsburg, Pennsylvania. We were working on the SNA link when Ron's beeper went off—it was his boss, who wanted to see him right away.

Ron said that I was doing a good job; he told me to continue, and that he'd be right back. Shortly after that, *my* beeper went off, so I phoned in. Now Ron's boss wanted to see me, too. I left my briefcase and toolbox at the data link, and headed for the boss's office in the main building.

The boss said he'd received notice to do an immediate reduction of the work-force, and he told me to turn in my badge and tools. I told him I'd left the tools over at the Communications Division on Beltline Road, along with my briefcase. He took my badge and told me to go clean out my desk. He said he'd call the Communications Division and have them let me in without a badge to retrieve my briefcase.

I cleaned out my desk, then picked up my briefcase and went over to Jo Jo's restaurant for a cup of coffee. I called Ron Sands at home, as he'd just been laid off, too. I said, "Hey Ron, I have a six-pack of beer—do you want some?"

He said to come on over. We had a few beers together. Ron's wife, Karen, also worked at Motorola Four-Phase, but she hadn't been laid off. By the time she came home from work, Ron and I had finished that six pack and then some, and we were pretty far gone. We had a few more beers; then Karen put us both to bed. I was way too drunk to drive.

Lufkin and East Texas

In spite of now being unemployed, I had a "nest egg" of substantial cash, so I started to look for a place to buy. I wanted to be near a lake where I could wind-surf, and I wanted it to have lots of trees. I got on my motorcycle and visited all the local lakes.

One place I visited was Lufkin. On my way there, I stopped in Athens for breakfast. The only place that was open was across the square from the court-house. It was called the B&B. I ordered breakfast and as I waited, I read through the pamphlet that was with the menus. It said B&B stood for Barrow and Bar-row, who had been the parents of Clyde Barrow—of the infamous Bonnie and Clyde. The couple had stopped by several times during their numerous runs from the law.

In Lufkin I stayed at the Sun-in-the-Pine Motel. I looked at a twenty-acre piece of property near Sam Rayburn Reservoir. The main problem was there were no jobs in the area, unless you wanted to work in the local sawmill.

I found another nice location in East Texas. It was called the Lake of the Pines. But it had the same drawback: no jobs.

Gun Barrel City

Next, I went to Gun Barrel City, Texas, which was located on Cedar Creek Lake. It was where my dad had been living when he'd died. The headland of the lake, near Kemp, Texas, was fifty miles from Main Street in Dallas, which put it right at the edge of my acceptable commuting distance. I figured I could relocate there and look for a job in Dallas.

I found a real estate broker and said I wanted to look at properties for less than $50,000, on the water. The broker practically laughed in my face. But she checked the listings anyway. She came up with a couple of places, but they were real dumps—old mobile homes set back from the lake in a slew.

Then the broker mentioned a lake cabin, a modified A-frame in the Loon Bay addition on the west end of Gun Barrel City. I'd loved the A-frames I'd seen in California. Unlike an apartment or house, with four colorless walls that closed in on me and made me feel like a prisoner, an A-frame had cozy pinewood walls on a gently sloping angle. This lake cabin was one lot away from the open water, on a tree-covered property near the North Park. It would be a great place from which to launch my windsurfer. The location stretched my commuting distance by another 15 miles, but I decided that was okay because there was only one signal light in Seven Points and I could make a right turn on red. The road was good, and the highway from Kemp to Dallas was excellent.

Nevertheless, I wasn't yet ready to commit, but wanted to look around some more. So I drove down to Lake Whitney, where my family had often gone fishing during the 1940s and 1950s. There was a lake cabin there, advertised as a "gingerbread, Hansel and Gretel" structure. I rode my motorcycle down and took a look. It was nice, but the area was barren. There were only a few evergreen cedar trees on a chalky, rocky ground.

I decided to go back and check out the A-frame again. As I rode down Highway 31, the landscape changed from barren to scattered scrub oaks when I crossed Interstate 45. I decided not to look at anything west of I-45. The closer I got to Cedar Creek Lake, the more oaks populated the landscape.

I looked at the A-frame sitting there among the trees, and I decided to make an offer.

The offer was accepted, and the owners were willing to carry the note themselves. When they asked about my ability to pay, I said, "No problem, I'm independently wealthy." Luckily, they accepted that story, as I was unemployed at the time! I did, however, have my nest egg, which was almost enough to pay cash for the place.

I moved in on November 20, 1984.

Friar Ockham's House

I named my little A-frame "Friar Ockham's House," after the fourteenth-century English logician, philosopher, and Franciscan friar, William of Ockham.

Like many Franciscans, William was a minimalist, idealizing a life of poverty. Friar Ockham (sometimes spelled Occam) is perhaps best known for "Occam's razor"—the theory that forms the basis of methodological reductionism. It's also called the principle of parsimony, or law of economy, and it states, *Entia non sunt multiplicanda praeter necessitatem*, or "Entities should not be multiplied beyond necessity." It basically says, "Keep it simple, stupid." The principle is that you use this theory, or "razor," to "shave off" unnecessary complications when considering an idea.

I decorated my A-frame in the Southwestern style, with a bright Native American rug and colorful Mexican blankets hanging over the loft edge. I added a couple of guitars, a set of maracas, and a pre-Columbian-style musical instrument called a "guiro." I made the loft into my office and built a small bookcase to hold some of my books. I put my desk up there, as well as my computer and all of my software. Downstairs there was a bathroom with a shower, a small bedroom, and a living room which contained the only source of heat: a wood-burning pot-belly stove.

There was also a small kitchen. From the ceiling in the kitchen, I hung a kitchen witch, along with a set of copper cooking pots and utensils, a salami, a cheese ball, some grapes, a long braid of garlic, and some drying wheat and barley plants. I filled the shelves with canning jars full of beans, rice, coffee, noodles, and various-colored pastas, as well as several cans of tea, and some herbs and spices. I added a find from the local flea market: a maroon basket with a wooden model of kneaded bread, a butter churn, and a rub board. I hung several colorful coffee cups from a shelf and put my Chicago cutlery block next to the cutting board on the counter top, along with my toaster, blender, copper-topped canisters, rice cooker, and Mr. Coffee machine. I set out my two cookbooks: one by Vincent Price and another called *Helen Corbitt Cooks for Looks*. It all made for a very colorful gourmet kitchen.

Next to the kitchen I hung some plates over a small cabinet to simulate a china closet. I would change out the plates with each season: bright hot colors for winter, cool blue china for summer, brown and orange plates plus a turkey platter for fall, and flowered plates for spring.

GBC

Shortly after moving into my little A-frame in Gun Barrel City (GBC), I went to see my aunt Saleta Jenson and her son Jimmy. When I told them I now lived in GBC they said that my great-grandfather was responsible for Gun Barrel City's unusual name, and they told the following story.

The father of my grandmother, Rena Bell Coupland, had once lived nearby in Troup, in Smith County, Texas. His name was Andrew A. Coupland (1841–1920), and he was a veteran of the Confederate Army. He was also a surveyor and had been hired by G. W. "Dodge" Mason and his partner Thomas H. Eubank to lay out a new town southeast of Kemp, Texas, on their ranch. This was in antici-pation of the extension of the Texas Railroad past Kemp. The new town was named Mabank, a contraction of the two partner's names. The extension of the Texas Railroad line became a reality in 1900; it followed the southeast border of Kaufman County and passed through the edge of the six-thousand acre Mason-Eubank ranch and continued into Henderson County.

The new town benefited from its proximity to the Tyler-Porter's Bluff Rail-road established in 1848, and from its nearness to some crude trails and wagon roads leading to Kemp, Prairieville, Whitehall, Phalba, Roddy Odom, Payne Springs, and Goshen, with Mabank as their focal point. All streets were drawn on the plat, followed by survey, aligning them 15 degrees southeasterly, parallel or perpendicular to the railroad tracks.

While surveying the fence next to the old Tyler-Porter's Bluff Road, from Mabank to Paine Springs, my great-grandfather, Andrew A. Coupland, com-mented that it was "straight as a gun barrel."

Soon the old wagon road became known as Gun Barrel Lane by the locals.

Gun Barrel Lane became Highway 198, as assigned by the Texas Department of Transportation, and the crossroad became Highway 85 (later reassigned as Highway 334).

The City of Gun Barrel began its growth at the intersection of Highway 198 and Highway 85 when Paul Hoskins erected his two-pump Texaco service sta-tion and grocery, the Gun Barrel Drive-In, in 1964. His attraction to this loca-tion was the nearby yet incomplete Cedar Creek Lake. Later a meeting was called to discuss the possibility of incorporation, on advice from Judge Reuben Lee.

In May 1969 by a vote of 37 to 21, the decision was made to incorporate the area, and Gun Barrel City was officially born.

Critters

There's a bumper sticker that says, "The more people I meet, the better I like my dog." I don't know that I'd go that far, but I've always been an animal lover.

When I first moved to Gun Barrel City, I would sit in my rocking chair on the front porch and drink my morning cup of coffee. Next to the porch was a tree, and often a friendly squirrel would climb down and just sit and look at me. I think she (I'd seen two babies, so I knew it was female) was begging for food and was probably sort of a local pet that neighbors down the street fed the way they did the neighborhood cats. I had a whole bowl full of pecans left over from Christmas, so I started to crack some and feed the squirrel. If I gave her an entire pecan, she'd run off and bury it, but if it were cracked she'd sit up and eat it.

Each day I'd put the cracked pecans closer and closer to my chair, until one day she climbed up my leg. As she sat on my knee, I fed her about ten pecans. For the next four months, every morning she'd come by and I'd let her sit on my knee and have breakfast with me. On any given morning, when she'd eaten all the pecans I had for her, she'd grab my hand with her paws and open it up to make sure I wasn't holding out. Once she was convinced my hand was empty, she'd rear up and investigate my shirt pocket, because sometimes I kept nuts in there. If the pocket was empty, she'd climb on my shoulder and look around behind my head and back. When she was satisfied there'd be no more breakfast that day, she'd jump down and go off about her business—whatever that was!

I named the friendly little squirrel "Momma," and her two offspring I named "Junior" and "J.R."

Her two babies followed her around, but she was trying to wean them. If they got too close to me, she'd jump down and run them off. I was *her* food source; they had to find their own now. But they were always nearby, watching her and learning how their mother survived.

One Sunday morning, I went down the street to visit my friends Dick and Doreen. While I was there, Jackie, my neighbor across the street, called and asked to speak to me. She said I'd just been robbed! Shocked, I asked what the robbers took. She said "Food, I think."

It turns out that in my absence, "Momma" could smell the pecans in the bowl on my coffee table. So she ate a hole in my window screen and went in and helped herself.

She'd entered through one window and had come out the other, bringing a pecan each time. This began occurring on a regular basis, and soon her youngster followed suit. The whole squirrel family had a regular routine: climb in through one window, pick up a pecan, and dash out through the other window.

The critters had ruined two window screens that I really needed to keep the mosquitoes out. I had to put a stop to their little venture. So I bought something called a "pet deterrent." It was a small transformer that converted 110 volts of alternating current into a couple of thousand volts but with very little current. It created an electric fence, like the ones used on a farm to keep cattle fenced in, only smaller. It would give a good jolt but would not hurt the animal.

I put some screen wire in the window flower box and connected the ground wire to it. I hooked the hot wire up to the screen. Then I plugged it in.

I went across the street to have coffee with my friends Jack and Jackie. We were sitting on their screened-in porch, waiting for the squirrel to start her rounds, when their cat headed over to my house and jumped onto the window box. She walked along the window for a few seconds—then made a dive for the tree in front of my house. Evidently, the cat had gotten zapped when she'd brushed up against the screen.

I never did see the squirrel again. She must have received a zap and decided not to come back stealing or begging for food.

The Nightly Parade

Meanwhile, there was a regular parade of critters across my backyard every night. A road behind my house led to a lakeside house in the Harbor Point addition, but other than that, there was nothing but woods. And a lot of critters lived there. A culvert went under the road and emptied out just inside my neighbor's area, and a drainage easement in back of my place emptied into the lake. Animals would go through the culvert and come out on my side of the road under the fence. They'd parade across my backyard, go up the street to the trashcans, and eat the dog food, cat food, and bird seed left out by the neighbors.

Skunk

When I first purchased the little A-frame, it had an air conditioner in the back bedroom window. I was used to sleeping with a window open, even in the freezing night air of Boston, so I took the air conditioner out, wrapped it in a black plastic trash bag, and put it in the garage. I installed one of those $19.95 fans in the upstairs window facing outward. It would pull the hot air out at night and draw cool air in through my open bedroom window. It functioned like the attic fan my dad had installed in our place in Rylie, Texas. This worked out well—until the night the skunk came to visit.

An enormous mother raccoon and her two babies came by every night, as did an armadillo, whom I named "Tank," and its babies. An opossum would often visit, as well as a skunk I named "Pepe Le Pew" after the cartoon character. I have a mild case of sleep apnea, and one night I must have snorted just as Pepe was under my open window, for he filled the air with his perfume. I came up out of bed in a hurry, gasping for breath, and dashed out my front door to get some fresh air.

Opossums

I kept my car in the driveway and used the garage mostly to store things. One Saturday I went into the garage to get something and heard a noise in one of the boxes stored there. I looked inside, and it was an opossum. She was building a nest. I closed the lid and carried the box across the street to Jack's house to show him. He was on his screened-in porch, so I dumped the contents onto his drive-way. Out came the opossum and a pile of leaves she'd put in the box. I figured that would teach her not to nest in my garage!

But then one night when I came home late, my headlights reflected off something under the garage door. When I opened the door, I saw three baby opos-sums running around in my garage. Mother had come back and finished her nest, and now she'd had her babies. I decided not to disturb the new family. She stayed in there for a couple of weeks until one night I saw her waddle across the back-yard with the three babies on her back.

I never saw them again.

Naming My Friends

My animal friends are aggressive, territorial, and fierce—even attacking each other, on occasion. But they don't lie or kill each other for emotional, political, or religious reasons, or for no reason at all. They've consumed nearly 100 percent of my garden, but I guess that's all part of some great plan.

In Genesis in the Bible, it says that Adam gave every animal a name. I fol-lowed suit. In addition to the animals mentioned above, there were three cotton-tail rabbits that I named "Thumper," "Brer Rabbit," and "Bugs Bunny." There was a small gecko lizard that I called "Jose"—after Jose Greco, the Flamenco dancer.

There was a blue jay that I named "J.B." for "Jay Bird," and another I named "B.J.," after the B. J. Honeycutt character in *MASH*. One mockingbird got christened "Chatterbox" and another "Bing" after Bing Crosby, because he was such a crooner. (I confess I almost had mockingbird stew several times, because each summer they'd peck and destroy every tomato in my garden.)

I gave the names "Morning Glory" and "Morning Star" to a pair of morning doves. The local cardinal became "Bernard Francis Law" after the Cardinal and Archbishop of Boston. But unlike his namesake, my cardinal had a mate, so she became "Mrs. Law." There were numerous robins: I called one "Bob," another "Bobbi," and a third "Robyn" after a lady I worked with named Robyn Casey.

Of course, the red-headed woodpecker was "Woody."

Hummingbirds began to fly around, so I put up a couple of feeders. I named them "HummVee," "H2," and "H3," after the vehicles. I put one feeder in front of the house and the other in the back. The hummingbirds would take turns monitoring each of the feeders and would chase off any other birds that tried to take advantage of this "virtual flower."

There are fourteen species of sparrow. I had a whole flock, and I think they were the Brewer's Sparrow variety. They were all chatterboxes, and there were far too many to name.

Another bird species that came around in abundance was the wren. This is a tiny little songbird with a rather thunderous song for its size. There are nine different types of wrens, and I think mine were from the Carolina family.

Other birds included a flock of crows, a little black-and-gray woodpecker, and a bunch of tufted titmouses—or are they called titmice? The tufted titmouse is similar to the plain titmouse, but it has rusty flanks and a black forehead or crest. They're found in central and southern Texas. There were also a bunch of similar-looking Mexican chickadees and mountain chickadees. Another bird found in abundance was the goldfinch. There were three kinds: the lesser goldfinch, Lawrence's goldfinch, and the American goldfinch.

Lake Birds

Egrets lived on two islands in the lake. They'd fly off during the day and stay with the cows in the local fields, which I guess is why they were nicknamed "cow birds." Also in the lake were mallards. A pair of them would come in for a landing, flaps down and gear down, right in front of my house. They'd walk up the street to a neighbor's house where they had a nest. The male would just peck around while the female sat on the nest. I named them "Donald" and "Daisy." There was another male mallard, too, so I named him "Daffy."

Every year white pelicans with black wing tips would land on the lake on their migratory flights to and from the north. There were also a number of geese living on the lake—I think the neighbors fed them.

Gardening

As someone once said, "I like Thanksgiving better than Christmas. It's cheaper giving thanks than giving presents."

One of the things I liked to do in my new A-frame was decorate it for fall and for the Thanksgiving season. I'd buy dry corn stalks and stack them around the front porch roof supports on each end. Then I'd put a couple of square bales of hay and several pumpkins near the corn stalks. I'd also purchase several ears of Indian corn at a dollar each and hang them on either side of the front door.

I'd heard that if you put seeds in the refrigerator they germinate better. I figured an outdoor frost was the same as refrigeration, so I put pumpkins in my garden in January and waited for the first frost. One night on the 10:00 PM news, the weatherman predicted a hard freeze. I ran to the kitchen, grabbed the biggest butcher knife, and dashed to the garden. Like a maniac from a slasher movie, I went into a frenzy, brutally stabbing at the pumpkins like a madman. It was cold, and I was in a hurry to split open the pumpkins, expose the seeds to the freeze, and get back into the warmth of my house.

I was just finishing my slaughter when a local police car cruised by. I guess he could see me violently stabbing something in the garden. He skidded to a stop and turned his spotlight on me. I'd just finished, so I stood up and walked back to the house, wiping my butcher knife with a rag.

He left without saying a word.

When spring came, I decided to plant a garden. The limb of a large tree hung over the place beside the road where I wanted to do the planting, and it shaded most of the area. To get more sunlight, but wanting to save as much of the tree as possible, I decided to cut the limb halfway off. I leaned a ladder against the tree near where I planned to cut. The ladder just reached the limb on the side toward the tree trunk. I climbed up and began sawing away. When the cut was finished, the big heavy end dropped off.

Having lost the extra weight, the remaining part of the limb sprang upward like a catapult.

Unfortunately, the ladder had been leaning against the limb that catapulted away, so the ladder dropped to the ground, taking me with it.

I managed to grab for the limb, and I held on for dear life. As I hung there, legs dangling, I heard my neighbor Jack Krueger laughing. He'd been watching the entire process. I yelled for him to come and help me. He was laughing so hard it took him almost five minutes to get me down safely.

Now that the area was sunny enough, I bought a used rototiller to turn over the soil, which was a mix of clay and hard tack. That old five-horsepower roto-tiller had a difficult time with such firm soil. It would bounce and jump around crazily trying to break up the clay and hard black "gumbo." After two hours, I'd made only one pass over the 15' x 40' area I'd chosen for the garden, and I'd turned over soil to a depth of only five inches. I refilled the gas tank and again cranked up the rototiller. The jumping and bumping caused gas to spill out the vent hole in the gas cap, adding to some gas I'd spilled during the refill. The con-tinued jumping and bumping caused some of the spill to fall on the hot exhaust pipe.

Next thing I knew, it burst into flame!

Again Jack came to my rescue, dashing across the street with a fire extinguisher he kept by his fireplace. As he came running across the street, he pulled the safety pin out and threw it away as if throwing a grenade. After he'd put out the fire, we went over to his screened-in porch and had a couple of beers and a lot of laughs. The next day I finally got the garden rototilled.

I decided to plant Indian corn, so I'd save the dollar-per-ear I was paying for my autumn decorations. I scraped the kernels off of a cob and planted several rows of corn: some red, some yellow, and some mixed with the pumpkin seeds from the night of the great pumpkin slaughter. Soon I had a wonderful stand of corn and lots of pumpkin vines. Every morning before work, I'd take a turkey feather out to the red corn and get some staminate (male pollen cells) from the tassels at the tops of the red stalks. Then I'd take it over to the pistillates (female cells) on the silk of the yellow corncobs. I was pollinating the yellow corn with red-corn pollen in hopes of getting corn with variegated kernels.

At work I told the lady in the next cube, who was a graduate of Texas A&M and knew about framing, that I'd been having sex with my corn in the garden. She roared with laughter.

I watched excitedly as the ears grew on the corn stalks. I was filled with eager anticipation. Would the ears indeed have both red and yellow kernels? Had my sexual experiments worked?

I never found out. One day I came home to find that Momma, Junior, and J.R., the supposedly friendly squirrels, had eaten every ear of corn in my garden.

Bill's Place on Winter Road

I took a trip to Bill Allen's place on Winter Road outside Columbus in Delaware, Ohio. Bill had bought his house from a guy named William Matix, who'd moved to Florida after his wife was murdered (probably by Matix himself). Not long after Bill Allen bought the house, William Matix made headlines after he and an army buddy named Michael Platt went on a bank-robbing spree in Florida. When the FBI cornered them, the two gunned down six FBI agents before they were finally killed. It was the worst gun battle in FBI history, and in 1988 it was made into a TV movie called *In the Line of Duty: The F.B.I. Murders.*

At Bill's house there was a glassed-in room for reading and watching the birds, which was what I did every day during my visit, as I drank Tennessee Tea (which is sort of like a whiskey sour plus Coca Cola). Bill had three pets: a cat named "Esme," another cat named "Big Dog," and a groundhog.

The groundhog had fallen into one of Bill's basement window wells as a baby, and it couldn't get out. So Bill fed the little critter and made friends with it before taking it out and releasing it. It was so tame that it would come up on the deck and take food from Bill while its siblings watched from a distance. Bill named it "Junior."

While at Bill's, we again talked about the possibility of going to live in the rain forest of Costa Rica, but I still wasn't convinced it was the right move for me.

Mexican Flower Pots

Bill wanted to take a trip to Mexico where he knew he could pick up some huge planters or flower pots for his deck, as well as some bright Mexican blankets for his sunroom and a couple of rugs. We decided to go by way of South Padre Island, Texas, and we took off on a road trip in Bill's big white Oldsmobile.

On the way down Bill spotted this decrepit-looking roadside restaurant in the middle of nowhere. It had a big weathered "Bar-B-Q" sign. He insisted that we try it. I was reluctant, but went along anyway. Turns out, that run-down hash-house served the best barbeque we'd ever tasted.

After we left the restaurant and got back on the road, the Oldsmobile started to overheat. As long as we kept it going about seventy, the temperature light would go out. But if we slowed down, it would heat up and that light would come on. We managed to make it to South Padre, where we checked into a room. Then we looked for a repair shop. We found one on the shore side, so we drove over and learned the fan had come off and there was no air flowing over the radiator unless we were going seventy.

We had the fan fixed and drove down to Brownsville the next day. We crossed over to Matamoras, Mexico, and Bill bought a couple of enormous pots that would barely fit into his trunk. On the way back to Ohio, I was driving when a couple of young kids kept getting in front of us and slowing down. I was ready to just ram them when Bill reminded me of the fragile pots in the trunk.

We managed to arrived in central Ohio with the pots intact, and Bill used them to brighten up his remodeled place on Winter Road.

Crappy Dunes

Bill and I spent several vacations in the Crappy Dunes.

During that first trip to South Padre Island, and several others, we stayed at a place with a very elegant name, the Capri Dunes. It was a dusty, dingy, moldy old fleabag of a motel, so we renamed it "The Crappy Dunes." We liked it because it had a deck that stuck way out onto the beach, so you had a great view down the length of the shoreline. Plus it cost only $25 per night.

I'd get up early and walk down the beach while Bill slept. When he awoke, he'd have coffee and work on whatever project he was writing at the time. Bill finished writing and editing his book of essays, *The Fire in the Birdbath and Other Disturbances,* at the Crappy Dunes. When Bill emerged from his work, we'd cruise South Padre. We usually wound up at Blackbeard's, a hamburger and sea-food place, for a late lunch. Later we'd shop in some of the stores and stop for happy hour at one of the bars.

Late one night we went to the Hilton to see the floor show and to try to meet some women. The entertainment was a guy doing 1950s-type rock songs. He was enormously talented. At the break, he came around the audience and talked to the customers. We invited him to sit at our table and chat. It turned out he was from Fort Worth, and he had a Saturday night show at the Summit Hotel, on LBJ and Josey Lane, across the street from where I worked in Dallas. I'd never seen his show, and I made a point to stop by the next Saturday when I got back.

COM Systems

Back in Dallas, I moved on to my next employment adventure.

I saw an ad in the newspaper for an engineering manager at a company called COM Systems, so I applied. It was a local PC shop set up to configure, install, and maintain Novell's NetWare networking system. The owner and manager was Larry Griffin. His wife, Jan, was the accountant and bookkeeper. For additional office staff, they had Larry's mother, plus a secretarial helper and receptionist named Shirley Ingram.

Larry's mom was getting up in years, and I think Larry had her on the payroll just to keep from having to support her for doing nothing. This way he could deduct her pay (or support) from his income tax. I believe she was becoming a little senile. For example, one day she showed up at work without a skirt. The long black slip she had on with her blouse looked kind of like a black skirt, but was clearly a slip made of silk with a decorative lace trim around the bottom.

Six COM Systems programmers were developing custom software to run on the Novell system for our customers. The company also had five technicians configuring, installing, and maintaining the systems. I was to be the "working manager" who would help with the workload and oversee the others.

The big problem—which did not show up until a few weeks later—was Larry.

He was prone to incredible fits of rage, the likes of which I'd never seen in a workplace. He'd get mad at someone and lose his temper, slam his fist into the wall, and fire that person on the spot. It got to the point where there was a monthly ritual of meeting at the Steak and Ale on Friday nights for a going-away party for whoever was fired that month.

This went on for two years. Finally, one day Larry got mad at his wife Jan and not only fired her, but divorced her! He wrote her a check on the spot for her part of the business and told her to get the hell out. I talked the situation over with the other employees, and they all said it was okay to call her and invite her to our monthly going-away party at the Steak and Ale. She became one of the regulars at our little parties. One time she showed up bringing her friend, Kay Massie, and Kay and I became good friends.

Meanwhile, Larry missed the deal of a lifetime because of his enormous ego.

Back in California, when I'd studied business, preparing to possibly start my own company, one of the things that all the books talked about was surviving the first five years. Few lenders would provide money to start-up businesses because so many fail. Rather than risk any money, they'd sit back and watch the start-ups to see which ones were successful. And if you were able to survive for five years, you could receive a lot of attention from the major player in the business world, because you'd obviously found a small niche and were able to make it pay off. After five years, if your niche was a segment of the market that a major player wanted to get into, that company might offer to buy your business and let you manage it for them. Usually your management obligations would last for a five-year period, and then you could bail if you wanted. Most of the time it was an ideal situation, with lots of potential profits in the form of stock, plus you'd receive a nice fat salary for five years.

Well, Novell was going to open an office in the Dallas market, and the company offered to buy COM Systems and use it as their Dallas location. They figured that with this approach, they could minimize their start-up costs, and only have to pay for expansion and advertising. They'd also get COM Systems' installed base.

But Larry the Idiot, not wanting to work for a large company, refused the deal.

Now that I think about it, it was probably not a bad decision on his part. He'd never have managed to work for more than a few weeks without losing his temper.

Anyway, Novell opened an office and warehouse just two blocks away, and took over most of Larry's customers.

Leaves and Lift-Offs

On my property and the one adjacent, there were a lot of leaves from all the trees. In the fall of 1985 I bought a riding lawnmower with a bag attachment to help pick up the leaves. One day as I was plowing along, the leaves began piling up in front of the mower because the rotor blades were not high enough for the enormous volume of leaves. As the pile grew, it reached the exhaust pipe that stuck out the front of the mower. The leaves caught fire, and the flames destroyed my riding lawnmower.

I decided to give up on motorized lawn-and-garden equipment.

My disaster with the lawnmower seemed to foreshadow several much larger disasters that made news several months later.

On January 24, 1986, I felt a sense of triumph as I watched *Voyager 2* on TV make its closest approach to Uranus. I remembered my childhood dreams about technology in general and space travel in particular. It seems that everything was coming true, and it wouldn't be long until civilians could travel in space.

Then, four days later, 74 seconds after lift-off, the *Challenger* Space Shuttle—with a civilian (a teacher from New Hampshire, Christa McAuliffe) aboard—exploded.

That disaster grounded the U.S. Space Shuttle fleet for several years.

Another tragedy occurred on April 26, 1986, when two explosions destroyed one of four nuclear reactors at Chernobyl, in the Ukraine. Among the many tons of radioactive material that were shot into the earth's atmosphere by the force of the blast were iodine-131 and caesium-137. The radioactive iodine has a half-life of only eight days, but the half-life of caesium-137 is thirty years. This means that as I write this in 2006, the risk to human health from the Chernobyl explosion is still present and will continue at least for another decade.

A Good-Bye and a New Beginning

In spite of all that was going on in my life and in the world around me, I found myself missing Leslie. I dreamed about her often. Then one night in December 1987, a few days before Christmas, she appeared to me in a dream and said "good bye" a final time. I still had occasional dreams about her—for example, about seven months later, I dreamed she was getting remarried—but after that December night in 1987, I finally accepted the fact that our relationship was truly over.

Meanwhile the new year of 1988 offered a new beginning. Science and technology continued to flourish with amazing new discoveries. For instance, medicine offered new areas of improved health and remedial surgery. A team of surgeons transplanted an entire human knee, while another doctor used a laser to clear a blocked coronary artery. These operations are performed fairly routinely now, but they seemed miraculous a mere two decades ago. In addition, the first successful five-organ transplant took place when a three-year-old girl received a new liver, pancreas, small intestine, and parts of the stomach and colon.

August eighth—8/8/88—was another Bonza Bottler Day. I thought back to that day so long ago, on May 5, 1955 (5/5/55) when the teacher had introduced us to Bonza Bottler Day. She'd encouraged us to ponder what we'd be doing and where we might be when it rolled around again. She must have taught us well, because in 1988 I thought about the next Bonza Bottler Day, 9/9/99, and wondered what I'd be doing and where I'd be.

Laredo Run—Or How I Narrowly Escaped Spending the Rest of My Life in a Mexican Prison

Bill Allen talked me into taking another road trip to Mexico. I figured he needed more colorful blankets or ceramic flowerpots to decorate his house in Ohio. So in

November I flew to Corpus Christi one Friday after work. Bill picked me up in a rented black Thunderbird, and we drove off to Laredo.

The road was straight with a few rolling flats through the desert. Bill had packed some snacks and sodas, and we cruised along at high speeds having a grand time. Bill stomped on the accelerator, and we did some "low flying" to see how fast that Thunderbird would go. It seemed we only touched the ground every now and then, and I don't think we passed another car all the way to Laredo.

Along the way, about sixty miles outside Laredo, I noticed several inspection areas where eighteen-wheeler trucks were being searched. Bill said the authorities were looking for contraband smuggled in from Mexico—things like drugs, weapons, and illegal immigrants. It didn't concern me, so I just shrugged and didn't think much more about it.

In Laredo, we checked into the La Posada Hotel. That night we went to the Tack Room bar for drinks. We had a wonderful evening discussing books, and all the places we wanted to go and see. Next morning we left the hotel, walked around the corner, and right there was the border-crossing into Mexico.

We walked through the Mexican border patrol building, but no one even glanced at our passports. People were coming and going as if there wasn't even a border. We crossed a bridge that spanned the Rio Grande River and strolled into a small plaza crowded with shops and people.

One of the shops had some colorful rugs hanging outside. I saw one that would look perfect in my A-frame, so I headed over for a closer look. As I was bartering for my rug, from out of nowhere this shady-looking Mexican appeared. A short swarthy guy with beady eyes, he had "crook" written all over him.

In a heavy accent he said his name was Tony, and he asked if we wanted any drugs.

I was shaking my head "no" when Bill enthusiastically responded, "Si!"

Tony hustled Bill down one of the dirty side streets without even looking around.

I was stunned. I'd had no idea we were on a drug run to Mexico. It took me a minute to regain my senses. By then they were almost a block away and disappearing into the crowd. I didn't want to be left there alone in a strange town in a foreign country, so reluctantly I headed off in their direction.

I should have turned, raced over that bridge, and gone back to the La Posada Hotel. But I was so shocked at the thought of Bill buying drugs that I just couldn't think straight.

Visions of the movie *Midnight Express* immediately popped into my head. Was Bill out of his mind? Would we end up in a filthy Mexican prison?

I caught up with them just as Tony was ushering Bill through a wooden door. The giant tooth painted on the nearby window was the tip-off that it was a dentist's office. As I entered, Tony was talking in Spanish to the receptionist. The reception area was crowded with shabbily dressed Mexicans. Some were rubbing their cheeks or moaning in obvious pain. But we were ushered past all of them. Tony gestured for Bill to follow him through a doorway into an examination room. Bill motioned for me to come along.

The poorly lit examination room contained some rusty metal tables and outdated dental equipment. As soon as we entered, another receptionist appeared. She looked from one of us to the other, and said something in Spanish to Tony, who translated:

"Señors, what drugs you want?"

I was sweating. My hands were trembling. I was too upset to speak. I didn't want any drugs. I didn't even want to be there. I just wanted to get across that bridge over the Rio Grande and back to the good ole U.S.A.

Before I could protest, Bill jerked his thumb at me and said, "He needs Valium."

It wasn't a lie. By now I was shaking so badly, I *did* need Valium or Librium or even Thorazine—anything to calm me down.

I watched as the receptionist wrote out two prescriptions for a hundred pills for each of us. Then she shoved a thick logbook in our direction.

In broken English, Tony explained that for legal reasons we'd have to write our names and addresses in the logbook. This was so they could record the prescription and have a record of our visit. Bill calmly filled out the lines, turned the page, and passed the book to me. With shaking hands, I carefully printed my name and address. I did not want the name on the prescription to be different than the one on my passport.

As the receptionist gave us our little slips of paper, *Midnight Express* popped back in my mind. I pictured myself spending the rest of my life in a Mexican prison. I envisioned the cruel jailers, and I could feel my asshole pucker as I was raped....

We each paid the receptionist twenty dollars. The entire transaction had taken less than five minutes, so it wasn't too much of an interruption for the poor Mexicans needing dental work.

Once we were outside the dentist's office, Bill asked, "Did you put your real name in that book?"

"My real name? Of course I did." It wasn't in my nature to lie.

Bill exploded in laughter. "You fool! You were supposed to give them a fake name!"

Oh, great, I thought as Tony led us down another dusty street to a pharmacy where we could get our prescriptions filled. I eyed Tony suspiciously, wondering if he, or the dental receptionist, or even the pharmacist would call the U.S. border guards and warn them that two Americans would soon be coming across with drugs.

Now, mind you, buying Valium in Mexico didn't seem like a criminal offence. We had real prescriptions—at least mine was real, since I'd used my actual name. But the prescriptions were Mexican prescriptions, and we hadn't even seen the dentist.

In my head, I began to rehearse the justification I would use at the border. "Officer, I have an extremely demanding job at Microsoft. You see I troubleshoot computer problems over the phone. I-It's incredibly stressful. Sometimes I even have nightmares. I need this Valium to relax!"

I didn't think the customs agents would buy my story.

Again, I flashed on *Midnight Express* and life in a Mexican prison, surrounded by rats and rapists and *cucarachas*.

Meanwhile, the pharmacy refused to fill our prescriptions.

Great, just great, I thought dismally. The pharmacist knows the prescriptions are bogus—what will the border guards think?

We shuffled along after Tony, who led us down some more unpaved alleys to another pharmacy a couple of blocks away, and then another and another. At last we found a store that was willing to sell us the drugs.

Once outside, pills in hand, we each paid Tony twenty dollars and bid him "Adios."

We trudged back to the bridge. I glanced hastily around us. I was sure any minute the *Federales* or Mexican police would appear with guns and handcuffs. I'd heard of rewards for drug dealers who informed on their customers. Tony didn't seem like the type who'd miss a chance to earn some extra *pesos*.

All the way to the bridge, my mind conjured up visions of the filthy, rat-infested Mexican prison that would be my home until I died.

The bridge, I thought, *if we can just reach that bridge!* Once we made it across the Rio Grande, we'd be in our own country. If there were any problems with the drugs, at least I'd spend life in a United States prison. I might still be raped and beaten, but I'd be raped and beaten by Americans, and on U.S. soil.

At the border crossing, I was sweating bullets. Bill went first, and as usual he strolled right through customs without even a look from the guard. Now it was my turn. Bill had put his bottle of pills in his pants pocket, but I was still carrying the little white sack from the pharmacy.

The guard wanted to know what was in the sack.

I panicked. I gulped. I could feel my eyes widen. What if I lied and they looked in the bag?

I stammered, "V-valium."

The guard gave a steely-eyed stare. Then he ordered me to report to the customs officer across the street. I looked over my shoulder as I crossed the road and saw him watching me, hand poised near his weapon, making sure I didn't make a run for it.

In the customs office, the official wanted to know what I had in the sack.

In a low voice, I confessed. "Valium."

He asked if it was for personal use.

"Yes." No kidding. At that point, I was shaking so badly I was ready to swallow a handful.

He shrugged, said "Fine," and "You can go now."

Barely believing my good fortune, I stumbled from the customs office into the daylight outside. I found Bill waiting calmly on the corner, smiling pleasantly.

I thought about killing him.

I wanted to choke the living hell out of him. What had he gotten me into?!

We went directly to the Tack Room bar where I had several stiff drinks in a row. That night, at the La Posada Hotel, I barely slept. I kept waiting for police to bash down the door and arrest us for drug trafficking.

Next morning, Bill dawdled over breakfast, but I had no appetite. I wanted to get as far out of Laredo as possible.

On the way home, I remembered the inspection stations I'd seen along the highway on our way down. We still weren't out of danger—they could pull us over at any one of these stations, and there we'd be, caught red-handed with illegal drugs smuggled from Mexico. That's always how it works—just when you think you're safe, they stop you for a search.

Once back home, I spent many nights expecting the DEA to break down my door at 4:00 AM, confiscate the Valium, and hustle me off to jail.

It took several weeks back in Dallas before I could get a good night's rest.

Rockport and Key Allegro

I made another trip to South Padre Island at Christmas, 1988, and again stayed at the Crappy Dunes. The new year, 1989, came, and on August 25 I watched on TV as *Voyager 2* made its closest approach to Neptune.

My friend Bill Allen had rented a condo near Corpus Christi on the bay side of Rockport, Texas. In November, I went down there to see Bill and our scientist friend, Eddie Greding. I stayed on Key Allegro, a tiny island adjacent to Rockport.

Across an inlet from Bill's condo was a beautiful Spanish-style house with a red tile roof. The owners had an obnoxious parrot that was up early every morning, squawking and making the worse racket imaginable. Sue Taylor was the real estate agent who had rented the condo to Bill. She told us she'd shown the home next to the Spanish-style house to a client, and he'd asked if the property would be $10,000 cheaper because of the noisy parrot.

The following year, Bill rented another condo on Key Allegro, and again I went down to visit him.

Amateur Radio License

The new decade brought exciting advances in technology. Surgeons at the University of Western Ontario in Canada performed the world's first bowel and liver graft transplants, enabling patients to consume normal diets. In addition, new computer and laser technologies were providing disabled people with greater mobility and communication.

Speaking of communication, in 1991 I decided to renew my amateur radio license. By now I owned a boat, and I wanted to have long-distance communications capability while sailing. I searched through the stuff in my garage and finally found the International Morse Code generator I'd built at Intel. I brushed off the dust and cobwebs, and sure enough, it worked just as designed!

I sat down and started practicing code again. It had been years since I'd used Morse code, and I had to look up several letters I'd forgotten. But I relearned my skills quickly. Soon I could copy five words per minute; then I picked up speed, and could do eight, then ten words per minute. After several months, I started for thirteen—but again, just as had happened years before, something about the spacing and grouping of characters and words just didn't connect for me at that speed.

I had a real problem at thirteen words per minute. Yet for some crazy reason, I decided to try fifteen. Believe it or not, fifteen words per minute was as easy for me as ten. So I jumped up to eighteen. Now I began to run into a problem with my printing speed. I couldn't block print at eighteen words per minute; I'd start to lose characters and get behind. So I decided to write the letters in script. I dropped back to five words per minute and wrote each character until using script became comfortable for me. After several months, I moved up to ten words per minute, and again I had problems with thirteen. As before, I went to fifteen words, and everything was easy from then on. Eighteen came fairly fast, then twenty. Boy, was I happy! I copied twenty words per minute for over a month, and decided to go a little faster. Twenty-two and even twenty-five were way easier than thirteen—go figure.

I contacted the local Cedar Creek Amateur Radio Club and scheduled a code test. The group met at KCKL radio station in Malakoff, Texas, which was where they administered the exam. Three witnesses and the code machine operator were there, and they were all very experienced hams. The operator asked if I wanted to start at five words per minute and work my way up.

"No," I said, "let's start at twenty and see if I can pass that speed."

He set everything up and the test began. I copied 100 percent, and then they gave me a test on what was sent and I scored one 100 on the test. Everyone was very excited—especially me. They next gave me the novice test, which I passed, but then I failed the technician test because I hadn't studied some of the rules.

Next month I came back after studying, and I passed the technician test, the general test, the advanced test, and, finally, the amateur extra test. I was issued the call code AB5MY.

On board ship, I needed a commercial license to work the equipment, so I began studying for the general radio telephone license. I passed it with the ship-board radar endorsement. That brought back memories from 1963, when I'd worked on the ASR-5 airport radar system.

Wang Labs

One day I received a call from a former coworker, Karen Sands. She'd left Motorola Four Phase and now worked for Wang Labs at the regional support center (RSC) in Irving. She said there were openings for good engineers in the telephone support group, and she wanted to know if I was be interested. You bet I was! So I went out and to see what they had to offer.

Wang had a great regional center in Los Colinas supporting the Wang VS system, which Dr. Wang had developed. This was a large microprocessor system designed for data processing and communications. The Los Colinas RSC was divided into several departments, offering support for the Wang VS system, communications, large systems, and a new PC group. I was offered a position in the PC group, which was headed by Morris Lyles. Wang had decided to have a token PC system, although all profits were derived from the VS system. The manufacturing cost of a VS was less than $5,000, but they sold for over $100,000.

Holly Neal became the PC manager, but I transferred to the communications department, headed by Stephen Schneider.

Wang was producing new products, about one every six to nine months. For training, the company often sent its analysts to Lowell, Massachusetts, north of Boston, where the home office was located. So I found myself going to Boston every six months or so.

Meanwhile, my mother was growing old and was having problems. All her sons were working and couldn't provide the level of supervision and care she needed, so my brothers and I decided to put her in a nursing home where she could have constant supervision, three nourishing meals, and accurate medication. When I wasn't in Boston for training, I visited her on Mondays, Wednesdays, and Fridays. My brothers Ralph and Raymond visited the other days. Ralph's wife, Fran, came as often as she could and did Mother's laundry and some housekeeping. Each time I had to go to Boston, I'd tell my mother where I was going and how long I'd be gone, so she wouldn't worry about me not showing up.

One day while I was in Boston, Mom asked Ray, "Is something wrong with Charles?"

Ray said, "No, mom. Why do you ask?"

Mom replied, "I can't understand why his company keeps sending him to Boston to be retrained."

Driving in Boston was always fun. In 1974, when the gas crisis was raging, to save gas the federal government had passed a fifty-five miles per hour speed-limit law. The law included a provision that all states had to allow right turns on red after a stop, except where posted. The key words were "except where posted." To get their share of federal highway funds, Massachusetts was forced to officially allow right turns on red; but to get around the law, Boston posted signs on every corner proclaiming, "No Right Turn On Red."

During one of my trips to Boston, I was with a female analyst named Robyn Casey. We decided to go to Boston's North End, the Italian neighborhood, for some great Italian food. Robyn didn't like driving in Boston so I drove the rental car. Before we reached the restaurant, I drove around the neighborhood so she could see what it was like. I went down a one-way street the wrong way, but no one cared. I turned onto a four-lane street and went a couple of blocks to the next one-way street going right. There was a "No Right Turn On Red" sign on the corner. While we waited for the light to change, seven cars ran the red light. I pointed this out to Robyn, saying, "Did you notice that not one of those cars made a right turn on red?!"

At Wang, I could see what lay ahead, and I tried to tell management that the PC would replace the VS system in a couple of years. But because of the high profits the VS brought in (again, the cost to manufacture a VS was less than $5,000, but the unit sold for over $100,000), the Wang folks ignored my advice. Then Dr. Wang retired, and his oldest son took over. The place nearly fell apart, so after a year, Dr. Wang came out of retirement and got the company back on its feet. But then he retired again. Eventually, Wang declared bankruptcy and started closing the RSCs.

Next thing I knew, I was laid off.

The Egg

By this time Bill was heading back to Ohio to teach. He wanted to know if I'd go along and help him drive, so I took another trip to Bill's place on Winter Road outside Columbus. When we got there, we talked about sailing the high seas. The plan was that I could take my $20,000 nest egg and buy a sailboat, and we could work our way around the globe. Bill could give writing classes from the boat to sailors who were writing travel stories for sailing magazines or to environmentalist who were writing articles and trying to save the world. I could repair computers and electronic equipment and do other odd jobs.

Next morning, I woke up scared as hell! It was a strange feeling—some kind of anxiety attack, I guess. But I told Bill it must be true that we were going around the world on a sailboat, because the idea had scared the hell out of me during the night. Bill put together an outline for the book he planned to write about our adventure. He called it *The Egg*.

Meanwhile, it was time for me to go back to Dallas and look for a job.

See America—Go Greyhound

My adventure with the big grey dog began uneventfully.

To look for a job in Dallas, I had to get back to Dallas. From Bill's place in Columbus, Ohio, the one-way plane fare was $336. But round-trip airfare was only $238. That made no sense, so I called Greyhound. A one-way bus ticket cost only $141, and the round-trip ticket cost $177. At least Greyhound had logical pricing. I could leave Columbus, Ohio, at 1:00 PM on the day of my choice, and arrive in Dallas the next day at 11:15 AM. The Greyhound agent said there was a forty-minute stop in Nashville, and a fifty-minute stop in Memphis. *Sounds good,* I thought. *It'll be an express from Memphis to Dallas, and I can sleep through that part of the trip.*

I packed my bags, and Bill drove me downtown to the bus station. The bus wasn't scheduled to leave for another hour, so Bill hung around and we went into the bus station's cocktail lounge. It was filled with weird-looking characters, reminding me of the bar scene in *Star Wars.* As we sipped our beers, the oddballs shuffled in and out, and some sat down at the tables beside us.

While I'd been staying at Bill's, he'd thrown a little party for me and invited some of his university colleagues. Almost everyone had asked me where I'd gone to college and what had been my major. The same thing happened with the patrons in the bus station bar, except the questions were a bit different: "What prison wuz you in, and what wuz you in fer?" Followed by "That will get you time every time."

We decided not to have a second beer.

By now, it was time to board the bus. I said good-bye to Bill, planning to see him again in Florida in January.

Little did I know what was in store for me over the next thirty hours.

I ambled onto the bus and found a seat next to a man who looked safer than the characters we'd seen in the bar. He was friendly enough, but he was quiet during most of our ride together, and I was thankful for that. We did make some small talk to begin with. He was going to Houston, Texas, he told me. I found out he'd paid $54 for his ticket. I asked how he'd gotten such a great deal, and he said that you had to give them the magic words—"no blanket"—to get the lowest rate. I never did find out what that meant, but I vowed to remember it in case I ever rode a Greyhound again.

The two-hour trip to Cincinnati was uneventful, a straight shot with no stops. I watched the terrain through the window as we cruised along, and admired the beautiful autumn scene as the afternoon sun cast its glow on the Ohio countryside. I learned there'd be a ten-minute stop in Cincinnati, for a smoke break, I guessed.

We arrived at about 3:15 PM. There was some confusion as we "debussed," and I noticed the bus driver having an animated conversation with someone I assumed was the supervisor. I could tell there would be a delay, so I sat down and ate the chicken sandwich I'd brought along. It turned out we were taking on a new group of passengers—a bus-full of migrant workers returning to Mexico. I also noticed one old guy who looked like a typical wino: sixty years old, chalk-white face, deep-set bloodshot eyes, and greasy yellow-white hair. His black suit was so dusty it looked as if he'd been tagged out after sliding into home plate. As he passed me, I noticed he had body odor that could take your breath away. After about forty minutes—instead of the promised ten—we got underway again.

We crossed the Ohio River and cruised into the river town of Covington, Kentucky. At that point, it was discovered that one of the passengers, a little old lady, had been left in Cincinnati. We had to stop and wait twenty minutes while the supervisor drove her over from Cincinnati in his own car. Then we got going again.

The trip from there to Louisville was okay. I'd brought along a book—*Surely You're Joking, Mr. Feynman* by Dr. Richard P. Feynman—and read for most of the way. The ten-minute stop in Louisville took about thirty minutes while unloading and loading took place. We changed drivers and picked up a real character. He began to give everyone a very bad time, especially the migrant workers. He bitched and carried on all the way to Nashville—which took some time, as this was a "local" with stops in every hamlet along the way. In Elizabethtown, Kentucky, we picked up an Amish lady and a little girl who were so typically Amish-looking they seemed to have just stepped out of a Norman Rockwell painting. The little girl had this angelic blank stare. The bus driver explained to no one in general that after the harvest, everyone went visiting and that's why he had to make so many local stops.

We left Interstate 65 at Munfordville and took back roads to Horse Cave and Cave City, where we collected more migrant workers. By now, people were crowding the aisle. It was like riding a bus in Mexico. All we needed were some chickens and goats.

Each time we stopped, the driver would say he was going to call Nashville to let them know we'd be late, and that we had thirty passengers who were going on to Memphis.

In Bowling Green, Kentucky, we picked up a couple of guys that I swear had just escaped from prison. The denim shirts they had on were exactly like the blue prison uniforms you see in the movies.

We finally arrived in Nashville at about 9:00 PM. Sure enough, twenty-five people headed for the bus to Memphis. By the time we lined up, there were at least fifty passengers. Some had been waiting an hour. After those from my bus had been standing for forty-five minutes, we were told we'd have to wait an additional hour because the buses for Memphis were late. We were also told that when the buses arrived, we'd have to wait still longer while all the passengers got off, and while the company cleaned up the vehicles.

By now the line had grown to about seventy passengers, so there'd be three buses to take all of us to Memphis. As we waited, a baby began screaming and wouldn't stop. I wondered to myself if she was practicing to be the next Whitney Houston or Aretha Franklin.

At last a bus arrived, and some passengers came into the terminal. I watched as a teenage girl saw a friend waiting for her and started to scream joyously. A young man standing in our line apparently liked the looks of this gal, so he started to scream, too. Then just for fun, half of the people in the terminal, including me, started to scream. It was a real circus. The girl saw another friend and started screaming again, so we all chimed in. This went on several times.

By now the two additional buses had arrived, but there was another sudden commotion. I saw, standing near me in line, an enormous black lady with a three-year-old child in tow. I hadn't noticed her before, but I figured she must have been standing there for an hour and a half. A group of people gathered around her, and a couple of guys helped her over to the office. On the floor where she'd been standing, there was about a quart of blood and goop. Someone whispered, "She just had a miscarriage." The guy next to me turned green, and I thought was going to throw up.

An announcement came over the public address system that there was a medical emergency and everyone should clear the ramp. An ambulance arrived and parked behind the buses marked "Memphis."

We were trapped.

After what seemed like an eternity, the woman and her three-year-old were loaded into the ambulance and whisked away. The cleaning people serviced the buses and mopped up the waiting room. We finally were allowed to board.

At 11:30 PM we were on our way. However, I'd been separated from my luggage. It was on one of the three buses, and I had no idea which one. But I was tired after all the standing, so I put it out of my mind and tried to get some sleep.

We reached Jackson at 1:00 AM, twelve hours after leaving Columbus. This was the half-way point, so I figured I'd get to Dallas by 1:00 PM the next day, at the latest. After a ten-minute stop, we again got under way. Most passengers were now asleep, and everything was finally going well.

It was the calm before the storm.

An hour outside Jackson and twenty miles from Memphis, a loud racket began underneath the bus. It sounded like the tires were coming off. The smell of burning rubber filled the bus, and we began veering and reeling and swerving all over the road. The driver finally gained control, and we limped to a stop at the side of the road.

The driver got off and set up flares and orange emergency triangles. Through the bus window, I could see him out there in the darkness with a flashlight, trying to find out what had happened. While he was investigating, the second bus went flying by on the inside lane, trying to pass a semi-truck. I was sure he hadn't seen us.

After fifteen minutes, our driver got back on and announced, "It blew a bag." Whatever that meant.

Now we'd have to transfer to another bus, assuming we could get information to the company to send one. The driver said he'd stop one of the other two buses and tell the driver to notify Greyhound of our plight. But I think the other two buses were already in Memphis. We waited for over an hour before another Greyhound came by. It stopped, and our driver sent word of our predicament. It was another half hour before our new bus arrived. We boarded after transferring all the cargo and bags. Finally, we were under way again.

At 5:00 AM we arrived in Memphis for a thirty-minute break. I went into the snack bar for a cup of coffee. The small cups cost eighty cents. What a rip off.

We went flying by Forest City at 7:35 AM and arrived in Little Rock at 9:00 AM. Things were going much better now, but as the sun was now shining into the windows, people began to heat up a little. We were all pretty ripe by this time. Then we were back on the road again, bound now for Texarkana. I'd read almost to the end of my book. In less than forty-five minutes we were off the Interstate, and at 11:00 AM we pulled into Hot Springs. No way would we make Dallas by 11:15 AM, as scheduled. But something positive happened: the smelly wino and the two prison guys got off the bus.

At 11:31 AM we stopped in Arkadelphia for a ten-minute smoke break. We arrived in Texarkana at 1:00 PM, which had been my second estimated time of arrival for Dallas. The quiet guy who'd been my seatmate throughout the trip got off to transfer to a bus bound for Houston, as did some of the migrant workers.

Now there were extra seats, so I stretched out and tried to sleep. The driver informed us that all the smokers would have to do without their nicotine as we were going to haul ass for Dallas.

That's when the rain began.

It poured all the way to Dallas—which slowed us down, of course. We arrived at 4:30 in the afternoon in the middle of a horrendous thunderstorm. Sure enough, there was an accident on I-30, as usual, slowing us down even more.

When we reached the bus terminal at last, I checked with baggage claim and was in luck. My bag had arrived ahead of me.

Thank God!

I caught a taxi to my car and reached home at 7:00 PM. I'd just spent a grueling thirty hours on a trip I could have made in couple of hours by plane. And all to save two hundred bucks.

So much for seeing America by Greyhound.

Microsoft

Microsoft opened its Product Support Services (PSS) center at 1423 Greenway Drive in Los Colinas in September, 1991, with thirty analysts and twenty managers and trainers. I saw the ad in the local paper and went for an interview in December. I aced the interview, was offered a position, and was told to report January 6, 1992.

I was being hired to do phone support for Microsoft's Windows 3.0.

We had a full month of training: eight hours per day in the classroom, plus a lot of self-study. Besides learning Windows 3.0, we also had to know DOS 5.0 and be familiar with WorkBench, CITS, TOMCAT, SRMAN, CompuServ, ABC Collector, the call-center information tracking systems, and Aspect, the telephone switch system. In addition, we were expected to write "knowledge base" articles. These articles would inform the other analysts about areas of confusion, questions, errors, and omissions, and would suggest procedures to identify and correct problems found during troubleshooting. We also wrote status reports that helped Microsoft determine support costs and that aided in accurate headcount projections for future support.

The building I was in was new. Only the first three of the five floors were finished. The fourth floor was under construction and would be used later for classes. The fifth floor was used for "team building."

Once a week, the regional manager would hold a "team building" meeting. We'd all get together in a circle. Then everyone would pass a basketball around. This was supposed to build trust between employees and managers.

It was fun playing basketball, I guess, but it seemed like a big waste of time when the phones were ringing off the hook from customers who needed support.

Three more classes went through the program in 1992, and four in 1993. By the end of 1992, we'd completely filled the building on Greenway Drive, so Microsoft leased another five-story building at 1200 Corporate Drive. That one, too, was soon filled.

At first, there was a beer party every Friday after work, sponsored by Microsoft. This was great fun! But it ended when our numbers reached over a thousand employees. At that point, I think the legal ramifications of company-sponsored beer binges finally occurred to management. Soon there were over three thousand employees working in the support center. We still had pizza parties once a month, but there was no beer or wine.

During the work day, Microsoft had free coffee and cola available on every floor, and there were always lots of free sweets, too: donuts, cookies, pies, cakes, and candies. At first I thought, "Gee, free snacks! What a great company this is!" Then it occurred to me that the goal was to keep the analysts wired with caffeine and sugar, so we'd work harder.

The most popular cola was called "Jolt"—it had triple the usual amounts of caffeine and sugar.

One Friday after work, someone suggested we meet over at D&Bs, a popular local hangout for young people. The average age of Microsoft employees was twenty-six, even with my fifty-three averaged in. Anyway, someone asked if I'd like to go with them to D&Bs.

Boy, would I! This was my kind of after-work social: shouting, hollering, playing a little pool, and chasing women. I jumped at the idea and said I'd meet them there. I dashed out to my car, jumped in, and sped over to D&Bs to get a head start with the women at the bar.

I bought a beer and started orbiting the bar and talking to a few of the ladies. Soon the "kids" arrived. I saw them come in, but then they all disappeared down the hall toward the restrooms. I didn't think much about it at first, but then I noticed they never came out. I finished my beer and decided to go to the restroom and see what was up.

Were they all in there using drugs?

Or were they in there having gay sex?

I carefully pushed open the door of the men's room, only to find it was empty. When I came out, I noticed a door at the end of the hall. So I opened it and went in.

I found myself in a huge video-game arcade. There were over a hundred video-game machines, plus all kinds of electronic car-steering and motorcycle-driving games, plus pinball, Pong, Dungeons and Dragons, and even virtual reality sets. I could not believe my eyes! D&B had modernized to entertain the young professionals.

The dress at Microsoft was not business casual, which usually means khaki slacks and a nice shirt. It was college casual—tee-shirts, jeans, and sneakers. The managers were a bit better dressed, with sports shirts and even a tie now and then, but I don't think I ever saw anyone wear a sports coat. The regional manager wore a suit and tie only on those rare occasions when he had to give business people tours of the facility.

Microsoft had an unwritten policy that they preferred to hire college grads right out of school who hadn't been tainted by other corporation's procedures. I think the strategy was that these kids wouldn't have a frame of reference to know they were being overworked and otherwise taken advantage of.

The company also hired new MBAs straight out of grad school. The MBAs were turned loose on the employees to try all the new techniques they'd just learned, with the idea of getting maximum productivity out of the workforce. These young zealots would have made the managers in *Brave New World* and *1984* proud. Indeed, they seemed to think they were starring in *THX1138*, George Lucas's film about the future dehumanization of workers. The MBAs used technology as it had never been used before—to micro-manage every employee's working time, hour by hour, minute by minute, down to the tenth of a second. All telephone conservations were monitored and recorded, supposedly "to ensure quality." Each manager had a computer configured for every employee he or she managed, and it was connected to the telephone switch. This provided "stealth access" to everything the employee was doing. Managers could intercept and read employee e-mail, and they had access to employees' computer systems and files.

Everything you did belonged to the company. Microsoft owned you, lock, stock and barrel. And if you didn't toe the line, you were out the door. Two new employees in my class were fired the first month, one for not fixing a customer's problem properly, and the other one for using a company computer to contact pop-culture guru Timothy Leary.

When we were first put on the phones, our first month's schedule was four hours of phone time and four hours of research. This was the same schedule as I'd had at Wang. The research time was essential, because we used it to find solutions to customer problems we hadn't resolved during the first phone call; we'd get the answer and call back to finish the job. But the second month, the goal was changed to five hours of phone time and three hours of research, and the third month, it was six hours phone time and two hours research. Six and two became the standard—until the volume of calls further increased. When this happened, management implemented a simple solution: they changed the schedule to seven hours of phone time and one hour of research. This meant you had a choice: you either had a dismal call-resolution record because you had no time to find solutions, or you worked an extra two hours each day, doing the research on your own time.

Paper Blizzard in the Paperless World

In theory, computer data was supposed to be replacing paper documents, because computerized data was easier to manage and store. The irony was that Microsoft employees were inundated with paperwork. There were daily and weekly phone stats, monthly stats, end-of-month reports, expense reports, call-coaching reports, development plans, performance goals, and bi-annual reviews with manager's feedback and peer feedback.

These latter reviews were supposed to be conducted in January and July. But because of the enormous amount of information to be gathered, input, and reviewed, the required data for the January review was submitted by the employee in December, and management didn't complete its rating until February—which may have been intentional, to delay the squeezing out of raises and bonuses. Similarly, data for the July review was submitted by employees in June, and the review by management was completed in August.

Grand-Slam Product Introduction

We'd been supporting Windows 3.0 for about three months when Microsoft released Windows 3.1. One of the bright new MBA marketing geniuses decided to release the new entire product on a single day. The intention was a "grand-slam product introduction and media-coverage blow out." So 400,000 copies of Windows 3.1 were shipped and preloaded to post office branches, Federal

Express offices, UPS warehouses, and DHL warehouses. All 400,000 copies of Windows 3.1 would be delivered the following Monday morning.

Bad idea.

That Monday morning was fairly quiet. We all expected some of the first Windows 3.1 calls to come in, but instead we had just the normal call volume of Windows 3.0 calls. We drank our coffee, ate our donuts, and did our work as usual.

Suddenly, in the afternoon, the switchboard nearly caught fire as 200,000 support calls came in. The call center switching system was set up to send a busy signal to incoming callers when the system was overloaded. After "grand-slam product introduction" Monday, that busy signal went out day after day without stopping.

This continued for two weeks. Then the MBA geniuses in call-center management decided to open the switch and let calls build up in the queues. The point was to get an idea of what the real call volumes were.

Another bad idea.

We weren't dealing with average people who would sit at their phones and use their fingers to dial the numbers, or who would maybe sit and push redial a few times, before giving up. Many of our customers were geeks—smart computer operators who had "dialer demon" technology that could call a phone number every few seconds, again and again, until they finally got through. So now we were more overworked than ever.

To deal with the call volumes, yet another group of MBA geniuses instituted what they called "meet the demand" days. The analysts taking the calls were expected to work as many hours as possible to keep up with the huge volumes of calls for Windows 3.1 support. We were given percentage scores for the number of hours worked, and each support analyst was given a ranking for each day. An available time score of 100 percent meant you'd only worked the normal eight-hour day (only!). My two heaviest "meet the demand" scores were 158.3 percent for April 19 (this meant my work day lasted nearly thirteen hours—making me the tenth highest in the company on that day) and 165 percent for April 20 (more than thirteen hours worked, and fourth highest in the company). My average for the month of April was 134 percent meaning that during that month I worked nearly 54 hours per week, every week that month.

There were other rankings, too, and I'm proud to say I was the only analyst who scored in the top ten of all three performance stats: available time (134 percent as I mentioned above), average calls handled per day (37), and average call time (11.2 minutes per call).

Work and Rewards

After "grand-slam product introduction Monday," I began to feel that working at Microsoft was like being in the movie *Ben-Hur.*

If you've seen the movie, you'll recall that Charlton Heston (Ben Hur) is sentenced to serve as a slave in a Roman war ship. All the slave oarsmen sit in the galley and row in time to a drumbeat, the signal that keeps the oarsmen in synch. Each beat means a pull on the oars. When the ship heads into battle, the drumbeat speed gradually increases—to fast, then faster, then faster still, until the pace is nearly impossible. And of course the slaves are chained to the oars, so there's no way they can stop.

That's what it was like to work at Microsoft.

To reward myself for all this labor, I went to William's Chevrolet dealership in Mabank and bought myself a new 1992 Geo convertible. Because I'd been driving my old Chevette for eleven years and banking what would have been car payments, I paid cash for the new car. The Geo was one of the great economy cars of the 1990s, and mine got fabulous gas mileage. I took a trip to visit Bill Allen in Rockport, Texas, in November, and on the way back, I filled up with gas about fifty miles northeast of Corpus Christi. I kept my speed up at seventy miles per hour, only slowing down to fifty for the traffic in Houston. Later, when I calculated my gas mileage, I found that I'd averaged 50.3 miles per gallon!

Meanwhile, the company rewarded me, too. I received a stock option of 280 shares of Microsoft stock at $68 per share.

I was moved into the group supporting Windows for Workgroup, which was Microsoft's first attempt at networking. There, my scores for available time, average calls handled per day, and average call time consistently exceeded the Windows for Workgroup performance averages in the entire state of Texas.

Microsoft introduced a new version of DOS, which I supported also. Again I consistently exceed performance averages for available time, average calls handled per day, and average call time. And again, there were rewards: the following year, I received another stock option for 180 shares of Microsoft stock at $74 per share.

I was working hard—as I'd been taught to do since childhood—and reaping the rewards.

I've never understood people who lack a work ethic. I remember that a couple of months before I got the second stock option (well deserved, since I was working my tail off every day at Microsoft), I stopped on the way home one day for a haircut at one of those unisex salons. As I sat there, I overhead a female customer talking to her hairdresser about her kids' graduation. Speaking of her own kids she said, "Their greatest ambition and goal in life is to either win the Texas Lottery or sue someone."

Hubble Fixed

On January 13, 1994, NASA declared the Hubble repair mission a complete success and showed the first of many much sharper images. The mission had been one of the most complex ever undertaken, involving five lengthy periods of extravehicular activity, and its resounding success was an enormous boon for NASA, as well as for the astronomers who now had a fully capable space telescope.

I was glued to JPL's website every night from July 16–22, 1994, watching the images of Shoemaker—Levy 9 comet impact Jupiter. That scared me, so I searched for near Earth objects (NEOs) with close approaches to Earth and found a list of potentially hazardous asteroids (PHAs), which is up to 831. The periodic comet 73P/Schwassmann-Wachmann 3, discovered on May 2, 1930, has broken into more than thirty different pieces as it approached the sun and could impact the Earth.

Windows NT

Microsoft introduced its first version of a networking and client-server system called Windows NT. Some rumors claimed this was an acronym for "New Technology," but Microsoft, fearing consumers would equate "new" with "unproven and error-prone," quickly squelched that rumor. I was assigned to be an analyst on the new NT system.

One of my duties was to help improve support-center productivity by identifying and creating "knowledge base" articles. These were information pages that other analysts could access to help them solve difficult customer problems, and the knowledge base helped us achieve team and unit objectives. The idea was to prevent analysts from having to "reinvent the wheel," that is, from doing redundant research to solve difficult problems for which somebody else had already found a solution. I submitted three knowledge-base articles plus one knowledge-base idea. I easily slashed my NT average call times, and in less than a month I had an average call time of 16.83 minutes, while overall average was 16.91 minutes.

Microsoft was big on demonstrating teamwork and positive peer relations, so I was assigned to work with team members to implement our own Windows NT AS domain (known as "Camelot"). The goal was to establish what we called "trusted domains" and create single domains, master domains, and multiple master domains in an effort to understand complex configurations that required several analysts' second computers. I contributed positively to my team and unit morale by cultivating and maintaining good relationships with those around me.

Meanwhile, a Windows NT MCP ("Microsoft Certified Professional") test was coming up. This test was something all our team members would have to take—and pass. So I started up a study group to help us all prepare.

By now, Microsoft's call volumes had exceeded the capabilities of the Aspect phone system, so the company began transferring a portion of incoming calls to the new system. But because the system was new, the company ran into a lot of start-up problems. During one month, the phones were down 10 percent of the time as a result of changing the routing program to balance call volumes. Other functionality suffered, too, that month: one function was down 13 percent of the time, and another was down 20 percent of the time. Additional distractions included the inability to update data information from the server. We were unable to access another system, too (one called "TOMCAT"), because of errors possibly resulting from problems on the network.

In spite of all these challenges, I remained confident that Microsoft liked my work. The tangible evidence of this was the fact that in July 21, 1994, I received another stock option—this time for 210 shares at $47.74 per share. The following year, there was yet another option, for 280 shares at $90.50 per share. And in 1996, I again received a stock option, this time for 100 shares $110.625 each.

Technical Account Manager

After learning NT, I became interested in becoming a technical account manager (TAM). TAMs supported Microsoft's large corporate customers: those with at least 20,000 desktop computers at their own companies. The TAMs helped these customers develop their own help desks, and supplied them with technical support CDs and access to our knowledge base.

To become a TAM meant I'd need to know how to support at least two applications. So I completed the Windows NT self-study program and the advanced server self-study program, and continued to study for my MCP certification. Meanwhile I attended courses in basic Microsoft Word and in Word usage triage.

I also attended seminars in TOMCAT, RAS, and TCP/IP triages and in Microsoft Support Network (MSN) international triage.

I was accepted as a technical account manager and passed three of six MCP tests toward my full MCP certification. This meant that I was ahead of the schedule I'd set for myself in my own development plan. I also completed all premier essentials training, including those for Windows 95, Microsoft PC Mail, and TCP/IP.

As a TAM, I was now more aware than ever that customer satisfaction was a Microsoft priority, especially satisfaction among large corporate customers. To measure customer approval among its largest clients, Microsoft used a paper-form survey called the "Premier Customer Satisfaction Survey." It wasn't easy to get busy corporate managers to fill out the forms. Nevertheless, I was able to achieve a 50 percent return on the surveys, despite the reluctance of some customers to even return phone calls, much less take time to fill out a survey. Microsoft wanted at least a 75 percent return on surveys, but this was practically impossible. In addition, the company wanted all the TAMs to write supportability reviews for each customer and product—a horrendous task, made all the more difficult by my lifetime problems with writing.

Microsoft sought to achieve Premier Customer Satisfaction Survey ratings of 100% "satisfied" and at least 85 percent "very satisfied." But some customers felt that a "very satisfied" rating was never justified. It was sort of like dealing with a teacher who gives grades of A minus but who never gives an A. There was always room for improvement, or so these customers thought. Unfortunately, this attitude could really put the screws to a TAM's performance review.

I succeeded, however, in maintaining an accurate customer profile and account plan for each of my customers, and I set the standard for how profiles and accounts were to be set up and maintained in TAM Central.

Meanwhile, because customer support was expensive, Microsoft began a campaign to recover the cost of supporting corporate customers. The company started tracking all time and materials used to support each customer. My analysis and action plans for all of my accounts achieved a rating of 115 percent for cost recovery, as well as a score of 100 percent for account renewal. I was also recognized for working effectively with customer account reps and customer account managers to facilitate global agreements, and to direct information on reseller/select questions.

$1,000 Bill

I'd continued to save up my cash, and as a tangible symbol of my success, I decided to get a $1,000 bill for myself.

I went to my bank and to several others, but each time I asked for one, I found that no one had a $1,000 bill.

Finally one bank's vice president informed me that the Federal Reserve Bank had been recalling all bills bigger than $100.

I phoned the Federal Reserve Bank in Dallas and learned that all existing bills over $100 were bank notes, designed to transfer large amounts of cash among banks. Since the implementation of electronic funds transfer technology, there was no need for the large bills. In addition, the Federal Reserve recalled the large bills to foil criminals, such as drug dealers, who typically deal in cash. Without the large bills, the crooks must tote around briefcases or even suitcases full of cash in order to transact business. The only place to get them was from numismatists—paper money collectors and coin collectors.

Confessions of a Weaponeer

I saw a PBS special, *Confession of a Weaponeer*, about a man named George B. Kistiakowsky who was President Eisenhower's science advisor. Eisenhower was afraid that the "military-industrial complex" in America had gained too much power as a result of all the money that was spent on World War II. One of his main concerns involved the strength of the Air Force's Strategic Air Command (SAC).

The Strategic Air Command had been formed right after the war as a strategic deterrent—so we could kick the ass of any country that tried to assault us.

The former post-war commander of the U.S. Air Force in Europe, General Curtis LeMay, commanded SAC and established its headquarters at Offutt Air Force Base, outside Omaha. In less than a decade, LeMay developed a powerful all-jet bomber force. LeMay eventually became Chief of Staff of the Air Force, and it was under LeMay that plans were begun for the intercontinental ballistic missile (ICBM) system.

Anyway, concerned about the power of SAC, Eisenhower sent Dr. Kistiakowsky to Omaha to learn more about LeMay's plans for the SAC bombers.

When Kistiakowsky arrived and started asking questions, he was told by LeMay he didn't have "need-to-know" status. LeMay added that civilians just could not be given the information Kistiakowsky was seeking. Kistiakowsky said,

"Fine, I'll just go back to Washington and tell President Eisenhower—who sent me here and who happens to be your *Commander-In-Chief*—that you don't think he has 'need-to-know' status."

Well, guess what? LeMay gave him his briefing.

But from this incident, it became clear that the military-industrial complex was extraordinarily powerful. Indeed, the people in this group had enough congressional support to get *anything* they wanted, and apparently no one—not even the President of the United States—could stop them.

According to the TV show, the military-industrial complex companies actually *created* the "Cold War." They did this to keep their World War II-era funding from running dry, to continue receive financial support from Congress to build new weapons. There was little the President or anyone could do to stop them.

In a number of instances, new weapons were developed that the military clearly did not need or want. But the new weapons generated revenue for the companies. So the weapons were built anyway.

One particular example was the F-111 fighter-bomber. It was advertised as an all-service aircraft; however, it was too heavy to land on aircraft carriers. One admiral called it a "tank," and he testified before Congress that there was not enough energy in the world "to make that tank a fighter."

Then there was an actual tank, the Sheridan Tank, produced between 1967 and 1970 and used in Vietnam. It intended as a lightly armored vehicle with heavy firepower. But like the F-111, it had severe limitations. The huge gun was too big for the light chassis, its firing interfered with the electronics system, and the tank jumped back forcefully whenever the gun was fired. As a result—surprise, surprise—the Sheridan was never widely deployed by the Army.

More recently, there is the Stryker. This is another light tank designed to be droppable from aircraft such as the C-130 Hercules. In 2000, the U.S. Army entered into a $4 billion contract with a consortium of companies, including General Motors, for production of approximately 2000 Strykers. The problem is, many feel the Stryker is too heavy for transport via the C-130—which is the main tactical aircraft used by the military. So in all likelihood, here's another waste of the taxpayers' money on a inefficient weapon.

When I look at this situation, I think, geez, given the military-industrial complex's abuse of power—plus the criminal behavior of companies such as Enron, WorldCom, Arthur Anderson, Xerox, etc.—perhaps we shouldn't be so worried about *foreign* enemies.

Our country is being destroying from within.

The Rise and Fall of Jimmy Ling

In the American business world, the 1980s and 1990s was the era of buyouts, takeovers, mergers, and acquisitions.

The founding father of this trend was Dallas entrepreneur Jimmy Ling.

Ling was a high-school drop out who established a tiny electrical company, Ling Electric Company, in Dallas in 1947 after he was discharged from the navy. He was so broke when he started that he was essentially homeless and lived in the back of his shop. After incorporating and taking his company public in 1955, Ling found innovative ways to market his stock, including door-to-door soliciting and selling from a booth at the State Fair of Texas. At first unsuccessful, he eventually discovered the magic of the merger. Over the next two decades, he began merging with other companies, primarily in the electronics and aerospace industries.

For example, one day Ling was reading the annual report for Wilson & Co. This business had started as a Chicago meatpacker and expanded into pharmaceuticals and sporting goods. Ling saw, in the annual report, that Wilson had substantial cash reserve. He slyly calculated how much it would cost, at the current price, to acquire the majority of Wilson's stock. Then he compared this figure to the reserve, and noticed that there was more cash in reserve than what it would cost to become majority stockholder—a lot more cash. He quietly started buying up shares of Wilson's stock until he indeed became majority stockholder. At this point, he marched in, called a board meeting, and voted a cash dividend to the stockholders. He made millions in profits.

That was in 1967. By 1970, his firm Ling-Temco-Vought (LTV) boasted billions in sales and a high place on the *Fortune* 500 list. Ling himself appeared on the cover of *Time.* But meanwhile, as a result of the bear market that began in 1969, investors had lost faith in large conglomerates: LTV stock plummeted from $167 to $11 per share. An anti-trust lawsuit was filed. Eventually, LTV's board of director's demoted Ling, and in 1975, he left the company he'd built. He still had Ling Electric Company to work with.

Undeterred, Ling continued his life as a corporate raider, doing both friendly and hostile takeovers, usually financing the deals by leveraging his other company assets. But he made some bad buys, and he'd built a house of cards. Ling formed several companies after his time at LTV, but none were as successful. Eventually the house collapsed.

A famous historian said, "Those who cannot remember the past are condemned to repeat it." The folks who got caught in the dot.com boom and bust of

1998 through 2001 would have done well to study Jimmy Ling before making their investments.

Dallas Revisited

Meanwhile, I sometimes revisited some of my old haunts in Dallas.

Downtown at the new Hyatt Regency Reunion Hotel, I rode the elevator to the top of Reunion Tower. There, the revolving Antares Restaurant was located in a rotating dome. As I rode around, it struck me as very ironic that I was above the flying red horse (Pegasus) on top of the Magnolia Building, and that I was going around and it was not.

Jefferson Boulevard

The Oak Cliff shopping district on Jefferson Boulevard had been second only to downtown Dallas in the early 1950s. Its decline began in the 1950s when the Baptists succeeded in getting Oak Cliff voted dry. A lot of the local businesses closed and moved across the river.

By the 1990s, the area was run-down and desolate—a shell without a kernel. I walked several miles from Marsalis to Willomet and back on both sides of the boulevard. A seedy Western Union office and a McDonald's were the only national franchises I passed. Skillerns, Sears, J.C. Penney, and all the other blue-chip concerns had decamped by the mid-seventies, leaving in their place marginal businesses operated by immigrants selling cheap goods produced by cheap labor in Hong Kong and Korea. There was no shortage of 99¢ stores, herberias, pawn-shops, TV rental outfits, and bridal-gown salons. The two most common signs were "Se Habla Español" and "Se Aceptan Estampillas."

In the decades since my childhood, Oak Cliff had become the slums for the "hooks and crooks."

The Texas Theater

In the early fifties, the Texas Theater on Jefferson Boulevard was the principal spot for allowable public pleasure in Oak Cliff. It was a polished place where Daddy took Mama to the show on Sundays. Already twenty years old by then, it was impeccably kept and was not even close to being run-down. But as Jefferson withered, the once-venerable movie house started falling to pieces also.

In 1990 to avert demolition, the nonprofit Texas Theater Historical Society (TTHS), with aid from the Oak Cliff Chamber of Commerce, bought the old landmark, pledging its restoration and development as a cultural arts center. To meet a $3,000 monthly mortgage, TTHS volunteers—many of them teenagers from the area—helped reopen the theater as a $2 rerun venue. But this proved unprofitable, and the theater closed its doors—again. It was threatened with destruction several more times after that, and somehow escaped the wrecking ball. But for several years in the mid 1990s, it stood vacant and pitiful-looking. And that's how I found it on that day when I visited my old neighborhood.

However, I'm pleased to announce that things have changed. The Texas Theater is now being developed as a performing arts center. In 2001, plans were announced to renovate and eventually restore the theater. Live performances will begin after renovation is completed, and restoration will be ongoing.

But on that dismal day in the 1990s, gazing upon the forlorn façade of the Texas Theater, I had no way of knowing what lay ahead. I only felt depressed.

The North Oak Cliff Library

The North Oak Cliff Library, at Tenth and Madison, had replaced the old Carnegie facility in 1987. Built in a trendy modular design, the new building called to mind an upscale shopping mall—all glass and reinforced concrete. It looked incongruous in a neighborhood given over to services for the homeless and the mentally handicapped.

Oak Cliff, I'd found, was rife with such surreal juxtapositions, a fever dream of the rust-covered and the shiny and new.

The local culture had completely changed in a single generation.

Lee Harvey Oswald's Grave

At the other end of Fort Worth Avenue—the Fort Worth end—Lee Harvey Oswald was buried somewhere in the eighty-seven-acre Rose Hill Cemetery. To discourage desecration of the grave by kooks, the exact location had always been kept secret. Oswald's "crime of the century" lay far away in the enigmatic past, and Oswald himself had since disappeared into a blur of dispute. His true role had been obscured by decades of the cloud cover of myth, legend, and suppression of facts.

According to the Warren report, he was a misfit driven to kill by resentment, envy, and even madness. That's not wholly inconceivable. And I'd known a dozen Lee Oswalds when I was growing up: quintessential South Oak Cliff losers mired in a hopeless system that denied them entry. The bottom line was always drawn just above their names. A large number of my graduating class were cops, and an equal number were crooks.

Dallas Reverie

Oddly, as I wandered past the relics of my childhood, instead of anger or disappointment, I felt a sense of almost Buddhistic complacency.

I looked at the rubble of gleaming childhood monuments not with anger or resentment, but with infinite composure and tremendous patience.

At over fifty years of age, I now had a keen awareness of time: I saw the passionate fullness of my life upon the wane, and I cast about for a new goal. Half a century had gone, and the best part of my life had passed. I felt, more than ever, the strangeness and loneliness of my little adventure upon the earth. I thought of my early childhood, of the chicken farm days, the high school days, the military years, the home-again years, the California years, the New England years....

I had a reason for quietude. For I had escaped my destiny. Somehow, I had come through disease, physical weakness, poverty, and the constant imminence of death and misery. I had emerged victorious over all of these.

As I lay in my bed one night, I watched through the window as a great star burned across the western quarter of the sky; I fancied it was climbing slowly toward heaven. And although I could not have said toward what point my own life was moving, I had a vision of a future freedom beyond anything I'd ever known. I had a conviction that I was being shaped to a purpose. I saw, more clearly than ever, that I was a stranger in a strange land, among people who would always be alien to me. Yet I was certain that my desire for possessions and wealth—a desire that was mixed inextinguishably with the current of my blood—would be fulfilled.

It had been a depressing day. I had seen the dilapidation and decay of things that had been bright, gleaming, hopeful beacons throughout my childhood. And now all was gone: the brightness, the hope, and—most painful of all—my youth.

The new millenium would soon arrive. I had no way to know what the year 2000 would bring for me.

But with familiar strength of the loner and dreamer I'd always been, I resolved it would be a beginning.

I lay there on my bed staring out into the darkness. "I'm going to make it," I thought. "I'm going to make it."

And the stars twinkled, as if in consent.

On the Road—in Dallas

The highway system in Dallas was a mess in the 1990s. I know this, because I commuted eighty-five miles to and from work every day during this period. I'd have sworn the highways were designed by frustrated amusement-park-ride engineers, and were managed by politically challenged (PC) politicians who wanted to spread around as much taxpayer money as possible to buy as many votes as possible, especially campaign contributors, in the name of "the greater good of the masses," meaning as many minority (voting blocks) groups as they can. Downtown Dallas was about two miles wide and one mile deep. It was bounded on the west by I-35E, on the east by I-45, on the south by I-30 and the north by Woodall Rogers. The highways cradle the city like a mote around a castle with sixty-story spires, and it's affectingly and appropriately named after an amusement park ride, "The Mixmaster."

My drive to work everyday was a combination of various types of scenery, roads, and drivers. It reminded me of diverse pieces of classical music.

Outside Dallas County, my drive was a beautiful tour through the country, with farms and ranches on either side of the road. Each morning I'd watch the sun rise over lovely scenery. One day I saw skies that looked like upside-down plowed fields. They were red with thin dark streaks, and they had patches turning to gold and peach on a background that ranged from baby blue to turquoise. Another time, along the horizon lay clouds so dark they were nearly black, while above them, the clouds were bright and flecked with gold.

Whenever I remember of this part of the drive, I find myself thinking of the "daybreak" segment from Suite No. 2 of Ravel's ballet *Daphnis et Chloe*. It begins with a muted musical depiction of dawn breaking over a sacred grove. I also think of Morning Mood from *Peer Gynt* by Greg.

I'd be jolted out of this pastoral mood as I crossed the Dallas County line. Demolition Derby time! Coming into town on Highway 175 from the southeast, the highway crossed the Trinity River bottomlands. The farms and the peaceful countryside ended abruptly. I'd drive up through Seagoville and Rylie (where I'd spent my chicken-farming days), through south Dallas (near my old home at 2210 Second Avenue), and across the raised highway near the Last Chance liquor store. A left turn through the Trinity River bottoms took me toward Cadillac Heights. As the road approached Highway 310 (South Central Expressway), it rose considerably to clear the train tracks. On the other side, I'd accelerate down that hill to encounter dead man's curve—a turning road flanked by a solid concrete wall, followed by a sharp 90-degree turn to the right. This was what we called a "come over darling" corner; it was like a roller coaster ride that slams you against whomever is in the seat beside you.

And the fun was just beginning! The three lanes of traffic now funneled to two lanes. Fortunately, the right lane became an exit, so if I didn't merge left quickly enough, at least I had an escape route. Not only were three lanes of traffic on Highway 310 bottlenecking into two, but the two lanes of Highway 175 merged with the two lanes of Highway 310, and the four lanes became three.

After everyone had settled territorial claims in the merge, the freeway began climbing and descending as it passed over the cross streets. This up-and-down process limited visibility to less than a couple of hundred yards, so I never knew what the traffic was like up ahead or what surprises lurked before me over the next rise. If the road was clogged and congested, I'd have less than ten seconds to slow down or stop after I topped the hill and saw the problem. This part of the drive was as hectic and hair-raising as charging into battle on a wild stallion, which is why it reminded me of the *Light Cavalry Overture* by Franz von Suppe.

To get onto Interstate 45 north, I'd exit and head uphill to the elevated freeway. And oh yes, just so none of us morning drivers became bored, there was an on ramp merging into this traffic from a side street. Now the good times really got rolling. I-45 north was one of the most heavily traveled freeways in Texas, coming all the way from Houston with traffic seldom slowing to below 70 miles per hour. (The east side of town was bounded by I-45, an elevated freeway under which the homeless had set up camp.) Because the Interstate 30 exit ramp lay less than a hundred yards from the Highway 75 on ramp, the art of merging became a daredevil skill. Half the cars on I-45 wanted to exit to I-30, while most of the traffic from Highways 75, 310, and 175 wanted to race straight ahead into town.

The long, eight-lane ditch (I-30) went under all the cross streets and was affectionately known as "the canyon." The canyon became a parking lot during rush hour. As a result, city planners built the Woodall Rogers Expressway to ease the traffic flow between I-35E and I-45, which was the elevated freeway to Houston. To cross town to I-35E, you had to exit right, climb up, and cross over to the left. It was just like riding an old wooden roller coaster. I always imagined hearing the clack, clack, clack, clack of the anti-roll-back apparatus on the wooden roller coaster at Fair Park climbing the first hill. At the top of the roller-coaster—I mean, at the top of the highway, I had to make a left turn, and again, like on a roller coaster, there would be a steep drop as my car sped into the canyon to go under the cross streets. And throughout this part of the drive that I've just described, I also imagined Wagner's *Ride of the Valkyries*. Then it was onward, like "The Charge of the Light Brigade": "half a league, half a league, half a league onward … canyon wall to the right, canyon wall to the left … into the jaws of death, into the mouth of Hell." The speed limit was fifty miles per hour, but pickups and other vehicles would whiz by, passing me as if I were standing still. They had to be going a hundred miles per hour.

There was no place to run and no place to hide. Once when I was coming down the hill at sixty-five miles per hour, the right three lanes all stopped because of a backup. Sure enough, some jerk pulled out of the stopped traffic into my lane. I slammed on my brakes and went screeching by in the breakdown lane with less than a foot to spare. The screeching sound was terrifying and deafening as it echoed off the canyon's walls. After I slid pass the jerk and saw I was clear, I down-shifted and continued on to work.

Another time, there was a backup and the three right lanes were slowed, so I opted to go to the end of the freeway and take Industrial Boulevard north.

The car would start the rise to go over I-35E, where there was a sign that read, "freeway ends ½ mile." Just on the other side of I-35E, another sign said, "stop ahead 800 feet," so I'd have to rocket downhill then brake hard to a dead stop.

When I was under the "stop ahead 800 feet" sign, I put on my brakes—and nothing happened. The pedal went to the metal. The traffic to Industrial Boulevard was backed up, and there was less than five hundred feet between me and the last stopped cars. I went through my James Bond emergency procedures, but the ejection seat didn't work. So I started down-shifting and yanking on the emergency brake for all I was worth. I steered the car to the curb, hoping the friction on the wheels would slow me down, and I managed to stop six inches from the rear of the car in front of me. My mouth tasted like it was full of new copper pennies. The light on Industrial changed about that time, so I put my car in gear and gently eased down to a nearby repair shop and had the brakes fixed.

Going home I-35E south (like I-310 and I-175) rose over the cross streets with limited visibility. Again, I never knew what thrills lay ahead over the next hill. To add to the fun, the Dallas North Tollway merged with I-35E right at the exit ramp for the Woodall Rogers Expressway. There was always a slowed lane of traffic on the right of southbound I-35E, waiting for traffic from the tollway to get onto I-35E. This caused some exciting moments as I frantically dodged the cars entering the highway. Just over the hill, traffic slowed even more, and because of the hill, I couldn't see the slowed vehicles until the exhilarating last second. Finally, I-35E went under Woodall Rogers Expressway only to rise again over Commerce Street (near where President Kennedy was assassinated). Then it snaked under Interstate 30 (the old toll road to Fort Worth), back up over the Trinity River, and south toward San Antonio.

After I managed to get on Woodall Rogers Expressway it became like the canyon, a long parking lot. Eventually I would get on the elevated I-45 south after fighting the south bound Central Expressway and the traffic trying to exit to I-30 east and west. It was a real challenge.

One day as I was driving home in my eleven-year-old Chevette with 407,000 miles on it, I had just passed Dead Man's Curve and accelerated to seventy-five to pull away from a pack of vehicles (they always seems to herd together). I was cruising along in the fast lane when I hit a pothole—hard. The left front ball joint failed, and the entire wheel came off. My car skidded over to the right, toward a five-foot ditch beside the road. As this all was happening, everything seemed to be moving in slow motion. But thank God and Sweet Baby Jesus, no other cars were around me! Somehow I controlled the skid and guided the car to a safe stop on the shoulder, a foot from the deep ditch. If the car had gone in the ditch it would have rolled over, and I probably wouldn't be sitting here writing this. After the car stopped I began to quiver, then to tremble, and then to shake. Somehow I got myself under control. I walked to the next exit, found the local police substation, and went in and called my sister-in-law Fran. She picked me up, and I spent the night at her place. After that, I decided to get rid of the eleven-year-old Chevette.

Going home at night, after crossing the Dallas County line it was like escaping from hell. As exonerated former death-row inmate Randall Dale Adams once said, "If there was ever a hell on earth it would be Dallas County." I felt I could relax after exiting Dallas County. The feeling is the same one I get when listening to Debussy's *Afternoon of a Faun.* I also think of the violin intermezzo or "meditation" from Jules Massenet's opera, *Thais.* Sometimes I'd watch the sunset and listen to one of my favorite radio programs, *Adventures in Good Music* with Dr. Karl Haas. That man could talk for a week about a single note. He'd explain why the composer chose that particular note, where else the note was used, and how many times the composer used the same note, etc. And somehow, Haas could make it all interesting.

The Basic A-Cord

Dr. Haas's radio program, *Adventures in Good Music* did a show called "Basic A-Cord," in which he outlined how to take three or four basic musical cords, repeat them in a given sequence, and then modify the sequence while adding a melody.

He started with a chant performed by a group of children, then he added some drums and other instruments. He also demonstrated the technique using Ravel's *Bolero* and Pachelbel's *Canon in D*.

Antonio Vivaldi's *Four Seasons* comes to mind when I think of the basic A-cord of fall to be the cool breeze blowing through the colored leaves and the smell of burning leaves. The melody seems to be the tap, tap, and tap of a hammer driving nails to finish a project before winter, or the chop, chop, chop of the fire wood getting cut, split, and stacked. This is punctuated by the bark of a dog, the call of a bird, or the honk of the geese flying south. Usually, there is the murmur of people discussing something just out of earshot.

The basic A-cord of winter seems to be the quiet, cold, white snow punctuated by the smell of hot cider and warm wood-burning fireplaces. The melody seems to be the jingle of sleigh bells and the scurrying around for Thanksgiving and Christmas. This is accented by Christmas music and carols. Then, there is the calm after New Year's.

The basic A-cord of spring is the warming of the weather, the first flowers, and the budding of the trees. The melody is the songs of all the birds and geese flying north, and the rush to get spring planting done and to freshen up the house and yard.

The basic A-cord of summer is the heat shining down from the sun and rising from everything on the ground. The melody is the tinkle of ice in the tea and the squeak of the porch swing or the rocking chair, as well as the occasional staccato beating of a grasshopper's wings in the still hot afternoon and the chorus of calls from cicadae on a late summer night.

Levelland Revisited—Close Encounters of the Second Kind

In 1997, on the fiftieth anniversary of the July 7, 1947, UFO crash in Roswell, I was out at Bill Allen's place in the Davis Mountains, home of the McDonald Observatory, when I said, "Let's go to Levelland!" It was where we had tried to go to in 1957, to get in on the UFO sightings there.

It didn't take much to convince Bill. So we took off. And this time, forty years later, we made it.

When we arrived we found a restaurant and began questioning some of the locals about their world famous UFO sightings. Everyone we asked looked at us like we were crazy. No one had the slightest idea what we were talking about.

We went to the local newspaper to see if they had any info. That newspaper hadn't even been in business in 1957, so we went to the library.

The library was closed.

Discouraged, I came up with a another idea: "Let's go to Roswell." It was only 150 miles away, not far from the state line in New Mexico. So we drove off, and once we got there, bingo! There was a UFO museum.

Inside we viewed the film the locals had made fifteen years after the 1947 incident, when the government finally stopped threatening them. I bought a tee-shirt and some other souvenirs. Then we went back to Bill's place in the Davis Mountains and began to read about the Levelland sightings in studies that had taken place since 1957. Bill got in touch with Irena Scott, a friend and professor at St. Bonaventure University, who had published numerous articles on UFOs. This is what she sent us:

The Levelland, Texas UFO Car-Stalling Incidents of 1957

This mystery began around 11:00 PM on November 2, 1957 on the outskirts of Levelland, Texas, only an hour after the Russians had launched a second Sputnik satellite (this time with a dog inside), but before the American public had yet been informed of the launching. Patrolman A. J. Fowler received a telephone call from one Pedro Saucedo who described how he and a man named Joe Salaz, traveling together on Route 116, about four miles west of Levelland, suddenly observed a brightly illuminated yellow, white torpedo-shaped object, that seemingly rose into the air from a nearby field. It was an estimated two hundred feet in length, and soon was flying at a high speed judged to be six hundred to eight hundred miles per hour. It passed right over the truck they were in, at which time there was a sound comparable to "thunder" with a "rush of wind." The truck actually rocked, there was "a lot of heat," and as it flew overhead the headlights went out and the engine died. As the UFO moved off into the distance the truck lights came back on, and then the truck started back up without any problem.

This was just the beginning of many similar reports. The next call came from a man who was driving about four miles east of Levelland (the direction Saucedo had reported the UFO had departed in) when he came upon a brilliantly lit egg-shaped object, again estimated at two hundred feet in length, sitting in the middle of the road. As he approached this, his engine failed and headlights went out. The object was lit up like a large neon light, and cast a bright glare over the entire area. This man stepped out of his vehicle, and when he did the UFO rose about two hundred feet into the air, and then blinked out. After this, he then had no difficulty in starting up his car.

Next, some eleven miles north of Levelland, another motorist described how he came upon a mysterious glowing object sitting in the road, and again how as he approached it the engine and lights failed, and again, when the UFO left, all worked well again.

A nineteen-year-old freshman of Texas Tech described how at 12:05 AM, approximately nine miles east of Levelland, his car failed, acting as if he had run out of gas. He got out and checked under the hood. Everything looked fine. When he put his hood down, he then noticed an oval-shaped object, with a flat underside, sitting on the road ahead. He estimated that it was about 125 feet in length, and reported that it appeared to be made of an aluminum-like material, while it glowed a bluish-green illumination. He got back behind the wheel and in futility tried to restart the car. Then he gave up and just sat there and watched it until after several minutes the object rose "almost straight up" and then disappeared "in a split instant." Naturally, after this the car operated fine.

Another caller reported that he was about nine miles north of Levelland when he came upon a UFO sitting on a dirt road. Again, as he approached it his lights went out and his motor quit. Soon the UFO rose vertically into the air, very quickly, until it reached an altitude of about three hundred feet, at which time its lights went out. After this, of course, the car lights came back on and the motor started back up with no problem. Then around 12:45 AM, a motorist driving just west of Levelland, close to the spot Saucedo and his friend had reported the initial sighting about two hours earlier, described how he had observed a large orange ball of fire. At first it appeared to be more than a mile away, but as he watched it continued to come closer. Then, at a distance estimated as about a quarter of a mile away, it softly landed on the highway. This motorist's truck engine "conked out" and his lights went out. A minute later the UFO rose vertically, and then everything mechanically returned to normal. When this UFO was initially seen in the air it was a red-orange color. When it landed it became a bluish-green, and when it rose back up into the air it changed back to a red-orange.

At 1:15 AM, Officer Fowler received yet another call. This time it was from a terrified truck driver from Waco. He described how northeast of Levelland he had come upon a brilliantly glowing egg-shaped object, within about two hundred feet of it, when it shot straight up into the air with a roar and then streaked away. His engine and lights had been affected, but again worked fine after the UFO left.

Around 1:30 AM, Sheriff Weir Clem and Deputy Pat McCulloch were out on the Oklahoma Flat Road, some four to five miles from Levelland, when they briefly glimpsed an oval-shaped object an estimated three hundred to four hundred yards south of their position. "It lit up the whole pavement in front of us for about two seconds," recalled the sheriff. Other patrolmen also glimpsed it or a similar object, but none of their vehicles were affected. Nonetheless, Levelland Fire Marshal Ray Jones joined in the search, and reported that his car's headlights dimmed and his engine sputtered as a mysterious "streak of light" passed him. In total, there had been at least seven separate UFO incidents reported wherein automotive vehicles were described being disabled and then rapidly and fully recovering upon the object's departure, and these all occurred in the same general area and over about a two hour period of time. Captain G. T. Gregory, then head of the Air Force's Project Blue Book, hastily proposed that the UFO was ball lightning, as there had reportedly been an electrical storm in the area at the time.

What we then got through Irena was a follow-up interpretation. This objective analysis helped shape our view of UFO sightings in general. It also made interesting points about human nature in general—specifically, about how unreliable our powers of observation can be. It said, in part:

The Levelland case is entirely dependent on eyewitness testimony. There were no physical traces on the ground, none of the affected vehicles were examined, and no photographs were taken. As a result, the witness testimony becomes the only way to evaluate the case. Unfortunately, many newspapers misquoted the witnesses or embellished the stories. Moreover, very few eyewitnesses were actually interviewed by the Air Force and journalists. In order to determine whether the seven reports were consistent, the source of the report and the manner in which it was reported must be reliable. In the Levelland case, witness credibility was not an issue with regard to the fact that something was seen. The Air Force and the journalists all agreed that something was seen on the evening of Nov. 2–3. Nevertheless, witness credibility was an issue with regard to the description and details of what was seen.

This follow-up story helped clarify the event, but given its parameters it wasn't able to ask some essential questions. In my opinion, one of the most important questions regards the issue of the events being a natural phenomenon, as suggested by the study's most reliable witness. The only natural phenomenon anyone could come up with had been ball lightning, and no studies in the past forty years have come up with any other possibility. The problem here was that ball lightning can't stall vehicles.

Newell E. Wright, the Texas Tech student in Irene's reprint, may have been the most reliable witness—but then he had ventured to call the incident a natural phenomenon, a position he has held to for forty years. Neither he nor anyone else can come up with a natural cause for how or why the vehicles stalled in the presence of the UFO then started again in its absence. This prompted Bill to ask a simple question: "What could have caused the vehicles to stall?"

Without even thinking about it, I said electromagnetic pulse (EMP), which I knew could neutralize electricity because it had done exactly that during the atomic tests of the 1940s and afterwards. EMP had the potential to be a powerful weapon and had caught the imagination of most people who knew about it. It had even entered the arena of popular entertainment. For instance, a form of EMP was used by an alien in the 1953 movie *The Day the Earth Stood Still* to halt most automated activity on the entire planet for thirty minutes, thus demonstrating his planet's technological superiority over ours. Then it was used again, with stunning effect, in the beginning of Spielberg's *Close Encounters of the Third Kind.* Spielberg had hired a team of scientists to research EMP's potential before he used it in the film. The scene involved stalled and then restarted vehicles in proximity to a UFO.

So could the stalled vehicles have been caused by electromagnetic pulse? The answer is an unequivocal yes.

Because our early atomic testing was done at White Sands and Alamogordo, New Mexico, close to Roswell and just over a hundred miles from Levelland, I became convinced that *the aliens are our own government,* a position held by many others. Following Friar Occam's (aka Ockham) dictate, I wound up convinced that if our government had used a form of illuminated aircraft utilizing EMP at Levelland, there would be no reason to suspect that the culprits were aliens from outer space. That's good, because the only alien invasions from space we know of for sure are the fossil remains of primitive microorganisms arriving in meteorites—along with the "building blocks of life," such as the carbon and amino acids present in passing comets.

One definition of "alien" is that it is something unknown and foreign to us. That fits the nature of our government as perceived by the general populace. Compared to the transparent and accountable governments in countries such as Denmark, Finland, and New Zealand, our government is similar to the former Soviet Union. It is secretive and utilizes disinformation in the name of national security. The end result is that our right hand (the average citizen) doesn't know what the left hand (the classified government programs) is doing.

In the Levelland case, my position prompts me to ask why White Sands, which after all is a research proving ground, would use civilian backyards to do its business. Such conduct compromises its policy of secrecy and wreaks havoc on the rest of us. (It also forces people like me to try to find out answers to otherwise simply explained events; in the case of Levelland this answer-seeking has been going on for almost half a century.) We have learned not to expect honesty where top secret projects are concerned. The lack of credibility within the Air Force's Project Bluebook has been common knowledge for years, but at the time of the Levelland sightings, it still was considered the most reliable source of explanation of UFO phenomena. The term "disinformation" was not in our 1957 vocabulary.

Given such practices, how has the U.S. been able to become the most powerful and technologically advanced country on the planet? The answer is science. Scientific achievements have been used to advance technology, which has been ineptly utilized by government. Government is still in the Stone Age, compared to science. If politicians could embrace the rigors and accountability of the scientific community, we would be living in a far different, more peaceful world.

Also, the average citizen is easy to fool. It's interesting that most UFO sightings began occurring around the dawn of the space age. Ever since, we have not only wondered about life out there—we seem to crave it. The reason may lie not only in human nature but also in nature itself. Ever since life began, it has tended toward what's called "clumping," the desire to bind or bond, making life more and more complex and empowered. Why do so many people say that they do not want to be alone in the universe? Do we want to clump with aliens? Could the psychological phenomenon of alien abduction, often involving sex (aka probing), be a fantastic projection of our need to clump? And is a need for a higher or supreme being also part of our preoccupation? Such deep-seated stirrings and concerns provide a convenient smoke screen for our government's monkey business.

My Search for Truth

I have worked all my life to find the truth. But when I find it and try to tell others what I've found, it seems they either don't want to hear it, or a few already know it and are trying to make a profit from it.

I've found that most of the people, approximately 60 percent, don't want to hear the truth I've learned. They just want to go on doing the same old stupid things they've always been doing.

I've also found that approximately 35 percent have no clue what I'm talking about. They're not interested, not aware, not looking for, and do not want to be bothered by anything outside their sphere.

I've found the remaining 5 percent are broken up into two categories: those who are looking for the truth but have not found it, and those few who have already found the truth. Most of those who have found it, do not know what to do with it—like me. The ones who know how to use it seem to be actively trying to keep everyone else from finding out, so they can continue to make a profit from it.

Ashleigh Brilliant said, "I have abandoned my search for truth, and am now looking for a good fantasy."

University Research in Twins Separated at Birth

In my search for the truth, I became fascinated by a TV show about a study of twins separated at birth. The study, which was conducted at the University of Minnesota, began with the Jim Twins. As the study continued, other recently reunited twins participated, as well. It was found that even though they'd been separated all their lives, the twins shared jewelry and clothing tastes, children's names, phobias (claustrophobia is common among twins—something about being crowded in the womb, maybe), fear of high elevations, thyroid disease, and speech defects. In one case, both twins were gay, but in another, only one was gay. One set of twins, Oskar Stohr of Germany and Jack Yufe of California, were separated after their birth in Trinidad and grew up in very different cultures. Yufe was brought up as a Jew, by his Jewish father in Trinidad, and Stohr was raised in occupied Czechoslovakia and went to a Nazi-run school. Nevertheless, they had certain things in common.

A lot of research is needed to determine what is controlled by DNA and what is controlled by social conditions. The show said that a sample of fifty was insufficient for drawing any firm conclusions.

Titan IV-B/Centaur

I and scientists from every corner of the globe watched in glorious anticipation as the *Titan IV-B/Centaur* launch vehicle carrying the Cassini-Huygens spacecraft was launched on the morning of October 15, 1997. The Cassini spacecraft will orbit Saturn for four years, making an extensive survey of the ringed planet and its moons. The Huygens probe is the first to land on a world in the outer solar system—on the surface of Titan, Saturn's largest moon. Data from Cassini and Huygens may offer clues about how life on earth began.

Redecorating

On December 9, 1994, a wonderful thing happened. Even today, it is hard to describe the great feeling I had. It was a rainy, cold, dreary day but my spirits were high. I was redecorating my home.

For ten years I had lived in what was essentially a hovel, and it looked great through my jaded eyes. But then, the interior director came by, put up drapes, and re-arranged the furniture a little. Suddenly, the house looked fabulous!

The redecorating was living up to my expectations. It wasn't exactly the way I'd imagined—the colors were different—but the charm was really coming out, as I knew it could.

For ten years, I'd never noticed all the dirt and spider webs. After all, the spiders were my friends: they ate mosquitoes, which were one of the major irritants in my life. I did not mind the mosquitoes taking a little blood now and then to survive, but the damn itching they caused resulted in a thousand times more damage to myself from myself than the mosquitoes had caused with the loss of a little blood. Then there was the buzzing. If it wasn't for hurting myself so bad, I could ignore the buzzing. But the buzzing triggered a frantic swatting and slapping that sometimes ended in hurting myself more.

I remember someone said that as long as you can hear the buzzing, you can be sure you're okay because nothing has landed yet. It's when the buzzing stops that you need to take action. I think my reaction to the buzzing is genetic and not a learned reaction. The medical implications seem to be minor, although I realize the medical part is really the most important.

Anyway, the little place always looked great to me. But now that she was getting a new dress, her inner beauty was really coming out. There was also a great satisfaction in knowing that the decorating was being paid for by interest I received on money I saved while living there.

Insecurity

For years, fear and insecurity drove me to longing for a house in the country, preferably near a lake. And it had to be paid for, so a loan company or rental agent could never evict me. I wanted a place to put my hand in the dirt and say, "This is mine." The image of Muley Graves in *Grapes of Wrath* had haunted me for years. I was obsessed with that scene where he lets the dirt sift through his fingers and says:

"My Grandpa took up the land, and he had to kill the Indians and drive them away. And Pa was born here, and he killed weeds and snakes. Then a bad year came and he had to borrow a little money. An' we were born here. There, in the door—our children born here. And Pa had to borrow money.... We were born on it, and we got killed on it, died on it. Even if it's no good, it's still ours. That's what makes it ours—being born on it, working it, dying on it. That's what makes ownership, not paper with numbers on it. And now they are throwing me off. I jus' ain't a-goin."

I had known poverty. And my childhood had been haunted by the specter of even more poverty. I could remember my mother telling me that we might someday have to live under a bridge. She was perpetually afraid of not having a place to live or a place to be buried. Later, she always reminded us about the grave plot in Rolins Cemetery, in Lancaster, Texas, that she had bought. She wanted to make sure we knew where she wanted to be buried. Well, she died on December 4, 1992, and she is safely buried there now.

I could also remember coming back to Texas after my last divorce, barely escaping bankruptcy. I remembered scoping out all the bridges in Dallas while I was looking for work, and choosing the one I liked best, and considering how I'd fix it up so no one could know I was there.

Now for the last ten years, I'd been free from those fears. I had my place in the country, near a lake, and it was paid for, and I had a garden where I could plant things, and I had dirt that I could call my own.

Now I was redecorating. Why? Because I had listed my place for sale.

I had decided to sell my house, buy a sail boat, and sail around the world.

It seemed like such a good idea. But that was from the security of my home. Then one night I awoke with the feeling of panic and a terror of eviction, and all those old haunting fears of homelessness reared up on their hind legs to frighten me. The next day I took it off of the market.

Losing My House

I had always received my tax bills in October, and I usually paid them right away to take advantage of the extra 1 or 2 percent deduction. One October day, when I hadn't received my tax bill, I decided to go to the school tax office in Kaufman and pay it anyway. So I stopped by the school tax office and said I was there to pay my tax bill.

The lady looked up my name and said I didn't owe any taxes.

I questioned her as to why not.

She said my name was not on the tax roles.

I told her I'd been paying taxes for the last fourteen years, so why not this year?

"What is your address and lot number?"

"The address is 154 Woodland Trail in Loon Bay, and the lots are 173 and 917."

She looked it up and said, "Those lots are owned by the First United Bank of Terrell, Texas."

"Oh no, they're not! I've owned that property free and clear for the last ten years, and I have the paid tax receipts to prove it."

"You'll have to go the county offices in Athens," she said, "to get this straightened out."

I jumped into my car and made a bee-line for Athens. There, a similar scenario took place as I explained that I owned that property, not the bank. I demanded an explanation of how the First United Bank of Terrell, Texas, got control of my land!

The lady said she would research it.

I said I would wait.

She came back in fifteen minutes and asked if my name was Charles Willingham.

"Yes."

"Do you live at 1432 Wallas Street, in Terrell, Texas?"

"No, ma'am, I don't. I live at 154 Woodland Trail and have for the last fourteen years."

"Charles Willingham of 1432 Wallas Street in Terrell had all of his assets put into a living trust for his kids, and that trust is administrated by First United Bank of Terrell."

"But 154 Woodland Trail is not owned by that Charles Willingham, because 154 Woodland Trail is owned by *this* Charles Willingham."

"Okay," she said, shaking her head. "Let me see what we can do."

"Fine. I'll be back in an hour."

When I came back, I learned that someone had mistakenly put all property owned by anyone named Charles Willingham into the First United Bank trust account. The clerk said the mistake had been corrected. But I learned how easy it was to lose your assets with a simple wrong keystroke on a computer's keyboard.

She gave me a copy of the record with my name as the owner, and I never had any more problems.

George Carlin had a skit where he talked about the unemployment rate. He said that it did not bother him that 6 or 7 percent of the people were unemployed. He said he believes that 20 percent of the population are idiots. The 7 percent of unemployed people does not bother him—it's the 13 percent of the idiots who are still working that bothers him. They are working in the DMV, your bank, your police department, and all the places where you need efficient work to take place for you to survive.

Well, to his list, I can add the county court houses around the nation where all records of our real estate assets are kept.

I have an update: the percentage of idiots has increased. The 2004 presidential election polls showed that roughly 50 percent of the population voted for Bush, and 50 percent voted for Kerry. One of those two 50 percents is made up of idiots.

And it is worse than that. Recently I was watching the New Year's Eve celebration on TV, and as the TV camera panned the crowd, every single person went crazy and acted like an idiot.

Meanwhile, Back at Microsoft

Microsoft had implemented a program called "Investing in our People." This meant better mentoring, tutoring, and training for all personnel. I mentored six new account managers, and tutored several others for their upcoming exams. As a result of this and other efforts, I received the "Teamwork" award for December 1997.

Also for my work in 1997, I received yet another stock option for 240 shares at just under $125 per share.

Then the company changed its customer-service goals, from a rating of "85 percent very satisfied" to the impossible target of "99 percent overall satisfied." I think Microsoft was trying to outdo IBM, which had a published a rating of "80 percent very satisfied." Whatever the reason, the push was on.

I accomplished a lot during my last few years with the company. For example, I exceeded my cost recovery goal by averaging 107.5 percent over eight accounts, with two global accounts, which made for an account ratio of nine. I achieved 100 percent renewal by my customers. Meanwhile, I completed training for Exchange 5.0 as well as professional development training, and I participated in regional recruiting efforts.

But regardless of how well I performed, somehow the figures were always jerked around before a review period. You could work really hard to get your call times down, only to learn that this counted for a mere 10 percent of your performance for the review.

Retirement—and How I Was a Millionaire for a Few Short Weeks

One thing I liked about the company, however, was Microsoft's employee stock purchase plan. Each employee could use 15 percent of his or her income to purchase Microsoft stock, and the company would match the purchase up to 10 percent. I made a point of always having the full amount withheld from my regular pay for stock purchase.

To retire, and to keep my income tax rate below 50 percent, I had to exercise one third of my options per year for three years. I also had to sell some stock to pay for the rest of the stock and to pay the tax. In addition, the company took out the maximum for social security and Medicare for those three years.

My taxes in 1997 were $29,588.37. By comparison, the taxes for my last three years were as follows:

    1998—$63,000
    1999—$65,000
    2000—$33,000

I retired at age fifty-nine on February 19, 1999, one day before my sixtieth birthday.

When I retired, I had a little over 10,000 shares of Microsoft stock left after taxes and paying for the stock. The price at the time was $119 per share. That was $1,190,000—so on paper, at least, I was a millionaire.

But as I once saw on a bumper sticker, "Ignorance can be cured. Stupid is forever." In April, 2000, the geniuses in Washington, headed by Allen Greenspan, decided to cool the economy by raising the interest rate.

Other geniuses in the Justice Department filed a lawsuit against Microsoft for alleged monopoly practices.

Microsoft stock fell $70 dollars per share, from $119 in February, to $49 in April.

On paper, within a matter of weeks of my last day of work, I lost $700,000 dollars—about 60 percent of my retirement.

Part Five:
The Sailing Years

Chapter Thirteen:

The Sailing Years (1973 to 2001)

The Beginnings

The sailing years really began in 1971 in California when my friend Ed Grady introduced me to windsurfing—his way of getting even with me for talking him into sky diving. Ed was a real athlete, and a few weeks after our skydiving escapade, he dislocated his shoulder while skiing. At the hospital he was put in a room with a guy named John who was also having his shoulder reset. John was an avid windsurfer, and he invited Ed to take a windsurfing class he was teaching in Redwood City.

When Ed's shoulder healed, he asked me to join him for the class. So up to Redwood City we went. I took one look at the windsurfers and thought, "This is a piece of cake! All you do is stand on a board and hold onto the sail." I soon learned that I was wrong—during the first hour, I spent fifty-five minutes in the water, four and half minutes standing on the board and less than thirty seconds doing anything close to sailing. Keeping your balance and holding the sail was a lot harder than it looked!

That little board has three degrees of freedom. Roll, pitch, and yaw. A ninety-eight pound, fourteen-year-old does not have much trouble standing up on the board, but a two hundred pound man exercises all three degrees of movement, like standing on a greased pig's back. Unless you're in the exact center of balance, you're in the water.

This sport is a worthy adversary. The first problem is getting the sail out of the water. A hundred square foot sail can hold a lot of water, and at a pound per pint ("a pint is a pound the world around") that is approximately eight hundred pounds of water in the sail. You cannot just yank the sail up. You have to slowly pull the mast out of the water and let the water run out of the sail. This is a non-linear process when it starts, and then it gets easier faster and faster. You are pulling with all your strength to get the sail out of the water and all of a sudden the mast comes up and hits you in the head, knocking you off the board backwards. All the while you are trying to get the sail up out of the water, the three degrees of freedom—roll, pitch, and yaw—are working against you. When you stand up on the board and find the center of balance, you have the uphaul to hold on to for extra balance.

After you finally get the sail up, and still able to stand up on the board, now the wind takes over and tries to whip the sail out of your hands. You have this tiger by the tail now, and it is trying to eat you. This flapping sail has to be thrown across your chest with one hand, in one quick motion, and the other hand must grasp the boom as you fall backwards toward the water. Then, you are in sailing position, and the wind catches the sail and holds you up and keeps you from falling into the water. Now, all you have to do is maintain this sailing position. But now you have five elements to contend with. Roll, pitch, yaw, the gusting wind, and the waves. I haven't even started to talk about maneuvering or tacking.

After three months, I finally was able to sail the board. After another three months of practice, in the training lake Ed and I decided to try the South End of San Francisco Bay.

We went to the Redwood City dock, where Charlie Brown's restaurant was, launched our windsurfers, and sailed out into south San Francisco Bay. It was great. The waves were higher than on the lake but it was good training. When we started back, we went up one of the side slews and there was this huge barge docked there. It had TV cameras all around, and they were watching us. When we sailed past the transom, it had "Glomar Explorer" painted on it. Oceanographers have long known that parts of the Pacific sea floor at depths between 14,000 feet and 17,000 feet are carpeted with so-called manganese nodules, potato-size chunks of manganese mixed with iron, nickel, cobalt and other useful metals. In the 1970s, Howard Hughes used the Deep Ocean Mining Project [DOMP] search for nodules, as a cover for building the ship *Glomar Explorer*. *Glomar Explorer* went to sea on June 20, 1974, found a sunken Russian sub, and began to bring a portion of it to the surface. An accident during the lifting operation caused the fragile hulk to break apart, resulting in the loss of a critical portion of the submarine, its nuclear missiles, and its crypto codes, said the government press release. However, according to other accounts, material recovered included three nuclear missiles, two nuclear torpedoes, the ship's code machine, and various codebooks. With the exception of the brief stint, the vessel has since been mothballed with the Naval Reserve Fleet in Suisun Bay, California, where it could be seen by cars crossing the Benicia Bridge on U.S. Highway 680 east of San Francisco.

After we docked and started to put our windsurfers, up there were a couple of fishermen fishing from the dock. To be friendly, I asked how the fishing was going. One of them pulled up a stringer and on it were about eight sharks three to five feet long.

He said, "Nothing but these darn sharks."

Wow! I wondered how many others there were out there where we were sailing, and if they were any bigger.

But—probably because I love a challenge—I became hooked on windsurfing.

After a year's practice, we found out that a lot of expert windsurfers would meet at the St. Francis Yacht Club in San Francisco on Thursday afternoons, and sail out under the Golden Gate Bridge.

Some of the expert windsurfers would sail across the Golden Gate to Sausalito. Afterward, they would have dinner and drinks at the Edwardian Club in the Marina district. I was not good enough to sail to Sausalito; I could just barely make it up to the closest Golden Gate Bridge pylon. There was always a thirty- to forty-knot wind blowing in through the Golden Gate, and depending on the tides, the waves were really great. If the tide was coming in the waves were just big four- to eight-foot rollers. If the tide was going out, it would create a three- to six-foot sawtooth wave by pushing the out flow water against the incoming waves the wind created. This created some great jumping ramps. You could lean way back against the thirty knot wind and jump clear of the water completely. Then you had to slack up on the sail as you went down behind the wave.

Windsurfing was fun, but it was also hard! It gave you a real workout, and it was worse than pumping iron for an hour. After one particularly grueling windsurfing session, I was so exhausted I thought I would have a heart attack. So I collapsed and tried to catch my breath on one of the nearby jetties.

While I was lying there gasping for breath, I gazed over at all the big sailboats in the Saint Francis Yacht Club's slips. I thought, "Sailing wouldn't be a bad sport if you didn't have to spend the whole time fighting that damned sail!"

After I rested up a little more, I said to myself, "Some day—I don't know when or how, but *some day*—I'm going to get me one of those big yachts and really do some sailing."

Learning

To prepare for my dream, I took every boating course I could find.

From the local Coast Guard auxiliary, I took six lessons in "Boating Skills and Seamanship." Then I signed up for "Piloting," "Marine Engines," "Inland Waterway Locks and Dams," "Radio/Telephone Equipment," and "Weather." Next, came my "Sailing and Seamanship" course, which included lessons in "Marlinspike Seamanship" and "Basic Sailing."

The more I learned about boating, the more I wanted to learn.

I took a class called "Coastal Navigation." This was really a great course that covered all the techniques of navigating along a coastline. We had several class meetings during which we did tide calculation around the San Francisco Bay. Our instructor pointed out that the ebb tide around Angel Island was really strong—sometimes as strong as eight knots. I learned that most small sailboats can't handle an ebb tide of this magnitude—they'd be swept out under the Golden Gate Bridge toward the Farallon Islands and would end up way off shore, out where the great white sharks go to feed. So when sailing in San Francisco Bay, you really had to watch the tides. Fortunately, most marine supply stores carried little folders with tide tables in them.

I took a course in "Advanced Coastal Navigation" and one in "Radio/Electronic Navigation." Then I enrolled at Foothill College in Los Altos Hills, where I took a class in "Celestial Navigation."

I learned that the terrestrial, celestial, and horizon coordinate systems are combined on the celestial sphere to form something known as the celestial triangle. When this triangle is related to the earth, it becomes the navigational triangle, the solution of which is the basis for celestial navigation. The calculations we did were fun and interesting; we used a resource called the *Sight Reduction Tables for Marine Navigation HO-229* (also called a "Nautical Almanac") to get the necessary data for our computations. This book contains precalculated solutions that can be used to plot the celestial line of position (LOP) by the altitude-intercept method.

To help me navigate, I bought a cheap Davis sextant, but it turned out I didn't need it. The navigation instructor gave us all the data we needed to do our calculations. Oh, well, the sextant looked good on my desk. It made me look like a real sailor!

All I needed now was a boat.

One of the engineers at work had a beautiful Catalina 26 sailboat moored at the Palo Alto Marina. I wasted no time becoming good friends with him, and soon I was one of his crewmembers. Whenever he sailed in San Francisco Bay, I'd tag along and help out and try to learn as much as possible. But he didn't go sailing often enough to suit me, so I sought out others with sailboats.

Ed White had a 26-foot Hobby Cat that he raced on a pretty regular basis, so I signed up to be ballast for him. We often raced around the bay, but the real fun was the offshore races at Santa Cruz. The racecourse there was huge, with long legs, so you could get a good feel for the wind and seas. The outer marker was the five-mile buoy, and there was always a good twenty-five- to thirty-knot wind. Sometimes I'd be hiked out on the trapeze, and Ed would get us up on one pontoon and we'd really scoot to and from the outer marker!

Sailing a Peacock in Buzzard's Bay

So I took a lot of sailing courses, and thanks to a couple of coworkers, I put my newly acquired knowledge to good use. Later, when I moved to New England, I immediately began hunting like a hungry shark for another friend with a sailboat. I found one—Dick Fow.

Dick was a programmer I'd hired to work with me at Alden Electronics. I can't remember if "Do you by any chance own a sailboat?" was one of the interview questions, but I confess it might have been.

Dick Fow had a gorgeous 38-foot Tayana sailboat berthed near Newport, Rhode Island in Falmouth, Massachusetts. The boat was named *The Peacock*. Dick and I would sail that bird into Buzzard's Bay and Rhode Island Sound, then out to Martha's Vineyard and back.

It was truly marvelous.

A Landlocked Texan Dreams of Sailing

As discussed in the previous chapter, life eventually brought me back to Texas and to Cedar Creek Lake outside Dallas.

I located the local Coast Guard auxiliary, and in 1988 I took another class in "Boating Skill and Seamanship."

I continued to dream of sailing—but nothing materialized. I was far from saltwater, and in spite of my best efforts, over the next few years no sailboating friends or coworkers appeared to satisfy my cravings. Nevertheless, to maintain my skills, in 1990 I took a refresher course in "Sailing and Seamanship." Meanwhile, I continued waiting and hoping to bump into someone with a nice boat for me to sail in.

But no boat-owners stepped up.

"Okay," I finally said to myself, "The heck with it! If I can't find a guy with a Tayana or a Catalina or even a Hobby Cat to help me out, I'll just buy a damned sailboat myself!"

I spent hours planning and scheming and dreaming about my future vessel and the voyages I'd take. Like a neurotic housewife clipping and saving coupons, I clipped and saved colorful magazine articles about all the places I'd visit. Soon I had not one but *two* boxes of file folders crammed full of articles. The articles in the alphabetically sorted and labeled folders were about Australia. And Barbados. And Bora Bora. And Easter Island, Fiji, Galapagos, Madagascar, New Guinea, New Zealand, Panama, Pitcairn, Polynesia, Samoa, South Africa, Tahiti, Tonga … and a host of other far-off and beautiful looking (in the glossy photos that accompanied the articles) places—many of which had exotic names that looked impossible to pronounce.

For the next decade, I fanatically clipped, filed, and hoarded every article in *Cruising World, Yacht World, Sail,* and *Seven Seas Cruising Association.* I meticulously recorded and filed the details of every conceivable port of call—including procedures, port restrictions, and quarantine areas, as well as the names, phone numbers (where available), and locations of officials and offices.

I made careful notes about the provisioning and repair capabilities of each site.

I read Earl Hinz's *Landfalls of Paradise,* and Marcia Davock's *Cruising Guide to Tahiti and the French Society Islands.* There are dozens of books about sailing in Polynesia, and I read nearly all of them. I studied Jimmy Cornell's books: *World Cruising Survey, World Cruising Routes,* and *World Cruising Handbook,* and I created my own world cruising handbook, based on his.

I even went so far as to write to "El Direccior Nacional Forestal Ministerio de Agricultura y Gandria." This was a head honcho in Quito, Ecuador, who oversaw the distribution of cruising permits for the Galapagos Islands. The permits were very hard to get, because in order to maintain the ecological balance of the area, the country restricts the number of cruising boats allowed to visit the island sanctuary each year. Miraculously, I received a permit, which I carefully filed in its proper place.

Each night I'd pour over my folders for hours. I'd read and reread the articles, obsess over the photos, and scrutinize my notes. Most importantly, I'd study and memorize tidbits of knowledge—knowledge I knew I'd need.

Someday.

When I bought my beautiful sailboat and set off around the world.

Saving

An article about sailing quoted the *Guinness Book of World Records* as follows: "Ninety-five percent of those who say they are going to sail around the world do not get their boats in the water, 95 percent of those who get their boats in the water do not get underway, and 95 percent of those who depart do not complete a circumnavigation."

Nevertheless, I started saving to buy a sailboat and sail around the world.

I was paid twice a month, so I'd live on one check and bank the other. My goal was $100,000.

I'd decided a 38- to 45-foot offshore sailboat would be perfect for my voyage. I took trips to Florida to shop for boats. I found some each time I went, but they were all over my budget by about $50,000.

When I first started looking, I had planned to spend $80,000 for the boat. That would leave $20,000 for the expenses of the cruise. Unfortunately, by the time I'd saved $80,000, prices had risen more than I'd expected, and the boats I wanted were up to $90,000. Then by the time I saved another $10,000 (about another year), prices had gone up to $100,000. So I diligently saved for another year, but by the time I had another $10,000, prices had gone to over $115,000.

I was getting nowhere.

Code

While I was working and saving for the boat, I made a list of all the things I should do to prepare for the trip. I needed long-range communication, so I decided to get my amateur radio license.

On board ship, I'd also need a commercial license to work on the equipment, so I began studying for the general radio telephone license exam. I passed it with the shipboard radar endorsement.

Regatta Del Sol and Veracruz Run

In spite of my success with communications, not only did I still lack a boat, I had little sailing experience.

The Regatta del Sol was a yacht race from Pensacola, Florida, to Mexico's Isla Mujeres on the Yucatan Peninsula. Sailors from the U.S. and Mexico enjoyed festivities at both the start and the end of the race. I tried to find a sailboat racer who'd let me join the crew for the Regatta del Sol, but most skippers were reluctant to take on someone with so little offshore experience.

The Veracruz Run was another race for which I tried to get a crew position, but again, with my lack of offshore experience, no one would take me on.

South Padre Island

Meanwhile, my childhood friend Bill Allen continued to visit Texas every year, and we'd often go to South Padre Island for the week between Christmas and New Year's. Most of the time we were there, we just partied and went to bars like Blackbeard's and Louie's Backyard. One time, we were in Louie's Backyard eying the waitress when a flashy young dude came in with a great looking lady. They sat at the table next to us, and Bill started chatting with them. Eventually, the talk got around to fast boats. The guy started bragging about how fast his offshore racer would go, and how often he made trips to Mexico and back. We sensed that the guy was a drug runner, so Bill launched into a story about the fast boats he drove in Florida for the DEA.

The guy turned pale as a sheet, and the conversation came to a standstill. Soon afterward, the couple left.

One year Bill stayed in a condo in Rockport, Texas, while he was writing "*Aransas: The Life of a Texas Coastal County,*" a book about the area. I visited him there, and we found another hangout called Tortuga Flats in nearby Port Aransas. We 'd drive over to Port A, take a five-mile walk on the beach, then go over to Tortuga Flats and talk about sailing while we drank beer.

A Good-Luck Omen

When I was a little boy, my mother said, "If you see a falling star, make a wish and your wish will come true."

One morning, my alarm sounded at its regular 5:40 AM time, and I jumped out of bed with excitement. That day there would be a falling star: the space shuttle would be streaking across the southern sky between 6:05 and 6:10 AM, on its way to a landing at the Kennedy Space Center in Florida. I'd heard it would make the horizon-to-horizon transit in less than two minutes.

I'd decided that if I managed to see it, I'd make a wish and it would be a good-luck omen. On this day I was scheduled to go to Houston to look at three boats, and I hoped one would be a winner.

I'd already checked out a very promising 36-foot sailboat called *Summer Time* when I'd been down in Houston with Bill Allen the previous Christmas. It had all the right stuff, but it seemed a bit too small. Size matters, and I was looking for something bigger. I'd initially set my limits in the 38- to 45-foot range, but I'd later adjusted them upward and wanted a 43- to 47-foot vessel.

However, as I began saving and struggling to keep up with rising prices, I realized my finances were limited. So I adjusted my expectations downward. I was now considering smaller vessels, between 36 and 41 feet, as well as boats that needed tender loving care (TLC)—a lot of elbow grease, rubbing, and fussing—to make the teak and brass shine. After that, I'd just have to buy some equipment and take the three shakedown cruises, then I'd be ready to take off on the greatest adventure of my life!

As I showered, shaved, and dressed, these words echoed in my head: "Just a little TLC, some cosmetic fixing-up, and it'll be ready to take you anywhere." That's what the boat salesman had said on the phone.

I was out the door by 6:00 AM. The temperature had dropped to 28 degrees, which put a spring in my step. I drove to What-a-Burger, grabbed two breakfast burgers and a large cup of coffee, and headed off to look at boats.

Meanwhile, I'd barely taken my eyes off the sky. The space shuttle was scheduled to pass over between 6:05 and 6:10 AM. As I left What-a-Burger, there were two con trails in the heavens. One running from north to south, so I didn't think that could be the space shuttle. And the east-to-west one was going the wrong way, so that probably wasn't it, either. Besides, I didn't think the shuttle would leave a con trail because it was up too high.

I never did see the space shuttle, so I didn't get my good-luck omen. *But after all*, I told myself, *not seeing doesn't mean* bad *luck*. And it was a beautiful morning anyway. *Thank you, Lord, for such a beautiful day*. Meanwhile, I knew the space shuttle was up there somewhere, falling through the sky as it sped toward Florida, so I made my wish on a falling star anyway.

I wished for just the right boat, and I hoped the wish would come true.

Boat Shopping

I arrived in Houston at 10:00 AM sharp. I was an hour early for my appointment with the salesman, so I decided to stop at Baker Lyman & Company to see what goodies they had.

Baker Lyman sold nautical charts for recreational boaters. Once inside, I was like a kid in a candy store. They had charts for the whole world! I drooled on one that charted a course from Panama to Peru. I found two more charts I liked: one for the Indian Ocean that traced a route from the Cocos (Keeling) Islands to St. Paul Island, and one from St. Paul to Rodriguez Island. These two charts had little spots in the opposite corners that indicated the islands—the rest of the maps showed only water. I bought a twenty-four-inch parallel rule for plotting, and an antique world map for my office, so I could show my co-workers where I'd be going for my date with destiny.

Outside, it was chilly, but the sun was shining and the air was calm. This meant it was a bad day for boat shopping. It's easy to buy the wrong boat on a beautiful day, but when it's raining and dreary, you can see the real condition of the boat far better. I'd hoped for an overcast sky and maybe a cold rain.

The previous Christmas, I'd fallen in love at first sight with the boat I'd looked at, a Union 36 sailboat called *Summer Time*. My salesman, Jerry, acknowledged that it was smaller than what I wanted. But we both agreed that otherwise it fit me perfectly.

Its construction was very similar to that of a Hans Christian sailboat. I later learned that it was, in fact, a spin-off from the Hans Christian—just like the Lord Nelson, another type of sailboat that I liked but couldn't afford.

Summer Time was fully equipped for world cruising: manual Aries wind vane, wind generator, repacked and recertified six-person Zodiac life raft, extra flares, water, Coast Guard safety package, 9,000 BTU air conditioning, new head and manual holding tank, remanufactured engine and transmission, new prop and shaft, Garmin global positioning system (GPS 45) navigation equipment, Auto Helm 3000 auto pilot, satellite navigation, single sideband (SSB) radio modified to provide weatherfax, ham radio antenna on the backstay, dinghy, Bimini top and dodger, and many more essentials. The only major equipment to be added were radar, an icemaker, and maybe a Gen-Set generator.

I thought about Bill, whom I'd already recruited as first-mate. He would want a generator. *He'll just have to make do with what I have,* I thought as I inspected the boat.

The look, the feel, the workmanship, the placement of everything was perfect. She was tailor-made for world cruising, which was exactly what she'd been doing for the past eighteen years. And that was the problem. As we say in Texas, she'd been rode hard and put up wet. She was worn out.

Although the teak was recently varnished and the color shone through giving off the original warm glow and feel, you could tell by the nicks and cracks around the mounting hardware that *Summer Time*'s grip on life was not what it used to be. The teak decks were spongy in a couple of places, indicating delaminating of the decks. There were lots of cracks around the deck at the hull-deck interface. The cabin roof had major rot areas around the dorades, and I spotted several places around the top corners where water had seeped in. If the delamination and rot were limited to the small areas that could be seen, these problems could probably be repaired for under $10,000. But if the damage was more extensive, a new deck and cabin top would be needed, which would cost some real money. And that was just for problems on top. If what I saw was any indication, the hull probably had suffered from neglect also.

I'd passed on the chance to make an offer on the Union 36 *Summer Time* in Houston last Christmas, but now I was back to look at three more boats.

The first was a Formosa 41.

The Formosa 41 was in worse shape than the *Summer Time*. It definitely needed a new deck and cabin top. The mast was cracked and needed replacing also. The six-volt electrical system would have to be converted to a twelve-volt system. I noted many more problems, and these were just the things that were easily visible. What was below the waterline, only a survey could tell. The boat was priced at $70,000, but the salesman said the owner was eager to sell, and I could probably pick it up for $35,000. It was a great price, but I was suspicious about the boat's condition. So it was on to the next candidate, the Endeavor 37.

The Endeavor 37, named *Dragon Slayer*, was a cheap production boat like the Morgan Out Islander that Bill and I had checked out in Rockport. When I looked in the storage areas at the hull-deck interface, I could see light filtering through the deck because it was so thin. The layout was for chartering, not world cruising, and the workmanship was of minimum quality. So it was on to the last candidate, a Yankee Clipper.

We had to go to the Lakewood Yacht Club to see the Yankee Clipper, named *Westphalia*. The boat was open and someone had clearly been working on it, but no one was there. This boat was in better shape than the other two, but nevertheless it needed a lot of work. Also, the layout was cramped and poorly organized. It did have GPS, radar, SSB, and Autohelm 3000 autopilot, but it would need about $35,000 worth of additional equipment to make it world-cruise ready.

Jerry, my salesman, could see I wasn't thrilled with any of the three boats. He suggested I make an offer on *Summer Time*, which was still for sale, and stipulate that the owner make repairs to correct the problems.

On the way back from the yacht club, I asked Jerry to drop me off at a coffee shop where I could sit and think.

Yet Another Boat

I sat in the Kettle Coffee Shop pondering what to do next. Should I make an offer on *Summer Time*? Should I keep boat shopping and hope for something better?

I decided to try and call David Cameron. He was another boat salesman, a friend of my acquaintance Glenda Carter. I wanted to see what he had to offer. I looked in my address book and saw that I had Glenda's phone numbers but not David's. This was in the days before cell phones were ubiquitous, so I went to the pay phone at the back of the coffee shop. I didn't have enough change to make a long-distance call, so I tried to call Glenda and charge it to my home phone. I'd done this many times in the past, but I learned the phone company no longer permitted this. I hung up, went back to my seat, and had another cup of coffee.

I thought, *I'll just dial Information and get David's number that way.* I changed two bucks and armed with my fistful of coins, I marched back to the phone and dialed 411.

"Operator, I need a listing for David Cameron."

"In what city, please?"

"I don't know the city, exactly. He lives in Texas."

There was a long pause.

"Do you know the area code?" the operator asked.

"I think it's 713."

"In the 713 area code, I'm showing five David Camerons, two of whom live on the same street."

Oh swell, I thought.

"Do you know his address?" she asked hopefully.

"No."

"Perhaps you know his middle name?'

"Unh-unh."

"Middle initial?"

"Nope."

I hung up.

I drank another cup of coffee and thought, Hell, I'll just call all five David Camerons and ask if they know Glenda Carter.

It worked! After plunking numerous quarters, dimes, and nickels into the pay phone, I connected with David's wife and she gave me his office number.

I dialed the number, but there was no connection—just a dead line.

So I tried again, this time dialing the 713 area code with the number.

I heard an ear-splitting tone, and then a recorded operator's voice said, "We're sorry, but we are not able to connect you with the number you dialed. Please check the number and try your call again."

I tried again, got the same ear-splitting tone and message.

Now I was down $1.75. With my last quarter, I phoned David's wife again and got direction to his office. She said she'd call and tell him I was coming.

David Cameron knew of a Tayana 37 that was for sale at the Lakewood Yacht Club. It was called *Liberty*. It was a lot like the 38-foot *Peacock* that I'd sailed with Dick Fow in Rhode Island Sound. However, this sailboat had a far different layout and was very cramped. David Cameron's spec sheet claimed it had a "master stateroom aft with head, shower, and queen-size bed," but it didn't. Granted, the boat was in good condition and had been well maintained, but its equipment was limited to bare essentials. I guessed it would cost forty grand to buy all the things I'd need for world cruising.

Disappointed, I went back to my original salesman, Jerry, to tell him it seemed like too much work and expense to fix up *Summer Time*. That boat was just too worn out.

As I turned into the Watergate area, I was greeted by multi-colored party balloons and a large sign for the Veracruz Run celebration: "Galveston to Veracruz Party Tonight."

It didn't make me feel festive. I was supposed to be at that party. But I was way behind schedule. By now, I should already have bought my boat and gotten it ready for my trip. The Veracruz Run was one of three shakedown cruises I'd intended to do that year. After that, I can't remember much about what happened. It is all in a fog. I kept getting these surrealist images flashing by.

I'd failed.

I can't remember much about what happened after that. I guess I went to a bar in Galveston and got drunk. Or maybe I fell asleep and had a series of dreams. Everything is a fog.

I seem to remember a highway—like the one Dustin Hoffman drives down in *The Graduate*. I'm driving along it, but I don't know if I'm going to Galveston or away from Galveston. I also have images of being crucified, again like Dustin Hoffman in *The Graduate*, when he's standing in the balcony of the church with his arms out stretched.

Lord, forgive me for I know not what I do.

I have a vision of being in this real shit-kicking bar listening to blaring country music, songs like "Hello, Walls." There's a horse there in the bar, but no, it seems to be a sea horse instead of a regular horse. I know I'm drinking Scotch and not beer. I'm sitting at the bar. My brother Rocky, who's been dead for five years, comes up to the bar looking very much alive. He's talking to the bartender and trying to help someone named Renee Summers, a waitress he knows at another bar down the street, get a better job. He keeps repeating, "She's honest."

There's the swank Flagship Hotel, built over the Gulf on pylons, with a fishing pier extending farther out. But right there beside it on the beach is an unemployment office and a welfare agency.

Then I'm in a restaurant, with a "fish market" sign on the wall, and I'm eating broiled mahi mahi with green string beans and corn and white rice.

Mom is there, and she's saying, "You have to eat your vegetables." So I eat the green beans and corn first. When I go to pay my bill, I don't have any money. All my cash is gone. I'm missing a fifty-dollar bill and several twenties. But I don't understand why. Did I get hit by a pick-pocket? Was I slipped a Mickey Finn at the bar?

Now I'm in the slums of Calcutta. I hear the twang of a sitar, the tinkling of tiny bells, and the "*Ommmmmm*" of mantras droning in the background. A Hindu night clerk stands before me at a hotel registration counter. He's saying he doesn't have any twenty-five-dollar rooms left.

The next thing I know, I'm in a cheap motel on Galveston Beach, with a blanket pulled over my head. I'm clutching my .380 automatic pistol. It's called the Mariner Motel, and it's as cheap-looking as the Crappy Dunes on South Padre Island. Except the walls are paper-thin, not cinder block. Ear-piercing rap music blares from the next room, and the stereo of a car parked in front of my door blasts more loud songs. The bass from the car is booming so loudly it feels like an earthquake is hitting. *Boom! Boom, Ka-Boom!* Grunts and screams are coming from the porno movie on TV. Through the thin curtains at the window I see the red neon sign outside blinking "*No Vacancy. No Vacancy. No Vacancy....*"

Suddenly I'm back at the boat place. *Summer Time* is there, and I go over to look at her once more. As I stroke her newly varnished toe rail, she rises out of the water and glares down at me:

"Worn out am I? You *loser!* I've been places you've only dreamed of. *You* are the one who is worn out—and you haven't even been anywhere. Well, you just missed the boat, bub."

Then the highway again.

The road is more clear now, I think, I'm on the highway north of Houston, heading toward home. No, I'm in Huntsville and I'm going to prison or worse to an old folks home. *Ignonimy!* Ayiiiiiiiiiiiiii!!!

"Dave … Dave … What are you doing, Dave? … I know, I've made some bad decisions lately. But I feel much better now, Dave.…"

"Daisy, Daisy, I'm half crazy.…"

From the movie *2001*.

The Second Coming

I awoke from my nightmare and found that I was, indeed, in a fleabag motel on Galveston Beach. Fortunately, in my stupor the night before I'd somehow had the presence of mind to hide my wallet, checkbook, and passport inside my car. Even more fortunately, I could now remember where I'd hidden them—inside the ice scraper glove and in the spare tire.

I retrieved my belongings and hightailed it back to my house in Cedar Creek Lake. From home I wrote Bill the following letter:

In light of recent events, I've decided to terminate my plans to sail around the world. Do you remember that story you told me about the second coming of Christ? Where Jesus came back to earth and they gave him electroshock treatments and now he is selling shoes in Sears? Well, I've decided to go into business.

I'm going to get me a Volkswagen bus, move down to Copano Bay, and sell stuff on the beach. I'll start out small—selling treasure maps, metal detectors, tee-shirts, and souvenirs. Next I'll open a storefront business and rent surfboards, wind surfers, and Jet Skis. As business picks up, I will expand according to the attached business plan. I will buy Blackbeard's on South Padre Island, then open a similar place on Galveston Island. Near Port A I'll open another place and call it Lafitte's, and then another place in Port O'Connor called LaSalle's. I will start a franchise chain of restaurants called Pirates of the Gulf Coast.

Please don't give away any of my business secrets.

Harvest Moon Regatta

Another shakedown cruise I'd intended to have completed by now was the Harvest Moon Regatta. Every October, on the weekend nearest to the full moon, Lakewood Yacht Club in Seabrook, Texas, organizes the Harvest Moon Regatta; the race sets off from the Flagship Hotel on Galveston Island and ends the next day at Port Aransas. Actually, for most participants, it's "overnight and over (the next) day." Apart from having to dodge a few oil-rigs, it's a straight shot south-west in the Gulf of Mexico, a race of 150 miles along the featureless Texas coast

452   IN MY TIME

with the boats on a reach in the prevailing southeasterly winds. It's sort of like a procession, perhaps, and not very exciting, following the faster boats ahead. So why do the number of entries keep increasing? What is the draw that pulls bay sailors out of the jetties, along with the hardened offshore racing crews? Maybe it's the call of the sea, an opportunity to create a dream and sail offshore, a chance to sample the cruising nirvana and achieve an ambition. For it's impossible to show me a sailor who hasn't dreamed of sailing the Caribbean, making landfall in Fiji, or tackling the Tasman Sea. Every Sunfish owner, windsurfer, and lake sailor has such secret goals, and the Harvest Moon Regatta is the first of many offshore passages to come.

RK Surgery

I recovered from my nightmare experience the night of the Veracruz Run, abandoned my grandiose business plan, and retrieved my dream of sailing around the world.

I began to think about things that could go wrong during my circumnavigation of the globe, and I realized I was far too dependent on my eyeglasses.

My eyesight was 20/400 in both eyes, so I wore thick glasses. During watersports, I'd always used special "sports holders" to keep my glasses on my face; nonetheless, I'd lost several pairs in Cedar Creek Lake while windsurfing. I'd fall off the windsurfer, water would wash over me and strip the sports holder over the top of my head, and the glasses would sink to the bottom.

I thought about sailing in the South Pacific. I didn't want to be wearing glasses there. In addition to the risk of losing them and being essentially blind, there was the problem that spray from heavy weather could collect on the glasses and obscure my vision so I couldn't see dangerous things around me.

I researched my options, which were radial keratotomy (RK) surgery and laser surgery. None of the laser surgery doctors could tell me much about the effect of the laser on my night vision. The remaining option was RK—the sharp stick in the eye.

My friend Dick Pool drove me to the Key Clinic in Dallas for the operation. When we arrived, they put some numbing drops in my eyes, and I sat in the lounge for about an hour waiting for the drops to take effect. Then they called me in and put me on a gurney and covered me with a sheet. They wheeled me into the operating room, where seven other patients were having the same operation. The assistant prepared me, and the doctor came over and put this clamp on my eyelid, so I couldn't blink. He told me to look at the point of light shining into my eye. They had taken some measurements during my previous visit, so they knew what template to use to correct my problem. The doctor pressed the template onto my eyeball, and it left an indention to indicate where he was to cut. Using a diamond scalpel he made four or five incisions where the template indicated. He moved to the other eye and repeated the process.

I was up and out of there in fifteen minutes.

They gave me some anti-bacterial drops and some pain pills, and sent me home with instructions to have dinner and go to bed. They told me to come back tomorrow, so they could check out their work. I could see just fine. Dick drove me home, and I had dinner and went to bed. I got up the next morning and could see great! I drove myself back to the clinic, and they checked my eyes. My right eye was 20/20 and my left eye was 20/25. Fantastic! I never had any kind of problem or pain.

Aqua Cat

When I'd worked at Wang Labs, I'd talked about sailing with my coworkers Allen McKensey and Irene Lacano.

Irene was an avid sailor and had a 14-foot day sailor she sailed twice a week at Grapevine Lake. Every Thursday night she'd participate in a race. On the weekends, if there were a race she'd be in it; otherwise, she'd just sail around the lake.

Allen McKensey, on the other hand, talked about how when he was a kid his dad had had a 26-foot sailboat, and they'd sailed it in Long Island Sound. His dad had bought him a catamaran that he sailed a lot, too, so we always swapped sailing stories. Allen kept saying he was going to bring his catamaran down to Texas the next time he went home.

Then Wang Labs closed the Dallas facility. Allen and I were laid off. Irene transferred to Wang's support center in Atlanta, Georgia. Later both Allen and I went to work for Microsoft in Irving. Naturally we still talked about sailing, so when Allen went home for Thanksgiving, he brought his catamaran back with him. He didn't have a trailer for it, so put it on top of his car. On the trip down, the catamaran wasn't secured very well and beat up both sides of his car. It was an older car and he wanted to get a new one anyway, so he did. Now he didn't want to put the catamaran on top of his new car, so he bought a new trailer for it.

When summer came, he towed it out to White Rock Lake and launched it. He sailed about halfway across the lake.

And then it sank.

All that banging around on top of his car had knocked holes in both pontoons, so they took on water in White Rock Lake. He managed to get it back to the dock and back onto his new trailer.

The following Monday morning, he came to work and asked me if I wanted his catamaran. Not only had it sunk on its maiden voyage in Texas, but he had another problem. Allen was fair-complected (a redhead with pale skin), and his doctor had told him not to spend too much time in the intense Texas sun. So the catamaran was becoming a big problem. He said he'd give it to me for free if I wanted it.

You bet I did!

I figured I could have some real fun on Cedar Creek Lake, where I lived. We set up the next weekend as the time for me to go and get it. I borrowed my neighbor's pickup, which had a trailer hitch, and went up and picked it up. It was a 14-foot Aqua Cat with a lateen sail. When I arrived back in Gun Barrel City, I didn't take it straight home. Instead, I stopped at Jerry's fiberglass repair shop on the highway near my house, where I had the owner take a look at the pontoon problem and give me an estimate. He said it would only be $200 to $250, depending on how much work was needed once he got started. Since the Aqua Cat hadn't cost me anything, I could afford that much in repairs—it was worth it to have a little fun. I left it with him. When I went to pick it up, I couldn't tell where he'd repaired it. He did an excellent job of matching the old faded-blue fiberglass.

I took it immediately down to the park and launched it. It was great fun sailing that little Aqua Cat. The only problem was, it didn't like to tack. Sometimes it would stall, going through the eye of the wind, and scoot backward for a little bit. But I soon learned how to get it to tack. I sailed it all summer and had some great times with it.

In September, we had a beautiful day with some good stiff winds and no thunderstorms. I wanted to see how fast I could make it go. I launched it and tacked out into the widest part of the lake, so I could have some good runs. I started out okay and kept sheeting in and going faster and faster. Finally, I got the upwind pontoon out of the water and was screaming across the lake. Well, for the pontoon still in the water it was hard to keep the nose up—it was being forced down into the water.

Sure enough, it took a dive.

I went over and "turned turtle" with the mast pointing straight down toward the bottom of the lake.

I'd read the manual and the instruction on how to right the craft if this should happen. So I took a break, took out a soda and drank it, and went over the procedure in my head. After I'd rested a little, I swam down and freed the main sheet to let the sail fall free so it wouldn't hold water. I climbed on the downwind pontoon and tried to pull the mast up on top of the water. The mast had a large Styrofoam ball on top—this was supposed to float the mast and hold it on top of the water until you could pull it up into the wind to help right the boat. It didn't budge. Not even two inches. I must have tried ten times to get the mast horizontal, and nothing worked.

Finally, exhausted after a couple of hours of trying, I decided it was time to get help. The next time I saw a boat coming near, I stood up and whistled and waved both arms in the distress signal. This guy and his wife came over, and I explained my predicament. They said they would help. I gave the guy the line that went to the top of the mast and asked him to motor out about twenty feet and just pull a little. I figured the mast should come horizontal with the big Styrofoam ball on top. He tried several times, and still the boat didn't budge. Finally, he tied the line to his transom and put his 150-horsepower motor at full throttle.

The boat did not budge.

But he'd managed to pull the mast out and get it horizontal. I asked him to tow me over to a nearby island. He said he would, so I lashed the mast to one of the pontoons, and he towed me. When we got to the island, we righted the boat by standing firmly on the ground in about two feet of water. But it was really a struggle. The water in the "trampoline" acted like a really good sea anchor and was tough to empty out. I could have never righted the boat out in the lake.

When at last we got the boat upright, he towed me over to where I had launched it. I put it on the trailer and took it home. I wanted to pay him for rescuing me, but he refused. So I just thanked him, and he left.

I sailed a few more times in September and November, but never again tried to fly at full speed. I'd just enjoy sailing around the lake in light winds.

"Boat Remodeling 101," as Chronicled in the Captain's Log

By now, like a good sailor, I'd begun keeping a diary or Captain's Log.

I'd been modifying my Aqua Cat to give it better downwind performance. On July 29, 1995 I wrote in my Captain's Log:

The mast is held up by an A-frame across the pontoons, with the mast creating another 45 degree A-frame along the centerline, bow to stern or front to back, by stepping (mounting) on a crossbar at the front of the pontoons. As a result, when sailing downwind and letting the sheet all the way out so the sail is abaft the beam (perpendicular to the centerline), the sail rubs against the A-frame creating two balloon-shaped blobs. Although the shape works and propels the craft well downwind, the sail rubbing against the A-frame creates chafe and it looks like an ugly duckling. I've decided to move the mast step back to the front trampoline crossbar support, thereby making the mast vertical. This will allow me to set the sheet all the way out, and the sail will fill without rubbing the A-frame and will give me better downwind performance. However, this has created the problem of supporting everything front to back. I will have to create a bowsprit and a backstay to hold the mast and A-frame up.

This was a chance to use all my engineering knowledge—right triangles, $\sin\theta$, and all that—to figure out how long the bowsprit should be and how long the forestay should be.

I wrote:

The height of the A-frame is eleven feet.

$$h^2 = a^2 + b^2$$

$$SinA = \frac{a}{c}$$

The bowsprit should be 6'4" and the forestay should be twelve feet, seven inches. That should create a sufficient right triangle to keep the mast and A-frame from falling backwards. A couple of backstays should keep it from falling forward.

I set about drilling the trampoline crossbar support and moving the mast mounting hardware back. The portable drill I'd purchased for this operation drilled the first half of the holes through the aluminum tubing. Then it died.

I should have taken this as a warning.

Undeterred, however, I returned the portable drill to the dealer and was given a replacement. This task completed, I tried to step the mast in its new position. The mast was too long. The arc circumscribed by the mast as it was currently attached to the A-frame went below the trampoline crossbar.

Curses! I'd have to modify the attachment of the mast to the A-frame to move the mast up or mount the mast step hardware horizontally on the trampoline crossbar. Mounting the mast step hardware horizontally would be the easiest but presented problems because I'd already drilled the trampoline crossbar vertically.

In my Captain's Log I wrote:

I've drilled the trampoline crossbar vertically, and drilling it again horizontally at the same place will weaken it. Also, transaxial tension loads across the horizontal screw holding the mast step could cause metal fatigue and fracture the mounting hardware, leading to a demasting. Using the sailboat only on weekends, and then only for a few hours, the fatigue failure probably will not occur in my lifetime. It might not occur in several lifetimes, but I do not want to risk it.

My attention shifted to the mast/A-frame interface. I disconnected the mast from the A-frame. By stepping the mast in the new step, and holding them up together, I could see that the hole in the mast was only a couple of inches above the A-frame mounting, but still within the frame head casting. All I had to do was drill another hole two inches above the current one. That done, the mounting of the mast to the A-frame at the new location went smoothly. I was ready to work on the bowsprit.

I'm ready to work on the bowsprit and have found a good 7-foot two-by-four. All that is needed is to drill through the aluminum tubing and the two-by-four the long, four-inch way. I had no drill bit that long, but the local hardware store solved that problem, and I have the long screws necessary to pass through the two-by-four and the tubing with washer to keep from pulling through the wood.

I drilled the holes and mounted the bowsprit. Now the time had arrived to erect the mast and secure it fore and aft with the forestay and backstay. In the Captain's Log, I wrote:

Secured the old backstay temporally, until I could figure out how to secure the forestay. The old rope chosen for the forestay was about twice the size of the backstay. There are going to be two backstays, so that should hold it. The thought that the forestay should be solid rod rigging or an aluminum tube has crossed my mind. It could take some of the compression load and remove some of the tension load off the backstay. The backstays could be left more to sail shape than to standing rigging. But that will probably be too expensive and can only be purchased in Dallas or Houston.

Raised the mast and while holding it vertically, took in the forestay and secured it to the old mast step bar across the front of the pontoons. All went well.

Now all that remained were some minor adjustments to correct the vertical. As I was stepping around the mast onto the new bowsprit, the backstay gave way. The whole rig came down around me. It trapped me across the bowsprit on my back, with the A-frame across my chest and legs, and lines everywhere.

I felt like Captain Ahab tangled up in the lines on Moby Dick's back.

But I probably looked more like Moby Dick tangled in the lines of the whaling ship bowsprit.

After extracting myself from the mess, I surveyed the damage and reported it in my Captain's Log:

Only minor bruises on my back. Could have been a lot worse. Ego suffered the most damage. No damage to the ship. This does illuminate the need for new lines, though.

So it was off to the local marine store for some new lines. With new lines and the mast raised, this time with two new backstays and a new forestay, it was time for the sea trial. I recorded it in the Captain's Log:

The breeze was blowing off of the weather shore about twenty knots. That meant I would have a good chance to check the downwind performance first. As the wind was offshore, it meant we were in the wind shadow of the land, so the winds would be light next to the dock and pick up as I crossed the lake. This would give me a very good chance to test the downwind characteristics in varying amounts of wind. I cast off, and let the sheet all the way out.

The sail filled beautifully and bulged like a spinnaker. The Aqua Cat took off like a shot in very light wind. The sail performed something like an asymmetrical chute. I was concerned that it might be overpowered when we got in the heavy twenty-knot winds. I could always turn into the wind and dump the air out of the sail if that started to happen.

This was great! Just the performance I'd hoped for. About halfway across the lake, I decided, *Enough of this downwind. Let's get on a screaming reach and do some really fast sailing.* I eased the tiller over and put her on a broad reach and started to sheet in. She began to come around, but instead of going faster she was slowing down. I sheeted in more, but the sail did not take the normal airplane wing shape. It kept the big bulge like a spinnaker. The more I sheeted in, the more the mast bent over, and the more the back stays became limp. But the big bulge in the sail did not change. Finally, the double sheet block was all the way down, and I couldn't sheet in anymore and she continued to slow down. About this time, I decided to tack and see if she would sail better on the other tack.

Big mistake.

As soon as I turned into the wind, she went into irons and stopped dead. Then she started to drift backward, downwind. Nothing I could do would force her to go anyway but downwind, backward. I could not even turn enough to get back on a downwind run, where I had some directional control and could maybe steer toward some beach or dock on the lee shore and avoid the rocks and stumps. As it was, I was headed up into the stumps and no man's land, toward the uninhabited island in the north shallows. There were no phones, and no roads, and nobody ever went up there because of the snakes. The only good part was that putting the tiller hard over, either way, would force the pontoons across the wind and slow the downwind drift. I had not considered that the sail would not create an air foil. Without an air foil, there was no power to do anything but go downwind. I had succeeded in optimizing downwind performance, to the exclusion of all other.

The Captain's Log says:

After an hour, swallowed my pride and let the sail luff, to keep the windage down as much as possible. Started looking for a rescue, before I drifted completely away from all boaters. A wandering pontoon boat on a cruise of Loon Bay came close enough to see me waving both arms in the distress signal and heard my whistles. Finally rescued, they towed me back to the dock where I started.

Plans are in the works to modify my windsurfer boom to hold the sail in air foil shape for the next attempt....

Flight to Annapolis

I saw an ad in the *Yacht Trader* for a Hans Christian 42 in Annapolis, so I called Interyacht, the yacht brokers specialists, and spoke with a yacht broker named Gillian L. Griffin, or Jill. She said there were a couple of Hans Christians and several similar yachts available in about the same size and price range, so I decided to fly up to Annapolis and check them out.

I called American Airlines. Round-trip airfare to Baltimore Washington International airport (BWI) was $189. I booked it for two weeks later. But just before I was to leave, Jill called and said the Hans Christian I was interested in was sold.

My ticket was non-refundable, so I decided to go anyway and see the other yachts.

Jill showed me a couple of other Hans Christians, but they were out of my price range. Then she took me down the coast to look at a Southern Cross 39.

I'd read several articles about the Southern Cross 39 and I thought it might do. When we went aboard, it was almost exactly what I was looking for—but a bit too small. The captain's cabin had only just enough room for a bunk and minimal storage space.

I said, "Let's look around a little more."

Jill had one older 44-foot boat, a LaFitte 44, that might suit me, so we went to see it. When I went aboard, I thought, "It's just right! If only I can get it for the right price."

I made an offer of $90,000 and flew back home.

I bought the LaFitte 44 in March 1996.

When the owner accepted the offer, I had to fly back up for the closing. I called the airlines, and I learned that a round-trip ticket to BWI now cost $245. I flew back and signed the papers and made arrangement to have the boat hauled and surveyed. The survey went well, and I arranged to have the boat brought over to Bert Jabin's Yacht Basin so I could work on it.

The plane ticket in April cost $378.

The next ticket, in May, cost $467.

I had to go back in June to sail the boat home. By now, the airfare was $1,457 to BWI. However, a ticket to Washington National, in downtown Washington D.C., was only $829. I flew into Washington National, and Captain Cunningham and his wife picked me up.

De-Naming

The LaFitte 44 is a cutter rig with a 63-foot mast, 130 percent roller furling head sail with a Yankee Stay sail and light wind flasher sail (a flat cut spinnaker). She has a blue hull and teak decks. She had not been sailed much by her previous owner and needed a lot of TLC. The problems were mostly the result of neglect and letting her become run down.

She was named *Quixotic*. The first thing I wanted to do was to change the name. I wrote the following for the de-naming ceremony:

Interdenominational De-Naming Ceremony

Oh, sovereign of the sea, King Neptune, and sovereign of the wind, King Aeolus, we are gathered here today to respectfully request your blessing and sanctification of the retiring of the good, loyal, and faithful name of *Quixotic* from your rolls. We ask that you strike the name *Quixotic* from all records, and we pray your indulgences in extending your goodwill and protection to this vessel in its new incarnated life.

I poured half a bottle of champagne over the name on the transom and drank the other half.

The Work Begins

I had to have all the hoses inside under the decks replaced by A&B Yachtsmen, Inc. I did the bottom job and put two coats of anti-fouling paint on the bottom. I put an ounce of cayenne pepper in each gallon of paint to prevent the spread of the zebra muscle.

There wasn't much in the way of equipment on the boat: only wind speed/direction, depth, heel angle, compass, VHF, and Loran. So I added a Raytheon radar and GPS. A&B Yachtsmen had to remove the mast to install the radar antenna on it. I also installed a Kenwood TS850 ham rig for HF SSB communications. I added a storm jib, and a storm tri sail, plus a sea anchor and drogue. I bought a Plastimo offshore life raft and a Trans-Ocean Pak offshore medical kit from West Marine.

I had all the seacocks, heads, holding tanks, and pumps fixed or replaced. A&B Yachtsmen were responsible for checking the fuel tanks for algae, cleaning the lines and tanks, and replacing the fuel filters. I made a point to ask them specifically three times if they had checked the fuel, and they said everything was in good shape. When they signed off on the survey, they made a note that the tanks did not need cleaning. A&B Yachtsmen were responsible for cleaning the fresh water tanks, also.

All of this was supposed to take three weeks, but it wound up taking two months. This was partly understandable as this was their busiest time of the year. I bought and had installed a Monitor self-steering wind vane to steer the boat, so the helmsman could go below to do things.

Renaming

On the next to the last trip to Annapolis in May, I had the new name, *Sir Charles*, painted on the transom. I had to have a christening ceremony. I wrote the following:

Interdenominational Christening Ceremony

Oh, sovereign of the sea, King Neptune, and sovereign of the wind, King Aeolus, we are gathered here today to respectfully request your blessing and sanctification of the good, loyal, and faithful yacht, *Sir Charles*. We ask and humbly pray for your indulgences in extending your goodwill and protection to this vessel in its new incarnated life.

I poured half a bottle of champagne over the new name on the transom, squirted champagne in all four directions of the compass, then bowed respectively and drank the rest.

Safety-at-Sea Seminar

Before I sailed the boat to Texas from Annapolis, I took a course in safety at sea. Prior to going out in the open ocean, I wanted to know what I was getting into and how to deal with problems.

I saw an ad for a safety-at-sea seminar offered on May 11, 1996, at Lake Lewisville, north of Dallas—so I signed up. It was a very informative day.

Throughout the seminar, the instructors sang the praises of something called a "life sling." It has been called "the most compact, economical, and effective solid lifesaver unit on the high seas." It's a bit like a standard donut-shaped life preserver, but the life sling is a horseshoe-shaped device, which, when combined with a hoist-and-tackle system, will, the literature says, "bring even your beefiest crew aboard safely, even if injured, unconscious, or in atrocious sea conditions."

The seminar included an in-water dockside demonstration of the life sling. One instructor jumped into the cold water, and the others went about rescuing him. They threw him the life-sling yoke, and he put it around his chest under his arms. Then they started to winch him up with the mainsheet block and tackle.

When they almost had him out of the water, except for his feet, the life sling ripped apart.

Apparently the stitching, at the point where the main line joined the harness, was insufficient for a soaking-wet two-hundred-pound person.

I decided not to purchase a life sling.

The Voyage Home

To help me get *Sir Charles* back to Texas, I'd hired an experienced sailor from Florida mentioned earlier, Captain Jim Cunningham. He arrived in Annapolis a day before I did, on Friday, June 7, 1996. On Saturday, he and his wife Hazel met me at Washington National Airport. They'd already made several stops for provisions and had their van full of these, plus the gear Jim used on his other deliveries. He often delivered new boats, which had no gear at all, so he had several "bags": a galley bag with pots, pans, plates and eating utensils; a sleeping bag with pillows, sheets, and blankets; a clothes bag with his personal things; and a ditty bag with some charts.

When we arrived at the boat, Robert LaTulippe, our first mate, was on board making ready. If ever there were a pirate who deserved the name Lafitte, Robert was the one: he fulfilled the classic image of a pirate on a sailing ship. With his slightly graying hair tied back into a ponytail, he quietly went about his duties making the ship ready. If he'd had a red bandana on his head and a patch over one eye, he could have stood next to Errol Flynn in any swashbuckler.

Preparations continued all day Saturday and Sunday. The size and seriousness of things began to come into focus for me as we picked up tools for the boat. Instead of normal tools, Captain Jim suggested we get the biggest bolt cutters, because the shrouds were oversized. Likewise, we needed the bigger hammer and crescent wrench. We also purchased foul weather gear, which was very expensive. We spent a lot of time at Fawcett Boat Supplies, West Marine, Boats U.S., and other marine and hardware places. We picked up a Garman handheld GPS as backup to the Raytheon GPS. Captain Jim and Robert both had handheld VHF radios, so we were okay for communications. We had no weather maps, though, so we picked up some charts and an East Coast chart kit.

On Monday, Captain Jim wanted to do a shakedown cruise to see how she handled and how the crew worked (i.e., what could this new owner—me—do?). We motored over to the fuel dock to get a fill up.

When Captain Jim had arrived on Friday, one of the first things he'd asked was, "Have the fuel tanks been checked?" He was told by the people in charge of the boat that everything was okay. I should have suspected something was wrong when we went to get fuel on Monday and couldn't get the caps off of the tanks. This was evidence that they hadn't been opened for a while. We must have spent 45 minutes trying to get the deck-plate access lids open. We bent several universal deck plate keys before we were able to take on diesel fuel.

Fueled up and motored out, we put up sail and cruised the bay for a couple of hours. After our shakedown cruise, Captain Jim decided to get under way early on Tuesday.

Sailing from Annapolis to Florida

Captain's Log:

June 11, 1996.

We shoved off down the Chesapeake Bay on Tuesday morning at 6:15 AM with partly cloudy skies and about ten knot winds out of the south. We motored out to Thomas Point light before we put up sail at 7:15 AM. At 9:15 AM we were abeam the "antenna prohibited" area. At 10:30 AM we were near the mouth of Choptan River, and it was clouding up. Around noon it started to rain on First Mate Robert's watch. Weather cleared up about 2:00 PM, and we passed abeam of No Point Lighthouse around 3:00 PM.

I took the helm for my watch at 3:36 PM abeam Point Lookout (N38° 03' 08' W076° 15' 33') bearing 089° heading 153°. Were near the mouth of the Potomac, about 4:00 PM, when we got a great air show from a couple of navy fighters diving on an anchored target ship.

We had sailed almost to the end of Chesapeake Bay, near Smith Point, when we lost radar at 5:05 PM. A light chop was developing from offshore. The batteries were going down, so we started the engine to charge them. The engine ran about a half hour and quit. It had not run long enough to charge the batteries up, so the radar still would not work. We spent the next four or five hours tacking back and forth, trying to get around Windmill Point shoal and up the Rappahannock River to Deltaville. We could see the lights of the marina, but the channel lights were lost in the lights ashore. Went aground about half mile out, trying to get into the narrow channel to the Deltaville Marina. Spent the first night there.
June 12, 1996.

First light showed, and we missed the channel entrance by about a quarter of a mile. We were headed directly for the first channel marker when we should have been going up the river, about another 30 to 45 degrees to starboard. A couple of fishing boats were coming out of the marina, so we watched their progress out past the last channel marker. Now we knew what to do to get into the channel.

Unfurled the big jenny to heel us over, raised the keel a few inches, turned the bow around. It slowly pulled us back out into the Rappahannock River. Worked our way up the narrow channel to Norton's fuel dock.

When we arrived at Norton's fuel dock, the engine mechanic was busy installing a pair of diesels in another powerboat so was unable to get to ours. Norton's called another Volvo Penta specialist from somewhere else in town.

This guy was a Cajun, and he was great! He took one look at the fuel and had it pumped out. After he cleaned the tanks and fuel lines and replaced the fuel filters, he found that there was an electric fuel pump installed in front of the fuel filters. He was puzzled by this arrangement, as the Volvo Penta engine had a perfectly good mechanical fuel pump built in.

We discussed the situation and decided that if the electric fuel pump was not needed, we'd just remove it, so there would be one less thing to break down. When the Cajun removed the electric fuel pump, it was completely stopped up with crud. He replaced it with a new line segment and then primed the fuel line. The engine sprang to life on the first hit of the starter and has been running perfectly ever since.

While we were waiting, we decided to get some breakfast. No food service was available at the marina, so the secretary called Taylor's restaurant in town for us. The owner drove her own Lincoln Town Car out to pick us up and take us to town. When we finished breakfast, we went across the street to a hardware and ship-supply store to get some parts. Then the owner of Taylor's restaurant drove us back to the marina. Great people, great service.

Robert went up the mast and mounted the new mechanical wind indicator and tried to repair the electronic datamarine instrument. While he was up there, he took some nice pictures.

Norfork and Newport News, Virginia

Captain's Log:

Got fuel lines fixed (algae build-up in tanks); got engine running. Replaced battery, and radar worked just fine after that.
June 13, 1996.

Got underway again at 6:00 AM with clear skies. At 8:45 AM we were near Wolf Trap light, and at 10:37 AM abeam Newport comfort. We rounded and came abeam of Fort Henry at 1:30 PM, entered Norfork channel, and proceeded down Battleship Row. We entered the intercoastal waterway (ICW) and started playing games with the drawbridges. Most bridges are on restrictions and cannot open except on the hour or half hour. Time and distance between them are such that a sailboat can't time the passage right and has to wait almost half an hour at every bridge. We were lucky though—the next-to-last bridge was a double: a car-and-train bridge. The train bridge got stuck down and we were delayed about an hour and fifteen minutes, but it finally went up at 6:00 PM and we got through.

After that, we heard on the VHF that it got stuck down again and would not be fixed until the next morning. Close call on that one.

Went through the lock at Great Bridge, VA, and stopped in Great Bridge at 7:50 PM for fuel. The dock man asked if we wanted some take out and a shower. We had a choice of pizza, fried chicken or barbeque. We ordered barbequed ribs, and they were outstanding! Seemed to be two racks of ribs; I saved half of my ribs for lunch the next day. We showered while waiting for the take out. Stopped at Coinjock Marina, NC, at 1:10 AM for the night.

June 14, 1996

Went down the Chowah River and out into Albemarte Sound, being chased by a thunderstorm. We missed the ICW entrance on the other side and nearly went up into a shallow slew. Corrected the error and got back on course.

We came down the Pamlico River and tried to cross into Pamlico Sound, but the corner was not marked very well and we ran aground several times. Finally, we were able to turn around and retrace our course back up the river a few hundred feet. We met a powerboat coming down. He seemed to know the correct route, so we followed. The captain of the powerboat hailed us on VHF and told us it was shallow a couple of hundred yards ahead where he was. Sure enough, we ran aground in the middle of the channel, but were able to bump along and get over the shoal and continue.

The navy gave us another aerial demonstration, as a couple of F-14 Tomcats kept flying very low over the area. I think we were near a navy or marine base.

Out in Pamlico Sound, something began to stink like shit.

The smell came up just as we got out into some big waves. It turned out that the maintenance people at A&B Yachtsmen had forgotten to put the inspection cover back on the forward head's holding tank, so its contents sloshed out in the big waves and ran into the bilge. As if this wasn't bad enough, the jib ripped, so we had to put into Oriental, North Carolina, at 1:30 AM in the morning—for shit removal and sail repair.

Oriental, North Carolina

Oriental is one of the nicest places anyone would ever want to visit or live. The people there were very friendly and helpful. The marina had an old 1975 Oldsmobile parked at the end of the boat dock, and this was available to any crewman who needed transportation around town. The keys were in the car all the time. It was an odd rusty orange color, and was known as the Orangemobile.

Less than a block away was a small grocery store, Harborside Grocery & Market, where we provisioned. The clerk carried the groceries to our boat and helped stow them. Oriental Sailmakers took time out from a previous job to come to the boat on Saturday to pick up the jib, repair it, and return it the same day. Extremely good service!

This is when we found that the roller furling sail guide was broken. If it was broken when I purchased the boat, A&B Yachtsmen should have notified me of the problem when they removed the mast to install the radar radome. That way, I could have had the problem corrected or replaced while the boat was hauled. However, I suspect that someone broke the guide when they re-stepped the mast, and they didn't want to take responsibility for breaking it. But a broken guide puts the ship and its crew in danger. If we had needed to remove the jib in a storm and were unable to get it down because of the broken guide, it could have caused us to be knocked down—and possibly, to sink.

While Robert was up the mast trying to free the jib from the broken guide he took some more great pictures. Captain Jim finally had to go up and get the jib free.

While the jib was being repaired, we worked on the forward head holding tank. We made a temporary repair by epoxying the inspection plate into place. The Inland Waterway Treasure store, across the street from the marina, would let you take a $100 item over to the boat to see if it would fit without paying for it. If it fit you came back and paid; if not you just returned it. I broke a lifeline termination and could not seem to get the new one on because of a barb on the broken screw. The salesman, who also did repairs on ships, came over and fixed the lifeline and didn't charge me for his work.

We went to the bar in the Oriental marina restaurant for dinner and had a great time. A woman named Mary was bartender, cocktail waitress, food waitress, and entertainment. She had a humorous comeback for any comment you made. Tit for tat. She was really entertaining. When we were ready to leave she hummed "Happy Trails to You" on a kazoo.

Word got around that we were headed for Texas, so a couple named Martin and John on a ship called *Magic Dolphin* called and wanted to know if we would mind them following along, as they were going to Galveston. This was great! Having another boat would give us some added security in case of more trouble. But when we were ready at 10:00 AM to leave, they were not quite ready. We waited until 11:00 AM, but they were still not ready and said for us to go on; they would be along shortly. They were two hours behind us when we started down the coast.

Beaufort, North Carolina, and Offshore

Captain's Log:

June 18, 1996

Went out into the Atlantic below Norfork, VA at Beaufort, NC, just below Cape Lookout, NC, 34° 37' 18" N 076° 31' 30" W. Spent half a day trying to get past Frying Pan Shoals buoy at 33° 29' 06" N, 077° 35' 24" W, fighting a head wind out of the south and the gulf stream flowing north.

Sailed all day and night.

June 19, 1996

Passed 80 nautical miles east of Charleston, SC. I took the helm at 3:00 AM and was enjoying a beautiful starry night as I followed my star. All of a sudden there was a big splash next to the boat cockpit.

When I heard the splash, I thought something had fallen off. After checking to make sure that Jim and Robert were still on board, I sat in the cockpit trying to figure out what had fallen. All of a sudden a dolphin jumped out of the water beside me! Then I saw another. They were having a great time riding the wake.

When dawn came and lit the sky, I could see a line of thunderstorms ahead of us. Captain Jim took the helm at 7:00 AM, and I got on the ham SSB and called several amateurs to get a weather report for Florida. Most of the people I reached were north of us and had no idea about the weather farther south. I kept trying to reach Florida, but the backstay antenna I'd installed acted like a directional antenna, with the mast as the reflector, so I couldn't talk to anyone in Florida.

I had breakfast and went to bed to get as much rest as possible, because I knew we were in for some heavy weather. Tropical Storm Arthur had closed in on the shore side, so we had to run halfway to Bermuda.

Captain's Log:

June 19, 1996

We put a double reef in the main and shortened the jib. Broke out the storm jib and storm tri sail in case we needed them.

Observation station 41002 is located at N32.3° W075.2°, and I think we sailed around it or at least near it. I took the helm at 3:00 PM in my foul-weather gear. Had to look up at the fourteen-foot waves, on top of ten foot seas, and had forty-knot winds, but made it okay through the first wave of thunderstorms. Captain Jim took my famous "Oh Shit picture."

Peeing in Heavy Sea

Sir Charles took the heavy weather with no difficulties at all. However, I had a little problem.

I will try to describe peeing in rough seas.

First, you're below deck. It's wet up on deck, so you decide to use the head (toilet), but you keep waiting because you've drunk too much coffee and peed just six minutes ago.

Finally, you can't wait any longer: you get up, then fall. You quickly get up again. You stagger forward, accidentally lunging for the towel shelf and pulling all the towels and washcloths to the floor. You fix the towel shelf. Now you gotta pee real bad.

You brace yourself over the head, yank down your foul-weather-gear pants, which pin your legs together so you can't balance, then you fall backward leaving a streak of pee. Because of the motion of the boat, the pee hangs in the air a moment longer than is comprehensible, from the toilet up to the hatch, about three feet over the wall and woodwork.

Of course, since you're falling, you can't work your "stop peeing" muscle, so you squeeze the end of your thingy and keep the pee in by sheer pinching pressure. You straighten up and tighten that muscle, which of course builds up just a tad too much pressure as you fall forward, so a thin squirt escapes at an odd angle—just as you try aiming for the bowl. But the lid has just slammed shut, nearly biting off your thing, so you pee on the lid, and maybe put your hand in it to stop yourself from hitting your head on the cabin top (which you do anyway).

Of course, which hand did you use? Exactly. So the rest of what's been waiting (between the pinched muscle and the pinched fingertips) dribbles down your leg into your foul-weather-gear pants, the ones that pinned your legs together so you couldn't balance.

Melbourne, Florida

Captain's Log:

Made it back to Charleston, SC, that night, before the main body of the tropical storm hit.

Magic Dolphin had to put in at Cape Fear, NC.
June 20, 1996.

Fueled up at Buzzard's Roost Marina on Johns Island, SC. Back out into the Atlantic and headed toward Cape Canaveral again, dodging the remaining thunderstorms. Was able to watch the Space Shuttle launch from offshore. Then had to duck into St. Mary's inlet, north of Jacksonville, FL, to avoid another line of thunderstorms.

Stopped at Pablo Creek Marina, Jacksonville, FL, for fuel and ice. Was able to call the Microsoft office and let my manager know where I was.

Went through old St Augustine, FL, 29° 51' N 081° 15' W, which has been restored to its grandeur.

We stopped at Harborside Inn at Palm Coast for dinner and to celebrate Captain Jim's birthday.

We were running all night until we ran aground in the fog about 3:10 AM at Nassau Coast Inlet near Titusville, Florida. We spent the rest of the night there. When we awoke in the morning and the fog had lifted, we could see we were in the middle of the channel but aground. We were able to motor over to a slightly deeper part of the channel.

We motored down the ICW on the inside of Cape Canaveral. We got hit by the daily thunderstorm about 11:00 AM, but we managed to round Dragon's Point and stop in Indian Harbor Marina in Melbourne, Florida—which was Captain Jim's and Robert's home port.

I made a lot of new friends during this stop. I met Robert's wife, Sue, and we all went to dinner at Conchy Joe's Seafood. I also met Nancy and her husband Chuck of Canvas Connection, Inc.—Hazel had them make a dodger for me from the rag that came with the boat. I met Ted Duay of Duay Marine and had him replace the roller furling.

Meanwhile, we had contacted the hurricane center. They said it was not a good time to cross the Gulf of Mexico: tropical storms were developing. To make things worse, Hurricane Boris in the Pacific was expected to come ashore in Mexico, cross into the Gulf of Mexico, and join the other tropical storms.

Fortunately I made yet another new friend, Loni Drum of First Class Travel Agency. She was able to book a great flight for me on a moment's notice. We had to pick up the ticket from the flower pot on her front porch.

I had a bimini made and installed by Canvas Connection. Robert installed some stereo equipment, tape deck and speakers.

Meanwhile, *Magic Dolphin* continued on down to Key West, where they spent a couple of weeks waiting for Hurricane Bertha to wear herself out, before they could cross the Gulf of Mexico.

Captain's Log:

June 22, 1996

Left boat in Melbourne, FL, and flew home to work for a couple of weeks while hurricanes clear out.

Sailing from Florida to Corpus Christi, Texas

After waiting out Hurricane Bertha in Texas, I flew back to Melbourne on July 13. Bill from Canvas Connection wanted to go with us to get some more offshore time. I said okay, but that I wouldn't be responsible for his expenses. This was okay with him.

Captain's Log:

July 14, 1996

Departed Indian Harbor marina at noon. Stopped in Varo Beach Marina on the ICW for ice at 5:23 PM before going to sea at 6:45 PM. We sailed all night.

July 15, 1996

Went into Ft Lauderdale at 12:38 PM to check the engine throttle and gear shift. Everything seemed okay, so went back out to sea and continued south. At 8:00 PM was abeam of Key Largo. We were hailed by another sailing vessel and found out that they had been struck by lighting and lost all navigation and wanted to follow us to Key West. We slowed down to five knots because that was their top speed. Passed Sombrero Key, FL 24° 37' 36" N 081° 06' 36" W, and turned into Moser Channel.

July 16, 1996

At 7:00 PM we went under the bridge at Moser Channel, leaving the Atlantic and entered the Gulf of Mexico. Stopped at Faro Blanco marina on Marathon Island at 1:00 PM to re-provision.

While we were re-provisioning, a 50-foot yacht stopped for the night. It had a hired crew of four, in white uniforms: a captain, two mates, and a lady cook. When they secured the yacht, the crew put down a mahogany gangplank with a handrail, then put out a green carpet from the gangplank to the shore end of the dock. Later the owner, his wife, and two kids came ashore and went somewhere. After they returned, about 8:00 PM, as I passed the yacht I could see the lady chef preparing dinner. The yacht had a beautiful dining room with a large mahogany table. The owner and family were served on the upper deck. It was a lovely evening. The next morning when they departed, the yacht left one of the crew on the dock. He went down a few slips and climbed aboard a 25-foot outboard fishing boat and followed the yacht out past the breakers. There he tossed one of the other crew a line and they tied the fishing boat to the back of the yacht and disappeared south.

Must be nice to have that much money.

Captain's Log:

July 17, 1996

Headed for Port Aransas, Texas, at 10:10 AM with southeast winds of fifteen knots. It should only take us five days at this rate. Working our way through East Bahia Honda Channel, which was unmarked and had sunken islands on both sides: Horseshoe Band and East Bahia on the port, Bethel Bank and Red Bay Bank on the starboard. At Honda Key, the Captain had to send First Mate Robert up the mast to the first spreaders so he could get a better view of the underwater layout.

Lots of dolphins playing around the boat's bow wave and wake.

We finally made it into the deep water of the gulf. We were following a thunderstorm front, so we should have had good winds.

Wrong.

The next day the wind died, and we had zero winds nearly all the way across the gulf. Robert went skinny-dipping in 2000 fathoms (12,000 ft) of water in the middle of the gulf. Then he called a passing ship on VHF to get a weather update. *Coastal New Yorker* had a lady communications officer named Linda. They were headed for Port Canaveral. The weather report showed high pressure over the gulf that would likely be there for a week.

With no wind, our progress was slow across the gulf. We ran out of drinking water on the seventh day, and ran out of bathing water the next day.

We feared we'd run out of diesel fuel halfway across and be stranded without food, water, or wind.

Thank heaven the fuel in the 110-gallon tank lasted to within one hundred miles of Port Aransas. And we still had the other 40-gallon tank as a reserve.

While we were about 250 nautical miles out, Captain Jim noticed that his mobile cell phone light was green, indicating it was connected to a cell somewhere. He called his wife to let her know we were okay. I tried to call work and let them know I'd be a couple of days late, but the phone wouldn't make that connection. I was able to call Bill Allen and let him know where we were and when we'd arrive in Port A, and he called Microsoft for me.

Then we lost the connection to the phone cell. Robert was able to pick up some good strong winds on his night watch and get us into Port Aransas at 8:00 AM on July 24, 1996. She was now safe in her new home.

We docked at Tortuga Flats and waited for Bill to arrive. The captain and mate secured the boat, and we started to walk over to the Seaside Cafe for some breakfast. Bill came flying into the parking lot as we walked across. He didn't recognize this motley crew. We were staggering on sea legs and looked to Bill like the local drunks after a binge!

After introductions we hurried to the cafe. We all needed a drink of water, as we'd run out of drinking water a day out from Port A. Some good food would help, also.

From Port Aransas to Bahia Marina

After breakfast, Glenda Carter arrived. Glenda had arranged a docking space for us in the Bahia Marina in Ingleside. Ingleside was about halfway up the Corpus Christi channel from Port A, so we proceeded up the channel.

Captain Jim decided to fly the colorful flasher sail so that Bill and Glenda could see it. It was quite an ordeal. He did not realize the wind had picked up and was too strong for the flasher's flying range. Down in the channel with the high banks on both sides, the wind was not blowing as strong as it was twenty-five feet up. When we set the flasher, it nearly turned the ship over. Then a tanker came toward us, down the channel, just when we needed to bring the flasher down. Normally, you'd turn the ship into the wind and dump the wind out of the flasher. But that would have taken us directly into the path of the oncoming tanker without any steerage and at a dead stop. We were left with trying to put the flasher back in its furling sock.

I was on the sock's furling line and jumping up and down with all my might and weight when a gust picked me up like a fly and almost tossed me over the side. I turned loose of the line about four feet above the deck. Thank heaven I landed on the deck. You don't want to be in the channel with a several-hundred-ton, nine-hundred-foot tanker bearing down on you.

I found out later that Bill was ecstatically videotaping this disaster from the safety of the cockpit. Finally the wind slacked, and several other crewmembers jumped on the furling line, and we managed to get the sock over the flasher just in time to get control of the ship as the tanker passed.

It was just the kind of drama that Bill liked.

We arrived at the Bahia Marina on a Wednesday, and learned it was closed on Wednesdays. Why on Wednesdays was a mystery, but it would turn out that at this marina, it was every sailor for himself. The channel into the marina was really shallow. The depth sounder went off and indicated the channel was just deep enough for us, but was getting shallower. We ran aground a couple of times, even with a five-foot, four-inch keel. We managed to dig close to the dock, just not close enough to the wharf to tie up. We were met by a small friendly beagle—the owners' pet and marina mascot. Seems he was the marina's Welcome Wagon.

Bill Allen's friend Sue Taylor arrived, and we finally found another ship owner who knew where our slip was. With some help, we freed the vessel and finally managed to get the *Sir Charles* in Slip 23 and tied up. After securing the boat, we sat in the cockpit and celebrated with drinks all around.

The channel into Bahia Marina had some really shallow water on the bay side. I mean *really* shallow. It was about three inches deep. Bill was sitting on the starboard side, looking out of the marina, when he noticed that the little beagle had jumped into the channel and was paddling across to the shallow water to start chasing the wild ducks wading over there. The water didn't even come up to his belly.

Bill looked at me and said, "This does not bode well."

As it turned out, Corpus Christi Bay is not all that deep anywhere—mostly ten to twelve feet. The Coast Guard pilot book gave a caution that when there is a stiff wind, the water in the bay on the upwind side drops about two to three feet, leaving only a seven-foot depth in some places. On the downwind side, the water rises two to three feet for a depth of fourteen to fifteen feet. If you have a seven-foot draft sailboat, and they are very common, don't sail on the upwind side of the bay.

Training on Corpus Christi Bay

During the following months, I commuted on weekends from Dallas to Corpus Christi. Deborah, a gal from Houston, showed up to sail with us. Every weekend, we'd sail out into Corpus Christi Bay, and I'd go over the training items for each weekend.

The first weekend I explained what and how long it took to turn or stop a sailboat.

Then I said, "Bill, I know you can't swim, but I need to know how long you can float if you fall overboard."

Bill said, "About fifteen, maybe twenty minutes."

I said, "That isn't long enough. It takes about thirty or forty-five minutes to turn a sailboat around and start searching for that one little bobbing head out in an endless sea of waves. I'll just have to say goodbye and call the Coast Guard to let them know where you fell off. Your body will float back up to the surface in a couple of days, if the sharks don't eat you."

Each weekend we'd practice tacking and maneuvering the boat under sail so we could get the feeling of how it would respond. I made everyone read the "rules of the road" for boaters, as we'd encounter many other boats in the channels and bay. Near the Ingleside naval station there was a huge offshore oil-rig platform manufacturing facility. These rigs were 15 stories high lying on their sides and three to five football fields long depending on how far out in the gulf they were located. A lot of 900-foot tankers came down the Corpus Christi channel to dock at the refineries around the bay. It was a very busy place to sail but good training. A couple of times we went down the channel to Port A and docked at Tortuga Flats to have drinks and dinner. It was a nice reward for a hard day of sailing.

Incidentally, the porpoises in Corpus Christi channel at Port A are always thrilling to watch. They work the ferries for food and delight in jumping the bow waves of large ships and swimming around passing sailboats.

First Offshore Trip

After some sail training, Bill was ready for his first offshore voyage. I read the following to Bill as we left the jetties at Port A: "Out there is the whole world just waiting for us to explore it, our new reality. Back there, our old reality and death. This is our birth, bursting fourth from the constraints of the womb into our new life."

Bill, a romantic, was visibly moved.

It was a beautiful day with wind at about twenty knots from the north, which made for a screaming reach sail. I kept sheeting in the jib and main sail sheet to make the boat go as fast as possible. It was great! The boat speed kept increasing until it was nearly ten knots, which is above the theoretical maximum speed. I think the knot meter was off a little. However, the boat healed over, almost putting the gunwales and safety line in the water, and the bow wave splashed into the bottom of the jib. This is the most exciting type of sailing. I told Bill to take his video camera over to the downwind side and tape the water splashing into the jib.

Bill wanted to know something about sail trim, a term he had heard.

I said, **"We don't trim sails."**

So Bill relaxed and did some videotaping, mostly of Deborah's legs.

Deborah cried out that this sailing was like an orgasm, so Bill started video taping her goose bumps.

I said, "Goose bumps and an orgasm all in the same weekend. We be sailing!"

While we were training in and around Corpus Christi Bay, Port A, and Port O'Connor, Bill filled me in on the local history and legends he'd learned while writing his book *Aransas*. I was especially interested in the story of the pirate Jean Lafitte, as my boat was a LaFitte. Here's Bill's story of the pirate Lafitte:

Jean Lafitte

Jean Lafitte was a "freebooter," a term that referred to men who lived beyond the law, who took matters into their own hands, and used their own power and cunning and wits to achieve desired effects. We generally tend to think of freebooters as pirates, but the word has a broader application; not all freebooters roamed the seas.

Aransas was perfect for Lafitte's purpose. Lafitte rarely boarded a ship at sea. It was more effective—and more fun—to tease a vessel into following him through a pass. Once Lafitte had his quarry in the bays, a game of cat-and-mouse began. He knew the reefs and shoals; his pursuers did not. With a series of deft moves, the freebooter could lead a tall Spanish ship to grief on a sand bar, then board her at leisure and take any treasure.

He could, when necessary, escape by sailing into a creek like Copano, Smuggler's, or Barkentine. Lafitte alone knew how to navigate Cedar Bayou. When Lafitte slipped into the bay from the open gulf, the Spanish hovered helplessly outside. In utter frustration, they watched the taunting flutter of Lafitte's topsail behind the dunes.

Lafitte established an operational base at Cedar Bayou, and another at the southwest end of San Jose Island. The two locations augmented a Galveston base Lafitte had taken from another privateer.

Lafitte built a house for Madam Frank, widow of one of his captains killed in service, on Blackjack Peninsula, now in the Aransas National Wildlife Refuge. Her home was on the point of land separating San Antonio and Espiritu Sanato bays. Lafitte made sure she always had plenty of everything, and never had to work for a living. He could go there when it became necessary.

"For three days and three night," Madam Frank said, the men labored to unload the booty of many years. Then, in a final speech, Lafitte released them from further obligation and divided the loot among them. Lafitte's share was, by right, far the largest. He decided to cache it back from Madam Frank's house, there on False Live Oak Point. She watched each trip they made, night after night, into the thick oak motts.

Lafitte told her, "There is enough treasure in those woods to ransom a nation!" When the last cask was buried, Madam Frank said Lafitte came out of the oaks alone; his bearers stayed forever buried with the booty.

"Dead men tell no tales."

I said to Bill, "When treasurers hunters read this, they'll be swarming the Wildlife Refuge with metal detectors, backhoes, and bulldozers. Damn the endangered whooping crane, there's gold in them blackjack trees."

First Trip to Port O'Connor

After several more weeks of training, and some more offshore time, I thought it was time for an overnighter to Port O'Connor, two days and one night. I took the ship out the channel, past the jetties and the number-six buoy, and headed northeast toward Port O'Connor. I had set up everything. I had the course of 85 degrees laid in, and I turned the helm over to Bill. Deborah was in the cockpit, so everything looked good. They could keep each other company and help control the ship.

I decided I'd better get some rest before my watch at 3:00 PM, so I went below and crawled into my bunk. The way I was lying down, the sun shown in on my hand and made it warm. I lay there a little while and then noticed that my hand turned cold. *The sun must have gone behind a cloud*, I thought.

Bill and Deborah were chatting away like magpies.

A little while later, I noticed my hand getting warm again. I didn't think much about it and tried to get some sleep. The next thing I knew, my hand turning cold again. Then I remembered that it was a cloudless sunny day.

Something was not right here.

I crawled out of my bunk and went topside to see what was up.

Bill and Deborah were still chatting away and not paying any attention to the compass.

I watched for a few minutes and noticed that the compass needle kept swinging around.

We had been sailing in circles.

I called off the trip and decided what they needed was some more offshore training about how to hold a course. I had taken too much for granted. It was not a natural instinct to these two. We spent the rest of the day sailing in straight lines offshore to hold a course.

Second Trip to Port O'Connor

Captain's Log:

Left Port A about 8:00 AM I turned the helm over to Bill and went below for a nap. We sailed all day, and I did the 3:00-to-7:00 PM watch. During that time, I lost the Garman handheld GPS. It was under the dodger, slid off the cabin roof, and was somewhere on the deck. Bill came up for his watch and asked where we were. I told him I didn't know because the Garman GPS had fallen off of the cabin.

Bill found the Garman. We were still on course, so I turned the helm over to him. The weather was nice, with winds out of the south, between ten and fifteen knots. I had full sails up, and we were making about seven to eight knots, which is about the maximum for a thirty-six-foot waterline.

I went below to get some sleep.

I hadn't been asleep very long when a violent jerk flung me out of my bunk onto the floor.

I ran topside to see what had happened.

The weather had freshened to about twenty-five knots, gusting to thirty-five, and we were way over-canvassed for that much wind. The boat had a very heavy weather helm and was turning into the wind. Bill would fight the helm and the boat would fall off sharply. That was what had thrown me out of my bunk.

Bill was fighting the weather helm with all his strength, but the boat was just too much for him. When I saw what was happening, I immediately ferruled the big 135 percent Genoa down to a very small one. That eased the weather helm considerably. I went forward and put a reef in the main sail. This put us back in control, and the boat settled down to normal sailing.

Bill later told me that he thought it was his finest hour, pitting himself against the elements and the ocean. Like something from *The Old Man and the Sea*. Bill saw the ordeal as a story from literature. Maybe he thought it would look good in a book he'd write. But he seemed to have no feel for the stress and strain he was putting on the ship and the danger it could cause.

At the time, I grumbled to myself, *Bill has no feel for sailing, and neither does Deborah*. Ultimately, though, the problem was that I hadn't yet learned to trim sails myself, much less teach Bill.

The next evening, as we approached Port O'Connor I couldn't see the entrance to the channel. I was sailing toward what I thought was the lighthouse near the entrance when Deborah cried out, "I can see the red channel markers off to our right!"

We headed for those channel markers. Sure enough, we were off course about two miles in toward shore and were headed into shallow water and the shoreline.

If it hadn't been for Deborah's great eyesight, we might have run aground on the beach.

La Salle

When we entered Matagorda Bay on our way to Port O'Connor, we passed the cofferdam where excavation of the *La Belle* was taking place. *La Belle* had been the flagship of the explorer La Salle; it had sunk in 1686, and then was rediscovered in 1995 in Matagorda Bay. According to the Texas Historical Commission's Web site:

The *La Belle* is one of the most important shipwrecks ever discovered in North America. The excavation, conducted in a cofferdam in Matagorda Bay, lasted almost a year and produced an amazing array of finds, including the hull of the ship, three bronze cannons, thousands of glass beads, bronze hawk bells, pottery, and even the skeleton of a crew member in the ship's bow, curled fetal style upon a coil of rope. The one million artifacts represent a kit for building a 17th-century European colony in the New World.

Bill knew all about this from the research he'd done for his book *Aransas*. He had even been a participating scholar and honorary crew member on the excavation project for the *La Belle*.

Here's his story of the explorer LaSalle:

La Salle was a visionary who had spent nine years as a Jesuit novice before becoming obsessed with discovery in the New World. He had developed the fur trade around Lake Ontario and built forts in the northwest. Then came his crowning achievement; LaSalle navigated the Mississippi River from the Ohio River valley to the Gulf of Mexico. He claimed French ownership of all territories in the river's watershed and along its tributaries. The next step was to establish a colony where the Mississippi emptied into the Gulf of Mexico.

King Louis enlisted LaSalle for the job and equipped him with four ships. LaSalle's flagship was the forty-five ton *La Belle*, his personal gift from the king, beautifully appointed. *La Belle* was fifty feet long and fourteen feet wide with two masts carrying square sails. Six bronze cannons were proof that *La Belle* was a very important vessel.

La Salle lost one of the four ships, a ketch, to Spanish pirates off Hispañola. When he reached the mainland, the mouth of the Mississippi was so hidden in a maze of bayous and sand flats that he never found it. He consulted his maps and found Rio Escondido, the Lost River, emptying into Matagorda Bay. LaSalle believed that it might be the Mississippi, but his transport ship *Amiable* ran aground only "a cannonshot from shore."

LaSalle reflected that the Spanish had named that bay and the island correctly. Matagorda means "starvation," or, literally, "reduce fat." It was an ill omen.

After they went ashore and set up camp the Karánkawas Indians crept in for a closer look. Posting a number of men to stay aboard *La Belle*, LaSalle commandeered canoes from the natives. The crew aboard *La Belle*, too fearful of the natives to go ashore, ran out of water. Some of the men tried to survive on the ship's store of brandy. Emboldened by it, they decided to sail to another part of the bay. A norther hit, and the drunken captain was unable to control the ship. *La Belle* ran aground. The crew salvaged what they could from her and carried it to the colony. The goods they brought hardly made up for the bad news: the hapless adventurers were stranded. LaSalle took his most able-bodied men and set off to seek help, hoping to find the Mississippi and follow it up stream to French outpost. Duhaut outlined a plan to shoot LaSalle from ambush, and when some of the others refused to go along, Duhaut shot him. A single bullet ripped through LaSalle's forehead. The man who had charted and changed a major portion of the North American continent fell dead.

"Whole Lot of Shaking Going On"

We started up the channel just after dark, and we were looking for the ICW cross channel when a fishing boat came in behind us and started to cut across the bay. I radioed the captain and asked if he was headed for Port O'Connor. Sure enough, he was. We fell in behind him. We never did see the ICW cross-channel markers, but we did get into Port O'Connor.

As we motored down the channel looking for St. Christopher Haven Marina, we stopped at a local dockside restaurant and asked the people where the marina was. They pointed farther down the channel, so we continued.

Sure enough, there was the St. Christopher Haven Marina, and we docked. No one seemed to be in the office, so we tied up for the night. Next morning, as we were preparing to leave, the marina operator came by and collected his fee. We sailed at about 8:00 AM and headed home. The two-day-and-one-night trip back was uneventful.

As we approached Port A, Bill put on a tape he'd made of rock-and-roll music. It started with Jerry Lee Lewis singing "Whole Lot of Shaking Going On" and continued an hour with all the great classic rock he could find. When one side was finished, the tape would switch to the other side and run for another hour, but Bill hadn't recorded anything on the other side, so there was only silence for an hour. We played the music at full volume as we sailed along. We sang and yelled and had a great time!

We arrived back at Bahia Marina late on Sunday night. We tied the boat up in the slip temporarily. I needed to catch a flight back home so I could be at work on Monday morning. Bill said he'd take me to Corpus Christi Airport, then come back in the morning and secure the boat. We left in a hurry, and I made the flight.

Next day Bill drove back to the marina. When he got out of his car, the little beagle dog growled at him and tried to bite his ankle. He was really mad at Bill for some reason. This was odd, because usually he was friendly. Bill ran to the boat and jumped aboard quickly to avoid getting bitten.

The dog sat on the dock, growling and barking.

Bill went about securing the boat and cleaning up. When he finished, he took a beer out of the refrigerator and went to the cockpit to relax. No sooner had he sat down than Jerry Lee Louis came on at full volume singing, "Whole Lot of Shaking Going On." It startled him at first, and then Bill began to laugh. The tape had been playing throughout the night, every other hour. The "live-aboards" in the marina must have been kept awake all night.

Now Bill knew what the little dog was so mad about. He, too, had been kept awake.

The Mayor of Ingleside

Bahia Marina, where I kept my boat, was in Ingleside, which was a small town right across the bay from Corpus Christi. The former mayor of Ingleside was featured on A&E one night.

He'd been a bright young man of promise who arrived out of nowhere, won over the town, and was elected mayor. But he turned out to be a sociopath who, while he was mayor, made money by running drugs into the little port town. Then he and a cohort killed one of their cronies who was ripping them off. They buried the body in a shallow grave—dug by a hired hand right on their property. The corpse soon began to reek and attracted the local dogs and, eventually, the police. Evidence on the body and from a statement given by the cohort led to the mayor, and he was charged with first-degree murder.

The A&E documentary suggested that Ingleside is one of the armpits of the world, full of scrub brush, muddy water, hot and humid mosquito-filled stagnant air—and populated by the stupidest people in the country, even for Texas.

I wanted out of Ingleside for a lot of reasons, and the Trout Street Marina in Port A was the closest and best place to move to.

Trout Street Marina

Bill was still writing his *Aransas* book. He asked to live on the boat because boat life would give him a feel for the area, as well as some atmosphere and local status. I said okay.

We agreed that the boat needed to be moved to Port A, where Bill could easily go to Tortuga Flats. We arranged for a slip at the Trout Street Yacht Basin Marina, and we motored the boat over.

Land Lines

Bill needed to have a phone line installed so he could get his e-mail and communicate with Ohio State University, where he was still teaching part of the year.

He talked to the Trout Street Marina management and, yes, there could be telephone service to the dock power post, but it was up to the boat owner to get the phone line from the power post to the yacht. Bill called the phone company and arranged for a phone line to be put into our power post. When the phone line was installed, Bill went about getting the adapter plug, the fifty-foot telephone line, and a cover tube to put around the adapter and phone line to keep the rain from shorting it out. He drove fifty miles to Corpus Christi to Boat U.S. Marine Supply, bought the supplies he needed, and drove back. As he was getting aboard, he heard this *ka-plush*. He looked down to see the $35 adapter plug slowly sinking into the water. He quickly put everything down, pulled off his shoes and socks, and jumped in after the plug, thinking he might be able to catch it before it went too deep.

Bill cannot swim.

He knew this, of course, but he thought he could handle the situation. As he hit the cold, slimy water, he abandoned the idea of saving the adapter.

He came up thrashing. Grasping frantically with his arms at full length, he was able to grab the two-by-six on the bottom of the dock with his fingernails.

He hung there for a few seconds, trying to figure what to do next.

His fingers began to slip, so he inched over to one of the posts holding the dock up, and wrapped his feet around it. Then he noticed a bowline on the boat in the next slip; it was hanging below the dock, just within reach. He inched himself over to that bow line and grabbed on.

Now he looked around to see how to get back up on the dock.

The only dock ladder was way down at the other end of the dock. No way could he make that distance. The other possibility was to try to climb up the sagging bowline. He threw one leg over the line and started to shinny up it. He was dangling on the underside, but he needed to get his soaking-wet body up on top of it. He managed to swing around and get on top of the two-inch thick line and straddle it, slowly inching his way up the last few inches up the swaying rope. He was able to keep the swaying down to a minimum by holding the dock with both hands wide apart and slowly pulling himself up to the dock.

Finally, he made it. He was on the dock. He lay there a few minutes to catch his breath.

He was about to get up and start back to the boat, when he noticed all this red stuff on the dock.

It was his blood.

When he'd put his feet around the post supporting the dock, it had barnacles all over it. They'd cut his feet to pieces. The barnacles were so sharp, and he was so desperate to get out of the water, that he hadn't even noticed.

I'd mentioned to Bill that I'd hoarded some little bottles of iodine to take to the South Pacific. I'd read that any little cut could turn pretty nasty in that tropical climate if not treated promptly, and I was going to use the iodine for barter. I had several bottles on board.

Bill made it back to the boat, went inside, took off his wet clothes, and doctored his cuts with iodine. Then he put on dry clothes and socks.

Now there was nothing for him to do but go back to Boat U.S. in Corpus Christi and get another adapter. He drove the fifty miles back, walked into the store, and grabbed another adapter. All the sales people were laughing when Bill told them his story. And he was pretty cheerful about it, too.

After paying the $35 for the new adapter, he drove the hour back to the dock. But when he tried to install the adapter, it turned out to be the wrong sex.

Damn!

It was getting dark and the store in Corpus Christi would be closing soon, so he rushed to the car and drove yet again back to Boat U.S. Talk about determination. However, the ordeal was taking its toll. This time when he walked in, no one laughed. They just looked at him with frightened eyes as he staggered over to where the adapters were. Sure enough, the adapter he bought had been on the wrong hook. Bill threw it down on the floor, grabbed a correct one, and stomped out.

The sales people stayed very, very quiet.

He drove back to the dock and put the adapter in. It fit perfectly.

Finally, something worked!

Then he picked up the cover tube and put the telephone wire in it. He measured how far it needed to go to protect the line, took out his Swiss Army knife, cut off the proper length, and taped it all up over the adapter to make it water tight. Then he threw his Swiss Army knife down in triumph and it stuck in the dock.

He danced around like Rocky, holding his arms over his head. Triumph over adversity!

Now to test it out to see if he could get his e-mail. He went below, hooked up his laptop, and got online. Everything worked. He was all set.

So he went back up on the dock to clean up his work area—only to trip over the Swiss Army knife he'd stuck in the dock. He broke the blade of his trusty knife and almost wound up back in the water.

If you ever go to the Trout Street Yacht Basin Marina in Port A, you can still find the tip of the Swiss Army knife in the dock near Slip 9.

Trip to Galveston

I needed to have some work done on the *Sir Charles*. The facilities to do the work weren't available in Port A, so we moved the boat to the Galveston and Kemah area near Houston to finish preparing for what we hoped would be a major world cruise.

The two-day-and-two-night trip from Port A to Galveston was uneventful, but the Houston ship channel was a real problem.

This was one of the world's busiest ports; it has very large tankers coming and going about thirty minutes apart, so we needed to stay out of their way. On the bayside of Galveston Island, the channel split in three directions. The far left, to port, was Galveston Island Channel, which went to the Port of Galveston on the inside of Galveston Island. Straight ahead was Bolivar Channel, which split, and the Houston Channel went north. It is a long channel up through Galveston Bay to Houston. It takes about four hours for a sailboat to get to buoy 39, which is the turnoff to Clear Lake and Kemah.

Offshore Training and Preparation

We continued our offshore training when we got to Kemah in Galveston Bay. It was really a stretch to get down the Houston ship channel and out past the jetties.

One of the items I needed to add to the *Sir Charles* was an autopilot. I chose the Raytheon ST80 as it was designed by the same manufacture as my GPS and radar system. After the installation, we studied the operation for a week, then took it out for a test. It worked well, but it wasn't user friendly.

The first problem was the changing from normal lighting to reverse lighting. At night, the bright light would hamper the night vision of the helmsman. To switch from daylight to reverse backlight was a twenty-seven-step process that took almost five minutes. It should have taken no more than one or two keystrokes, but it didn't.

The next major problem was making changes. There was a built-in five-second timer that would reset to a default setting if nothing had been pressed in time. While trying to learn the system, I would read the manual and activate an event, then I'd go back to the manual to read the next step. Before I could enter the next setting, the unit had reset itself back to the default setting. So I had to memorize all the steps in a sequence to make any changes.

This was very discouraging.

I had to enter every waypoint manually, and there were a lot of them. There should have been a way to download them from a computer. A number of other operations should have been much easier, too.

ST80 Design Change Request

I called Rayetheon's support line a number of times and received little help. Finally I logged onto the Rayetheon's Web site and started submitting design change requests:

During the set up, the default "sail boat" mode enables a number of options not available in Power mode such as VMG to wind page. The automatic advance to next waypoint option could be created on the navigate or route page and default to current implementation of helmsman intervention ("press track to accept") to advance to next waypoint. Add an option for the captain or helmsman to select automatic advance to next waypoint. The same alarms should be maintained: short beeps 0.1 kilometer from waypoint and a long beep at the waypoint. If automatic advance is selected at the end of the long beep at the waypoint, and if there is no helmsman intervention prior, then the system should automatically advance to next heading and to the next waypoint.

This is critical when single-handling or when a course change is needed at the waypoint. The helmsman can prepare for the course change by freeing the windward sheets and preparing the leeward sheets while keeping positive control on the lines and not worrying about pressing the tack button to accept the course change.

In sailboat mode only, create an option for the captain or helmsman to select/un-select automatic advance to next waypoint without helmsman intervention on the navigate or route menu.

I submitted an ST80 critical design change request in January to be able to upload and download waypoints and routes from the captain's computer program to the masterview with the navigate keypad. I just had the instruments replaced to get your latest software upgrades and lost all of my waypoints and routes. I have all the waypoints and routes stored in my laptop computer, and it would save me a day of entering all those lost data points by hand if I could just load them from the laptop.

Cap'n Voyager

I purchased a laptop computer from SeaTec Marine and installed Cap'n Voyager software with Maptech full color, precisely calibrated digital marine charts from CD-ROM.

Each region includes GPS navigation information, and makes route planning as simple as clicking a mouse. Click, and you've put a waypoint on the chart. Click again, and another waypoint to create a route.

The software turns my laptop into a chartplotter displaying my real-time vessel position and key navigation information when connected to my GPS system.

I plotted all the cruising routes I created based on Jimmy Cornell's book *World Cruising Routes.*

Hitting a Boat and a Marina Wall

As I said before, we had gone into the Galveston and Kemah area to check the engine throttle and gear shift, but then it started acting up again.

We'd just returned from a day of offshore training, and I was letting Bill dock the boat. A crosswind made it very difficult to turn the boat sharply enough and quickly enough without going into the slip too fast. We needed to get the boat lined up with the slip before the wind blew the back of the boat sideways.

After the wind blew the boat almost crossways, I told Bill to back out and try again. It had to be done quickly before the side of the boat rammed the walk and knocked a hole in her fiberglass hull. Bill made it out of the slip okay, but the crosswind blew us all the way across the channel toward the boats docked there. When Bill tried to shift from reverse to forward, the gearshift slipped and stayed in reverse.

As a consequence, we were blown into the back of another ship docked on the other side of the channel. The crash broke one of the davits holding the ship's dingy. There was no damage to the *Sir Charles*, but our life lines were hung up in the other ship's dingy rigging. The owner and guests were having drinks in their cockpit and were understandably upset.

They helped get us apart, though—and then we were blown still backward into the seawall, crushing *Sir Charles's* monitor wind vane, because the shifter was still stuck in reverse. While we were pinned against the seawall by the wind, I was able to get the gearshift out of reverse and move the ship back up to the slip and into it.

Bill Walking The Plank

I had to swing the boat quickly and enter the slip fast to prevent another disaster. I told Bill to stand on the bow outside the safety lines and jump onto the dock. I needed for him to hold the boat to prevent it from ramming the dock.

Just as we approached the walkway, some five feet away, I yelled, "*Jump!*"

Bill jumped.

Trouble was that the jump need to be timed by the jumper, not by someone screaming at him.

Bill hit the walkway at a bad angle, damaging his foot and ankle.

He was able to prevent the boat from ramming the dock, but I haven't heard the end of the story of how I made him "walk the plank."

Chuck Finally Learns to Trim Sails

Deborah had an acquaintance named Captain Firestone who was an experienced sailor. She talked him into coming along with us for a training session. Captain Firestone taught us how to trim sails and many other valuable skills.

I was relieved to have overcome my ignorance of such a basic skill of sailing.

But the experience made me realize that I still had a lot to learn.

Shakedown Cruise

At last, I was ready for the shakedown cruise.

My crew consisted of Bill Allen, Deborah Russo, Dana Greshner, and Cheryl Reed.

In addition, we planned to pick up Bill's friend Juliet Wenger, a seventy-six-year-old journalist, at our Port Isabel destination. Juliet had wanted to go on the complete cruise with us, but her husband, Chris, said no. So she could only go from Port Isabel back to Port A, just for an overnighter. Chris didn't want to be without his cook for any longer than that.

I loaded up my little Geo convertible with all the new gear my expanded crew would need: extra harnesses with built-in life vests, lifelines, foul-weather gear, and some other stuff. I'd just passed Huntsville, Texas, and was cruising at about seventy when I reached New Waverly, about thirty miles north of Houston. Suddenly, the timing belt broke, and the little three-cylinder engine quit. I coasted into a Texaco station and called Triple A's road service. This was at about 4:00 PM. Triple A said it would be two hours before they could dispatch a tow truck to take me to the nearest Chevrolet dealer. The problem was that the Chevrolet dealers would all be closed by them.

The attendant at the Texaco station overheard me talking to Triple A, and when I hung up he said, "I know the Triple A tow-truck driver here in New

Waverly. He's just a few blocks away, on the other side of the highway. You want me to call him?"

"Yes, please!"

The attendant called the tow-truck driver, who came right over. When he arrived, we discussed the best place to take the Geo. He noticed I had Triple A Plus towing coverage, which has a one-hundred-mile towing radius, so he said he could take me to any dealer in the Houston area. There was one on the south side, near the League City turnoff that went to Kemah. I figured the taxi ride from there to the boat wouldn't cost too much, so asked the tow truck driver to go there.

As we passed Houston and approached the Chevrolet dealer, I asked the driver how many miles we'd come. He said, "Seventy-five miles." That left twenty-five miles out of the one hundred that were covered, so I asked him just to take me to Lafayette Landing Marina, where the boat was berthed and everyone was waiting on me. I could deal with the Geo problem when I returned from the shakedown cruise.

The tow truck pulled into the marina, and I went to get Bill. I asked him to come to the car and help me with the gear. When he saw the Geo hanging from the tow truck, he broke up laughing.

Well, I'd made it on time.

The tow truck dropped the little Geo and left. Bill and I took the supplies in and secured everything, and then we went to dinner at Joe's Crab Shack.

My crewmate Deborah Russo, who knew the Houston area very well, said there was a good repair shop called Top Notch just two blocks from the marina. I figured I'd call them from Port Isabel to see if they could fix the Geo while we were gone.

We came back from Joe's Crab Shack and started to bed down for the night. Cheryl was going to sleep in the U-shaped settee bunk we'd prepared behind the folding dining-room table. I'd made a plywood support spacer to fill the leg space and make a nice wide bunk. But when she put her weight on one knee to climb into the bunk, the plywood slipped off and dumped Cheryl on the floor with a loud bang and some screams. She was stuck behind the folding table with her ankle pinned between the table support and the plywood spacer.

I heard the screams and thought, Well, we've just had our first accident.

It was quite a scene trying to free Cheryl and get her back up on the settee. There was very little space in which to get a good hold on her and lift her out of the tight spot she was squeezed into. Finally she was extracted, but while she was sitting on the settee, it became evident that her ankle might be broken. It was swelling up quickly and had a good two-inch knot on one side. We offered to call an ambulance and have her taken to a hospital and X-rayed, but we said we'd have to winch her out of the cabin, as the paramedics couldn't get her up the companionway on a stretcher. Also, it might take a cherry-picker to move her from the boat to the dock.

She refused.

Suddenly her ankle wasn't hurting all that much, she said. She decided to let it recover overnight.

Next morning the swelling had gone down, and she thought it would be okay, so we shoved off around 6:00 AM and motored out to the Houston ship channel.

I'd programmed the autopilot with the waypoints of all the ship-channels buoys, and it followed them perfectly. We motored as we headed out toward the ends of the jetties, following the channel markers with care until we were past marker 7A and clear of the shoal that it marked to our port. From the marker, I suspected that not all skippers had read their charts for this area and would be oblivious to the dangers until it was too late. After we passed 7A, we raised the sails and cut the engine.

It was as fine a sailing day as could be, with winds out of the north about fifteen to twenty knots—perfect for where we wanted to go. I set the autopilot and entered the next way point, WP20 (29º 08.30' N 094º 26.30' W) out the safety fairway about fifty kilometers away, to get us out past all the oil rigs and into the clear open ocean. Then we turned south by southwest and headed for Port Isabel with the sun shining brightly and a good, constant fifteen-knot wind. All we had to do now was sit back, drink coffee or soda, and watch the new autopilot guide the ship to each waypoint. It worked perfectly.

After my daytime watch, I went to bed to get some rest before my evening watch. While I was asleep, a sparrow came aboard and flew down into the cabin where I was asleep. It rested on my leg, and Bill made a video of it sniffing my butt. Other than that, the trip was uneventful. We sailed all night under a beautiful full moon and starry skies.

The next day was a repeat of the first day and night, with more good sailing weather. We arrived outside the Brazos Santiago Pass (Port Isabel Channel) about noon on the third day and motored up the channel. The Sea Ranch Marina was just inside the south end of South Padre Island, but we had trouble finding the channel up to the marina. It was really close to the island, and I was afraid we'd run aground getting that close. Indeed, we did run aground several times in the shallow muddy bay looking for the channel. Finally, a powerboat came in, and we followed him up close to the island, where the channel was, and got to the Sea Ranch Marina, where we secured a slip.

I called the information operator and was given the Top Notch's phone number. I called and asked if they could pick up the Geo at Lafayette Landing Marina and replace the timing gear. I'd left my car keys behind the visor for them. They said they would.

It was time for dinner, so we called Island Cabs and had a Mexican cabbie named André drive us up to Blackbeard's for dinner. Bill sat in the front seat with André, and they proceeded to discuss Einstein's Theory of Relativity. André asked Bill what he thought of the concept of time. Bill said he thought it was a construct created by man, and that it really did not exist.

This was just the preliminaries, and soon they really got deep into Relativity Theory. I think Bill was sad to have to get out of the cab to do something so low as eat when he'd found someone who could intelligently discuss Relativity. After dinner, we returned to the ship for a night's rest.

Next day, we rose early and walked down the beach in front of the Hilton Hotel and all the way down to the Crappy Dunes, where Bill and I had stayed several times when he'd come down on Christmas break from Ohio State. He videotaped the whole walk. On the beach near the Crappy Dunes there was some sandcastle artwork, and Bill and Deborah gushed over this for almost an hour.

That night Juliet Wenger arrived, and we went to dinner at Scampi's Restaurant on the bayside, because our old favorite Louie's Backyard was closed. As we had dinner, we sat on the end of Scampi's pier and watched one of the most beautiful sunsets ever.

We'd planned to sail back the next day, Thursday, but a storm blew in. It was raining with really gusty winds, so I said, "No way am I going out in that kind of weather with an inexperienced crew." We just piddled around Thursday.

Friday morning, the wind in the marina was about ten knots out of the south, and the weather looked like it was clearing, so I decided we should go for it. We secured the ship and left Sea Ranch Marina and went out past the jetties. The wind was directly out of the south, 180 degrees at twenty to twenty-five knots—perfect sailing weather. However, the waves were still a little rough from the storm the previous day. The sea had eight- to ten-foot swells left over from the storm, and some eight- to ten-foot waves on top of that.

It was nothing the boat or I couldn't handle, but Deborah Russo and Cheryl Reed got seasick. Deborah stayed in her bunk all day and night, and Cheryl set on the lee side of the cockpit where she occasionally "called the shark York," which is slang for barfing. We sailed out to the 97 degree meridian and turned north. The 97 degree meridian would take us directly into the mouth of the Aransas Pass Channel.

Dana and I rigged the spinnaker pole out on the starboard side and put the main sail out on the port side. We were configured for wing-on-wing sailing. It was beautiful, and the autopilot handled it perfectly. I watched the GPS, and we didn't vary more than .01 degree from our course all day.

As night began to fall, I told Dana we should take the spinnaker pole down so if the wind freshened during the night, all we'd have to do to shorten sail would be to roll up the roller faring jib. We wouldn't have to worry about the spinnaker pole.

A problem developed when we tried to stow the pole.

The swivel joint on the mast end was bent and wouldn't allow us to secure the bottom of the pole to the mast. I had to climb up the mast a few feet to catch the line that stowed the pole. When I reached for the line with my right hand, I lost my left-hand grip on the mast and fell backward to the deck. I still had the line in my right hand, and when I reached the end, I felt a sharp stab of pain in my biceps. I got back up, and my arm ached, but I could still use my arm with very little pain and no apparent limitation of movement or strength. We ended by putting a cushion under the spinnaker pole to keep it from tearing up the teak deck, and we secured it with several lines to keep it from swinging around.

At the time, the pain in my arm didn't seem to be a big problem. But a month later, when I finally went to a doctor, I learned about all the damage. My biceps tendons had been pulled away from my forearm bone. The doctor said there wasn't anything he could do because by then, the muscle had already atrophied, and reattaching it wouldn't help.

We arrived in Port A about two in the morning and tied up at the fuel dock. Bill and Juliet went up to the phone and called Juliet's husband, Chris, to come get her. Meanwhile, I listened to the weather report, and it said there was fog along the coast. I wasn't sure I wanted to start dodging oilrigs in the fog. While I was considering the issue, Chris arrived for Juliet.

Bill said he'd continue with me if I chose to go in spite of the fog, but otherwise he was jumping ship and going with Juliet and Chris. Meanwhile, Cheryl and Deborah were too sick to continue. Bill and Cheryl wound up going home with Juliet. Dana lived in Port A, so she was home, but her car was in Kemah. I decided to motor over to the transit slips, tie up, and leave the boat there until I could find another crew.

Next day, after we secured a transit slip for a month, Dana, Deborah, and I caught a Southwest Airline flight to Hobby airport in south Houston. We took a taxi to Kemah and picked up Dana's car, and I went over to Top Notch and paid for the repairs on my Geo.

What a trip that was!

Returning to Galveston with Irene

The boat was still down in Port A, and when I got back to work at Microsoft, I phoned an old sailing friend I'd worked with at Wang Labs: Irene Lacano. As previously mentioned, she had a small sailboat and went sailing a couple of times a week at Grapevine Lake. We'd always talked about doing some offshore sailing, and when she heard I had a LaFitte 44, she lost no time calling because she wanted to do some offshore sailing on a big boat. I'd been putting her off, but now that I needed an experienced sailor as a crew member, she'd have her chance.

She was ready to go—just let her know where and when. We scheduled the trip for Thanksgiving Weekend, only a couple of weeks away, so she wouldn't have to miss any work. She'd fly into Hobby airport in south Houston, and I'd pick her up and drive her to Port A. When we got back, I could drive her to get her return flight.

Bill was available for the trip back from Port A, also. Friday night, Bill and I picked up Irene at the airport, and we drove to Port A. Bill left his car in Port A because Eddie, our friend, would pick him up at the Corpus Christi airport and drive him to Port A to get his car.

Saturday morning when we started to leave, the boat's engine wouldn't start. I called around, and no one was available to work on marine engines on Saturday. While we were sitting in the cockpit, I saw a mechanic go down our dock to another boat. I ran up and asked what he was doing, and he said he was working on an engine. I asked if he'd stop on his way back and take a look at my engine. He said he would.

In about thirty minutes, here he came. He ran a couple of tests and said it sounded like one of the cylinders was full of water. He removed an injector, and he was right—the engine had taken on some water. I asked if he could fix it. He said he'd try. He removed all the injectors and turned the engine over with the starter. Water squirted out all of the injector ports.

Well, that cleared all the water out of the engine.

He put the injectors back in and primed the fuel lines. The engine started after a couple of tries, and we were ready to go.

I'd planned to use the return waypoints out past the oilrigs, in case fog developed. We sailed all day and night without any problems. The weather was great, but because of the delay, we were running a little late for Irene's airline schedule. We came in at WP20 (29° 08.30' N 094° 26.30' W) out at the end of the safety fairway about 4:00 PM. We saw two tankers going into Houston, and I told Bill and Irene to follow them, because they knew where the channel was. Wouldn't you know it, the wind was dead on our nose trying to go down the safety fairway, so I had them roll up the jib, start the motor, and motor in.

No sooner had we entered the Galveston Channel, between the jetties, when a big 900-foot tanker behind us gave out with five blasts on his horn.

That meant, "Get out of my way!"

Sailboats have the right of way, except in restricted channels where the tankers cannot maneuver; then, the smaller vessel is burdened to get out of the way. Unfortunately, there was another big tanker coming out of the channel. I couldn't get into the outbound lane, and I couldn't get close to the jetties for fear of running into them and knocking a hole in the hull. I held my course until the last minute and the outbound tanker passed, then I swung the *Sir Charles* around to follow the outbound tanker and let the inbound tanker go by.

That was close.

When the inbound tanker passed, I did another U-turn and started back up the inbound lane. No sooner had I settled in the inbound lane, than another tanker with only three little lights on gave me the five blasts again.

Another one? Darn!

I decided I'd make a run for the anchorage up ahead.

Irene was getting plenty nervous, and her eyes were wide open with fear. She did not like playing tag with nine-hundred-foot tankers.

Bill said, "Relax Irene. You don't want to die all tensed up. You want to look good in the casket."

I said, "Bill, you don't understand. When one of these big tankers runs you down and crushes your ship, the big props of the tanker suck everything through the props, which chews you up into little bitty pieces making chum for the fishes to eat. There won't be any need for a casket."

Poor Irene. She was pale white by now.

We made it to the anchorage and got out of the way of the second tanker. We continued on into the Galveston Channel and docked at the Galveston Yacht Basin Marina. It was about 8:00 PM. We found a telephone and called a taxi to take Irene to the airport. She'd missed her 7:00 PM flight, but she said she would rather stay at the airport and catch the next plane to Atlanta.

After that experience, she never called and asked to go sailing again.

Next morning we left the marina. As soon as we were out in the Galveston Channel and before we reached the intersection of the Bolivar and Houston Channels, the engine started to overheat. I shut it down and turned the boat around. I told Bill to steer while I handled the sails. We sailed back into the marina.

I told Bill to turn on the engine and put it in reverse to keep us from crashing into the dock.

We managed the docking just fine.

I walked over to the Galveston Yacht Service and asked the manager if someone could come over to the *Sir Charles* and check out the overheating problem. They sent a repairman over. He found that the engine-water input seacock was closed and the impeller was destroyed trying to suck cooling water in through the closed seacock. He replaced the impeller, and we were off again in about half of an hour.

Voyage to Chicxulub and Progresso, Mexico

Bill wanted to go to Mexico—specifically, to Progresso, Mexico. The center of the Chicxulub meteor impact crater was just fifty kilometers offshore from Progresso. Bill thought a description of it would make a great introduction to the book he was going to write about our trips. He envisioned an introduction that would start at the end of the dinosaur era and the beginning of mankind.

So we hired Captain Walter Grimes and his first mate, Katherine, to help sail us to Progresso, Mexico, on the Yucatán Peninsula. The planned trip would begin on August 17, 1998, and would end on August 21.

The Chicxulub Meteor Impact

My own research on the Chicxulub meteor impact crater led me to a Web site with information from the American Geophysical Union. I found an article called "Chicxulub Impact Crater Provides Clues to Earth's History" from a journal called *Earth in Space*:

"In 1980, Luis Alvarez and his geologist son, Walter, proposed that a giant asteroid or comet struck the earth approximately sixty-five million years ago and caused the mass extinctions of the dinosaurs, and over 70 percent of all life on earth. This bold proposition resulted from their discovery, near the medieval town of Gubbio, Italy, of a centimeter-thick clay layer among limestone depos-

ited on the earth's surface at the time of the extinction event, between two geologic time intervals, the cretaceous (K) and tertiary (T). The limestone directly
beneath the clay layer abounds with planktic formaniferids of latest cretaceous
age, while the tertiary limestone unit immediately above the clay layer, showed
only rare and poorly formed fossils. Thus the clay layer itself, the scientists reasoned, must hold clues to the duration and nature of the mysterious KT extinction event, one of the most dramatic calamities to afflict earth's biosphere since
the development of complex life over a billion years ago. They found that the clay
contained high concentrations of the element iridium, extremely rare in earth's
crustal rocks but quite abundant in certain meteorites, and proposed that this
clay was the altered remains of the dust cloud blasted around the world when a
ten-kilometer-wide asteroid or comet struck the earth. Fifteen years of research
has upheld this idea, and, now, all indications are that the source crater has been
found.

The collision occurred on the Yucatán platform and is centered near the port
city of Progresso, Mexico. The 200- to 300-kilometer wide crater lies buried
beneath 1100 meters of limestone laid down in the intervening years and few
clues of its presence remain at the surface. Yet prominent circular anomalies in
geophysical data gained the interest of Petroleos Mexicanos and in the early
1950s, they began an exploration campaign that included deep drilling to recover
samples of the subsurface rocks. The buried feature became known as the Chicxulub structure, named for the first well located near the Mayan village by the
same name. Pemex drilling continued throughout the early 1970s, and by that
time, Mexican scientists realized that the Chicxulub structure was quite unusual.
Three wells near the center had recovered silicate rocks with igneous textures, initially mistaken for volcanic rocks, and others, located between 140 kilometers
and 210 kilometers from ground zero recovered breccia deposits hundreds of
meters thick, indicating catastrophic or explosive conditions. By 1980, at least
one scientist at Pemex felt that the evidence pointed to impact.

Beginning in 1990, however, samples from the Pemex wells were located in
Mexico City and teams of scientists from the United States and Mexico quickly
developed an impressive case that the Chicxulub structure was indeed the KT
"smoking gun." Mineral evidence of shock metamorphism, requiring pressures
and strain rates considerably higher than those produced by terrestrial processes,
indicated that the crystalline rocks within the basin were melt rocks formed by an
impact event and not by volcanism. Biostratigraphic information indicates that
the structure was formed in uppermost cretaceous rocks, consistent with a KT
age. Argon and uranium-lead age determinations reveal that the melt rocks and

the associated breccias are the same age as the tiny spherules of impact glass found within KT boundary deposits in Haiti and Mexico and the unmelted granitic fragments found in KT boundary exposures throughout western North America. Isotopic analyses demonstrate that the Chicxulub melt rocks and the ejecta spherules originated from the same source rocks. Consequently, there is a clear chemical, as well as, temporal link between the Chicxulub structure and the KT boundary deposits."

Brazos River Iridium

Dr. W. F. Hillebrand of the University of Texas at Austin wrote a book titled *T-Rex and the Asteroid of Doom*. In this book, he theorized that the Chicxulub asteroid was the cause of the dinosaur's extinction, and that after the dinosaur was gone, mammals were free to evolve. These ideas fascinated Bill and me.

I saw a PBS special that showed Dr. Hillebrand on the bank of the Brazos River, near Waco, Texas, where the KT line, or iridium layer, was deposited by the Chicxulub impact. The iridium layer is a thin, greenish band that encircles the earth, and nothing like it has been seen before or since. It is one of the most important geologic events in evolutionary history.

From the film and Internet pictures we'd seen, Bill and I got the idea of seeing the green KT line first-hand, so we drove over to the Brazos River and crossed it three times. Finally we found the particular bridge shown in the PBS special.

We parked and went down to the river's edge, south of the bridge and along the west bank. We clawed through the brush and brambles, and poked in the dirt until we found the excavation site. Since the site had been restored to its original condition, it took us days to clear the vegetation and dig down several feet to where the layer was supposed to be, according to the pictures.

When we knew for sure that we looking at it, we jumped up and down and hooped and hollered, "Eureka! We found it!"

Bill put his hand on it, having me do the video filming for a change. Touching the iridium sent some sort of cosmic energy through him, he said. He felt that he had experienced the passage of sixty-five million years in an instant.

So it seemed to fit that we wanted our international sailing trips to start at the very spot where the asteroid of doom struck earth and changed the history of the world.

The Voyage Begins

We left Galveston with Captain Walter and Katherine and sailed due south with perfect sailing weather for the first two days. Then we ran out of wind. Dead calm.

We started the engine, and it worked for about four hours. But our diesel fuel contained algae that had built up in it. We put poison in the fuel tanks to kill the algae, but the dead algae just settled to the bottom of the fuel tank. That can be okay as long as you stay in calm waters, but when you get to sea, it sloshes around and goes into the fuel filters and then into the fuel pump. We cleaned the pump, but it still did not work. It was late, 9:00 PM, so we decided to wait until morning to work on it.

Had I been a superstitious person, I would have said the gods were telling us to go home.

Anyway, it was a beautiful starry night, so we had cocktails. We were sailing on a heading of 151 degrees. While we were having drinks, I watched the compass. It would swing around to 330 degrees (180 degrees from the direction we wanted to go) and send us north at one knot. Then it would swing around again to 150 degrees, but would not stop there. We were in one of the gulf stream's eddy currents. We were just going around in circles. Our sails were at maximum, so we must have been stirring the atmosphere in a "butterfly effect."

Next day we repaired the fuel pump and continued to Progresso, Mexico. We could see a line of thunderstorms building behind us.

On the way down, Katherine had a fishing line over the transom and was trolling a fuchsia lure. She snagged a twenty-pound yellowfin tuna and hauled it aboard, but no one wanted to clean it, so she threw it back in the gulf. The evening before we arrived in Progresso, a flight of about twenty-five little "Huey birds" landed on the ship. One landed on my head and another one on my hand, which was on the helm. The rest made themselves at home on the boom and various parts of the boat. They were about two inches long and did not weigh over a half an ounce. There are not a lot of places for little birds to land and rest when they're 120 miles out at sea, so they take advantage of any opportunity that becomes available.

I called them "Huey birds" because I'd heard they eat the hot Mexican chili peppers off the vine, then fly backward into the wind going, "Whooeeee!"

The next morning, when we were fifty kilometers from Progresso, I woke Bill up. We opened a bottle of champagne, and we all toasted the center of the Chicxulub meteor impact crater, the death of the dinosaur, the birth of human-kind—and the beginning of our travels.

Arrival in Progresso

When we arrived in Progresso, we cleared customs and Captain Grimes introduced us to Señor Lance, the port manager at Progresso. He also introduced us to Judy and Juan Manuel Mier, the owners and editors of *Yucatan Today*.

We heard on the radio that the first gulf hurricane of the season—named, ironically enough, Hurricane Charlie—had developed exactly where we were slowly turning around.

Now I know how hurricanes get started: by a sailboat caught in a gulf-stream eddy stirring the air around in a circle with its sails.

We couldn't sail back just yet, so we decided to do some sightseeing. We went to Mérida and rented a VW bug. On the way back to the boat, we ran out of gas, so Señor Lance, the port manager at Progresso who had taken us to Mérida, brought us some gas.

The next day we started off for Chichén Itzá. The little VW bug didn't have air-conditioning, so we had all the windows down and the little vent windows up front all the way open. About a half hour into the trip, a large tour bus passed us. We decided to draft it and get a little better gas mileage and some extra speed.

Well, this was okay for about ten minutes, then someone on the tour bus flushed the toilet and a great cloud of stuff exploded all around the VW and the vent windows sucked a lot of it inside.

First lesson learned: while traveling in third world countries do *not* follow tour buses. They have toilets on them and don't use holding tanks.

Chichen Itza

We finally arrived at Chichen Itza at about 10:00 AM. Tourism is one of the largest income-producing enterprises in Mexico, so they try to make things as pleasant as possible for the gringos. The Chichen Itza site included a large reception area, a dining room, gift shops, and toilets. We paid our admission and started the tour.

We climbed the famous Temple of the Sun and the Temple of the Moon. We toured the whole place. There was the ball court where the losers lost their heads. There was a carved wall of stone heads, and Bill took my picture with my head among the others.

About 4:00 PM we started back. When we arrived at the boat, we heard that Hurricane Charlie was in full blow, so we weren't going to sail back that week. Bill, Katherine, and I decided to fly back. Captain Grimes drove us to Mérida, where we caught the plane to Houston. We'd come back down a week later for the boat.

Smuggling Illegal Weapons into Mexico

While we were back in the states I e-mailed Judy Mier, owner and editor of *Yucatan Today,* who also ran a tourist agency. I asked her to book us a hotel in Mérida and a rental car. She did and, in return, she asked us to bring down some pepper spray and mace for her daughter who was starting college. These are illegal in Mexico, but she didn't want the girl to be completely defenseless.

Bill and I went to a couple of local gun stores and picked up two small cans of pepper spray and two cans of mace. I wasn't willing to have them in my baggage while going through customs, so Bill said he'd do it.

His heart rate is normally only about forty, so he's usually calm under pressure. I think he could lie to a polygraph and get away with it. Nothing gets his heart rate up except sex and violent exercise.

When we went through customs, sure enough, his "random selection" light came on, so he had to talk with the customs official. She asked what the can of mace was. He smiled and said it was shaving cream. She did not question him further.

Sightseeing

In Merida, we picked up the rental car and found the little hotel. It was a very nice one in downtown Merida. We drove out to Judy Mier's house and delivered the pepper spray and mace.

We spent the night in Mérida and the next day drove to Cancún. On the way, we stopped at another Mayan ruin, Coba, and visited it for a while.

Coba is a very large remote Maya site on the shore of five lakes. It is still largely unexcavated, and thrived between 600 and 800 A.D., when Chichen Itza became dominant. The pyramid El Castillo is one of the tallest in the Yucatan. It is not as well restored as many of the more famous ones, and the steps are rather treacherous, without any ropes to help you along. But the view of the jungle and the surrounding lakes from the top is spectacular.

I spent about thirty minutes under one of the steps in a small cave until the rain in the jungle stopped.

I could hear the song "You Belong to Me" going through my head:

See the Pyramids along the Nile,
See the sunrise on a tropical isle,
Just remember all the while
You belong to me.

Fly the ocean in a silver plane,
See the jungle when it's wet with rain,
Just remember till you're home again,
You belong to me.

The 42-square-mile site has lots of small roads and paths through the jungle, and it's easy to get lost. There were lots of birds and butterfly species there, as well as many statues in various states of preservation.

We drove on to Tulúm, and then up to Cancun.

Cancun is split into two parts. There is the Isla Cancun, which is a strip along the Caribbean shore with expensive hotels, shopping centers, and water activities. We stayed on the mainland in the older part of the city, Ciudad Cancun, at the Howard Johnson's, and we visited all the sites on the island.

I was really disappointed. It looked like any mall or tourist trap you find in the U.S. There was a Hard Rock Cafe and lots of fast-food places and department stores, just like in the U.S.

The only new entertainment was the Coco Bongo. It was advertised as one of the three best dance places in the world. Bill and I thought we might go and see what it was all about. Only problem was, it did not even open until one in the morning, and the cover charge was $35. I thought $10 was a lot at the totally nude places in Houston!

We left Coco Bongo to the jet setters and ecstasy dancers.

The next day, we took a ferry from Puerto Juarez to Isla Mujeres. This was one of the places I'd wanted to sail to in the Regatta Del Sol. Puerto Juarez is a small town on the mainland about five miles from the lighthouse on the south end of Isla Mujeres. The town serves as a terminal for ferries servicing Isla Mujeres. A long pier extends from the shore abreast of the town.

Isla Mujeres is about four miles from Isla Cancun. It's about four miles long, low, narrow, and wooded. The south part is slightly elevated and has trees about twenty-five meters above the sea. We visited the ruins of a square watchtower, which stands near the southern part of the island. The east side of the island is composed of fairly steep, rocky shelves which terminate at the north end in El Yungue (Anvil Rock)—square, black, and about two meters high. Some white cliffs rise about midway along the east side of the island. A large, square, white hotel stands on the north extremity of the island and has been reported to be a good radar target.

In 1517, Isla Mujeres was discovered by Spaniards led by Francisco Hernandez de Cordoba, who arrived at this abandoned site and discovered monuments shaped like women. So the Spaniards named it Isla Mujeres, which means the island of women. The island was later used as a pirate refuge by famous men such as Jean and Pierre Lafitte because many Spanish ships sailed very close to its shores.

Future Plans

While we did our sightseeing, I was planning our future trips.

At Isla Mujeres I thought, Now that I've seen the place and know about the harbor and the docks, maybe we can stop by on our way to the Panama Canal.

We also planned a possible voyage to Australia and had decided to stop by the Galapagos Islands on the way. Bill wanted to include a reference to Charles Darwin in his book about our travels. He wanted to investigate what Darwin had seen that helped him formulate his theory on natural selection and evolution, and write his famous books *The Origin of the Species, The Voyage of the Beagle,* and *The Descent of Man.*

We also wanted to visit places that Captain James Cook visited on *HMS Endeavour Bark* in 1768 during his discovery of the southern hemisphere. We wanted to go to Otaheit where he observed the transit of Venus. We also wanted to see the Society Islands of French Polynesia, Tahiti, Moorea, Bora Bora, Huahine, Raiatea, Tahaa, and Maupiti. Then we'd visit the Cook Islands of Aitutaka, Atiu, Mauke, Rarotonga, and Magaia.

Next we wanted to go to the Samoan Islands before stopping in Tonga, then Fiji, and then heading on toward New Zealand.

I wanted to see the place Joshua Slocum, the first solo circumnavigator, visited, where the native chief wanted to know where his crew was. When he said he was alone, the chief said, "You had crew, you ate'm."

I wanted to spend some time sailing around New Zealand, one of the best sailing places on earth, before crossing over to Australia.

Bill wanted to see the places that Jack London and his wife Charmain saw aboard his sailboat *Snark* and that he talked about in his book *The Cruise Of The Snark*. Bill had also read sailing books by Melville, Robert Louis Stevenson, and Joseph Conrad.

The Threat of El Niño

During my research, the TV newscasts kept referring to the El Niño cycle, which they thought would last for several years and affect shipping in the South Pacific.

I knew we didn't want to be in the South Pacific between December and May, for that was the typhoon season—but this El Niño was a year-round phenomenon.

There is a regular annual transit of about eighty boats that sail from New Zealand to Fiji in June and back again in November. In late October 1991, there was a freak typhoon in the South Pacific dubbed the Queen's Day Storm or "Killer Storm." It sustained winds in excess of sixty-five knots, gusts to over eighty knots, and seas at least forty-five feet. Eight boats were sunk, and twenty others were disabled. One boat was lost with all hands. Neighboring countries sent search planes out to locate the EPIRBs (emergency position-indicating radio beacons) that were activated, and to assist as best as they could until surface ships arrived.

I began to think it might be dangerous to try to sail around the world while El Niño was active.

Return to Texas

After visiting and sightseeing in Mexico, we headed for home. The cruise back to Texas was fast and smooth. Bill went back to Ohio State in September. It had been a near-perfect sail, better than the shakedown cruise. I felt more confident about our skills than ever. Only Zeus himself could stop us now, which would turn out to be exactly what happened by the time we were ready for the South Pacific voyage.

Part Six:
The End of an Era

Chapter Fourteen:

The Retirement Years (1999 to 2007)

Retiring

As mentioned earlier, I retired from Microsoft on February 19, 1999, one day before my sixtieth birthday. It wasn't long before my childhood friends, Bill Allen and Eddie Greding, were retired, too. A few months after my retirement came Bonza Bottler Day, and a few months after that, there was a whole new millennium to contend with.

Another Bonza

September 9, 1999. 9/9/99.
    It was Bonza Bottler Day again.
    And I was depressed.
    I was not where I wanted to be.
    I was not on my sailboat.
    I was not in the South Pacific, as I'd hoped.
    I was just sitting in my A frame—which Bill called "the badger hole"—writing these memoirs.

New Millennium

But then the new millennium dawned.
    I came out of my funk. This was fantastic—more exciting, even, than Bonza Bottler Day!

It was the best time, ever, to be alive—the greatest century in the history of man had just been completed. The new century was a new time, a pregnant time, full of promise and expectations, as though time were young and fearless. There were no limits, no boundaries at all, to what the future might hold. We now found ourselves in that magical time that only occurs between major ages in human history. We were poised either to step boldly into an unknowable future or to tumble down the slopes of science and pseudo-democracy into the wells of chaos.

Yes, it was a new and exciting time, but what had been lost to the past?

Ladies were not ladies any more.

You couldn't trust a gentleman's word.

And tomatoes didn't taste as good as they used to.

The pillars of civilization—had they vanished?

You couldn't trust any one whose job depends on a vote.

The last hundred years had been crowded with fantastic advances and discoveries—but it was also a banged-up century, some said, one of cheating, of murder, of rot and decay, of scrambling for public and private lands and damn well getting them by any means at all.

Our little nation at the dawn of the century had been torn with complexities and was too big for its britches. Things got going for us during World War II, and afterward, the Soviets took us on. We beat them, of course, but unfortunately, we wound up with a monstrous national debt and tens of thousands on the public welfare roles. Next thing we knew, our soldiers went to Iraq, and no one seemed to know why. There were some stories about weapons of mass destruction, but I suspected the real reasons were covered up. The war seems to be doing two things: It provides income for our military/industrial complex, a test of the new Force Transformation doctrine and it's a training ground for our generals.

But I've jumped ahead of myself. First I have to go to France.

France, November 19, 2000

I'd remained friends with my ex-wife, Joan, the one with whom I'd moved from Texas to California decades before.

I had always suspected that Joan had some kind of supernatural powers. As I said earlier, she had this strange intuitive perception ability. She didn't measure things analytically the way I always did. Instead, she just knew instinctively what needed to be done. She could observe a situation and instantly know what to do, while I'd have to analyze the situation, which might take me weeks. For example, she has never balanced her checkbook. She just says, "When I feel I need to make a deposit, I will."

She'd remarried by this time and was living in Pebble Beach. She and her husband, Jerry Winters, had sold the three houses in Carmel and bought a house in Pebble Beach. They also opened the Winters Fine Art Gallery in Carmel and were making a fortune. They opened five boutiques in Decatur, Illinois, to help her daughter Martha. They flew to France on vacation several times and eventually bought a little farmhouse in the tiny hamlet of Bréves. They vacationed four times a year there for a month each time. They invited me to come visit, and if I liked the place, I could stay on and house-sit—or rather, farm-sit—while they were away back in the U.S.

So I went to France November 19, 2000, and stayed until February 27, 2001.

The flight to France was awful. To save money, I'd bought a coach-class ticket, and I was cramped for hours in a small seat in the crowded cabin. We finally arrived in Paris at Charles de Gaulle airport, and I somehow managed to pry myself out of the tiny seat and walk from the plane into the airport. After clearing customs, I caught the shuttle to the Hotel Mercure, and checked in. Joan and Jerry Winters always spent the night there to recover from jet lag. I checked in and went down to the coffee shop and waited for Joan and Jerry. We spent the night at the hotel and took off early the next morning after breakfast. They had rented a car to drive to their house.

The trip to their home was complicated. I was glad they drove, because I'd never have found their place on my own. We took a highway called A1 (Paris) out of CDG, drove a short distance to A3, and headed south to A104-E15. We turned off on N104 (Marne de Vallee) and continued for about thirty minutes to A6 Sud (Lyon). We continued on A6 Sud for over an hour and exited at Auxerre. There we picked up N151 to Clamecy. In Clamecy, N151 turned off to Nevers, so we turned onto D985 to Corbigny.

Bréves

Their village, Bréves, lies about halfway between Clamecy and Corbigny, almost a three-hour drive from Paris.

The village has a population of about three hundred (I told you it was tiny!) and is located on the Yonne River and Canal du Nivernais, in the Bourgogne, or Burgundy, region of France—where they make the famous wines.

Joan and Jerry lived on a half acre of land in a three-bedroom, one-bath house. It had a living room with a fireplace, a dining room, a kitchen, an attached one-car garage, and a full basement with a special oven for bread baking. There was also a full attic, in which the previous farmers had stored hay. Attached to the garage was the livestock feeding room. There was also a shed for the sheep, named Bill, Hillary, and Chelsea. There was a beautiful patio out back, and a veranda over the garage and feeding room.

Across the street was an old church—and I mean *old*. It was built in 1150.

Before I left home, I'd logged onto the Earthlink website to find how to get my e-mail in France. Earthlink showed phone numbers for their servers in Auxerre and Dijon. I tried to use the Auxerre number but had some problems with it, so I used the Dijon server, and it worked fine. I was able to get my e-mail and do everything on the Internet that I could do in the States. It worked out very well for me while I was away.

First Side Trips

The village of Bréves is too small to have supermarkets; those are found in neighboring towns. The Atac supermarket is in Corbigny, and the Auchan supermarket is in Clamecy. There was also a hardware store in Corbigny, called Maxibrico. ("Bricolage" means "do-it-yourself" in French.)

I drove down to the capital of the region, Nevers, to have a look around. While I was cruising Nevers, I saw a store called Brico Depot, so went in to see what they had. It was exactly like a Home Depot in the States, right down to the check-out stands, the warehouse aisles, and the banners identifying the aisles. Everything was the same, except it was all in French.

"Brocante" means "second-hand dealer," and it's a word the French use for "flea market." Almost every town in France has one, and the larger cities have several, some of which are enormous. We shopped at several in the area. I found and bought a six-foot wrought-iron cross at one, and some adzes (wood-working axes) at some others.

Vézelay

As I became more comfortable being in France, I began to make other side trips. One of the first of these was to Vézelay, a mountain-top village near Bréves. Like most small French towns, it had an enormous church—this one was in the Romanesque style and dated back to the twelfth century. For a long time it was thought to contain the relics of Mary Magdalene, and so it's called the Basilica of St. Magdalene.

According to legend, at least two crusades began in that very church. Apparently a priest called for the Second Crusade during an Easter sermon in 1146 in front of the French king. A generation later, Richard I of England and Philip II of France met there before leaving for the Third Crusade.

Auxerre

I went into Auxerre with Joan and Jerry to meet their banker, Thierry Corniot, at the Crédit Mutuel, where I opened a bank account. With a population of nearly 40,000, Auxerre is the closest city to Bréves, so we did a lot of shopping in there.

Clamecy

I already mentioned that in Clamecy, we often did our grocery shopping at the Auchan. Clamecy also contains a wonderful restaurant, called L'Opaline, which is housed in an old church. France really seems to have a lot of old churches, so I guess they didn't need this one. Anyway, the food was fantastic and equal to that of any four-star restaurant in the US.

I thought Joan and Jerry's place needed some colorful flowers to brighten things up, so I was delighted when I located a wonderful plant nursery in Clamecy. I bought several planters, three rose bushes, some bulbs, and four giant sunflowers. I planted the bulbs in the planters, and had Gilbert Billiat (Joan and Jerry's neighbor) plant the roses. I put the four giant sunflowers on the patio table.

Corbigny

We also grocery shopped at the Atac in Corbigny. When Joan and Jerry first came to France, they stayed in Corbigny. La Marode was a wonderful restaurant, where Ceryl was the waiter and bartender. We had several lunches and dinner there. Yves, a chef at La Marode, also ran a pastry (Patisserie) shop, with his partner Crystal.

Road Trip—Vallèe de la Dordogne

We drove down to the Dordogne River valley to visit some places that Joan and Jerry liked. On the way, we ran into snow going over the mountaintop, but we didn't have any real problems.

Sarlat

Wikipedia on the Internet calls Sarlat "one of the most attractive and alluring towns in southwest France." We stayed overnight here, at the Auberge Bourmande hotel.

Sarlat is famous as the town where pâté de foie gras (goose-liver paste) is created by force-feeding geese to make their livers larger; the enlarged livers are puréed to make the pâté. There are several large foie gras factories and numerous goose farms, not to mention shops where you can purchase the local delicacies. We went to a place called Saveurs du Terroir—or Flavors of the Earth ("terroir" means "terrain," not "terror"!) and bought some pâté de foie gras. I sent several gift packages back to the States.

Beynac

Next Jerry drove us to Beynac, which has been called one of the most beautiful villages in France ("l'un des plus beaux villages de France"). Joan and Jerry stayed there in a castle on their first visit to France. Jerry drove us up this really narrow street, just wide enough for the little car, to show me where they had stayed. He had to back all the way down the street to get us out, because there was no room or place to turn around. Even if there had been, the road was so steep that going down it head first would have been really scary. It was scary enough just backing down.

On the way home we stopped at a brocante, and I bought some wood-working planes.

Train to Paris

It was about time for Joan and Jerry to take off on their return trip back to the U.S., so a few days in advance, we decided to take the train to Paris to see how difficult it would be to get to Charles de Gaulle airport by train. We boarded in Clamecy and left the car at the station. The train from Corbigny took us to the

Gare-de-Lyon in Paris. The Gare-de-Lyon is the station for the trains running between Paris and Lyon.

In the station we went down three levels to take the subway, or Metro, two stops to Châtelet Les Halles. Then we had to go up two levels to another Metro line to get to Charles de Gaulle airport. On the way back, we took the 31 line from the airport to Châtelet Les Halles, and went down two levels to catch the Metro to Gare-de-Lyon.

We boarded the train to Dijon, but had to change in Laroche-Migennes to get to Clamecy. It all went very smoothly, so Joan and Jerry decided to take the train to the airport when they left on their trip back to the U.S.

Home Alone in France

They departed in January and left me alone for nearly two months.

They didn't have a TV, so I listened to the radio. There were several very good classical stations, and I enjoyed a lot of classical music and a number of operas. One station played a lot of American jazz that I hadn't heard in years. "Le Jazz Hot" is very popular in Europe.

Back to the Airport

In late February, my long stay in France had come to an end. Joan and Jerry were in California, so an American friend of theirs named John Broadbent drove me to Paris to catch the plane back to Texas. John and Doreen Broadbent were commanders of the Salvation Army in Monterey, California, where Joan and Jerry had been members. John and Doreen had retired to Dompierre-sur-Héry, a town just south of Corbigny.

On the way to the airport, I smelled diesel fumes, and when we stopped at the gas station for fuel, John's car wouldn't start. When John raised the hood, we saw that one of the fuel-injector lines had broken off of the injector. John called a tow truck, and we were taken to a garage in a nearby town. The tow-truck driver went around the garage, which was full of cars in for repair, and borrowed a fuel-line injector tube from a similar car. He replaced John's broken line and charged us only twenty euros.

Then we were off to Paris. I'd planned the trip so that I arrived the day before my departure. I would spend the night at Joan and Jerry's favorite Paris hotel, the Hotel Mercure, and go to the airport the next day. Being late to the hotel was no problem. That night, I caught the shuttle to the airport, went to the ticket counter, and upgraded my seat to business class.

It was a much better flight going home than coming over.

Carmel

A few months later Joan and Jerry called and invited me to come visit them. It was summer, and the temperature in Carmel was somewhere between 40 and 60 degrees. In Texas, the temperature was somewhere between *hot* and *hotter*.

I said, "I'm on my way."

I didn't want to drive through Death Valley in the summer's heat—I feared I might wind up as one of those piles of bleached bones in the sand that you see in the pictures—so I took the cooler northern route through Denver, Salt Lake City, and Reno into San Francisco.

Northern route, my ass—the heat was blistering. My Geo's air conditioning went out in Provo, Utah. I found a Chevy dealer and had the mechanic take a look at it. With dollar signs in his eyes, he gave me the bad news: it would be $1,100 to repair, and it would take three days because they'd have to order a new compressor.

I said no thanks, found a hotel room, and took a long nap. I left around 1:00 AM and crossed the Great Salt Lake at night. I didn't see much of Salt Lake City because I passed through at 2:00 AM. One nice thing was that the speed limit was seventy-five per hour, and my little Geo wouldn't go that fast unless it was heading downhill and had a tail wind. So I didn't have to worry about getting a speeding ticket. I prayed I wouldn't have any more problems with the car. It was the middle of the night, I didn't see any cell phone towers, and I didn't think my phone would've worked.

I put the radio on "search mode" but couldn't find any stations on FM or AM. Not even XERF Del Rio, Texas. Back in the 1940s and 1950s, you could *always* pick up XERF Del Rio, Texas, anywhere in the good old USA. The station was memorable because it offered autographed pictures of Jesus.

I'd brought along my Radio Shack metal detector because I'd planned to stop in "gold country" and do some panning for gold. In Auburn, California, I bought a flake of gold from Pioneer Mining Supplies to calibrate my Radio Shack metal detector. I took a room at the Gold Country Inn in Placerville, California. Then I hooked up my new gold metal loop and tried to calibrate the metal detector. Nothing worked. So I put the regular loop on the machine, and it would find anything *but* gold.

Damn! It was 102 degrees outside, I'd expected the temperature in the mountains to be much cooler. Not in the summer. I decided to skip the gold prospecting until my metal detector was fixed.

Finally I arrived in Carmel-by-the-Sea. It was good to be back in California, and I had a great time. Unfortunately, I had to be in Houston on August 3 because I had to move my boat to a new slip. Also, I had some doctor's appointments in August at the VA Hospital in Dallas. They wanted me to come back soon, but I said I didn't think I'd make it until next year.

I didn't want to trust my Geo to make the trip back, so Joan, Jerry, and I discussed train travel. We all agreed that it was great fun. They especially liked riding the *California Zephyr* from San Francisco to Chicago. They said it was "top of the line." To insure that I'd return sooner, they bought me a first-class round-trip ticket, for passage plus a fancy sleeper compartment, to Chicago and then to Dallas.

It sounded *great*. I could remember, from fifty years earlier, riding the train with its click-clack, click-clack, and the gentle swaying back and forth. I recalled how relaxing it was, and how it sure put me to sleep.

See America by Amtrak

The Amtrak "Route Guide" describes the *California Zephyr* this way: "Experienced travelers say the *California Zephyr* is one of the most beautiful train trips in all of North America.... Certainly, it's the most comfortable way to travel between Chicago and the great cities of the West.... On board, you will experience the comfort and relaxation of train travel while witnessing some spectacular scenery. We are happy to have you aboard today, and we want to ensure your trip is everything you want it to be."

I was ready for comfort and relaxation.

I was excited about the scenery.

I was *not* prepared for the violent slamming down and crashing up of train travel in the new millennium.

It was as if some sections of the track were an inch lower than the other sections. Then there was the violent *whipping* from side to side as if other sections of track were misaligned by two inches. As if this weren't bad enough, there was the jerking forward and jamming backward, like being on a stretched rubber band. All of this happened every ten or twenty minutes, all day and all night. I thought the train would derail at any moment. (As it turns out, this thought was prophetic.) It was especially bad near Salt Lake City and Provo, Utah. I thought we were off the track several times.

Whatever happened to that click-clack and gentle swaying from the fifties? I later learned that the tracks haven't been very well maintained in the past fifty years, and this has destroyed the relaxation of train travel. It's also made train travel more dangerous, as we shall see.

Like a cheap stripper, the *California Zephyr* bumped and grinded its way across the country. In the California mountains we averaged about thirty miles per hour. We followed the American River to Tahoe, and the Truckee River to Reno. I'll admit there was some breathtaking scenery. And I could even see it fairly well through the filthy windows.

While we were on the flats in Nevada, I noticed a truck on the highway, about a mile away, doing seventy to seventy-five miles per hour, and the train kept even with the truck. What speed!

A surly black lady was in charge of making the announcements. She'd come on the P.A. and bark commands about dining times, and orders, and the service hours of the snack bar and lounge car.

I went to bed the first night just after Reno, and woke up somewhere near Provo, Utah, where the train was rocking and rolling like crazy because the tracks were so bad.

We followed the Colorado River, and I saw a lot of rafters and kayaks shooting the rapids. When the rafters beached for a rest, they almost always mooned the train as it passed.

I took lots of pictures, especially of Brice Canyon, through the filthy windows. In Denver, a guy with a squeegee came along and washed the coach windows, but by now we'd passed all the beautiful scenery so it was too late for the really great picture-taking. Most of my photos were taken between Grand Junction and Denver through dirty windows. The window washing should have been in Grand Junction. There wasn't much to photograph in Kansas and Nebraska except corn and wheat, and besides, we passed through there at night.

I went to sleep the second night outside Denver, not long after the window-washing, and I woke up in Omaha. I needed to transfer to the *Texas Eagle*, which runs between Chicago and San Antonio, so I left the *Zephyr* several miles southwest of Chicago at Galesburg, Illinois, and was bussed to Springfield where I boarded the *Eagle* at 7:00 PM.

Amtrak describes the *Texas Eagle* this way: "Ready for a real travel adventure? Hop aboard the *Texas Eagle*."

"Travel adventure" was right.

I'd had dinner in the diner, gone to bed, and was sleeping soundly when I was jolted awake. I'll never forget the date: Monday, July 30, 2001.

At 2:15 AM, the train left the track and crashed into a gully. The two engines landed on their sides and slid about fifty feet in the muddy rain-swollen gully.

The first car was in the gully. The second car flew sideways and almost went onto the highway about fifty feet away.

I was in the third car, the sleeper car, and we went bumpty, bumpty, bumpty, bumpty as the wheels hit the railroad ties and careened off the track at about a 30-degree angle. We tilted into the gully in about four feet of muddy water. Thank God for the trees, because they were what kept us from rocketing down the gully full of water.

We were in the middle of a terrible thunderstorm. Thunder was crashing, and lightning lit up the scene dramatically every few minutes. Other than that, it was really dark. Plus, none of the emergency equipment worked, including the emergency lighting system.

One big fat ugly black lady made it to the door *screaming* in absolute hysterical panic before the car stopped moving. She continued to scream as everyone else surveyed the problem.

A porter arrived and asked if anyone was injured. He tried to get everyone to stay still until he could determine if the car was going to remain upright or turn over into the river and drown everyone.

Swell.

A real "travel adventure."

We were still in almost total darkness, so the porter issued each of us a bioluminescent light. I could see cars passing on the road about fifty feet away. Finally, some of them stopped and pointed their headlights toward the train.

I could see water running under my window, but I didn't know how deep it was. Finally one of the truck drivers waded across the gully clutching onto the trees to keep from being washed away. The water was raging down the gully, but it was only about waist deep.

There were emergency procedures posted in my compartments, and they listed three emergency exits: 1) the door you came in, 2) specially marked doors, 3) your window. Well, the train was leaning against the ground on the door we came in on, so that one was blocked. The specially marked doors were jammed shut (that was where the big fat ugly hysterical lady was still screaming from). That left the window.

I didn't want to climb out the window, but it was better than being in the train when it turned over into the river and drowned us.

I prepared to escape through he window.

I yanked on the little red handle.

It came off in my hand.

It was supposed to pull the rubber out from around the glass so you could remove it. Obviously it didn't work. Luckily, the man in the next berth was able to get his glass out.

By now, some rescue workers had arrived. They put a ladder up to the window, which was good because we were on the second level and it was quite a drop to the ground. We "detrained" into ankle-deep mud.

By now half the town of Annapolis, Missouri, (population 310) had turned out. They made a human chain across the gully to help us escape. Several ambulances stood ready to take the injured to local hospitals.

Water was still raging down the gully, and that's what had caused the accident. The downpour had washed away some dirt from around the railroad ties, so they couldn't support the two engines pulling the train. The engines fell over on their sides, leaving the rest of the train to jackknife back and forth across the track.

We all got out okay. The train didn't fall over, and no one drowned.

The local sheriff and his deputies were right there on the spot—along with a couple of preachers. The sheriff took our names, and we were bussed to a school where the local people gave us bath towels, hair driers, and blankets. They had set up tables with hot coffee, orange juice, milk, donuts, cakes, and pies in the cafeteria. One of the locals opened his pizza business, and in no time we had pizza.

Later, at 4:00 AM, the locals broke out their emergency rations of potted meat, potato chips, Coke, Pepsi, 7-Up, and I don't remember what all. We were very well-fed. At about 6:00 AM they served breakfast, with sausage, eggs, biscuits, gravy, English muffins, Eggos, and fresh-baked donuts.

Finally, the Red Cross arrived. They brought stretchers and army blankets, but not much else. The state police arrived and took charge, but all they did was take names and dates of births.

Then the local media descended upon the scene. A couple of reporters brought TV cameras and started interviewing passengers. Several passengers complained that none of the emergency procedures had worked and that we'd had problems escaping from the train. When they interviewed me, I told them my whole story, just as I've told it here.

Interestingly, none of these interviews ever made it to the national news. I think the Amtrak people saw to that—they didn't want the negative publicity. There was only a small report about "a derailment."

By 9:00 AM the Amtrak people had begun to arrive. Their first priority was to get everyone, or as many as possible, on busses to their destinations. They said they'd send the luggage some time later.

I asked, "What do you mean some time later?"

Well, they said, they'd have to clear up the wreck site. Then a hazardous-materials team would have to contain the diesel-fuel spill to keep it out of the river. This would all take about a week. Then they'd have to retrieve, survey, and identify the luggage—which would take another week.

"Well," I said, "I'm not going anywhere without my luggage. My house and car keys, my checkbooks, and my ID are in my briefcase."

They said they'd deal with me later that afternoon.

June Reeves of Amtrak had all of us write our names and addresses on a sheet of paper, and describe our luggage in detail to help the survey team identify it. One lady was worried about her luggage in the baggage car. I told her the luggage was racing down the river and was probably halfway to Memphis.

They put most of the passengers on buses to Chicago, Little Rock, Dallas, Houston, or Austin. But about ten of us insisted on waiting for our luggage. Judy put us up in a Days Inn in Farmington, Missouri. She took me to Wal-Mart and said to pick out anything I wanted. So I got a new outfit: shoes, socks, underwear, pants, and shirt; plus toiletries such as a toothbrush and toothpaste.

Meanwhile back in Dallas, my neighbor Jack had driven fifty miles to Union Station to meet my train and drive me back to Gun Barrel City. He waited in the station for six hours, then asked the clerk if the train from Chicago was delayed.

They said it wasn't delayed, it was derailed.

Jack was really mad, because they could have made an announcement on the radio or TV, or at least at the station. He'd driven fifty miles each way for nothing.

I phoned him from the hotel and said I'd let him know when I'd be in.

Next morning, Amtrak had managed to retrieve all my luggage. Judy sent me to St. Louis in a taxi, and I boarded a TWA flight to Dallas/Fort Worth, where another taxi was waiting to drive me to Gun Barrel City. So Jack didn't have to come back a second time to pick me up.

I later learned the well-hidden (by Amtrak) fact that there's a train wreck in the U.S. almost weekly. Granted, most of these are freight trains, but it does indicate how bad the tracks and the systems are. In January 2005 there were three derailings in a single day: one in South Carolina, one near Seabrook, Texas, and one near Reno.

In South Carolina five Avondale Mills workers are known to have died after chlorine leaked from several railcars involved in an early morning accident. The crash derailed fourteen cars, including three that were carrying ninety tons of chlorine and another car carrying sodium hydroxide. About 240 other people became ill. About 5,400 residents were unable to return to their homes. Unsafe levels of chlorine were recorded Saturday near the scene of the accident, despite forty tons of lime dropped on the area in an attempt to contain the fumes.

In Seabrook, Texas, the freight train derailed and one of the tanker cars leaked a large amount of hydrogen peroxide. There were no serious personal injuries.

The Reno derailing was of the *California Zephyr* on the route I took to Chicago.

Glutton for Punishment

I don't know if it was because I didn't want to hurt Joan and Jerry's feelings by not using the return ticket they'd bought for me, or if I'm just a glutton for punishment, but within a months I was ready to get back on Amtrak and head for California.

I was scheduled to catch the *Texas Eagle* to Chicago on September 15 at 11:00 AM, and then transfer to the *California Zephyr* for my return to California.

September 11th changed everything.

I received a message on my answering machine saying the train had been cancelled and giving me an 800 number to call for more information. I called this number, and an ignorant-sounding black woman on the other end said, "There's no problem. All trains are running on schedule."

So I asked my neighbor, Jack, to drive me down to Union Station on Houston Street. When I went up to the ticket counter, the guy there said that the *Texas Eagle* had been cancelled, and that he'd called me and left a message on my answering machine. I told him I'd called the 800 number and a woman there said all the trains were running on schedule.

Then I asked, "When will the next train be going?"

"Don't know. I can give you a refund for this half of your ticket, or I can book you on the southern route via San Antonio through Los Angeles."

"Okay. When can you get me on the route through LA?"

"The next available seat is on a train that leaves the day after tomorrow. By the way, it's a seat in coach. The sleepers are all booked."

"I'll take it."

So Jack and I went back home. On Friday, Jack (who was probably sick by now of driving the fifty miles each way to Union Station) picked me up at 9:00 AM so I could catch the train at 11:00 AM. When we got to the station, we learned the train was delayed until 4:20 PM.

I told Jack to go back home; I'd just wait in the train station.

At 1:00 PM, they posted a message on the electric sign saying the 4:20 PM train to San Antonio was delayed until 5:30 PM. Then at 3:00 PM they posted another message saying the train wouldn't arrive until 6:30 PM. Then at 4:30 PM they posted yet another message saying the train would be in at 8:00 PM. At 7:00 PM, I'd been sitting in the station waiting for my train for nine hours when they posted another message saying the train would arrive at 8:30 PM. But it was not going to San Antonio. It was only going to Fort Worth; from there, they'd bus us to San Antonio.

I said, "Why can't they just bus us to San Antonio from here?"

The response was, "It's easier to do it this way."

"Will there be dinner service on the train to Fort Worth?"

"Yes. The dining car will still be open."

"Okay," I said, "but if they bus us to San Antonio, will there be vacant seats on the train down there?"

"Yes. They have two cars on the siding now, and you'll get the seat you've booked."

We finally boarded the train to Fort Worth at 9:00 PM. The dining car was closed, but we could get a free sandwich in the snack bar. We all lined up to get into the snack bar. I hadn't eaten all day. I received what they called a "cheeseburger," a Coke, and some potato chips. The so-called cheeseburger was a microwave-heated biscuit with a gray-looking patty inside and a squirt of Cheez Whiz.

When we arrived in Fort Worth, sure enough, they had several buses waiting for us. We all piled in and started off. Somewhere after Waco the buses stopped at a Burger King on the highway. We stumbled out and eagerly wolfed down some *real* fast food, complete with heartburn.

When we arrived in San Antonio at 1:00 AM, the stationmaster informed us that the train from New Orleans was running late. Big surprise. When it arrived at 3:15 AM, they had to hook up the two cars from the siding. We weren't allowed on while they were hooking them up, but finally they lined us up to board. All the families with children got to board first. The rest of us had to fill in wherever there were empty seats.

I said, "What do you mean fill in? I'm supposed to have a reserved seat."

"No, you'll have to fill in."

But I had a seat assignment, and I insisted that it be honored.

While I was waiting to board, I had a conversation with the conductor. He said, "You know, Amtrak's running out of money, and it might fold in a couple of months."

I replied, "Yippiieee! Then maybe we'll get a real train system, like the ones they have in Europe."

The conductor snatched my seat-assignment ticket from my hands and made me go to the car with the families. To "fill in."

I should have kept my mouth shut.

At 4:00 AM, we finally started for LA. We bumped and banged along all day. When we stopped in Del Rio, Texas, I went into the train station and asked if XERF was still on the air. The two young men behind the counter had never heard of XERF. Well, that answered that question of why I hadn't been able to pick up their signal on the road several months earlier. Autographed pictures of Jesus Christ must have not sold very well.

The XERF station was actually in Mexico, so I figured that maybe it made money after all, and the Mexican government nationalized it, then ran it into bankruptcy. That's what happened to a lot of money-making businesses in Mexico during the 1980s. It was socialist policies such as these that ran the currency-exchange rate up to 4200 pesos to the dollar.

I took photos of Bill Allen's place—or, at least, the mountain his house is on—in Fort Davis as the train passed through Alpine, Texas.

We continued through El Paso and on to Phoenix. I went to sleep in my uncomfortable seat and woke up the next morning in Palm Springs. We arrived in LA at about 9:00 AM. Then I had to wait until 11:00 for the *Coast Starlight* to San Francisco. Believe it or not, that was a fairly nice train, and I had a great trip along the coast. I arrived in Salinas at 6:30 PM, and Joan and Jerry were waiting for me.

I spent two weeks visiting my friends; then it was time to go home.

I limped along in the Geo. Thank God I had a cell phone and knew I could call Triple A if the Geo broke down in some God-awful place.

I'd bought a case of thirty-weight oil before I left, and I'd gone through six quarts by the time I reached Bakersfield.

I filled up with gas and oil in Barstow before crossing the desert. It was getting dark, so I decided to go for it, and planned to spend the night in Kingman, Arizona, if I made it that far. The Geo limped on, and I made it to Needles, gassed and oiled up again, then pressed on toward Kingman. Next morning, I made

Flagstaff, where I found a Wal-Mart and bought another case of thirty-weight oil. Thank God I did, as I really needed it.

I started up the Continental Divide in Arizona, and the car just would not go. Luckily, I had a set of Champion sparkplugs. I swapped out the plugs—the old ones were fouled from burning oil—and the Geo took off like a striped-ass ape! We made it all the way to Amarillo, where I pulled up to a Motel 6 and turned off the engine.

The Geo wouldn't start the next morning.

I used the cell phone to call Triple A. I figured the car needed a little ether starter juice. The mechanic tried everything he knew, but it refused to start. I had it towed to the local Chevrolet dealer, and $286 later they were still having problems getting it to run. This was after replacing the timing belt and some other belts.

I asked if they could just get it started. They did. It smoked a lot, but it ran pretty well after it got warmed up. I said, "Don't turn it off—I'll take it just the way it is." I paid the bill, jumped into the smoking thing, and took off. I filled up twice before I got home, without turning off the engine. Somehow I made it.

It used twenty-four quarts of oil between Carmel, California and Gun Barrel City, Texas.

Back to France

I was still aching to go to Australia, but I wound up back in France.

As mentioned previously, I'd done some house-sitting in France early in 2001 for Joan and Jerry while they were in the U.S. They had figured it would be okay to leave the house unoccupied, so I'd come back to the U.S. But late in 2001, Joan called and said someone had broken into their place in France. Something had been stolen, and she was worried the thieves might return.

Joan asked, "Would you be interested in returning to France and house-sitting until we can return?"

"You bet! I'd love to spend another three months in France."

I bought a ticket for December 29, 2001, and Jack drove me to the airport. I flew to Chicago, the United Airlines hub, and changed planes. On the previous trip I'd flown "cattle car" (coach) on a Boeing 777, and my knees had stayed pressed against the seat in front of me all the way to Paris. I couldn't stretch out, and my knees were bruised from rubbing against the seat back. This time I asked at the ticket counter to use my frequent flier miles to upgrade to business class.

Being in business class made a world of difference on that long flight. I was in the front row and could see into first class. I could hardly tell the difference between what they had and what I was getting, except for the champagne service.

I landed at Charles de Gaulle airport and rented a car to drive to Bréves. By now, I was confident that I could find my way to Joan and Jerry's house on my own. Unfortunately, I missed the first turn-off from A1 to A3. I figured I'd just get off at the next exit and make a U-turn.

Wrong.

The freeways in France are very different from those in the U.S. Instead of just heading down the exit ramp and up the entrance ramp to get back on, I had to drive through a small town and search for the road back to the freeway. I found it, but then I had to get onto A3 from the opposite direction, which was also difficult. By some miracle, I wound up on the right freeway going in the right direction, but then I missed the N104 (Marne de Vallee) turn-off. After driving around lost for an hour, I saw a sign that said "A6 Sud"—so I took it. But this didn't look like the right road, either, so I stopped at the tollbooth and asked. The attendant said that the *autoroute* (toll way) would take me to Auxerre, which was near where I needed to go.

I continued and got off at the "N6 Auxerre" exit. Great! I was home free. I made it to the house and talked the neighbors, Monsieur Gilbert and Madame Joellen Billiat, into letting me in the house, and into taking me to return the rental car. Joan had a 1985 Renault 5GT turbo that was a kick to drive, especially when the turbo kicked in. I almost lost it a couple of times when this happened as I was passing another car—it felt like the Renault was ready to jump out from under me.

Memoirs and Self-Taught French Lessons

Alone in France. Now was the perfect time to make real headway on my memoirs. Every morning I'd get up, go into the dining room where I'd set up my computer, and write, write, write.

Meanwhile, my French was extremely weak, and I wanted to improve it. So when I was tired of writing, I'd practice with my flash cards.

I'd bought a box of blank Avery business cards in Auxerre, from the stationary store that was across the street from my bank. Using phrases and words in a Barron's phrase book and a dictionary for travelers, I'd written words, numbers, and common phrases in French on one side of each card, and in English on the other side. I would read the French and try to remember what it meant in English.

During my stay, I was able to memorize all the numbers to one hundred.

Before going out, I'd research the words and phrases I might need for that expedition—words like "stamps" if I was going to the poste office, for example. Then I'd make more flash cards with those words and phrases. When I went out, I carried my flash cards with me. At the post office or anywhere else, I would try to pronounce the French. If the people didn't understand what I was trying to say, I could just show them the typed version on the card. This worked out very well.

Happy Hour

I would type on the computer until 4:00 PM each day, which was "break time"—also known as "my solitary happy hour."

The area where I was living was known for its fine wines, and they were cheap. You could get a nice bottle of table wine for about eighty-eight cents. I purchased a Beauvillon table wine and a 1999 Beaujolais. The area also had some good cheeses. Each day at 4:00 PM, I'd have a wedge of cheese and a glass of wine.

It was very nice visiting France. There were several great classical music stations, so I often listened to classical music or operas.

At night before I went to bed, I'd have a glass of milk and a piece of baguette—that long French bread.

Dijon

One of the side trips I took was to Dijon. This is the ancient capital of the French province of Burgundy. It has lots of historic monuments as well as—need I say it?—old churches.

Of course, Dijon is known for its strong and tasty mustard, so I went to the visitors' center to get directions to a store that specialized in this delicacy.

I did not know there were over fourteen different versions. I purchased several souvenir boxes of assorted mustards to send home.

Escargots

I stayed in France until March 28, 2002. Joan and Jerry's friend John Broadbent drove me to the airport again. On the way we stopped in Auxerre. At the big Auchan supermarket, I bought a couple of jars of escargots to take home. There were over six-dozen snails in each jar. I also bought a couple of cooking plates especially designed for snails, and eight little forks for snail-eating.

Flight Home

The flight home was on a United 777. In the cockpit were a couple of new lady pilots in training. I asked one if she had any navigation charts of Europe that she didn't need. I told her my wife was a corporate pilot and updated her Jeppesen charts regularly. She said she'd look.

Later she came back and told me the captain on this flight was Danny Fitch. Danny had been a passenger on United's Flight 232 from Denver to Chicago on July 19, 1989. This flight was famous because the DC-10 blew an engine and lost all hydraulic power. At that time, Danny was a United training and check pilot with over three thousand hours on the DC-10. Danny had assisted the flight crew with the emergency, and it's lucky he was on board because as a flight instructor, he was an expert in the necessary emergency procedures. He maintained power to the engines and was able to steer the plane to a crash landing at Sioux City airport in Iowa. The plane broke apart on impact and exploded, but the majority of those aboard survived.

Thank You

Once I got home, I wanted to send a nice gift to Joan's neighbors, Monsieur Gilbert and Madame Joellen Billiat, because they'd been so helpful to me during my stay in France. I thought it would be great to send them something uniquely American, and uniquely Texan, too. Knowing the French are always interested in good food, I bought two types of onions, the Texas 10/15 Supersweet, and the Noonday.

The 10/15 was developed by horticulturists at Texas A&M University who reduced the *pyruvate*—the chemical that makes you cry when you peel onions—resulting in a mild and sweet taste. The name 10/15 comes from the fact that in Texas, for best results you should plant these onions on October 15th, or 10/15.

The other onion variety, Noonday, is named after a small town south of Tyler in Smith County. The onions there are famous for their sweetness. There's even a Noonday Onion Championship each May

I found information about these two onion varieties on the Internet, copied it, and put it in the package that I shipped via FedEx to Monsieur Gilbert and Madame Billiat.

The Da Vinci Code

Almost a year to the day after I returned from France, a blockbuster best-selling novel called *The Da Vinci Code* hit the stands. Reading it confirmed everything I'd imagined about my ex-wife Joan.

I'd often suspected she had some kind of inside track to a higher power. I've already written about how she had this incredible intuition that defined science, mathematics, logic, and reason. Well, when I read in *The Da Vinci Code* about the secret society Priory of Sion, I believed it. According to the novel, Mary Magdalene was actually the Holy Grail (the bearer of Jesus' blood); in the book, the Holy Grail is not a physical chalice, but a woman, namely Mary Magdalene, who gave birth to Christ's descendants and thus carried the bloodline of Christ.

The existence of the bloodline was a secret contained in documents discovered by the Crusaders after they conquered Jerusalem in 1099. The Priory of Sion and the Knights Templar were organized to keep the secret, as the bloodline flowed through the ages down to the present time. The character Sophie Neveu was a desendent of Jesus in the story.

As I mentioned earlier, not far from Joan's town of Bréves is the mountain-top village of Vézelay. In this village is an enormous Romanesque church, the Basilica of St. Magdalene, that was said to contain the relics of Mary Magdalene.

The Old French expression for the Holy Grail, *San Gréal*, actually is a play on *sang réal*, which means "royal blood" in Old French. In the book, the Grail relics of Mary Magdalene were hidden by the Priory of Sion in a secret crypt—perhaps in the Basilica of St. Magdalene? That must be what brought Joan to Bréves. The "Rose Line" as it's called in the book, is a meridian line that's marked in the floor of Saint-Sulpice (another old church) in Paris and that has an important, mystical, and divine significance: it leads to the Holy Grail.

Well, guess what? The Rose Line passes right through Joan's house in France. Like Sophie Neveu, Joan does not know she is a desendent of Jesus.

I guess I should call her Saint Joan.

Australia

The threat of El Niño continued to loom, however, disrupting our sailing plans in the immediate future—nonetheless Australia's siren call beckoned with urgency.

I was still tremendously gung ho for the South Pacific and Australia.

I told Bill, "I'm out of my rut."

I continued, "The American dream is an A-number one stick it to 'em and twist fleece-job! The corporate executives are raping the workers, taking all the cash out of the corporation and dumping the pension plans. All the middle class jobs are going over seas, to Mexico, China and India. The American economic population is going to be an hourglass with no middle class."

I preached on.

"The individual in America is dead, my friend. As for as I know, we're the last two left. And sometimes, I have my doubts about you. It's the last days of the Roman Empire, I'm telling you true. The same thing happened to Greece, France, England, the Mayans, and the Egyptians. You've got violence, drugs, per-version, and greed. You've got the poor versus the rich, the haves with their hi-tech lifestyles versus the have-nots who still can't afford indoor plumbing. We're rotting from the inside out. One of these days, Bill—sooner than most people think—it's all going to implode. It's all going to collapse in on itself. Just like the maggoty piece of fruit it is. The handwriting is on the wall."

I concluded, "I worked my ass off and for what, Bill? What did it get me?"

Bill, "A generous benefit package?"

"No! Well it wasn't because I didn't keep up. I kept up, damn it. I read, I grew. It was like learning a whole new business every other week, but I did it, brother, and I did it well. I kept up with all those twenties geeks and was the only one in the top ten of all three performance measures. For eight years I put in ten hours a day at Microsoft. Plus overtime, weekends, on call around the clock. I know computers better than … better than they know themselves. I've been a computer, damn it. I dreamed in binary! But hey, why pay a six-figure salary to someone who knows computers from as far back as the Cretaceous/Tertiary Period, when for half that much they can get some kid who thinks the Cuban Missile Crisis is an old movie his parents talk about but he hasn't seen? All those young punks, Bill … those virtual-reality techno-nerds. Who was it who made them possible, answer me that? Who was it who ushered in the golden age of computers?"

Bill asked, "So what will you do now?"

"To hell with El Niño," I finally replied, "and to hell with lousy weather as long as the planes are still flying!"

"Let's fly to Australia then," Bill said. "Now?"

We stopped by an expediting passport services place that advertised U.S. passports in as little as twenty-four hours to get Bill's passport renewed. The place was run by Iranians. No wonder terrorists can enter this country so easily. They run the passport offices and can create all they need.

We loaded some of Bill's things into his car and headed toward Texas. We stopped at the Econo Lodge in Forest City, Arkansas, that night. It, too, was run by Iranians. The next day we made it to my place in Gun Barrel City where I picked up a few things also and we rested overnight. I thought we should go to Port A and have a toast to Australia and say goodbye to Eddie and Marcia and visit our former hangouts.

Within a week, we'd booked a flight to Sydney.

We arrived at Sydney International on September 26, 2002.

After years of thinking about it, of planning it, of scheming how to make it happen, I had at last gotten myself to what was, for me, the "Magical Land of Oz."

Having learned in childhood about the duck-billed platypus, Bill and I were now finally making our long-talked-about pilgrimage to this Noah's Ark of fantastic creatures. Wombats! Rabbit-eared bandicoots! Death adders! Two-headed lizards! Tasmanian devils!

It was our own version of "lions and tigers and bears, oh, my!"

But why was there so much bizarre life down there at the bottom of the world? Why were the creatures so similar to, yet different from, other animals on earth?

I liked the notion of Australia as a lost continent, a time capsule, a place out of the pages of H. G. Wells, Jules Verne, and Conan Doyle. The reality of the place, however, is even more fascinating.

Geologists know that Australia is a vestige of the great supercontinent, Pangaea ("all lands" in Greek), which was heaved out of the sea more than three billion years ago. The southern section, Gondwanaland, eventually broke away, taking with it seven continents, including Australia. Australia then separated from Gondwanaland and drifted northwest, where it escaped the geological upheavals that led to the more familiar animal species elsewhere. The rainforest regions along the east coast and the Great Dividing Range are all that remain of a once lush, tropical continent.

Apartment in Sydney

Bill and I had rented an apartment in advance. We planned to be in Australia for three months, and we figured an apartment would be much cheaper than a hotel. I let Bill handle all the details of the rental, and he assured me it was very lavish. He told me it was in an area called "King's Cross." That certainly sounded luxurious.

So we took a taxi to our waiting apartment, where I soon realized that King's Cross is the red-light district of Sydney.

Bill had chosen that area intentionally.

King's Cross

More interesting than the animals of Australia is the human wildlife. King's Cross is the nightlife center for the whole city of over five million people. King's Cross never sleeps. The clubs rock till dawn. Prostitution is legal, of course, and it's hard to tell the straight girls from the hookers. They're all gorgeous—especially at night and after a couple of highly alcoholic Australia beers.

Not only is it hard to tell the straight girls from the working girls, it's also sometimes difficult to tell the boys from the girls. This is a true zone of tolerance, where almost anything goes. As with other resigned countries in the world, Australia has learned that it's better not to impose laws on Mother Nature.

Back to the girls. At least seven prostitutes stood nightly on the corner across the street from our third-floor apartment. Four others were constantly lounging under our window. All night long cars would stop, and the girls would vie for customers. Young males would drive past and shout obscenities at the girls.

I'd stare at this nightly action from our window. Some of the girls outside just didn't look quite right to me. One evening I said something about this to Bill. He came over, craned his neck to look, and said, "Yeah. They're either transvestites or transsexuals." Transsexuals (trannies) are a lot harder to spot than transvestites—and many are far more beautiful than your ordinary street girl. The majority seemed to be from Thailand where, for some reason, men and women behave less differently from each other than they do in other parts of the world.

Bill decided to do an investigation. He said it was important for him to do some research into this subject, for he was planning a book on prostitution. He trolled the area almost every night during our first week, then came back to report that the only difference between a trannie and an ordinary girl is that a trannie has a pronounced Adam's apple.

A district called Darlinghurst was just up the hill from our apartment. It's the main drag for topless bars, strip joints, and porn stores. It's world famous, and hardly anybody goes to Australia without cruising the half-mile-long crush of bizarre and scantily-clad humanity. It's something like the New Orleans Marti Gras, but goes year-round and has fewer parades. Prince Charles was there when we were, and he was, no doubt, in the same live sex shows and cathouses that Bill was.

Our apartment was on Williams Street, about two blocks down from Darlinghurst. Williams was one of the main six-lane highways into downtown Sydney from the west, so every morning bumper-to-bumper traffic poured into downtown. An 18-wheel rock hauler would come all the way down the hill every morning with his screeching brakes on and wake me up.

Bill's a late riser, so I'd get up quietly, dress, and walk up the hill to a cafe called "William's on Williams," or to the Holiday Inn restaurant, and get some breakfast or coffee. William's on Williams served a great breakfast for under five bucks. It was also an Internet cafe. For about a dollar, you could use an online computer for an hour. These Internet cafes were everywhere, but that one was the least expensive. Sometimes I'd turn right on Victoria Street and go down to a German restaurant called Una's. I liked sitting at these sidewalk cafes and watching the human Australian wildlife.

One day I walked through an area called Woolloomooloo (yes, that's really the name) to Potts Point and to Woolloomooloo Bay. Another day, I walked down to Hyde Park. Williams Street turns into Park Street at Hyde Park. Several bus lines use Williams to get to downtown, so there's a bus about every five minutes. It cost only twenty-five cents to the Circular Quay ferry terminal, where all the bus lines terminate. From there, you can transfer to any bus and go anywhere in the city.

Blog Diary Entrees

By now I'd begun writing a blog. Here are some of my entries from the Australia trip:

October 1

Took the ferry to Manly Beach and had fish and chips, but the stand didn't have salt or ketchup. As we left the Circular Quay ferry terminal, I took almost a whole roll of pictures of the famous Sydney Opera House and the Sydney Harbor Bridge. The ferry ride lasted almost an hour and covered a lot of Sydney's harbor.

532   IN MY TIME

We spent most of the day at Manly Beach, where Bill photographed topless sun-bathers.

October 2

Went to Chinatown and had dinner. From downtown, we took the Monorail around Darling Harbor to Haymarket and Dixon Street.

October 4

Took the bus to Bondi Beach. Spent most of the day at Bondi Beach and had lunch at Charlie's Place.

October 5

Had breakfast at the Customs House at the Circular Quay, which had an exhibit of Salvador Dali's sculptures. We didn't even know he did sculptures.

October 6

Went to Coogee Beach and had fish and chips at the Beach Pit. Twelve of the best beaches on earth circle Sydney.

October 7

Walked four blocks down Williams Street, found the Hard Rock Cafe, and had a hamburger. The place was totally identical to the ones in Dallas and Cancun. Went to the Central Train Station and picked up literature for train schedules.

October 10

Went to the Sydney Observatory and found out there is going to be a total eclipse of the sun in two months, on December 4. Bill and I discussed it, and decided we should go to the town of Ceduna in South Australia to view the eclipse.

October 11

Went to the Queensland Travel Center, booked trip to Brisbane and Cairns for two weeks, to scuba dive off the Great Barrier Reef. Booked a five-day diving-certification course at the Pro Dive shop. Booked a four-day, three-night Cod Hole outer reef extended dive by Taka. Bill has no interest in diving, so I'll be going alone.

October 12

Went to the Royal Botanic Gardens and saw the Opera House and the Governor's House from the opposite side.

Bats

The Royal Botanic Gardens are fantastic. They cover seventy-five acres right there by the water, and they showcase the lush and gorgeous vegetation of the region. While we were there, we noticed some strange-looking things hanging in the trees.

Bats!

And these were giant ones with four foot wingspans.

We went back at sundown so Bill could photograph the giant bats feeding in the gardens. As the sun went down, even more of the mysterious megabats—known as flying foxes or fruit bats—arrived from their roost across the harbor in North Sydney. The "bat chatter" in the trees around us grew louder.

It was a little creepy.

Like most people, I'd grown up fearing and loathing bats, thinking of them as creatures from hell. I had cut my teeth on Batman and had seen most of the Dracula movies—some of them multiple times. Bill was attracted to their legends and mythology, to their power as metaphor—but I didn't want them in my hair any more than the next person.

Now I was about to see them in the wild: hundreds or maybe even thousands of them in this park in the middle of Sydney.

There were forty or fifty flying foxes in a single tree, chirping, squeaking, and screeching, engrossed in their all-important territorial concerns. They were clambering upside down, hanging from clawed toes, pulling themselves forward with hooked fingers on their four-foot-wide, raincoat-like wings.

As they flew from tree to tree with loud, drum-like wing beats, I noticed they were much larger than I'd expected.

In a hushed, excited tone, Bill said, "Once you learn about them, you get obsessed. The way they fly! In the evening in summer, they'll catch the rising thermals and spiral up like hawks or eagles. It's just—just wonderful!"

The flying fox is nothing like the small, echolocating, snub-nosed microbats that some of us actively dislike in the states—in fact, they may not even be related.

John Pettigrew, a neuroanatomist at the University of Queensland in Australia, has made a case that flying foxes are primates that evolved independently from the microbats that they're commonly confused with. Pettigrew, an expert on the brain's system for processing visual information, stumbled onto his hypothesis when he looked at a flying fox's brain tissue and saw that it had visual pathway traits seen only in creatures like ourselves. Unlike microbats, flying foxes hunt with eyesight, not echolocation, and their visual acuity is among the best of any mammal.

More Blog Entries

October 14

Went to the top of the Sydney Tower and took pictures of Sydney in all directions. Took bus 324 to Watson's Bay Ferry Wharf to have dinner at Doyles on the Beach. Doyles is one of the oldest restaurants in Sydney; it was established in the 1800s.

On the bus Bill met a very nice lady, a marathoner from Canada. I asked Bill to find out from her where Doyles was located. Several passengers on the bus quickly spoke up and said it was at the end of the line. The people of Sydney are very helpful and friendly. On numerous occasions, people were extremely helpful with directions.

October 15

Cleaned apartment. Bill got laid. Went to the Bourbon Steak House for dinner.

October 16

Went to the South Australian Travel Center, Mezzanine Level, 247 Pitt Street, and rode the monorail to Chinatown for dinner again.

October 17

Almost time to go to Brisbane for diving. Did laundry and packed. Took the Captain Cook Cruse of Sydney.

Brisbane and The Great Barrier Reef

On Friday, October 18, I caught the train by myself to Brisbane at 8:00 AM.

The plan was to dive the Great Barrier Reef, a lifetime desire of mine. The terrain between Sydney and Brisbane reminded me of East Texas, Louisiana, and Mississippi. There were a lot of small livestock farms and a lot of trees. I felt right at home seeing all the cattle on the little farms. Almost to Brisbane, the train stopped at a street crossing, and the conductor said that some lady had run her car into the engine.

I'm starting to think that when it comes to train travel, I'm accident-prone.

We sat there for about an hour as the local police investigated the accident. Then we had to wait another hour while the railroad did its own investigation. As a result, we were three hours behind schedule. The train stopped 150 kilometers short of Brisbane, and we were bussed to town and dropped off at the train station. I'd planned to pick up some postcards in the train station and write them to my friends during dinner, but by the time I arrived, everything was closed.

I went across the street and had a subway sandwich.

I went back to the train station to spend the night so I could catch the first train to Cairns the next morning, but the security guards ran me out at 1:30 AM so they could close the train station. I said, "But where will I sleep?" They said, "Don't know, mate. Maybe the steps out front?"

I found a bus stop and slept on the bench.

All my life I'd worried that some day I'd be so poor I'd have to sleep on a park bench, and here it had finally happened.

At 5:00 AM the station was unlocked. I went back in and waited for the gift shop to open at 6:00 AM. I bought post cards and wrote to my friends while I had breakfast.

More Blogs

October 19

Boarded the Sunlancer train to Carins at 8:55 AM. Most of this area between Brisbane and Carins is tropical. The farms are a little larger, and grow sugarcane, pineapples, and bananas. Saw a couple of kangaroos, plus some sheep and chickens.

October 21

Went to the local hospital for the dive physical and learned, to my great dismay, that my lungs lack the capacity for scuba-dive certification. The doctor said I needed a score of 70, but I could only muster a 59 to 61.

So I failed the dive physical.

I've failed a lot of tests in my life, especially as a kid in school, but this has got to be the worst.

As consolation, I went to a Japanese restaurant, the Cherry Blossom, and had some good sukiyaki.

Pro Dive

I don't give up easily. If I couldn't scuba dive, at least I could snorkle.

On Tuesday I was picked up by the people from Pro Dive at 6:00 AM and boarded the dive boat. When they assigned the rooms, I found I was bunking with a nice woman named Sally Carroll. For a snorkeling partner, I was paired with Lorraine from Switzerland. Snorkeling seemed to be a great way to get to spend time with females—I should have tried it decades earlier! The dive master,

Dennis, was from Holland, and he was six feet eight inches tall. He always had to bend over going through the hatches.

On Wednesday we went out again, and I snorkeled at a place called "the whale."

On Thursday, we snorkeled at the Flower Garden in the morning. When Dennis had certified all the new divers, he asked me if I'd like an introduction dive. He said he could hold onto my hand, and we could go down eight or ten meters (about twenty-five feet) for a few minutes.

I said, "You bet!"

I suited up, and he went through a few instructions. When going down, I had to be able to clear my ears to align the air pressure inside with the water pressure outside. I had to be able to clear the water from my mask by exhaling through my nose. I had to be able to remove my mouthpiece and then clear it when I put it back in. Then, in case I lost my mouthpiece, I had to be able to lean over to the right so the hose would hang out to the side, then rotate my right arm to bring the hose and mouthpiece back around in the front, and put it back in my mouth.

Blog Entries

These are the entries I wrote about the end of the dive trip:

October 24
    Went on an introduction dive with Dennis, and it was *wonderful!*
    I cannot express how really terrific it is to fly under water. I took a few pictures with my $12 throw-away underwater camera, but they didn't come out well.
October 26
    Did not snorkel or dive today. After we arrived back, the crew had a dinner for us at the Chapel Restaurant.
October 27
    Did laundry, and went to Global Gossip to check the value of my Microsoft stock. It was at $51.52, so I placed a sell order for five hundred shares at $53 per share. Went back to the hotel and had chicken korma.
October 28
    At the Cairns Yacht Club I drank a Cuba libre. Then I went to the Cairns Hotel and Casino and had a look around.

An Ad about the Reef

Here's an ad I found about the Great Barrier Reef. It describes it perfectly:

3-Day 2-Night Dive to Adventure for Certified Divers and Snorkellers

The Great Barrier Reef is renowned as the eighth Wonder of the World and a declared World Heritage Site. It encompasses 350,000 square kilometers of the South Pacific Ocean, extends for 2,000 kilometers, and is made up of 2,900 individual reefs and 71 coral cays. Its Ecosystem supports the greatest concentration of life on this planet including 1,500 species of fish, 350 different kinds of coral, 4,000 species of mollusks, and 10,000 species of sponges. Water temperatures range from 22 to 28 degrees Celsius, allowing comfortable diving all year round.

Dive Sites

Milln Reef—perfect introduction to the outer reef, relaxed and varied, sleeping turtles

Flynn Reef—excellent visibility, a photographic haven

"Gordons"—intriguing coral mazes which end at "Clown Fish City"

"Coral Gardens"—magnificent coral enlivened by a myriad of marine life

Pellowe Reef—exclusive pristine dive destination (weather permitting)

Thetford Reef—outlying bommies with swim throughs (weather permitting)

Weather permitting, we never dive the same dive site more than twice.

Extended Seven-Day Dive

Next I took a Taka Dive out of Cairn which was an extended liveaboard dive trip of seven days to remote regions of the outer reef of Northern Great Barrier Reef with the Giant Potato Cod and Maori Wrasse at the Cod Hole.

We would dive at the Cod Hole with the giant potato cod named George. He was a regular pet who always waited for the divers to come and feed him. There was a professional diver and photographer on the trip who made beautiful underwater pictures, so I bought his CD with the photos.

End of Dive Trip

After we returned, I caught a Virgin Blue flight back to Sydney for $89. I wasn't about to press my luck with another train adventure. The airline was modeled after Southwest Airlines out of Love Field. They used the same 737 aircraft and had the same no frills passage.

Total Eclipse

Bill and I were both excited about seeing the total eclipse. On December 3, for $176 each (Australian dollars), we caught a Virgin Blue flight to Adelaide. We spent the night in a hotel room that was over a bar. We were on the top floor, and I swear I could hear footsteps on our ceiling.

The next day we took a commuter flight to Ceduna. When we landed, we wanted to take a taxi, but there were only a couple of them and they were taken. The place was packed with people there for the eclipse. While we were waiting, a bus arrived and Bill hopped aboard. I was in back of the bus line waiting to board when a taxi arrived. One of the guys I'd been talking to in the bus line asked if I wanted to go with him in the taxi. I said yes, but I needed to tell Bill where to meet us. Bill and I had talked about where the best viewing spot would be, and we'd decided to try to find the farthest point west. I went around the bus and found the window where Bill was sitting and told him to meet me at the visitor's information center. I'd get there first and see if they had any last-minute room cancellations.

He nodded, so I jumped into the taxi and went to the visitor's center. Sure enough, they had some last-minute cancellations. The guy with me took the single room, and I took the double room. The rooms were two blocks from the visitor's center so we both ran down to secure our rooms. I dropped off my bags and ran back to the visitor's center to wait for Bill.

I didn't see him, so I went into the store next door and bought some postcards. Then I sat out in front of the tourist information office and addressed my postcards. After waiting for several hours, I decided to walk over to the post office a block away and mail my postcards.

When I returned, Bill was still not there. I went into the sandwich shop across the street, bought dinner, and ate at a table right across from the visitor's center. It was almost time for the eclipse to start, so I went back to the room to get my camera.

As I walked down the street, through my sunglasses I could see the moon just starting to cover the sun. I took a picture with my dark glasses over the camera lens.

I walked down to the jetty in the middle of town, where several TV camera-crew trucks were set up. The jetty was blocked off by the police, so I stood around the trucks and watched the eclipse on their TV monitors. Whenever I saw a good shot, I'd step around the truck, put my dark glasses over my camera lens, and snap a picture with my throw-away camera.

The wind was blowing at about twenty knots and coming off of the water, so the chill factor was plenty high. The truck blocked the wind, and the heat from the running engine kept me warm. When the total eclipse came, I was able to get three pictures of the thirty-two-second total eclipse. After totality, I stayed around for about three more pictures as the moon passed by.

I went back to the cafe and bought a snack and drink and waited another hour for Bill. He never showed. As I went back to the room and bedded down for the night, I wondered where he would sleep?

Next morning I caught a bus to the airport for the flight back to Adelaide. There was Bill. He and several others had spent the night in the airport. He told me that he found the farthest point west, near Thevenard, and sat there in the wind the whole time, chilled to the bone.

He was thrilled, however, because he'd gotten even better pictures than he'd hoped for.

In fact, they were published on the *Astronomy Magazine* Web site within seventy-two hours after he took them.

Adelaide

We caught our commuter flight back to Adelaide and returned to the hotel on Rundle Street, over the bar. I told one of the waitresses about the footsteps I'd heard the first time we'd stayed there. She said there'd been stories of ghosts in the building.

We had dinner at an Indian restaurant the next night, and at a French restaurant, La Guillotine, the night after that.

One day I went to the Cleland Wildlife Reserve in the Mount Lofty Range on the side of Mount Lofty, just twenty minutes from Adelaide. It was set in seventy acres of shady bushland, and it was teeming with native birds and animals. It had a spectacular panoramic view of Adelaide. I enjoyed the many walking trails. I walked among the kangaroos and got to see some cuddly koalas. I had a staring match with a kangaroo as we both lay on the ground about five feet apart.

We rented a car from Thrifty, a Mitsubishi, to drive the Great Ocean Road to Melbourne—a trip of six hundred miles. We planned to stop along the way and spend the night and take some pictures.

The Great Ocean Road

We spent the first night in Portland at the Admella Motel on Otway Court. I was really disappointed in the drive all the way to Port Campbell. It was just like driving down Highway 287 from Armadillo to Fort Worth. But after Port Campbell, things began to pick up. We stopped at Loch Ard Gorge and the Twelve Apostles, and the terrain began to look like Big Sur in California.

The Loch Ard Gorge is named after one of the area's most famous shipwrecks, that of the *Loch Ard,* which killed fifty-two people there in the 1800s. The area is gradually eroding away, so you see a sandy beach and a beautiful sea at the foot of the towering cliffs.

Five miles away, the Twelve Apostles is the name of a group of limestone rocks jutting out of the water. It used to be called Sow and Piglets, but in the 1950s, wanting to attract tourists, they changed the name to the more appealing "Twelve Apostles"—even though there are only nine jutting rocks.

The next 120 miles to Torquay was really beautiful. Even more beautiful than Big Sur. I think it has to do with the angle of the sun's rays. The water seemed to be bluer and the greens looked greener.

Apollo Bay

On the second night we stayed at the Beachcomber Motel and Apartments. We had breakfast at Kafe Kaos.

Ballarat

Bill wanted to go to Ballarat, where there was a wildlife park that he knew about from talking to Jeremy Hogarth, an Aussie naturalist and filmmaker. Greg Parker, the owner, was not there, so his sons Stuart and Christopher showed us around the park.

Before we'd left the states, our biologist friend, Eddie Greding, had warned us of the dangers waiting in Australia.

"Watch out," this ordinarily fearless researcher had said, "for the saltwater crocodile, the Sydney funnel web spider, the box jellyfish—and for God's sake, watch out for the snakes."

I knew Australia had the deadliest spider and jellyfish on earth, but this warning about snakes puzzled me, especially coming from a guy who has spent his entire life collecting fer-de-lances, diamondback rattlers, and copperheads. Eddie said that 75 percent of Australia's snakes are poisonous, and at least four of the six most poisonous snakes on earth live here. The two most lethal snakes outside Australia are cobras. Eddie said, "To put them in toxic perspective, consider this: the Australian inland taipan's venom is forty times more powerful than any cobra's."

Wow, I thought. But why does Australia have such a lethal snake? I found out the answer from Bill's Aussie friend Jeremy Hogarth. It turns out that, in addition to Australia's predominantly bleak, dismal landscape, it has the most unpredictable, unreliable, and variable climate anywhere. Hogarth, like many scientists, believes that because of such adverse conditions, Australia can support its amazing diversity of plant and animal life.

Here, more than anywhere else in the world, animals and plants have survived by adapting in extraordinary ways. They are opportunistic. They must live in desert conditions, yet be ready to exploit the inevitable floods that give rise to new or dormant life. Plant life has evolved to the point where both fire and water stimulate growth. The red kangaroo is so adaptable that it can simply suspend development of any growing embryo until conditions are right; it can provide different milk to joeys of different ages; and it can abort all of its offspring at will and wait for better breeding conditions.

For snakes, adapting to this extreme country means making every bite count. To really know this wild land, you should get to know its snakes. By the time I left, I think I knew even more than Eddie about the inland taipan. It's believed to have the most toxic venom produced by any land snake.

Six dangerous snake species are commonly encountered in Australia. The order of deadliness is taipan, tiger snake, death adder, king brown snake, brown snake, and copperhead. Stuart and Christopher showed us all of them at Ballarat wildlife park. Australia's deadly snakes have neurotoxins as the principal component of their venom; this means they attack the voluntary muscles and the nervous system. Death is usually the result of suffocation caused by blockage of the respiratory passage.

Inside a display case in the snake house, an aggressive Australian tiger snake came right at Bill with its head flattened in an attack posture. Without the protective glass between us, it would have gotten Bill for sure.

The Australian tiger snake is the only one that actively preys on its poisonous comrades, having somehow developed an antivenin that can neutralize their poison. If antibodies for human use could be made from their molecular structure, we might finally have an antivenin to protect against bites from multiple snakes.

Greg's sons saved their best—or worst—snake for last. "Here we have the inland taipan, also called the 'fierce snake,'" Stuart said. "This is probably the most venomous land snake investigated in the world so far. It's six hundred times as venomous as your western diamond-back rattler."

The display case was crawling with inland taipans. A half-dozen were plopped together safely within the glass. Their small, dark-brown, slithery heads were somehow scarier than the big heads of American snakes that I'd come to know back home in Texas. I backed away as he unlocked the inland taipan's window and raised it wide open, teaching me the meaning of true fear.

"There," he said. "Now you can get a better picture."

Bill did get a good picture with his camcorder, albeit a shaky one. An inland taipan came part way out of the case, not three feet from his lens.

Bill said later, "I don't know what someone else would have done, but I kept on filming. Maybe I trusted Stuart's judgment. Maybe I was just incredibly stupid."

Stuart next introduced us to the largest crocodile I ever saw. He demonstrated how agile the big salt-water croc was by holding a rat on a stick about six feet above its mouth. The croc eyed it for a few seconds then jumped up and grabbed the rat. It was all the way out of the water except for the end of its tail.

Scared the hell out of Bill and me.

We also saw koalas, wombats, goannas, giant tortoises, wedge-tailed eagles, and wallabies—as well as the duck-billed platypus, the missing link!

In the eighth grade, Bill's science teacher introduced him to the duck-billed platypus. Part reptile and part mammal, the odd looking, web-footed animal lays eggs and nurses its young.

"You have to hold them by the tail," Stuart said, "Otherwise, they wiggle free—and you don't want to get stuck by those spurs on the hind legs."

These seldom-used spurs are the platypus's only known defense, with enough toxin to cause agonizing pain to humans, or to kill an animal the size of a dog.

The platypus at Ballarat wildlife park was fully mature (almost twelve, the average life span), and smaller than I expected—about sixteen inches long. This was a long-awaited moment for me; I savored it, studying the creature as though it were under a magnifying glass.

It was dark brown on top with a white, almost silvery belly.

I was especially taken by its fur. Platypuses are better insulated than polar bears, with an upper layer of long fur and beneath that, a fluffy undercoat so dense that it can't be parted to find skin. On top of its tail, the hair is coarse as steel wool; everywhere else, it is the thickest, softest fur of any animal on earth.

The platypus bill is its most striking feature, but it's nothing like a duck's bill; it's made of strong, rigid bone covered with rubbery skin, like that of human lips. The bill's one million electro-receptor nerve endings make the platypus the only higher vertebrate that can sense electrical fields. When a yabbie (a freshwater shrimp) swims, its muscles emit electrical impulses. So the platypus, by using its bill, can catch its prey even with its eyes, ears, and nose closed.

If a scientist puts a small alkaline battery at the bottom of a platypus pool, the platypus will attack it repeatedly because it gets more of a charge from the battery than from a yabbie.

Bill didn't scoff at the notion of platypus-as-missing-link. He'd researched them, and he said they share an amazing number of characteristics with reptiles and mammals. They lay soft eggs like reptiles, have internal testes and a cloaca like reptiles, and have bone structures that are known only in reptiles and in no other mammals.

The big puzzle to scientists is that the platypus is practically one of a kind. It's part of an ancient group called monotremes; the short- and long-beaked echnidas, or spiny anteaters, are the only other members. Theoretically, there should be other species in this group, but there is no fossil evidence to support this theory. More recently, the idea of the platypus as a missing link has begun to give way to the idea that it's actually a mammal that developed independently of all others. Australia used to have many large land-dwelling mammals; the platypus stayed small, learned to live in water, and now exists on the only part of the continent that has the water it needs.

Another interesting creature we saw in Ballarat was the bilby. Large water-dependent carnivores are scarce in Australia. But the country has evolved its own unique meat-eaters, such as the bilby. With kangaroo-like hind legs and big ears, bilbys resemble rabbits—but they have long sharp teeth and relish live mice. There's also a rat-sized carnivore marsupial called the brown antechannus; the males live only one year, but they have an extended mating frenzy before they drop dead.

544   IN MY TIME

The best known Australian carnivore is the Tasmanian devil, and we saw these, too. I knew the creature as the amusingly thick-witted whirling dervish in Bugs Bunny cartoons. Having now seen several of the real things, I can state with authority that there's nothing amusing about them.

The early English settlers of Tasmania named them, quaking when they heard their demonic screaming and howling after dark. To see a group of devils feeding and fighting over carrion is to witness a scene out of perdition. When they saw me watching them, the adults tried to stare me down. Their dry noses turned wet, and they opened their wide mouths menacingly so I could see into their gaping jaws.

Their ugliness doesn't stop with their jaws. They have bear-like heads, and grayish-pink snouts, like rats. Their bodies, also rat-like and up to two or three feet long, are black with a few white marks here and there. They have little whippy tails and stumpy, rat-like legs.

Devils will eat any animal, dead or alive, fresh or rotten. And they're totally unconcerned about what body part they're devouring. Their great jaws are seven times stronger than a Doberman's. They crush bones as easily as meat, and they'll even eat feathers.

In mating, the females are dominant. They pick their mates, and have a pecking order to which they adhere until all the females have copulated. Once successfully paired up, the female ovulates about a hundred eggs, has her mate fertilize about sixty of those, and gives birth to thirty or forty young. Then comes the big race, with all the babies trying to be the first to latch onto one of only four nipples. All but the lucky four soon die, becoming a nice snack for Mom.

Tasmanian devils are not especially happy creatures, and with good reason. It's not like they've struck some demonic deal to gain temporary bliss on this mortal coil. They live to be eight or so, but their vigorous sex life lasts only a few years. They have the highest incidence of cancer of any known mammal, perhaps because of their diet. And they stink horribly, exuding a gaseous odor similar to sulfur. Their backbones literally dissolve as they reach old age, and their hair falls out, especially around the rump. When they're old and sick enough to be noticed as such by other devils, they're ganged up on and devoured.

Melbourne

After seeing the snakes, the crocodile, and the Tasmanian devils, I was ready to escape to Melbourne.

We made our way into the city, drove downtown, located the Thrifty car rental office, and followed the tram tracks until we found the Arden Motel, where we secured a room and dropped off our bags. Then we drove back downtown to Thrifty and turned in the car.

We walked around town for a while and stopped at a sidewalk cafe, Softbelly, for dinner, then caught a tram and went back to the motel.

The tram goes through Queen Victoria Market, with peddlers selling sweet fruit and crisp vegetables, and the meat hall selling everything from still-flapping fish to chops and chooks and even goat and crocodile.

Here's some information about Melbourne that I found in a travel brochure: "The 'upper market' is a bargain hunter's bounty of homewares, fashion, CDs, plants, and handcrafts where stall after stall is filled with 'must-have' and only the strong willed leave empty handed."

The next day, we went back downtown and visited Chinatown. Two grand red arches mark the two ends of Chinatown. The spicy, aromatic air invites you to eat. We had a great Chinese dinner. Shops and businesses sell everything from Chinoiserie to fungi, and some still stock the same products they did over a hundred years ago. The Chinese Museum tells the Chinese-Australia story over five levels of exhibition space.

We did some more looking around and walked over the hill to the Fitzroy district. Fitzroy is the home of the city's Bohemian culture. Brunswick Street is very "alternative lifestyle," while around the corner is Johnston Street, with Spanish culture and tapas and flamenco bars.

We visited some of the city's seven magnificent arcades. The honeycomb of arcades is the extensive, and the opulent Block Arcade is the crowning glory. It was the shopping mall of its day (1891) and features imported Italian mosaic floors and a central octagonal glass dome. The Royal Arcade is Melbourne's oldest (dating from 1869); it features Gog and Magog, two mechanized folklore figures who strike the hour. The center of town is Bourke Street Mall. The Walk, Australia on Collins, and the Galleria Shopping Plaza all sit within a block of each other off Little Collins Street. Majestic and tree-lined, Collins Street is a classy street of chic designer shopping, soaring steeples, five-star hotels, exclusive jewelers, and private clubs; it's also the city's premier commercial streetscape and is home to some big corporations.

Back to Sydney and Back to Texas

On the flight from Melbourne to Sydney I reflected that Melbourne is proletariat, artistic, and intellectual, while Sydney is cosmopolitan. There are only twenty million people in the whole country. Sydney has five and a half million while Melbourne has five million. Sydney thinks of itself as Americanized, while Melbourne thinks of itself as European. Half the population of the country lives in these two cities, with the rest living around the edge, mostly on the east coast. The center of the continent, the Outback, is desolate and barren.

Australia is a great country overall, full of beauty with a natural world unlike any other. For Americans at least, it's the last Wild West for those of us who grew up thinking, "Go West, young man."

I'd arrived in Australia on September 26, 2002, and I flew home on December 23, 2002.

Now that my travels were over for the time being, I decided to buy a Celestron 8 telescope and do some star gazing. Bill, Eddie and I were always excited about viewing the stars and now I had some time.

Jewel of the Solar System

Cassini-Huygens entered orbit on June 30, 2004, around Saturn.

Saturn's beautiful rings are what set it apart from the other planets in our solar system. It is the most extensive and complex ring system in our solar system, extending hundreds of thousands of miles from the planet. Made up of billions of particles of ice and rock, ranging in size from grains of sugar to houses, the rings orbit Saturn at varying speeds.

Many Moons

Saturn's thirty-four known moons are equally mysterious, especially Titan. Bigger than the planets Mercury and Pluto, Titan is of particular interest to scientists because it is one of the few moons in our solar system with its own atmosphere. The moon is cloaked in a thick, smog-like haze that scientists believe may be very similar to earth's before life began more than 3.8 billion years ago. Further study of this moon promises to reveal much about planetary formation and, perhaps, about the early days of earth as well.

The European Space Agency's Huygens Probe dove into Titan's thick atmosphere in January 2005. The Huygens Probe was named after Christiaan Huygens, a Dutch astronomer who in 1655 discovered Titan, Saturn's largest moon. The 319-kilogram (703-pound) Huygens probe separated from the Cassini orbiter in December of 2004, and began a twenty-day coast phase toward Titan. The Huygens probe plunged into a planetary atmosphere farther away from earth than any other deep space probe has gone before. It was the first spacecraft to land on a moon in the outer solar system.

New Horizons

The New Horizons spacecraft Pluto probe was launched January 19, 2006.

It is expected to fly by Pluto in 2015. It is planned for New Horizons to fly within about six thousand miles of Pluto. New Horizons will come as close as 16,800 miles to Charon, Pluto's moon, although these parameters may be changed during flight. After passing by Pluto, New Horizons will continue further into the Kuiper Belt. Mission planners are now searching for one or more Kuiper Belt Objects (KBOs) on the order of thirty to sixty miles in diameter for flybys similar to the spacecraft's Plutonian encounter. As maneuvering capability is limited, this phase of the mission is contingent on finding suitable KBOs close to New Horizons' flight path, ruling out any possibility for a planned flyby of 2003 UB313, a trans-Neptuinan object larger than Pluto.

Economic Indicators

Once back in the U.S., I finally was forced to turn away from the stars and face the reality of the economic collapse and the erosion of my retirement fund.

When I was working, I sometimes felt that I was the only one working and paying taxes. It seemed like all around me were armies of slackers on various assistance programs, such as fake disability, WIC (women, infants, and children), TANF (temporary assistance for needy families), food stamps, unemployment, welfare, etc., etc., etc.

The following is from a *USA Today* article published on March 14, 2006:

Federal Aid Programs Expand at Record Rate

A sweeping expansion of social programs since 2000 has sparked a record increase in the number of Americans receiving federal government benefits such as college aid, food stamps, and health care.

A *USA Today* analysis of twenty-five major government programs found that enrollment increased an average of 17 percent in the programs from 2000 to 2005. The nation's population grew 5 percent during that time.

It was the largest five-year expansion of the federal safety net since the Great Society created programs such as Medicare and Medicaid in the 1960s.

Spending on these social programs was $1.3 trillion in 2005, up an inflation-adjusted 22 percent since 2000 and accounting for more than half of federal spending. Enrollment growth was responsible for three-fourths of the spending increase, according to *USA Today*'s analysis of federal enrollment and spending data. Higher benefits accounted for the rest.

Programs that grew over the past five years are aimed at the under-65 population.

It almost seemed that if I stopped working, the economy would come crashing down.

And sure enough when I retired it did just that.

I'd thought the economy and stock market should have crashed in 1990, but there was not even a major slowdown. I worried about this for some time. Then I heard on TV that the Dow would keep going up because baby boomers had figured out that there wouldn't be any retirement for them. Corporation were slowly beating them out of their pensions, and Social Security was underfunded and would be gone when they retired. So the baby boomers were investing in the stock market to create their own retirement plans.

Do not believe anything a politician says; you know he's lying if his lips are moving.

Do not believe the so-called financial experts, either, because like the politicians, the experts have hidden agendas.

Do not believe anything in the media, either.

Advertisements are the real barometer of the economy, the true economic indicators. When you keep seeing ads for "Zero Down!" "Zero (or near zero) Interest!" "No payments for a year!"—you know the economy is in trouble.

It's especially bad when the U.S. car manufacturers run such ads.

"What's good for General Motors is good for the country." So goes an old saying. If U.S. car manufactures are dumping excess inventory by giving everyone employee discounts, then our economy is in the toilet.

One ad I saw said, "zero down, zero interest for five years." That means they think the current depression will last longer than five years.

Eddie's Retirement Party

Bill had taken early retirement from teaching English at Ohio State. Now my other high-school friend, Eddie, retired from his job as a college professor in the natural sciences department at Del Mar College in Corpus Christi. Bill and I attended his retirement party.

He'd invited a number of colleagues from school. They were stuffy academic types and were all having a boring time making nice polite toasts to Eddie for his work and his teaching. After all, he was the elder statesman of the college. It was really boring.

Bill and I decided to liven things up.

So we told the story of that night in high school when I hid in the basement of an abandoned house and pretended to be a corpse, while Bill lured the poor and gullible and unsuspecting Eddie inside in search of bats. Bill and I each told our part, playing off of each other, and cracked up the partygoers. It broke the ice, and the party was great fun after that, with dancing, funny stories, and other carryings on.

It turned out to be the only party in Eddie's life that he'd ever enjoyed.

Retirement

Someone should tell the best-kept secret about retirement: it's not what it's talked up to be.

First your body starts to go. Sex is usually first—why do you think Viagra, Cialis, and Levitra are such big sellers?

Then the other organs began to weaken. Hair and teeth fall out, so you need hairpieces and dentures.

Then the hearing goes, and you need hearing aids.

Eyesight deteriorates in a steady progression: first you need glasses, then bifocals, then trifocals. After that, if you live long enough, you go blind.

Your joints begin to hurt. Arthritis sets in. So do all sorts of other ailments.

You get gas. There is a reason they call us "Old Farts."

Then you get high: high blood pressure, high cholesterol, and high blood sugar.

By now, to alleviate these conditions you're taking about ten different medicines. The drugs have side effects. Your lower back begins to hurt and you have diarrhea and constipation and trouble urinating. So you add Metamucil, and Benefiber, and Fibercon to try to keep yourself regular. Then your sinuses get blocked up and you have sinus headaches. Next the sinus starts to drain and you are always trying to clear your throat and you have trouble swallowing. You get lots of exercise by pumping up the pharmaceutical companies' bottom lines and painting them black.

Remember what I said about the Tasmanian devils? How their sex life lasts only a few years, they have a high incidence of cancer, they exude a gaseous odor similar to sulfur, and their hair falls out? I'm not sure aging humans are much better off.

Next your mind starts to go. Is it, uh, what do they call that disease again? Oh yeah, Alzheimer's? You start with short-term memory loss. Soon then you can't remember your own address, or your friends' faces, or your relatives' names. But the latter is probably a good thing, because your relatives—especially your kids, if you have them—are up to their asses in debt and working two and three jobs to pay their bills, and they don't have time to come visit anyway and when they do, they try to borrow money. You start to hope *they* get Alzheimer's and forget where you live.

But look on the bright side—it's not all bad. With Alzheimer's, you wake up to a new world every day.

While all this is happening, the stock market tumbles and inflation hits. Your retirement fund goes down the tubes, your money is worth less, your finances bottom out, and you can't pay your own bills, let alone lend your relatives money to pay theirs.

The only thing close to golden in your *golden years* is your urine.

If you want to avoid all this, I think sixty is a good time to die. I retired twenty-four hours before I turned sixty.

Gonza Bonza Bottler Day

Early in April of 2006 I looked at the calendar and realized, "On Wednesday at two minutes and three seconds after 1:00 in the morning and afternoon, the time and date will be: 01:02:03 04/05/06. Wow! That's a Gonza Bonza Bottler Day!!!"

I thought back to that Bonza Bottler Day in 1955, on 5/5/55 when the teacher wrote the date on the board. On that day I wondered what I'd be doing on June 6, 1966, which seemed an eternity away. And now I thought of all the Bonza Bottler Days since then.

Deep Space

In mid-2006, *Voyager 1* will be more than nine billion miles from the sun. *Voyager 2* will be at a distance of seven billion miles. Both craft are healthy and continue to send data back to earth. In the next ten years, scientists expect the *Voyagers* to cross the heliopause, the edge of the bubble created by the sun's supersonic wind, and become the first craft to reach interstellar space.

The Solar System Shrunk

If you woke up Thursday morning on August 24, 2006, and sensed something was different about the world around you, you're absolutely right. Pluto is no longer a planet. The International Astronomical Union resolved one of the most hotly-debated topics in the cosmos by approving a specific definition that gives our solar system eight planets, instead of the nine most of us grew up memorizing. The Astronomical Union has designated as the prototype for a new class of celestial objects, to be called "dwarf planets." The International Astronomical Union has decided that, to be called a planet, an object must have three traits. It must orbit the sun, be massive enough that its own gravity pulls it into a nearly round shape, and be dominant enough to clear away objects in its neighborhood.

The dwarf planet family also includes 2003 UB313, nicknamed "Xena." In addition to Pluto, Ceres and 2003 UB313, the Astronomical Union has a dozen potential "dwarf planets" on its watchlist.

The organization has decided that most asteroids, comets and other small objects will be called "small solar-system bodies."

A Life Full of Miracles

Reflecting back on my life now, it seems full of miracles. I've seen a man walk on the moon. I've seen total strangers on opposite sides of the globe who've never met become connected by the Internet.

I've seen the rise of medical technology, mapping of the human genome, development of the Internet, and amazing advancements in cosmology or study of the universe, including quantum physics.

I was born poor and by all rights, I should have stayed poor. I'd barely graduated from high-school, and my learning disability should have doomed me to ignorance. Plus I'd had serious health problems as a child and should never have lived into adulthood.

Yet somehow, in spite of my background, I beat the odds. I'd risen to become, among other things: a Cold-War spy, an early Mary Kay entrepreneur, a Silicon Valley computer pioneer, a weather-satellite engineer, and a Microsoft support guru. I conquered my family's lack of culture and education and become a classical music enthusiast, an appreciator of art, a student of world religions and philosophy and science, and a French wine connoisseur. I'd never been taught about investing, yet I'd made money in real estate in California, New England, and Texas, and I'd made money, lost money, and made money again in the stock market.

I've lived in Germany, San Francisco, Carmel, Los Angeles, Boston, France, and Sydney. I've crossed the U.S. by train, plane, car, and bus. I've hobnobbed with generals, corporate executives, and multi-millionaires, and I've slept on a park bench. I've traveled to Canada, Mexico, Chile, Peru, England, Germany, France, Spain, Italy, Turkey, and Australia. The sickly little boy from Texas became a professional dance instructor, a sky-diver, an expert swimmer, a hiker, a windsurfer, a pilot, and a sailor with my own 44-foot sailboat. And I survived near-plane crashes, train accidents, and storms at sea.

I sometimes wonder how I managed come up out of the shit-pile of life, smelling like a rose. Was it divine intervention? Was it my own determination and work ethic? Or had I just been lucky enough to be in the right place at the right time?

Chapter Fifteen:

Revelations (Epilogue)

Revelations

I'm going to conclude my story by summing up what I have experienced and learned in my time.

Lies

First of all, I have been lied to by almost everyone and everything: friends, relatives, employers, the church, the government, on and on. I would estimate that 98 percent of everything I've been told during my lifetime was wrong. The problem is that the lies they have told me are the lies that they themselves have been told. They believed these lies and never questioned them. What truth I have learned was in science.

Science is generally true and repeatable with the information available at a given time, but as new theories arise, some of the accepted foundations are eroded. New theories have led to new frontiers. Newton's Laws are still valid for most physical systems, but these laws have been superceeded by relativity theory and quantum physics. If we can land a man on the moon and a probe on Saturn's moon Titan, then we have to be doing something right, something that is part of the scientific method, and part of consilience—an old concept made current by scientist Edward O. Wilson.

Consilience, says Wilson, is the concurrence of generalizations from separate classes of facts in logical inductions, so that one set of inductive laws is found to be in accord with another set of distinct derivation. It suggests a coming together of knowledge, and it is rooted in the Western concept of an intrinsic orderliness that governs our cosmos. Consilience is very much at odds, however, with the newer concept of chaos theory in quantum physics.

Meanwhile, our political system has gone in the direction of chaos, greed and corruption. Without consilience in the social sciences, the wrongs that men do live after them, and errors are compounded exponentially. Not all disciplines, systems, and areas of knowledge have checks and balances comparable to the scientific method, and until they do we'll continue to fall into the abyss of chaos.

Some things happening today need repeating. They are all from another earlier tragedy. Polonius's speech to Laertes from William Shakespeare's *Hamlet* Act I Scene III and Hamlet's "To be, or not to be, that is the question" soliloquy in Act III Scene I.

"Beware of entrance to a quarrel [and that warning includes war.]
Costly thy habit as thy purse can buy,
Neither a borrower nor a lender be;
For loan oft loses both itself and friend,
there's the rub!
the law's delayed,
The insolence of office, and the spurns
That patient merit of the unworthy takes,"

Riding the Wave

Bill, Eddie, and I managed to ride the economic tsunami of the good times created by baby boomers as they plowed through the decades. Like good surfers, we managed to keep our heads down and stay in what surfers call "the tube" of a very long-lasting wave, while others wiped out or were impaled on reefs and eaten by sharks. Being in the right place at the right time was the key to our success. This country was full of promise after World War II, and everyone was very optimistic. Now however, I see the high-water mark where the wave broke and started to recede. It has left a whole generation (perhaps a whole country) high and dry in the devastation.

The U.S. has been a war-based economy ever since World War II. As it says in the Bible, in both Matthew and Mark, there will be wars and rumors of war. These benefit large corporations in the military-industrial complex, enhance their bottom lines, and divert attention from the real cancer in the system.

The U.S. is now a debtor nation. As of April 14, 2005, the national debt was $7,786,560,972,566.27. The estimated population of the United States at that time was 296,574,802, so each citizen's share is $26,506.14. The national debt has increased an average of $1.64 billion per day since September 30, 2004!

On March 16, 2006, faced with a potential government shutdown, the Senate voted to raise the nation's debt limit for the fourth time in five years. The bill passed by a 52-to-48 vote, increasing the ceiling to $9 trillion. On March 17, 2006, the debt was up to $8,270,880,983,362.78. Your family's share is $132,235—and this does not include state, local, and city property taxes, school taxes, and sales tax. There's also the increasing costs of home mortgages, auto loans, insurance, health care, utilities, furnishings, groceries, and entertainment. Plus there's the (ever-increasing) cost of $300,000 per child just to get the kid through high school.

General Motors announced it would lay off 30,000 employees and close twelve assembly plants; then Ford Motor Company announced it would lay off 30,000 employees and close fourteen plants; then Chrysler announced a large lay-off; (together totaling 105,000 and equaling 25 percent of the workforce and 12 percent of American white-collar workers); then Kraft, the largest food manufacturer, announced 8,000 job cuts. This does not include all the suppliers that are affected. With public schools now teaching Chinese, you can bet that the economy is in real trouble. China's Geely Automobile Holding Group unveiled its newest family sedan, the compact Ziyoujian (or "Free Cruiser"), at a recent North American International Auto Show in Detroit. It is scheduled to sell for less than $10,000. At about the same time, GM announced a reduction of 1 to 3 percent on prices for 80 percent of its automobiles. Soon additional cheap Chinese goods will replace expensive American products, and our economy will become worse than during the Great Depression of the 1930s. Unable to pay our national debt, we'll be pressured by foreign countries to do their bidding.

The progression of American history looks like this:

From bondage to liberty
From liberty to abundance
From abundance to selfishness
From selfishness to complacency
From complacency to apathy
From apathy to dependence
From dependence back again to bondage

As we become more dependent on foreign oil and foreign products, we go back into bondage, and we won't be able to win, we won't be able to break even, and we won't be able to get out of the game.

The U.S. is no longer following the natural law of the survival of the fittest; the U.S. is following the law of the survival of the unfittest—because the unfittest can vote.

Politicians are so busy kissing everybody's ass to get re-elected that they cannot even see the problems that need to be corrected, much less formulate a plan to fix them.

How long can this nation survive supporting the mass of the unproductive lower class?

The good are destroyed while the evil survive and prosper.

This Country Is Screwed Up

Ever since World War II, the U.S. economy has been controlled by the military-industrial companies, aka the defense industry, which even frightened President Eisenhower. In a famous speech president Eisenhower warned us to never get into a ground war in the Middle East. Government contractors have figured out that a product should be manufactured in as many states as possible, so more politicians will vote for the project to gain jobs—and votes—back home.

The only thing close to eternal is a government program.

Now we're at war in the Middle East. Did anyone on President Bush's staff ever read *The Seven Pillars of Wisdom* by T.E. Lawrence (Lawrence of Arabia)? Lawrence showed that the people of the Middle East are incapable of governing themselves and cannot maintain a democratic system or nation; 4,000 years of tribal strife plus the legendary bickering of Iranian rug merchants makes the Middle East the least likely place for a democracy.

Have you ever heard of the comedy George of the Jungle? This is a story of *didus ineptus* George of Arabia—or should it be *didus ineptus* George of *Iraq*?

The intelligence offered by the hawks in Washington has led our president down a rosy garden path to war. In my opinion, we went to war with Iraq for all the wrong reasons. The U.S. sent 160,000 troops to Iraq and hired nearly 120,000 civilian contractors to support the troops. The majority of the contractors are from Vice-President Cheney's former company, Halliburton/KBR.

U.S. Business: Utterly Corrupted

This country is slowly destroying itself from within. Witness the criminal behavior of corporate executives, such as Equity Funding board chairman Stanley Goldblum, as well as the following Enron executives: Jeffrey K. Skilling and Kenneth L. Lay; chief financial officer Andrew Fastow and his wife, Lea; senior accountant Wesley H. Colwell; treasurer Ben F. Glisan Jr.; Jon C. Doyle, Robert C. Kelly, H. David Ramm and Thomas White. Then there's WorldCom's chief financial officer Scott Sullivan and controller David Myers. And executive Dennis Kozlowski and finance chief Mark Swartz of Tyco International Ltd. Cable TV, with the help from Arthur Anderson's CEO Jerry Kent, and CFO Kalwarf. Let's not forget JP Morgan Chase and Citigroup, plus four Merrill Lynch executives: Robert S. Furst, Schuyler M. Tilney, Daniel H. Bayly and Thomas W. Davis.

Dishonesty, Greed and Hypocrisy in Corporate America

My title for this section comes from the title of an article by Huck Gutman, published in *The Statesman* July 14, 2002.

There have always been "robber barons" throughout history, and this last century was no different.

The former CEO of Halliburton, who was in charge when those accounting practices were introduced, is Dick Cheney, currently Vice President of the United States.

President Bush and Vice President Cheney, ever mindful of campaign contributions from rich and powerful corporate executives, ever mindful of their circle of friends the wheelers and dealers and "captains of industry," ever mindful of their own past practices, are themselves in no hurry to see significant changes made.

The mergers of large corporations are creating a near monopoly. Soon there will only be three or four banks, loan companies, insurance companies and all the major businesses. Like Wal-Mart, they will drive the small companies out of business, and when they control the market they can demand whatever price they want to—everyone will have to pay or do without. Then they will lobby for laws that you have to have their products to be a resident of the state, like auto insurance, trash pick up, connections to municipal water and sewer services, and that you must, of course, pay property and school tax. The 20 percent loan sharks of the 1920s are now presidents and CEOs of Banks, savings and loans (S&Ls), and credit card companies.

Governmental Watchdog Groups Are Foxes in the Hen House

The U.S. Food and Drug Administration (FDA) is influenced by high-paid pharmaceutical lobbyists, political action committees (PACs), and trade association's PACS. There are 287 pharmaceutical lobbying firms, and there are more pharmaceutical lobbyists than all the members of Congress combined. The sum exceeds actual number of registered pharmaceutical lobbyists because many work for more than one type of company. PhRMA is one of the biggest drug lobbies.

The operation of the Pharmaceutical Benefits Scheme (PBS) has substantially increased the costs of medicinal drugs. High-paid pharmaceutical lobbyists who are "helping" the legislature actually write legislation. The natural purpose and driving force of the pharmaceutical industry is to increase sales of pharmaceutical drugs for ongoing diseases and to find new diseases to market existing drugs. By this very nature, the pharmaceutical industry has no interest in curing diseases.

The eradication of any disease inevitably destroys a multi-billion dollar market of prescription drugs as a source of revenues. Therefore, pharmaceutical drugs are primarily developed to relieve symptoms, but not to cure.

Clinical studies and research are funded by the pharmaceutical industry to ensure the results, thus prompting one university professor to say to his graduate research students, "You are not going to find a cure for cancer on my research grant money."

It Gets Worse

Every year, the Centers for Disease Control and Prevention says 5,000 deaths, 325,000 hospitalizations and 76 million illnesses are caused by food poisoning. The frequency of serious gastrointestinal illness, a common gauge of food poisoning, is 34 percent above what it was in 1948. Why, in an age of technologies that protect food, is food poisoning at least as common as it was a half-century ago? The variety of foods available has expanded considerably faster than the FDA's ability to inspect them. In the last decade, grocery stores have doubled the number of items they stock, from every corner of the world, some carrying new organisms that scientists still cannot identify, much less treat. In fact, the amount of contaminated food that reaches store shelves only to be recalled for posing health risks has reached its highest level in more than a decade. "We are the canaries in the coal mines," said Dickson Despommier, a professor at the Mailman School of Public Health at Columbia University. "We do have a real problem," said Joe

Levitt, food safety director for the FDA. Amid the proliferation of foods, the F.D.A.'s resources to scrutinize them have scarcely changed, often making consumers the first to test a product's safety. Food safety focuses on meat and poultry, by the Department of Agriculture (USDA), especially beef; the General Accounting Office estimates that 85 percent of food poisoning comes from the fruits, vegetables, seafood and cheeses that are regulated by the FDA and claim a larger share of the American diet each year. The FDA has less than a tenth of the inspectors of the Department of Agriculture. The pharmaceutical companies have no interest in solving this problem. It is another supermarket avenue for their drugs.

EPA

The United States Environmental Protection Agency (EPA) is biased by the energy companies.

There are 601 lobby firms for Environmental and Superfunds, 300 lobby firms for fuel, gas, and oil; 362 lobby firms for clean air and water; 646 lobby firms for nuclear energy; and 285 lobby firms for utilities.

A reporter on the TV show *Sixty Minutes* interviewed a NOAA-named researcher Bob Corell. Corell led 300 scientists from eight nations in the "Arctic Climate Impact assessment" study. A transcript at the CBS.com Web site points out, "Back in 1987, President Reagan asked Corell to look into climate change. He's been at it ever since…. 'Right now, the entire planet is out of balance,' says Bob Corell."

But the real problem is that the White House has been censoring Corell's reports to throw a better light—i.e., put a better spin—on the issue to deflect the argument that oil and gas supporters are causing the problem. Under the cover of the war, the White House edits reports and great pains are taken to deceive the public.

SEC

The U.S. Securities and Exchange Commission (SEC) is in bed with the finance industry.

There are 449 lobby firms for the finance industry.

Shortly after the stock market crash of 1929, a regulatory body called the Securities & Exchange Commission (SEC) was born. Its stated goal was to restore investor confidence and faith in a financial sector that was notorious for fraudulent activities (and still is), easy credit and hazardous investments. Two significant proposals by the U.S. Congress, the Securities Act of 1933 and the Securities Exchange Act of 1934, led the way to the formation of the SEC and, ultimately, a structured financial industry under government supervision. (That really makes me feel good.) The aim of both of these acts was suppose to protect "investors" and large financial companies from any indiscretions that could arise from dishonest and unscrupulous individuals dealing in the securities markets.

In other words, "We're here to protect you."

Banks and S&Ls

There are 367 lobby firms for the banking industry.

Banks and S&Ls are not in the business of protecting your money. The FDIC and the FSLIC are. Banks take your money and invest it to make more money. If their investments fail, it is up to the FDIC and FSLIC to cover your money. Banks are now charging you a fee to take your money out. But don't expect to get your money out if money gets tight. You can only get a few one-hundred dollar bills, just enough to last a couple of weeks. There are no more $500 or $1,000 bank notes. Electronic funds transfers took care of those. You have to give the bank two weeks' notice to get any large amount of your hard-earned money out, and if too many people—three or four plus their friends—want their money, the banks will simply close their doors as they did in the 1930s. Your money is just a bunch of ones and zeros in a computer somewhere, and with one push of the "delete" key, it is gone.

Mr. Charles Keating, a lawyer, hired none other than the present chairman of the Federal Reserve Board, Mr. Alan Greenspan, to lobby government to allow diversification from home loans into direct equity investments. In fact, Mr. Greenspan praised Lincoln as "a financially strong institution that presents no foreseeable risk to the Federal Savings and Loan Insurance Corporation." Mr. Greenspan was successful. Mr. Keating was virtually free to speculate directly. Mr. Greenspan's lobbying had in effect pulled the cork. By October, 1986, when San Francisco regulators first investigated Lincoln, the S&L already had *unreported losses* of $135 million.

In all, the cost to taxpayers for the failure of this S&L, alone, is estimated at from $2 billion to 2.5 billion.

Few articles have conveyed the favors for favors mechanism underlying American politics better than the article entitled "An Amazing Tale" in the August 26, 1989, issue of *The Economist*. It deals with the Lincoln Savings and Loan debacle.

If you next read "High-Rolling Texas: The State That Ate FSLIC," you probably won't need to read much more about how Congress operates.

Insurance

There are 328 lobby firms for the insurance industry.

The insurance industry is not in the business to pay large claims. Small, pocket-change claims are okay. They are in the business of collecting as much money from as many sources as they can, investing it to make more money, and fighting tooth-and-nail to limit large claims. They are like HMOs; they make money by denying service. That is why there are so many exclusions in the small print.

Insurance companies are a second- or a third-generation spawn of protection rackets.

Government Protection Rackets

There are all sorts of protection rackets. The most important one: protecting the public from ever hearing the truth.

The people in charge tell the rest of us that they will protect us from sundry evils, including burglars, serial killers, polluters, bigots, greedy corporations, uncaring employers, and even "global warming"—and all we have to do is ante up the cash to pay for it. Should anyone refuse the government's protection, he or she will shortly find himself in the deep trouble for failing to pay his "fair share."

Criminal Justice System

Crime does not pay, but traffic citations do. If a criminal commits a crime and the police catch him or her, it will cost the city and state a million dollars to prosecute them. If the police write a million traffic tickets, they can receive 40 million dollars income.

Do the math.

"Follow the money," Deep Throat said.

Where would you put your resources?

There was one convenience-store robber caught on a surveillance camera who left his date book planner with a picture ID, address, telephone number, and where he was going to be the next week. The police could not catch him and asked anyone who knew him to turn him in.

One community was so fed up with the local police's inability to control crime they brought in a few nuns from the Sisters of Charity order to pray for their community and talk with the criminals in hopes it would help.

The criminal justice system (CJS) has grown so big that it now has a life of its own, creating jobs and not reducing crime. If it reduces crime it puts itself out of business. The CJS consumes large amounts of federal, state, and local funds and has more revolving doors than all Macy's department stores combined. Sentencing a criminal to life in prison is cruel and unusual punishment—for the taxpayer. Because of political appointments and more congressional reps going to jail, the Supreme Court will declare that punishing political criminals violates their civil rights—the ones which they passed for themselves, such as their pay raises. The Congressional Immunity Act infers that certain criminal acts are normal congressional activities. A review by the House Ethics Committee review of the six-term Republican Congressman from Florida, Mark Foley, scandal once again proved congress has no ethics.

The Prime Directive of Congress appears to be gain power, retain it at any cost and, then get re-elected. Wasting time debating and voting on a non-binding resolution is congressional masturbation and shows how impotent they are and unable to perform.

The worst crime of the last century was the government investigation and cover-up of the shooting of President Kennedy.

TV Nightly News

The old meaning for "nightly news" was "a report of recent events: new information."

The new name for TV news programming should be SAP: shock, awe, and propaganda.

Shock: 1) a violent shake or jar, 2) a sudden or violent disturbance in the mental or emotional faculties, 3) a sense of outrage to one's convictions especially of morality or propriety, something that causes outrage, horror, stupefaction, or disturbance or agitation in a person, or organized system.

Awe: veneration and latent fear inspired by something sacred, mysterious, or morally impressive.

Propaganda: dissemination of ideas or information: spread by deliberate effort through any medium of communication to further one's cause.

The TV media leaves the report of new information to the last minute of air-time. A news bite is 1/30 of a thirty-minute news cast, or 1/120 of the total news. A TV news cast puts the one-minute of actual news at the end of the hour to drag you through all the SAP.

One year, eggs are bad for you, then the next year beef is bad for you. Then for some reason, eggs and beef are okay, but vitamins are bad. Then soft drinks are bad, and coffee is good. Then the opposite is reported.

I suspect that certain lobbyists did not donate enough money to the right political campaign or the right candidate, and that's how these reports got generated. The lobbyists probably respond by paying sympathetic researchers to write up and publish an opposing viewpoint. I am sure journals of medicine and other scientific-paper-mill tabloids check, recheck, and check again with a minimum of three independent researchers every report they print, just like the *National Enquirer* does.

Right.

You cannot believe anything you see or hear on the TV. This was proven in the 1960s when the game show programs were exposed as rigged. The TV Network executives made it clear in their congressional testimonies that they consider anything on TV as entertainment and that it should not be taken seriously.

The nightly news reports how well the economy is doing. The unemployment rate is at an all time low of 5 percent, and there are 10,000 new jobs created. However, there were 150,000 workers laid off, and there are 50,000 new college graduates. That means 200,000 workers are applying for the 10,000 new jobs just created—a minimum of twenty people for every job, and that does not include all the workers who have fallen off the radar by exceeding their unemployment benefits, not to mention all the illegal aliens.

Television Is an Advertising Medium

TV is the most powerful advertising medium in history.

In the seven or so years following the end of World War II, the fledgling upstart medium of television advertising would irrevocably alter the social, economic, and political landscape of the United States.

Like the beginnings of most new technologies, the first era of commercial television was a wild and wooly period fueled by an entrepreneurial spirit, gold rush mentality, and corporate interests.

Single Sponsorship

The Revlon cosmetics firm was the sponsor of *The Revlon Mirror Theater*.

There was the *Colgate Comedy Hour*, with Eddie Cantor, Frank Sinatra, Eddie Fisher, Abbott and Costello, Spike Jones and the City Slickers, and Wictor Borge.

Remember "Halo Everybody Halo"—"The shampoo that glorifies your hair"? How about these:

The Dinah Shore Chevy Show—"See the USA in your Chevrolet!"
General Electric Theater—hosted by Ronald Reagan
Mobil Masterpiece Theater
Texaco Star Theater—with "Mr. Television"
General Foods' Kraft TV Theater
Westinghouse Electric Corporation's *Studio One*
The Elgin Hour
Playhouse 90, with The Miracle Worker, The Comedian, The Helen Morgan Story, Requiem and Judgment at Nuremberg
Philco TV Playhouse
Cavalcade of America—sponsored by DuPont Co.
Camel's Man Against Crime and Camel Newsreels
The Chase and Sanborn Hour
Maxwell House Concert
Pabst Blue Ribbon Bout
Chesterfield Supper Club
Bell Telephone Hour
Sylvania's Beat the Clock
Geritol's Twenty-One

And don't forget those advertising icons:

Speedy Alka-Seltzer
Kellogg's Tony the Tiger
Maytag's Lonely Repairman
And last but not least: The Marlboro Cowboy

No programming format mesmerized televiewers of the 1950s with more hypnotic intensity than the "big money" quiz show, one of the most popular and ill-fated genres in U.S. television history. Single sponsorship virtually ended after the quiz show scandal. It was replaced by scatter plan (or scatter shot) advertising: a series of spots, throughout the day and evening. With five or more advertisers for each show, no one would have ultimate power. Prices varied throughout the day for ad spots.

Demographics

By the early 1960s, demographic information began to dominate advertising, and the buying and selling of thirty-and sixty-second slots.

Nielsen's rating data—in terms of sex, age, economic and educational status, urban or rural location, and other factors—were especially valued by advertisers.

Today we have:

Prison Break
America's Most Wanted
Cops
The Sopranos
Law & Order
Conviction
CSI, CSI NY and CSI Miami and NCIS
Desperate Housewives
Wife Swap
Criminal Minds
Drowned Alive
Blowing Up
Cold Case
Desperation (Stephen King)

What demographic are they targeting?

Quiz shows became game shows, shifting focus from knowledge to puzzles, word games, and reality shows. Today these include:

Wheel of Fortune
Survivor
Deal or No Deal
Fear Factor

Awareness

Every TV news program should start with a Gaia report on the health of the planet.

If nothing new is being done to save the world and our species, it should be headline news until something is done. If Congress reduces the emissions requirements for the EPA, it should be headline news and should cause strikes and protest marches.

I was amazed at the information about pollution available on the Internet that is NOT being reported in the news. There are hundreds of Web sites keeping data every day. One called AirData creates a detailed map by country, state, city, and even zip code. The Gaia report should be like the weather report. It should start with an air-quality summary and a pollution report on things such as the pollutant count, with the totals of each pollutant for the day, totals for the year, and a list of the types and percentages of each. There should be full-sphere picture of the earth like the GOES satellite image, showing the major pollution distribution area, like the water-vapor display on the weather report. Then there should be a list of the ten biggest polluters in each country, similar to the ten most active stocks on the PBS *Wall Street Week* report. Mankind should be put on the endangered species list, not because there are too few of us, but because there are too many, causing loss of habitat.

Education

Every elementary school, middle school, high school, and college should require a minimum of three courses in environmental sciences for graduation. If bilingual and gay studies are taught, then environmental sciences should be also. There is an Ethical, Legal, and Social Issues in Science (ELSI) Project sponsored by the Lawrence Berkeley National Laboratory. The project advocates air pollution software and curricula for use in networked classrooms. The art of building effective Internet tools for science learning is being developed in partnership with the Knowledge Integration Environment (KIE) Project of the University of California School of Education.

A high-school biology teacher and a physicist, Dean Rockwell and Tony Hansen, have developed an instrument capable of measuring graphitic carbon aerosol (a suspension of airborne carbon particles better known as soot). This instrument can be easily understood, built, and operated by students. It is inexpensive, so teachers can afford to assemble it on their small science budgets.

Rockwell and Hansen developed a procedure that uses simple, commonly available materials, such as facial tissue, a vacuum cleaner, a large garbage bag, a light bulb, plastic cups, and a $2.40 photocell that could be attached to an ampmeter. This low-tech procedure produces data extremely similar to those provided by the best high-tech equipment!

There should be a Rachael Carson award given, like the Pulitzer Prize, to the author of the best environmental article, paper, or book each year, and there should be a Nobel Prize in Planet Quality awarded to the corporation that does the most to eliminate pollution. There should be a Freddy Krueger award for the worst polluter, with a one million dollar fine to be put in the superfund for cleanup. But such a program would never get past the congressional male whores in Washington. Awareness and early education are the keys to saving the environment.

IRS

In 1862, during the Civil War, President Lincoln and Congress created the office of Commissioner of Internal Revenue and enacted an income tax to pay war expenses (the Revenue Act of 1862).

The income tax was repealed ten years later. In 1894, Congress revived the income tax, but the following year the United States Supreme Court ruled, in *Pollock v. Farmers' Loan & Trust Co.,* that taxes on capital gains, dividends, interest, rents, and the like were direct taxes on property, and that the statute in question was unconstitutional because it had not apportioned the direct taxes among the states according to population. In 1913, however, the states ratified the Sixteenth Amendment, which removed the requirement that income taxes (whether considered direct or indirect taxes) be apportioned by population.

Since then the following taxes have been enacted:

Accounts Receivable Tax
Building Permit Tax
Capital Gains Tax
CDL License Tax
Cigarette Tax
Corporate Income Tax
Court Fines (indirect taxes)
Dog License Tax
Federal Income Tax
Federal Unemployment Tax (FUTA)

Fishing License Tax
Food License Tax
Fuel Permit Tax
Gasoline Tax (42 cents per gallon)
Hunting License Tax
Inheritance Tax Interest Expense (tax on the money)
Inventory Tax IRS Interest Charges (tax on top of tax)
IRS Penalties (tax on top of tax)
Liquor Tax
Local Income Tax
Luxury Taxes
Marriage License Tax
Medicare Tax
Property Tax
Real Estate Tax
Septic Permit Tax
Service Charge Taxes
Social Security Tax
Road Usage Taxes (Truckers)
Sales Taxes
Recreational Vehicle Tax
Road Toll Booth Taxes
School Tax
State Income Tax
State Unemployment Tax (SUTA)
Telephone Federal Excise Tax
Telephone Federal Universal Service Fee Tax
Telephone Federal, State, and Local Surcharge Taxes
Telephone Minimum Usage Surcharge Tax
Telephone Recurring and Non-Recurring Charges Tax
Telephone State and Local Tax
Telephone Usage Charge Tax
Toll Bridge Taxes
Toll Tunnel Taxes
Traffic Fines (indirect taxation)
Trailer Registration Tax
Utility Taxes
Vehicle License Registration Tax
Vehicle Sales Tax
Watercraft Registration Tax
Well Permit Tax
Workers Compensation Tax

Not one of these taxes existed a hundred years ago, and our nation was the most prosperous in the world, had absolutely no national debt, and had the largest middle class in the world—and only one parent had to work to support a family.

"Tax freedom day" once occurred sometime in September or October. That's the day when your employer stopped deducting tax from your earnings and sending it to the government, and you got to keep what you earned until the end of that year. Then they passed a law raising the taxable income withholding, so now it is impossible to get to the "tax freedom day" for the average citizen.

The federal income tax is a pay-as-you-go tax.

From the Internal Revenue Service United States Department of the Treasury Tax Information for Individuals:

These frequently asked questions and answers are provided for general information only and should not be cited as any type of legal authority. They are designed to provide the user with information required to respond to general inquiries. Due to the uniqueness and complexities of Federal tax law, it is imperative to ensure a full understanding of the specific question presented, and to perform the requisite research to ensure a correct response is provided.

The IRS is not even responsible for what it says.

All of Washington is hooked on money, power, or drugs—or all three. Money and power are far more addictive, pervasive, and insidious. Money to a congressional rep is what black tar is to a heroin addict. Jack Abramoff was one of the biggest dealers. Now that he is naming names and pointing fingers, and the White House and a lot of congressional reps are returning contributions or donating them to their favorite charities. As if that makes it okay and absolves them of all guilt of taking a bribe. That's like being just a little bit pregnant.

The pain and suffering caused by U.S. business criminals and politicians is not usually apparent, is easy to ignore, and is hard to estimate and imagine. How can anyone swindle a widow and orphans? The stock market, upon which so many Americans, wisely or unwisely, depend, keeps going down from lack of confidence in the government's ability to do anything about corporate crime and accounting scandals.

Homeland Security

Homeland Security is just a bureaucratic smokescreen to punish us into thinking we are more secure today than on 9/11/01. The yellow stripe (or line in the sand) painted around our border and the multicolored threat board (crime watch on patrol) signs are not working.

We now have a living memorial (TV news) to the 9/11 hijackers' martyrdom. Every year on 9/11, TV news celebrates their martyrdom by displaying their names and photos.

I don't fear terrorists; I fear our federal government more. A terrorists cell may kill a few thousand but our government can kill a lot more.

Being a Politician Is Not Easy!

Bending over and kissing everyone's ass, licking all of the rich campaign contributors' boots, hiding billions of dollars, carrying all that money to your friends, and keeping all of your lies straight is not an easy job. Also shaking all those hands and kissing all the babies while smiling, and while jousting with the press and avoiding all the reporters' questions is quite a workout.

Playing all those games is hard, too, like jumping through all the hoops for everybody and playing hide and seek.

It's expensive, also, to hire a staff of spin doctors to keep the right words in your head while paying off all the writers and reporters to say good things about you.

All the expense is not necessarily monetary. You have to sell your soul and you have to make pacts with the other devils.

It requires sacrifices: you have to sacrifice your integrity and honesty, your family, parents, wife, and children.

You must be educated in the ways of politics: you must learn brainwashing techniques and how to double speak. You must master how to allude, insinuate, seduce, arouse, evoke, suggest, entice, intimate, hint, theorize, inspire, offer as a possibility for consideration, and allude to the conception of an idea without ever saying what it is.

You must learn to defend yourself and your position despite all the facts and evidence against you until your dying day. And never admit you did anything wrong. Which will be true in your mind, anyway. California representative Randy "Duke" Cunningham broke this cardinal rule. Politicians are like diapers; they should be changed often, and for the same reason.

The value of the U.S. dollar has dropped to an all-time low against the Euro and the yen, apparently for reasons similar to those affecting the stock market. A weakening dollar makes foreign investors nervous, and it is feared by financial experts that a sharp decline in the dollar might lead to its collapse by causing a sell off "as everybody tries to get out the door at the same time."

The U.S. economy depends on foreign investment money to finance our enormous negative trade balance. That is, as a nation we spend more than we earn. "Private foreign purchases of American securities ran at an annual pace of $260 billion for the first quarter of 2004 compared to $400 billion for year 2003, a 35% decline," according to the *Taipei Times* on June 22, 2002. A run on the dollar could have dire consequences for the world economy.

It is difficult to have much confidence in our government to regulate business, let alone punish business criminals and deter their type of crimes. Regulation has been in disrepute since the Reagan administration twenty years ago—despite the disastrous effect deregulation had on Californians and the economy of California in 2001.

Chaos and the Decline of Civilization

John Hanchette, in the *Niagara Falls Reporter*, observes, "Scientists and mathematicians frequently advance the Chaos Theory to explain unanticipated consequences. It states that tiny variations in beginning conditions can trigger mammoth, lasting transformations in the end results. Its basic principle is sometimes referred to as the Butterfly Effect—as in, say, a hurricane being whipped up by ever-increasing wind vectors over a period of days and weeks after being catalyzed by the tiniest imperceptible change in air current when a butterfly somewhere flaps its pretty little wings."

This theory of seemingly random cascading events is scoffed at by some academics, but I'm starting to believe in it.

Next they will say a man can't have a house. Who is it that wants to take my property? Great public thieves have come along and picked the pockets of everyone who had a pocket.

The Supreme Court ruled that local governments may force property owners to sell out and make way for private economic development when officials decide it would benefit the public (for the good of the community for development of a shopping mall, football stadium or any other purpose), even if the property is not blighted and the new project's success is not guaranteed. Advocates argue that forcibly shifting land from one private owner to another, even with fair compensation, violates the Fifth Amendment to the Constitution, which prohibits the taking of property by government except for "public use."

As John Steinbeck said in *East of Eden*, "There are monstrous changes taking place in the world, forces shaping a future whose face we do not know. Some of these forces are evil to us, perhaps in themselves but because their tendency is to eliminate other things we hold good…. There is great tension in the world, tension toward a breaking point, and men are unhappy and confused."

As alluded to earlier, the simplified laws of thermodynamics are: you cannot win, you cannot break even, and you cannot get out of the game.

The good are destroyed while the evil survive and prosper.

This is nothing new for the great Western civilizations. Mesopotamia, Babylonia, ancient Greece, ancient Egypt, the Roman empire, the Byzantine empire, the Maya, Charlemagne's empire, the Sumerian civilization—all these had their rise and fall.

Why did the civilizations of Greece and Rome collapse? The real answer is greed and corruption. It seems that the U.S. has succumbed to the same weaknesses. Like the English in India in the nineteenth century, the U.S. politicians are out to get what they can, taking part in the administration if it pays them, and contemptuous of the traditional culture, except insofar as it provides more wealth. The idea of keeping to the ideals of the founding fathers never enters their heads. There have always been "robber barons" in all civilizations.

The Gaia theory is ultimately in control.

Kids.Net.Au, at http://encyclopedia.kids.net.au/page/ga/Gaia_theory, explains the Gaia theory this way:

The Gaia theory is a broadly inclusive name for a group of ideas that living organisms on a planet modify the nature of the planet to make it more suitable for life. This set of theories holds that all organisms on a planet regulate the biosphere to the benefit of the whole. Because they may choose to do so selfishly for only the benefit of their own species, perhaps as parasites, these theories are also very significant in green politics. While there were a number of precursors to Gaia theory, the first scientific form of this idea was proposed as the Gaia Hypothesis by James Lovelock, a UK chemist, in 1970. While controversial at first, various forms of this idea became accepted to some degree by many within the scientific community.

The greatest century of scientific development in the history of man will be followed by the greatest fall of a civilization in the history of man. It will be a long, slow and painful fall. This is probably inevitable but a necessary thing. This is Gaia's re-establishing the way things work. What goes up must come down.

The far east civilizations have had 4,000 years of nearly continuous growth. Although stymied by a class system that did not promote development, they did survive many internal wars and political changes, and now they are challenging Western Civilization.

Congress will curtail funding for the manned shuttle and international space station, just like it did for the Super Collier, in favor of unmanned probes such as those we've already sent to the planets. There is not enough excitement about the manned-space-flight program to get politician re-elected. The next manned space program will be lead by the Chinese, and the first men on Mars will be Chinese.

Dr. James Burke, in his book and TV show, *Connections*, pointed out how interdependent our civilization is on all of the interconnected systems we have developed. The question is what will be the trigger that will start the change and fall of our civilization, and start the total collapse of our society? Is it the rising national debt, greed and corruption, the war in Iraq, high energy costs, the World Trade Center attack on 9/11, bank credit-card debt and late fees, or some other catastrophe?

Will the trigger be Hurricane Katrina, or a terrorist attack? The recent hurricanes highlighted the fact that this country is not ready for a terrorist attack with a weapon of mass destruction such as chemical, biological, or radiological weapon. The Federal Emergency Management Agency (FEMA) was the main obstacle to help and recovery, and George Bush commended the head of FEMA, Michael D. Brown, for doing a "good job."

A good job?!

The system is flexible enough to absorb all of these problems, but there will be one that will eventually trigger the change and bring the system down.

Edward O. Wilson's book *Consilience* says that the greatest enterprise of the mind has always been and always will be the attempted linkage of the sciences and humanities. But what we should try to do is save the sciences. The Eighteenth-century Enlightenment believed we could clear away the debris of millennia, including all the myths and false cosmologies that encumbered humanity's self-image. People are innate romantics, they desperately need myth and dogma, and scientists can not explain why people have this need. The propositions of the original Enlightenment are increasingly favored by objective evidence, especially from the natural sciences. Consilience is a test of the truth of the theory in which it occurs.

Democracy has always been prone to corruption. Free enterprise and free markets seem to be the most susceptible to greed and corruption. Because we were so successful with the free enterprise system and capitalism, the world has tried to be like us. However, one does not have to have a democracy to have capitalism. A benevolent dictatorship or a constitutional monarchy is more efficient. The religious right has led us to believe that God was on our side and we could do no wrong.

Government of the people, by the people, for the people, as Abraham Lincoln said in his Gettysburg Address, is no longer true. The U.S. government has become a government of large corporations, by the executives, and for the very rich.

Big Brother Is Really Watching You

George Orwell missed the date by only twenty years. Instead of 1984, what he predicted came to pass in 2004. With all the surveillance cameras being installed around the country, and warrant-less wiretaps, Big Brother is really watching you.

"Your call may be monitored to ensure quality."

You can get an aerial view of your own street, house and yard. You can view cars parked in someone's driveway and with infrared photos; you can even view people inside a house and your bedroom, or cars parked in anyone's garage. You can get satellite aerial photos of your state, city, town, or street, acquired within the last few months. Photos are available for most locations in North America and for specific locations in fifty-six other countries. Satellite surveillance of earth is a fact of life. Intelligence agencies have been using satellite images for decades. But a new program called "the National Map" is being conducted by the U.S. Geological Survey. Its focus is on detailed images of urban areas, or "knowledge needed by the nation." Media reports over the past few months show that the Pentagon has been monitoring anti-war, anti-government, and anti-military activity for the TALON, Threat and Local Observation Notice program.

Your Copier Is Spying on You

Every time you print a document, it automatically includes a secret code that identifies the printer's serial number, date, and time. In an effort to identify counterfeiters, the U.S. government has succeeded in persuading laser-printer manufacturers to encode each page with this identifying information.

The DocuColor series prints a rectangular grid of 15 by eight miniscule yellow dots on every color page. The printer serial number is a binary-coded-decimal, with two digits per byte (constant for each individual printer). The code shows the year (without the century, 2005 is coded as five), month, day, hour (which may be a UTC time zone, or it may be set inaccurately within the printer), and minute that page was printed.

Systems

Whole systems are slowly coming apart. First there was the bus system, then the education system, and then the trains. Next, airplanes will begin to fall out of the skies as, to compensate for higher fuel costs, airlines cut maintenance cost instead of cutting executive salaries and bonuses.

The military-industrial complex, the oil and gas industry giants, the auto industry, and the church are the biggest gorillas in the world. They're so big they make King Kong and Mighty Joe Young look like pygmy marmosets.

Moron's (Mobil/Exxon's) $65 billion in profits is enough to buy all the congressional reps, lobbyists, and lawyers needed to keep any improvements to the environment and consumers from interfering with their bottom line.

George W. Bush said in a televised speech, "Oil and gas profits from Iraq should go to the Iraqi people." If that is the case, shouldn't the profits for U.S. oil and gas go the American people instead of to corporate executives?

What happened to the thousands of electric cars that were built in the 1990s? They didn't use gas or oil, need a tune-up, or need oil changes, spark plugs, oil, or air filters—so the auto industry ground them up to keep anyone from copying them. Now they are introducing the Hybrid car, which is worse for consumers but good for the auto manufactures.

In the 1974 fuel shortage, the automobile industry picked the worse solution possible for the consumer but most profitable to the auto industry: the catalytic converter. To reduce gas consumption and reduce emissions, all they needed to do was limit internal combustion engines to four cylinders.

My Geo has only three cylinders and gets 50.3 mpg and does a very good job at that.

History Lesson

On December 17, 1903, at Kitty Hawk, North Carolina, the Wright Flyer became the first powered, heavier-than-air machine to achieve controlled, sustained flight with a pilot aboard.

Here are some U.S. statistics for 1906, found in an e-mail widely circulated on the Internet:

The average life expectancy in the U.S. was 47 years.

Only fourteen percent of the homes in the U.S. had a bathtub.

Only eight percent of the homes had a telephone.

A three-minute call from Denver to New York City cost $11.

There were only 8,000 cars in the U.S., and only 144 miles of paved roads.

The maximum speed limit in most cities was 10miles per hour.

Alabama, Mississippi, Iowa, and Tennessee were each more heavily populated than California.

With a mere 1.4 million people, California was only the twenty-first most populous state in the Union.

The average wage in the US was twenty two cents per hour.

The average U.S. worker made between $200 and $400 per year.

A competent accountant could expect to earn $2000 per year, a dentist $2,500 per year, a veterinarian $1,500–$4,000 per year, and a mechanical engineer about $5,000 per year.

More than ninety-five percent of all births in the U.S. took place at *home*.

Ninety percent of all U.S. doctors had *no college education*! Instead, they attended so-called medical schools, many of which were condemned in the press *and* by the government as "substandard."

Sugar cost four cents a pound.

Eggs were 14¢ a dozen.

Coffee was 15¢ a pound.

Most women washed their hair only once a month, and used borax or egg yolks for shampoo.

Five leading causes of death in the U.S. were:

1.   Pneumonia and influenza

2.   Tuberculosis

3.   Diarrhea

4.   Heart disease

5.   Stroke

The American flag had forty-five stars. Arizona, Oklahoma, New Mexico, Hawaii, and Alaska hadn't been admitted to the Union yet.
The population of Las Vegas, Nevada, was only thirty!!!!
Crossword puzzles, canned beer, and ice tea hadn't been invented yet.
There was no Mother's Day or Father's Day.
Two out of every ten U.S. adults couldn't read or write.
Only six percent of all Americans had graduated from high school.
Marijuana, heroin, and morphine were all available over the counter at the local corner drugstores. Back then pharmacists said, "Heroin clears the complexion, gives buoyancy to the mind, regulates the stomach and bowels, and is, in fact, a perfect guardian of health." (Shocking? *Duh*!)
There were about 230 reported murders in the *entire U.S.A.*!
Cost of living—prices of items in 1950, in 2006, and projections for 2050:

	1950	2006	Will cost in 2050
Car:	$1,750	$17,500	$87,500–$175,000
Gasoline:	$0.27/gal	$3.00/gal	$15.00–$330/gal
House:	$14,500	$145,000	$725,000–$1,450,000
Bread:	$0.14/loaf	$2.62/loaf	$13.10–$432/loaf
Milk:	$0.82/gal	$4.36/gal	$21.80–$232/gal
Postage Stamp:	$0.03 cents	$0.39	$1.95–$4.00
Average Annual Salary:	$3,800	$50,388	$251,940–$670,160
Minimum Wage (per hour)	$0.75	$6.65	$26.60–$53.20

All projections into the future have been wrong, and I hope these are, also.

Religion

I have been to many churches and Bible study groups, and have found that religion is the result of the master of ceremonies. The ability to inspire and motivate is done by the speaker.

There's a film called *The Singer Not the Song*. It's essentially a story about religion and whether it is the church or the clergy that makes religion work. The film tells the story of a priest and a gunfighter. The priest sees the gunfighter as a soul to be converted, and the gunfighter is impressed that the priest is not afraid of him. During the course of the film, they have several philosophical discussions, and at one point the gunfighter asks, "Is it the song or the singer who makes the music?" Just as the priest is about to win over the gunfighter, the gunfighter is killed in a gunfight. The priest administers last rights, and the gunfighter, with his dying breath, asks again, "Is it the song or the singer who makes the music?"

Like a doubting Thomas, I don't believe in miracles. I have seen too much flimflam, fraud, cons, prestidigitation, hoaxes, and "hoc est corpus" to believe anything I see or hear especially by religious groups. The word "hoax" came from the magic incantation "hocus pocus." "Hocus pocus," in turn, is commonly believed to be a distortion of "hoc est corpus" ("this is the body") from the Latin Mass. However, many etymologists (perhaps intimidated by the church) dispute this claim.

What I have suspected for many years and thought was too good to be true is now being voiced. I have always thought some of the miracles in the Bible were a fantasy or Christian mythology. Now the tribulation is here. Mary was not a virgin. Jesus' father is suspected to be a nineteen-year-old Roman soldier. Jesus had a girlfriend, and maybe even a wife who had a child. Judas was not a bad guy after all; he was just doing what Jesus wanted him to do. There was no resurrection, just a plot to perpetuate Jesus' image.

Jesus seemed to be just another unemployed carpenter/soothsayer protesting in the streets until he turned over the moneychanger's table in the temple. He had interfered with the great pagan god, Money, who still rules the earth today, and that sealed his fate. He interfered with commerce, where government officials collected taxes and negotiated deals to keep their power, just like they do today. That would be equivalent of creating a computer virus that interfered with the electronic funds transfer (EFT) of the Federal Reserve banking system and disrupting campaign contributions and bribes to congressional representatives today. Why do you think politicians go to church? Not to pray or worship—they could do that at home. It is to advance their political agenda and create the illusion that they are good, kind, and trustworthy, and to solidify their voting base and hold on to power.

You can do anything you want to, but don't step on my green felt money-changing table.

Jesus was hauled before a kangaroo court similar to those of our criminal justice system. If you're rich or have political influence and have enough money, you get a slap on the wrist or a pocket-change fine. The worst you will get is to be sent to a club fed or put under house arrest. Otherwise, you get the needle after twelve years of Orwellian TV appearances and courtroom shows with ACLU lawyers' songs and dances. The jury in Pontius Pilate's court was probably paid by lobbyists to convict Jesus and let their friend—a rapist or murderer—go free.

If Jesus was the son of God, he failed miserably against the pagan god Money. And his church today is not doing any better. In fact his followers seem to have made alliances with the pagan god Money and have built mega-churches to honor him. Jesus never built any churches. He walked the land and told his disciples, "Go, and do thou likewise." Jesus said to the rich man, "Go and sell that thou hast, and give to the poor." Then said Jesus to his disciples, "I say unto you, that a rich man shall hardly enter into the kingdom of heaven. It is easier for a camel to go through the eye of a needle, than for a rich man to enter into the kingdom of God."

Jesus went on a vision quest like shamans had done for thousands of years. For forty days and nights of sensory deprivation he experienced his inner alternate state of consciousness.

The light to which he so often made reference was the light of truth—inner truth—and his use of the idea of repentance was identical with the modern psychological technique of integration, which teaches you to go back over your life and discover in it your real motives, real fantasies, real purposes, urges, instincts, escapes, lies, cheatings, and vanity. You can't know yourself unless you know all the negative and inferior aspect of yourself. That's a thoroughly reasonable and scientific principle. A man who did not know himself could not in any way trust what he thought about other men or the world. Jesus' premise was that an individual man is able, through integrity alone, to follow the elements of his subjective nature to their outermost boundary. He concluded that any man who did so would find the boundary infinite and immortal. The subjective disciplines that he tried to explain were precisely analogous to the scientific method toward objective fact now used in the modern world. The identity of the thinker is as much a part of what he or she is thinking about as the thought itself; just as, in relativistic physics, the nature and position of an observer is as much a part of the conclusion as measurements of the observation. This idea has been repeated many times in many ways by scholars and psychologists like Carl Jung and Sigmund Freud.

But is was best said by William Shakespeare:

This above all; to thine own self be true,
And it must follow, as the night the day,
Thou canst not then be false to any man.

If Jesus comes again, the modern church will see that He will be given electroshock treatments and put to work selling shoes in Sears. Oh, we get our reward later (sounds like our government full of empty promises) in heaven, and we get "beamed up" at the Rapture the church promises. This sounds like the pilot episode of *Star Trek*. If you believe that, I have a bridge in Brooklyn that has been sold a few billion times.

The church today, which is supposed to be for good and human rights, should have led a crusade into the Jihad in Iraq and Afghanistan, and should have led the charge for the tsunami and hurricane relief efforts.

Where are all the Knights of Columbus and Masons when they are needed?

The preachers of my childhood in Oak Cliff made it all sound too much like the threat-and-rescue method of psychological introduction, or brainwashing.

Bill Allen says there is a God gene, or as I like to call it "the witch doctor" gene. If you were not afraid of or scared by the witch doctor, you had an accident or did not receive the witch doctor's medicine when you needed it, so you died before you could reproduce. Thus, the witch doctor kept the "fear of God" gene in the gene pool.

Religion has impeded man's development since the beginning of time.

Name ten scientific breakthroughs to advance mankind that religion has created.

The Worst Proposal in Sixty-Five Million Years

In Texas, our Republican Governor Rick Perry's solution to the energy situation involves building twelve new electric generating systems, to be powered by Texas coal. Adolf Hitler's final solution (Endlösung) killed six million. Rick Perry wants to do Hitler one better. (Everything in Texas has to be bigger.) Perry wants to kill everyone. His proposal will screw everyone, even the whole world, to line the pockets of his friends and campaign contributors.

He says his proposal will reduce our dependence on foreign oil and help reduce electric rates. It might reduce our dependence on foreign oil at the expense of polluting the atmosphere and killing us all. It will do nothing to reduce the electric rates. Electric rates in Texas are set by the price of natural gas. Even that is fixed at a high rate, set by Governor Perry last year and even if natural gas costs go down, electric rates won't. The coal-fired electric generation will only reduce costs for the energy producers and increase their profits. It will *not* reduce costs to the consumer. It will increase pollution and speed global warming, leading to the elimination of all life on the planet. It is the worst proposal in sixty-five million years—even worse than Adolf Hitler's final solution.

The only reasonable solution would be if the twelve plants were powered by hydrogen.

The Sixth Extinction Event

The Sixth Extinction Event, aka the "Final Solution," is a race between weapons of mass destruction, over-population, the destruction of the rain forest, pollution-caused global warming, changes coming in molecular and cellular research, or another asteroid of doom. Cornell professor Frank Drake's equation is that $N = N_* \, f_p \, N_e \, f_l \, f_i \, f_c \, f_L$, where:

N represents the number of stars in the Milky Way Galaxy

fp represents the fraction of stars with a planetary system

Ne represents the number of planets in a given system that are ecologically suitable for life

fl represents the fraction of otherwise suitable planets which life forms of life evolve

fi represents the fraction of inhabited planets on which intelligence forms of life evolves

fc represents the fraction of planets inhabited by intelligent beings on which a communicative technical civilization develops, and

f_L represents the fraction of planetary lifetime graced by a technical civilization

N might be as small as 1, because civilization tend to destroy themselves soon after reaching technological phase. As Carl Sagan says in *Cosmos,* there might be no one for us to talk with but ourselves.

I heard that everyone has one book in him, and I have spoken my piece.

The Handwriting Is on the Wall

Alas poor America, I knew her well: a lady of infinite jest, opportunity, compassion, and generosity, all of which led to her demise, and thus she is no more.

—End—

About the Author

In spite of my background, I beat the odds. I'd risen to become, among other things: a Cold-War spy, an early Mary Kay entrepreneur, a Silicon Valley computer pioneer, a weather-satellite engineer, and a Microsoft support guru. I conquered my family's lack of culture and education and become a classical music enthusiast, an appreciator of art, a student of world religions and philosophy and science, and a French wine connoisseur. I'd never been taught about investing, yet I'd made money in real estate in California, New England, and Texas, and I'd made money, lost money, and made money again in the stock market.

I've lived in Germany, San Francisco, Carmel, Los Angeles, Boston, France, and Sydney. I've crossed the U.S. by train, plane, car, and bus. I've hobnobbed with generals, corporate executives, and multi-millionaires, and I've slept on a park bench. I've traveled to Canada, Mexico, Chile, Peru, England, Germany, France, Spain, Italy, Turkey, and Australia. The sickly little boy from Texas became a professional dance instructor, a sky-diver, an expert swimmer, a hiker, a windsurfer, a pilot, and a sailor with my own 44-foot sailboat. And I survived near-plane crashes, train accidents, and storms at sea.

I sometimes wonder how I managed come up out of the shit-pile of life, smelling like a rose.

978-0-595-68196-9
0-595-68196-4

Printed in the United States
109563LV00003B/1-48/A

9 780595 681969